AF356226

Maritime Security and Risk Assessments

Maritime Security and Risk Assessments

Editors

Marko Perkovic
Lucjan Gucma
Sebastian Feuerstack

Basel • Beijing • Wuhan • Barcelona • Belgrade • Novi Sad • Cluj • Manchester

Editors

Marko Perkovic
University of Ljubljana
Portorož
Slovenia

Lucjan Gucma
Maritime University of
Szczecin
Szczecin
Poland

Sebastian Feuerstack
Institute of Systems
Engineering for Future
Mobility
Oldenburg
Germany

Editorial Office
MDPI AG
Grosspeteranlage 5
4052 Basel, Switzerland

This is a reprint of articles from the Special Issue published online in the open access journal *Journal of Marine Science and Engineering* (ISSN 2077-1312) (available at: https://www.mdpi.com/journal/jmse/special_issues/KF5VBPY2VE).

For citation purposes, cite each article independently as indicated on the article page online and as indicated below:

Lastname, A.A.; Lastname, B.B. Article Title. *Journal Name* **Year**, *Volume Number*, Page Range.

ISBN 978-3-7258-1677-4 (Hbk)
ISBN 978-3-7258-1678-1 (PDF)
doi.org/10.3390/books978-3-7258-1678-1

Cover image courtesy of Marko Perkovič

Contents

About the Editors

Marko Perkovic

Marko Perkovic is a maritime expert with a PhD in Maritime Studies from the Maritime University of Szczecin and a master's in Transportation Engineering from the University of Ljubljana. He currently serves as Vice Dean for Research and Development and Assistant Professor at the Faculty of Maritime and Transport, University of Ljubljana. With extensive experience in simulation-based education, ship safety, port operations, and environmental protection, Marko has led over 30 industry projects and published hundreds of scientific and technical articles. He previously served as a National Maritime Expert at the EC Joint European Research Institute and is an active member of several international maritime associations.

Lucjan Gucma

Lucjan Gucma is a professor in the Navigational Faculty at the Maritime University of Szczecin, Poland. His main scientific interests focus on probabilistic methods for assessing navigation safety and ship simulation methods. He has published four books and over 100 papers in various journals and conference proceedings. He has also participated in numerous national and international projects related to the safety of navigation and maritime transportation.

Sebastian Feuerstack

Head of the Safe Automation of Maritime Systems Department at the Institute of Systems Engineering for Future Mobility at the German Aerospace Centre. His research focuses on systems engineering and, in particular, on safety methods to ensure the trustworthiness of highly automated and autonomous vessels at sea, on waterways and automated systems in ports. To this end, the Institute operates a testbed, which couples digital twins of port infrastructure, vessels and reference waterways in simulation with a physical testbed of highly automated research vessels, VTS operators and remote control operations setups that extends along the German Bight and includes several ports to test and evaluate highly automated maritime systems.

Preface

This reprint focuses on maritime operations and safety, covering topics such as vessel traffic analysis, collision risk assessment, and the use of AIS data to improve maritime safety and surveillance. Our primary goal is to contribute to maritime safety and the protection of the marine environment by providing innovative solutions to the challenges facing the industry. This work is aimed at scientists, shipping professionals, port authorities, non-governmental organisations, and maritime authorities, providing valuable insights and strategies to improve maritime safety.

Marko Perkovic, Lucjan Gucma, and Sebastian Feuerstack
Editors

Journal of
*Marine Science
and Engineering*

Editorial

Maritime Security and Risk Assessments

Marko Perkovič [1,*], Lucjan Gucma [2] and Sebastian Feuerstack [3]

[1]   Faculty of Maritime Studies and Transport, University of Ljubljana, 6320 Portorož, Slovenia
[2]   Faculty of Navigation, Maritime University of Szczecin, 70-500 Szczecin, Poland; l.gucma@am.szczecin.pl
[3]   German Aerospace Center (DLR), Institute of Systems Engineering for Future Mobility, Escherweg 2, 26121 Oldenburg, Germany; sebastian.feuerstack@dlr.de
*   Correspondence: marko.perkovic@fpp.uni-lj.si

Abstract: The main topics of the presented papers focus on various aspects of maritime operations and security, including anomaly detection in maritime traffic, collision risk assessment, and the use of Automatic Identification System (AIS) data for enhancing maritime safety and surveillance. These papers cover a wide range of subjects within the maritime domain, such as trajectory clustering, kinematic behaviour analysis, Bayesian networks for risk assessment, resilience analysis of shipping networks, and the development of novel methods for detecting abnormal maritime behaviour. The emphasis is on using data-driven approaches, statistical methodologies, and advanced technologies to improve maritime operations and security.

Citation: Perkovič, M.; Gucma, L.; Feuerstack, S. Maritime Security and Risk Assessments. *J. Mar. Sci. Eng.* **2024**, *12*, 988. https://doi.org/10.3390/jmse12060988

Received: 28 May 2024
Accepted: 29 May 2024
Published: 12 June 2024

1. Introduction

Shipping is relatively safe and clean, but maintaining this state of affairs is an intensive and expensive business. As technology advances, operating systems on board commercial vessels are becoming more specialised and complex. The development of automated systems to monitor, analyse, and regulate the various operations or services on board relies on computer applications to centralise and optimise decision making. Such systems are highly vulnerable to cyber attacks. In addition, crew reductions and the general trend towards reducing the number of people on board ships are being implemented on a large scale. In some ports, pilotage services are already being provided remotely. The first LNG ship recently sailed autonomously across the ocean, and several maritime universities are already preparing for the new era of seafarers who will remotely monitor and control ships' navigation and propulsion elements. There are many technical and regulatory challenges, such as the robustness and resilience of autonomous navigation technology, on-board systems, communications, land-based traffic management, piracy, and cyber security; in addition, ports are a key element of the maritime transport chain and are also vulnerable to cyber attacks. In addition, larger ships and increased port traffic can lead to increased risks at the ship level.

This Special Issue on "Maritime Security and Risk Assessments" presents a comprehensive exploration of key maritime operations and safety issues. Its collection of 16 manuscripts covers various topics, including vessel traffic analysis, collision risk assessment, and vessel traffic service (VTS) operations. Researchers have looked at predicting vessel traffic density using advanced time-series models, mapping fishing activity using Automatic Identification System (AIS) data and developing frameworks for anomaly detection and route prediction based on vessel patterns. The importance of AIS in enhancing maritime security is a recurring theme throughout the manuscripts, focusing on technological advances and efforts to reduce piracy. In addition, the papers highlight the metrics and provider-based results for satellite-based AIS services, emphasising the importance of completeness and temporal resolution in maritime surveillance. The real-time identification of anchorage collision risks based on AIS data is a critical aspect discussed in several manuscripts, highlighting the need for advanced risk assessment models in complex maritime environments. Overall, this collection of manuscripts provides valuable insights into ensuring navigational safety, mitigating maritime risks, and improving maritime governance through innovative technologies and analytical approaches.

2. Contemporary Challenges in Maritime Safety and Security

In the dynamic landscape of maritime security and risk assessments, sophisticated tools, devices, and sensors are pivotal in fortifying safety measures and bolstering surveillance capabilities [1]. The integration of the Automatic Identification System (AIS) for real-time vessel tracking and collision avoidance as well as radar systems, which are adept at detecting vessels and assessing risks in diverse weather conditions, forms a robust foundation for maritime safety protocols. Complementing these technologies, Closed-Circuit Television (CCTV) cameras offer visual surveillance of critical maritime infrastructure, while cameras, including near-shore and satellite imaging, provide invaluable insights into vessel behaviours and anomaly detection [2]. Leveraging remote sensing technologies, such as satellite imagery, enhances the monitoring of maritime traffic and facilitates the identification of illicit activities. Vessel Monitoring Systems (VMSs) ensure regulatory compliance among fishing vessels [3], while Global Navigation Satellite System (GNSS) technology enables precise navigation and efficient route planning. The seamless communication facilitated by VHF radios fosters effective coordination between vessels and maritime authorities, optimising operational efficiency in busy waterways. By synergising these tools, researchers and stakeholders in the maritime domain can elevate their capabilities in monitoring, analysing, and responding to maritime incidents, thereby promoting a safer and more secure maritime environment even when conventional and autonomous ships meet [4].

The security issue is still relevant in maritime piracy [5], a problem recently modelled by Contribution 12, who developed a Bayesian network model to assess the risk of pirate hijacking. By incorporating various factors, such as geographic location, vessel characteristics, security measures and historical piracy data, the prediction of the likelihood and impact of hijacking incidents with surveillance data, including AIS information, is of great importance in improving the model's accuracy. It demonstrates the effectiveness of Bayesian networks in maritime security and provides valuable insights for improving preventive measures and decision-making processes to ensure safer maritime operations; also, it explores strategies for predicting and preventing maritime piracy. Ref. [6] propose several policy recommendations to enhance maritime security, including the implementation of international cooperation frameworks, the deployment of naval patrols in high-risk areas, and the adoption of the best management practices by shipping companies.

Recently, however, we have been confronted with entirely new threats for which we do not yet have the right answers. These include attacks on ships by militant groups [7] and severe jamming of GNSS receivers, which can have a significant impact on ship safety. The security threat posed by Houthi attacks on ships in the Red Sea has been the subject of academic and political debates. These attacks, which began in earnest in late 2023, have significantly disrupted global shipping lanes and caused economic ripples worldwide. The Houthis have targeted various commercial vessels, prompting major shipping companies to reroute their voyages around the Cape of Good Hope, significantly increasing transit times and costs. This can add up to two weeks to a shipment's journey. Figure 1 shows the diversion of ships around the Cape of Good Hope [8].

The attacks have led to increased insurance premiums and operating costs for shipping companies, with notable disruptions in the supply chain. Some ships have experienced extreme delays, affecting the trade between Asia and Europe as well as increasing shipping costs. The international community, including the United States and NATO allies, has responded with military operations and sanctions to mitigate the threat, but the situation remains tense and unresolved. Traffic through the Suez Canal has fallen by 70% in tonnage. These developments highlight the urgent need for a coordinated international effort to secure vital sea lanes and address the wider implications of such disruptions for global trade and economic stability. Furthermore, interference with GNSS systems is also very threatening, either through jamming or, even worse, spoofing [9]. Recently, positioning systems have been jammed in the Black Sea, the Eastern Mediterranean, and the Suez Canal. Pilots already face the challenge of navigating large ships through a narrow and winding canal [10] and berthing MGX class container vessels to the quay, hopefully not contacting the STS crane [11]. The situation becomes even more complex when a large ship, exposed to strong winds, heavily drifts in a narrow canal. To make matters worse, the crew suddenly loses confidence in the positioning system (failure of GNSS incidentally also affects radar stabilisation). The extent of such spoofing and jamming was also discussed at the last CHENS meeting (Chiefs of European Navies), where, among others, Androjna and Perkovič [12] presented some recent examples of spoofing, such as the sudden jump of a ship in Suez, which is shown in the Figure 2.

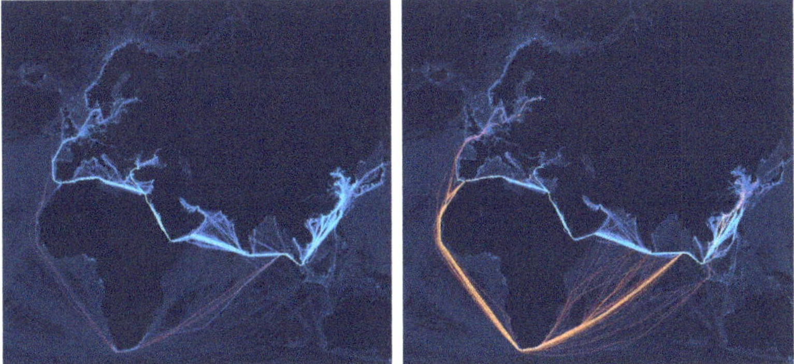

Figure 1. The increase in trade voyages between China and Europe via the Cape of Good Hope has been mapped using monthly AIS data (October 2023 vs. January 2024) [8].

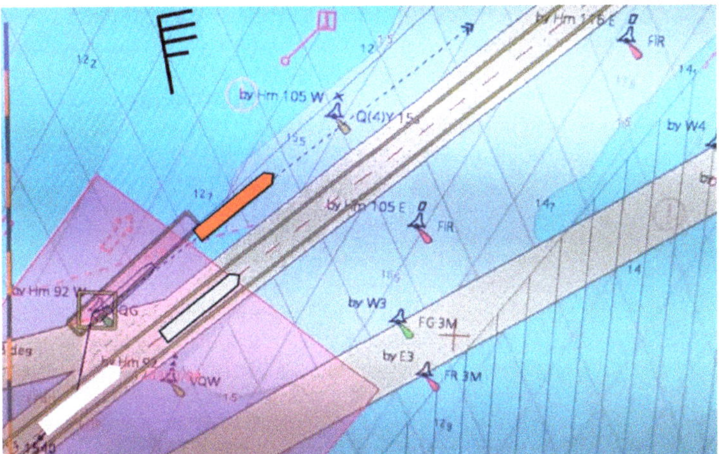

Figure 2. When you lose confidence in your vessel's position at a critical moment, as happened with the box ship Maria Elena recently at the exit of the Suez Canal in strong winds (35 knots), the consequences can be dire. This loss of confidence can stem from purely technical reasons. While major position shifts clearly indicate spoofing, even small deviations, which are harder to detect, can be extremely dangerous [12].

Such deliberate shifts or jumps in position due to intentional acts or technical failures pose a challenge to the ship's crew, marine pilots, and surveillance/control centres that monitor shipping traffic. Intelligent ship control is increasingly available, but it must reconstruct the trajectory correctly under uncertain conditions [13] and, when needed (in case of collision avoidance), predict vessel prediction or turning [14], which is extremely challenging when data are malicious. It should be noted that four levels of ship autonomy are currently defined, the first three of which require direct human involvement. Full autonomy also requires more accurate ship models, but even the most accurate model's prediction of a ship's behaviour will still deviate from reality. This is particularly important when it comes to enabling autonomous docking, where the positioning, communication and control system must be flawless, and a model must be included to represent the work of the ship's pilot [15].

3. Published Articles

This Special Issue encapsulates a diverse array of cutting-edge research endeavours to enhance navigational safety, mitigate risks, and bolster security measures in the maritime domain. The 16 manuscripts within it delve into pivotal themes such as collision risk assessment, anomaly detection, vessel traffic service (VTS) operation optimization, and utilising Automatic Identification System (AIS) data for maritime surveillance. Drawing on a rich tapestry of methodologies and tools, the researchers explore the complexities of maritime traffic patterns, the identification of abnormal behaviours, and the development of advanced risk assessment models.

By integrating microscopic and macroscopic factors, the scholars aim to comprehensively understand the collision risks between ships and in specific water areas. Furthermore, the significance of satellite-based AIS services underscores the importance of data completeness and temporal resolution in maritime surveillance. The manuscripts also shed light on the potential of emerging technologies, such as artificial intelligence and machine learning, to revolutionise anomaly detection and route prediction. Moreover, optimising VTS operations is identified as a critical area for future research, emphasising the need for advanced decision support systems to manage maritime traffic effectively. Researchers aim to pave the way for a safer and more secure maritime environment by addressing governance challenges and exploring innovative solutions. Through a multidisciplinary approach and a synthesis of diverse methodologies, these manuscripts collectively contribute to advancing maritime security and risk assessments, offering valuable insights and paving the way for future research endeavours in this dynamic and critical field.

In particular, the authors delve into utilising Bayesian networks and Petri nets to address maritime piracy confrontations, emphasising the importance of AIS and surveillance cameras for data collection. Another investigation focuses on the risk assessment of pirate hijacking, employing Bayesian networks and the N-K model, relying on surveillance cameras and risk assessment tools for a comprehensive analysis. Furthermore, the dynamic irregular grid approach for navigation safety evaluation, utilising GPS systems and navigation tools to enhance maritime operations, is explored. Additionally, hierarchical analysis, fault trees, and Bayesian networks are employed to conduct a ship collision causation analysis, highlighting the significance of vessel tracking systems and collision detection tools in understanding collision scenarios. Developing a ship collision risk evaluation model and a multi-ship encounter risk assessment model underscores the importance of radar systems and collision risk assessment tools in enhancing maritime safety protocols. Another study adopts a Bayesian network approach for the resilience analysis of maritime accidents, emphasising the role of data logging devices and sensor networks in assessing accident scenarios. The application of movement extraction, data aggregation, data analysis, computer vision, and deep learning techniques showcases the significance of AIS, surveillance drones, and image processing tools in maritime piracy analysis. Additionally, the utilisation of AHP and TOPSIS for multi-criteria decision making highlights the importance of decision support systems and data analytics tools in enhancing decision processes.

Through these comprehensive studies, this Special Issue significantly advances our understanding and implementation of maritime security and risk assessment strategies. By leveraging the latest advancements in technology, data analytics, and risk assessment frameworks, researchers are poised to shape a safer and more secure maritime environment, setting the stage for continued innovation and progress in this critical field.

Author Contributions

Liu et al. (Contribution 1) investigated and addressed the increasing complexity of vessel traffic in waterways due to increasing vessel traffic, varying environmental conditions, and the need for effective collision risk prediction and management strategies. The research problem focuses on the need for advanced tools and methods to assess and manage the risks associated with complex vessel traffic situations to improve safety and efficiency on busy waterways. The researchers proposed a model based on a molecular dynamics approach to assess the complexity of vessel traffic from both temporal and spatial perspectives, considering vessel motion parameters and spatial distribution characteristics. The proposed model was shown to accurately identify and quantify vessel traffic in a waterway using real AIS and simulated data. By objectively assessing the complexity of vessel traffic, the model can help maritime surveillance operators to monitor and organise vessel traffic more effectively, especially in complex traffic scenarios. Prospects for future research in modelling vessel traffic complexity include extending model parameters, adapting coefficients to different water environments, improving validation with real data, improving risk assessment methods, and integrating different surveillance technologies to increase safety and efficiency in managing complex vessel traffic scenarios.

Paper 2 addresses the problem or challenge of decentralised documentation of maritime incidents and presents a novel approach to improve incident investigation in the maritime industry. The study focuses on developing a system that allows multiple entities to contribute and document data in a trusted, decentralised, and tamper-proof manner to support conflict resolution. Existing approaches focus primarily on data from one's own vessel, neglecting valuable information from external sources that could contribute to a better understanding of critical traffic situations. The process of aggregating data from different actors equipped with sensors, which is then signed by a time-stamping authority to ensure data integrity, is proposed. The results highlight the importance of collecting data from multiple sources to fully understand critical traffic situations, especially when traditional investigation methods are limited. By proposing a method that guarantees non-repudiation of recordings and ensures data authenticity, the research addresses the trust issues that can arise between stakeholders during incident analysis. The conclusion highlights the importance of establishing a decentralised and trustworthy documentation system for maritime incidents, emphasising the need for a reliable database that includes contextual information beyond ship-related data.

The research by Ma et al. (Contribution 3) examines the identification and analysis of human errors in maritime safety operations to improve safety measures on board ships. Among the key factors influencing human error in maritime safety identified in the study are the error-producing conditions (EPCs) present in specific tasks on board ships. These EPCs are critical in determining the likelihood of human error in maritime operations. Factors such as inadequate training, lack of teamwork, inadequate supervision, poor communication, and lack of safety awareness are significant contributors to maritime safety errors. The study uses a novel hybrid approach that combines the SOHRA model, the entropy weighting method, and the TOPSIS model to calculate the probability of human error. By incorporating these models, the research aims to minimise the subjectivity of expert judgements and improve the accuracy of human error assessments. The methodology involves applying the proposed approach to a case study of cargo loading operations on a container ship, involving tasks that pose risks to crew members, shore-side workers, cargo, equipment, port facilities, and the environment. Through expert judgement and calculations based on the hybrid approach, the study identifies specific error-prone tasks and associated mitigation measures. The results show significant differences between the scenarios, with one scenario showing better results regarding the likelihood of human error. The study concludes that while it may be impossible to prevent all human error, it can be minimised through effective analysis and mitigation strategies.

Paper 4 presents a study on route evaluation and similarity measurement of container shipping fleets between the ports of two shipping lines. The study used Automatic Identification System (AIS) data to evaluate route characteristics for container vessels of a single fleet calling at the ports of Savannah and Charleston on the East Coast of the US. The methodology included AIS data pre-processing, route generation, and similarity measurement using Discrete Fréchet Distance (DFD). The research focused on assessing route length, duration, speed, and similarity between different trips, highlighting these parameters' statistical distributions and skewness. The results showed moderately positive correlations for trip length, negative correlations for duration, and highly positively skewed speed distributions. The study found that routes of the same vessels showed the highest similarity. In contrast, routes of different vessels showed greater dissimilarity, with DFD providing a valuable measure for route interpretation. The authors emphasized the importance of route analysis for fleet management, safety considerations, and decision making in the maritime industry. There is potential for future research to extend the analysis to include longer time periods, additional ports, and alternative measures of similarity to improve the reproducibility and applicability of the methodology for wider maritime studies and operational improvements.

The selection of appropriate anchorages is essential to ensure the safety of ships, crew, and passengers during navigation. By selecting appropriate anchorages, the risks of collision, impact, and other hazards can be minimised, with the welfare of all parties involved being the primary consideration. In addition, optimum nautical anchorages contribute to the efficient use of maritime space by providing ample room for ships to manoeuvre, ensuring smooth access for arriving and departing ships. Beyond safety and operational considerations, strategically selecting anchorages can also enhance the attractiveness of coastal areas for nautical tourism, attracting visitors and supporting local economic activities. In addition, carefully siting anchorages plays a crucial role in environmental protection by preventing damage to marine and underwater ecosystems. By carefully considering various criteria and applying multi-criteria decision analysis methods, stakeholders can make informed decisions that balance safety, economic viability, environmental sustainability, and user expectations, ultimately establishing well-planned and effective nautical anchorages. A case study of Split-Dalmatia County is presented in paper 5, which examines the optimisation of mooring locations using multi-criteria decision-making (MCDM) methods, specifically the Analytical Hierarchy Process (AHP) and Technique for Order of Preference by Similarity to Ideal Solution (TOPSIS). The research methodology gathered users' opinions through a questionnaire to determine the most important criteria for nautical anchorages, focusing on safety. Based on the weighted criteria, the study identified 15 optimal locations out of 86 possible variants.

With China importing a significant amount of bauxite from Guinea and Australia, the maritime logistics involved in this process face challenges such as long distances, complex marine environments, and the unique property of bauxite to fluidise. To ensure the stable and continuous import of bauxite and the development of China's aluminium industry, assessing and mitigating the risks associated with bauxite shipping logistics is crucial. Paper 6 discusses the importance of risk assessment in maritime logistics and introduces new parameters and methods to effectively prioritise and analyse risks. Traditional methods such as Failure Mode, Effects, and Criticality Analysis are commonly used in risk assessment, but they have limitations in dealing with the complexity and uncertainty of maritime logistics risks. Therefore, this paper aimed to introduce an improved risk assessment model that combines FMECA, fuzzy Bayesian networks, and evidence reasoning to provide a more accurate and comprehensive assessment of risks in bauxite maritime logistics. A risk parameter characterisation system was proposed to introduce sub-parameters to measure the severity after risk occurrence in maritime logistics. To capture the relationships between the antecedent attributes (input parameters) and the criticality levels (output) representing the severity of consequences in the bauxite maritime logistics system, 125 fuzzy rules were created. Finally, a sensitivity analysis was performed to ensure that the model responded appropriately to variations in the input parameters and maintained consistency in the risk assessment results.

Today, ships rely heavily on the availability of GNSS for accurate positioning, navigation, and timing (PNT). However, civil GNSS currently lacks sufficient security measures, and even cryptographically authenticated GNSS signals have recently been shown to remain vulnerable to spoofing attacks; spoofing attackers can replay authentic GPS signals in near real time to break receiver lock. Combining secured GNSS information with nautical data and other positioning techniques may be able to prevent position spoofing. The paper by Spravel et al. (Contribution 7) proposes MAMA, a software-based GPS spoofing detection framework based on anomaly detection. Unlike other approaches, it is software-based, relies solely on network traffic monitoring, and can be retrofitted. The authors generated a comprehensive tagged dataset using a maritime simulator based on real-world data. It includes several legitimate and spoofed samples and was used to evaluate their approach, and it is also available to others for benchmarking. In their future work, the authors plan to extend their evaluation and explore the potential of an intelligent ensemble of individual detection methods. They aim to improve detection capabilities and address deficiencies in detecting GNSS spoofing attacks in maritime systems.

Future maritime autonomous surface ships (MASSs) will rely heavily on machine learning algorithms and computer vision systems to navigate, detect, and avoid obstacles. However, the maritime industry faces new AI-related threats, such as adversarial attacks. In their paper, C. Lee and S. Lee (Paper 8) investigated the vulnerability of object detection and classification algorithms. They highlighted the importance of considering the security threats of deep neural network (DNN)-based object classification algorithms to various attack methods, such as FGSM, I-FGSM, MI-FGSM, and PGD. In particular, they investigated a very popular approach, the "you only look once version 5" (YOLOv5) algorithm, by training it on the Singapore Maritime Dataset (SMD-Plus) and then challenging it with adversarial attacks based on perturbed images generated by six pre-trained DNN algorithms and four adversarial attack methods. The results show that all algorithms and adversarial attack methods significantly reduce accuracy and pose significant security threats to AI systems during training. The study also acknowledges limitations in the experimental design. It highlights the need for further research into the vulnerability of AI systems to adversarial attacks and the development of mitigation strategies to enhance cybersecurity in the maritime industry.

Liu et al. elaborate a safety analysis method (paper 9) based on a network model of ship collisions, which relates causal factors with their inter-relationships to quantify the importance of each factor and thus identify key causal factors of ship collisions. The study proposed a successive safety evolution process based on the network model, where the safety protection strength of each causal factor was quantified, and a safety evolution process was initiated to control the proliferation of ship collisions. Based on the study of 300 ship collisions in Chinese waters, they extracted 98 causal factors. Unlike the existing studies that directly use the number of accidents as an important indicator of a particular causal factor, they propose to examine network efficiency to measure the importance of accident causation. Using this approach, their results are consistent with previous studies that have identified, for example, poor lookout as a major cause of accidents. However, the network analysis also identified those factors for which the probability of an accident is lower than for others, while at the same time having a higher degree of influence on the incidence of ship collision accidents.

A comprehensive review by Stache et al. (Contribution 10) provides valuable insights into the advances and challenges in maritime anomaly detection to improve the safety and efficiency of vessel traffic services. The review highlights various publications from 2017 to

2022, focusing on AIS-based anomaly detection techniques, statistical analysis, clustering, and neural networks. It highlights the importance of descriptive statistics and compares the results of different anomaly detection techniques in specific scenarios. The Automatic Identification System was the primary data source for anomaly detection due to its ubiquity, standardised structure, and coverage of relevant maritime traffic information. In terms of detection techniques combined with anomaly types, the anomaly types of route deviation and kinematic deviation are covered by the commonly used techniques of descriptive statistics, clustering, or NN-based classification. It is worth mentioning that NN-based detection in ship behaviour has shown promising results. However, a key concern is the lack of explainability in the decision-making process of NN-based systems, which may hinder trust and acceptance by VTS operators. The authors also discuss future work required to improve the utility of decision support tools (DSTs) in daily VTS operations, including integrating VHF data, generating annotated training data, and implementing realistic operator test beds.

The authors of paper 11 focus on improving Maritime Domain Awareness (MDA) in Brazil through a data-driven framework (CV-MDA) using computer vision technology. The research addresses the challenges of monitoring vessel activity in territorial waters and highlights the importance of accurate and reliable data for effective surveillance. By integrating data sources such as Automatic Identification Systems and Long-Range Identification and Tracking, the framework aims to consolidate vessel movement data to improve MDA. The proposed CV-MDA framework consists of three layers—Data Acquisition, Integration, and Detection/Classification—using a fine-tuned transfer learning approach for vessel detection. The study highlights the importance of data integration in increasing awareness. In addition, the research contributes to the advancement of maritime surveillance by addressing the challenges of monitoring vessel activity and enhancing the capabilities of surveillance systems. By leveraging computer vision technology and integrating multiple data sources, the framework improves situational awareness, supports effective decision making, and strengthens maritime surveillance worldwide to address security risks and illegal activities in territorial waters.

Through a systematic risk analysis approach, paper 12 provides valuable insights into the complex dynamics of pirate attacks along the Maritime Silk Road to enhance maritime security in the region. The paper addresses the critical issue of maritime piracy and its impact on shipping. The study identifies key risk factors that affect the vulnerability of ships to pirate attacks, highlighting the interaction of hazard, mitigation capacity, and vulnerability/exposure as crucial components in assessing the risk of attacks. By applying dynamic Bayesian network analysis, the research provides a comprehensive understanding of the pirate attack process, enabling stakeholders to better understand and mitigate the associated risks, highlighting the impact of natural and man-made hazards. DBN extends traditional Bayesian networks by incorporating temporal dependencies, enabling dynamic behaviour modelling over time. This enables the analysis of how different factors such as waves, visibility, number of pirates, weapons, political and economic situations, and naval support interact and evolve to influence the risk of pirate attacks. By considering these dynamic variables, DBN enables a more scientific, intuitive, and accurate assessment of the process risk of pirate attacks on ships. The study highlights the importance of crew training, anti-piracy drills, self-defence equipment, and surveillance intensity to improve ships' resilience against pirate threats.

In paper 14, Liu et al. present an analytical model for the real-time identification of anchorage collision risk based on AIS data to address the increasing challenges of maritime safety in congested port environments. The model incorporates three key aspects: microscopic collision risk, macroscopic collision risk, and traffic complexity. The microscopic aspect focuses on the relative motion between vessels to assess the collision risk of individual vessels. In contrast, the macroscopic aspect evaluates the characteristics of the anchorage area, considering safe navigable waters within variable boundaries. The model was developed to consider traffic complexity by analysing the spatial compactness of vessels in the anchorage. By integrating these aspects, the model provides a comprehensive approach to assessing collision risks in anchorage areas, enabling stakeholders to make informed decisions and implement targeted safety measures. The model's real-time monitoring capabilities using AIS data enable the continuous risk assessment and proactive intervention to improve safety and operational efficiency in ports worldwide. Experimental case studies have validated the effectiveness of the model in accurately identifying collision risks. The model also considers vessel traffic density and spatial complexity for a more accurate risk assessment.

Brcko and Luin (paper 14) present a novel decision support system that uses fuzzy logic to improve situational awareness and assist the navigator in collision avoidance in multi-ship encounters. The study begins by highlighting the conventional approach of using a simple radar circle model with a radius of up to one nautical mile for ship-safe areas. The methodology section outlines the research process, including a literature review focusing on emerging collision avoidance decision models, the development of algorithms to calculate collision risk and avoidance manoeuvres, the use of fuzzy logic for decision making, and the conduct of Monte Carlo simulations to evaluate the performance of the proposed model. The integration of the International Regulations for the Prevention of Collisions at Sea (COLREGs) with artificial intelligence techniques forms the model's backbone. Simulation tests show the effectiveness of the fuzzy-based decision model in two-vessel scenarios, although it has limitations in complex multi-vessel situations. Decision support systems are an important step towards integrating autonomous ships, as the management of these ships in the transition phase will mainly depend on the experience and competence of the seafarers. In addition, the paper highlights relevant issues for further study, particularly concerning revising the COLREGs on collision avoidance at sea, especially in multi-ship situations. A particularly important area in this respect is the study of the feasibility of determining the right of way from a ship's perspective. Spatially based route planning is a promising solution to these challenges and should be the subject of thorough investigation and research.

Vessel traffic service providers need efficient, context-sensitive tools to detect anomalous vessel behaviour. Current approaches, such as deep neural network models for detecting anomalous vessel behaviour, lack interpretability. Olesen et al. present a novel method for detecting anomalous vessel trajectories based on two-stage clustering that considers providing contextual decision support to VTS operators (paper 15). Their basic idea is to relate predictions to behavioural clusters and consider kinematic similarity measures (i.e., changes in speed and course) to assist the operator in accepting or rejecting the algorithm's prediction. For the evaluation, they created and published two large hand-annotated AIS traffic datasets containing more than 30,000 trajectories from 11 vessel types and marked anomalies covering a full day, including a collision accident, search and rescue activity, and anomalous commercial traffic. The detection performance of their proposed method is lower than that reported for using variational recurrent neural networks (VRNNs). However, it is much faster to compute and provides contextual decision support.

Another article (Contribution 16) examines the landscape of Global Navigation Satellite System (GNSS) spoofing. It is well known that Automatic Identification System (AIS) spoofing can be used for electronic warfare to conceal military activities in sensitive maritime areas. However, recent events suggest a similar interest in spoofing AIS signals for commercial purposes. The shipping industry is experiencing unprecedented fraudulent practices by tanker operators seeking to evade sanctions. This article addresses the issue of AIS manipulation, particularly in the context of the illegal oil trade, which is fuelled by the ongoing war between Ukraine and the Russian Federation. It highlights a major gap in the nexus between law, economics, science, and policy. Repeated cases of discrepancies between AIS position reports and satellite radar images have led to the detection and documentation of AIS position falsification by tankers carrying Russian crude oil in closed ship-to-ship (STS) transfers and by unofficially registered tankers ("ghost ships"), with direct implications for maritime safety. The risk posed by STS transfers depends on the boundaries of the transfer area, environmental restrictions, weather conditions, sea state, traffic density, good anchorage conditions, ship characteristics, and compliance with international and local regulations. These false ship positions underline the need for effective tools and strategies to ensure the reliability and robustness of AIS.

4. Outlook Research

The outlook for maritime safety and risk assessment research shows a promising trajectory with several key focus areas and associated challenges. Critical research directions are improving collision risk assessment models by integrating microscopic and macroscopic factors, refining anomaly detection techniques to identify abnormal vessel behaviour, and optimising vessel traffic service (VTS) operations through advanced technologies. In addition, overcoming challenges related to data completeness and temporal resolution in satellite-based AIS services, exploring the potential of emerging technologies such as artificial intelligence and machine learning, and addressing governance issues such as piracy prevention and environmental protection are essential to advancing maritime safety and security. In ref. [16], several possibilities for the future development of maritime safety are identified. In particular, further developments in automation and remote control with autonomous ships and unmanned navigation systems will reduce operating costs and increase safety. Integrating artificial intelligence and machine learning will create smarter navigation systems for better hazard prediction and optimised shipping routes. Cybersecurity measures must be strengthened to protect digital technologies and communication systems from cyberattacks. Integrated maritime traffic management systems combining satellite data, AIS, and radar will improve traffic management and reduce collision risks. Education and training programs must be interdisciplinary in their focus on technological advances and environmental issues. Finally, international cooperation is essential for harmonising global regulations and standards to ensure shipping safety worldwide.

To achieve these goals, researchers can use data analysis, machine learning, geospatial analysis, simulation and modelling, network analysis, risk assessment frameworks, and remote sensing technologies to enhance their studies and develop innovative solutions.

Author Contributions: Conceptualisation, M.P., L.G. and S.F.; writing—original draft preparation, M.P.; writing—review and editing; All authors have read and agreed to the published version of the manuscript.

Funding: The publication of the paper is supported by the research project (L7-1847; Developing a sustainable model for the growth of the "green port") and the research group (P2-0394; Modelling and simulations in traffic and maritime engineering) at the Faculty of Maritime Studies and Transport, financed by the Slovenian National Research Agency.

Conflicts of Interest: The authors declare no conflicts of interest.

List of Contributions:

1. Liu, Z.; Wu, Z.; Zheng, Z.; Yu, X. A Molecular Dynamics Approach to Identify the Marine Traffic Complexity in a Waterway. *J. Mar. Sci. Eng.* **2022**, *10*, 1678. https://doi.org/10.3390/jmse10111678.
2. Jankowski, D.; Möller, J.; Wiards, H.; Hahn, A. Decentralized Documentation of Maritime Traffic Incidents to Support Conflict Resolution. *J. Mar. Sci. Eng.* **2022**, *10*, 2011. https://doi.org/10.3390/jmse10122011.
3. Ma, X.; Shi, G.; Li, W.; Shi, J. Identifying the Most Probable Human Errors Influencing Maritime Safety. *J. Mar. Sci. Eng.* **2023**, *11*, 14. https://doi.org/10.3390/jmse11010014.
4. Šakan, D.; Žuškin, S.; Rudan, I.; Brčić, D. Container Ship Fleet Route Evaluation and Similarity Measurement between Two Shipping Line Ports. *J. Mar. Sci. Eng.* **2023**, *11*, 400. https://doi.org/10.3390/jmse11020400.
5. Pušić, D.; Lušić, Z. Multi-Criteria Decision Analysis for Nautical Anchorage Selection. *J. Mar. Sci. Eng.* **2023**, *11*, 728. https://doi.org/10.3390/jmse11040728.
6. Sun, J.; Wang, H.; Wang, M. Risk Assessment of Bauxite Maritime Logistics Based on Improved FMECA and Fuzzy Bayesian Network. *J. Mar. Sci. Eng.* **2023**, *11*, 755. https://doi.org/10.3390/jmse11040755.

7.   Spravil, J.; Hemminghaus, C.; von Rechenberg, M.; Padilla, E.; Bauer, J. Detecting Maritime GPS Spoofing Attacks Based on NMEA Sentence Integrity Monitoring. *J. Mar. Sci. Eng.* **2023**, *11*, 928. https://doi.org/10.3390/jmse11050928.

8.   Lee, C.; Lee, S. Evaluating the Vulnerability of YOLOv5 to Adversarial Attacks for Enhanced Cybersecurity in MASS. *J. Mar. Sci. Eng.* **2023**, *11*, 947. https://doi.org/10.3390/jmse11050947

9.   Liu, J.; Zhu, H.; Yang, C.; Chai, T. A Network Model for Identifying Key Causal Factors of Ship Collision. *J. Mar. Sci. Eng.* **2023**, *11*, 982. https://doi.org/10.3390/jmse11050982.

10.   Stach, T.; Kinkel, Y.; Constapel, M.; Burmeister, H.-C. Maritime Anomaly Detection for Vessel Traffic Services: A Survey. *J. Mar. Sci. Eng.* **2023**, *11*, 1174. https://doi.org/10.3390/jmse11061174.

11.   Emerick de Magalhães, M.; Barbosa, C.E.; Cordeiro, K.d.F.; Isidorio, D.K.M.; Souza, J.M.d. Improving Maritime Domain Awareness in Brazil through Computer Vision Technology. *J. Mar. Sci. Eng.* **2023**, *11*, 1272. https://doi.org/10.3390/jmse11071272.

12.   Hu, X.; Xia, H.; Xuan, S.; Hu, S. Exploring the Pirate Attack Process Risk along the Maritime Silk Road via Dynamic Bayesian Network Analysis. *J. Mar. Sci. Eng.* **2023**, *11*, 1430. https://doi.org/10.3390/jmse11071430.

13.   Liu, Z.; Zhou, D.; Zheng, Z.; Wu, Z.; Gang, L. An Analytic Model for Identifying Real-Time Anchorage Collision Risk Based on AIS Data. *J. Mar. Sci. Eng.* **2023**, *11*, 1553. https://doi.org/10.3390/jmse11081553.

14.   Brcko, T.; Luin, B. A Decision Support System Using Fuzzy Logic for Collision Avoidance in Multi-Vessel Situations at Sea. *J. Mar. Sci. Eng.* **2023**, *11*, 1819. https://doi.org/10.3390/jmse11091819.

15.   Olesen, K.V.; Boubekki, A.; Kampffmeyer, M.C.; Jenssen, R.; Christensen, A.N.; Hørlück, S.; Clemmensen, L.H. A Contextually Supported Abnormality Detector for Maritime Trajectories. *J. Mar. Sci. Eng.* **2023**, *11*, 2085. https://doi.org/10.3390/jmse11112085.

16.   Androjna, A.; Pavić, I.; Gucma, L.; Vidmar, P.; Perković, M. AIS Data Manipulation in the Illicit Global Oil Trade. *J. Mar. Sci. Eng.* **2024**, *12*, 6. https://doi.org/10.3390/jmse12010006.

References

1.   Goudossis-Goudosis, A.-A.; Katsikas, S.K. Towards a secure automatic identification system (AIS). *J. Mar. Sci. Technol.* **2018**, *24*, 410–423. [CrossRef]

2.   Pallotta, G.; Vespe, M.; Bryan, K. Vessel Pattern Knowledge Discovery from AIS Data: A Framework for Anomaly Detection and Route Prediction. *Entropy* **2013**, *15*, 2218–2245. [CrossRef]

3.   Natale, F.; Gibin, M.; Alessandrini, A.; Vespe, M.; Paulrud, A. Mapping fishing effort through AIS data. *PLoS ONE* **2015**, *10*, e0130746. [CrossRef] [PubMed]

4.   Baldauf, M.; Kitada, M.; Froholdt, L.L. The Role of VTS Operators in New Maritime Safety Situations: When Conventional and Autonomous Ships Meet. *Appl. Emerg. Technol.* **2023**, *115*, 413–423.

5.   Fenton, A.J. Preventing Catastrophic Cyber–Physical Attacks on the Global Maritime Transportation System: A Case Study of Hybrid Maritime Security in the Straits of Malacca and Singapore. *J. Mar. Sci. Eng.* **2024**, *12*, 510. [CrossRef]

6.   Jin, M.; Shi, W.; Lin, K.-C.; Li, K.X. Marine piracy prediction and prevention: Policy implications. *Mar Policy* **2019**, *108*, 103528. [CrossRef]

7.   Pedrozo, R.P. Protecting the Free Flow of Commerce from Houthi Attacks off the Arabian Peninsula. *Int. Law Stud.* **2024**, *103*, 2.

8.   Batavier Gregory The Red Sea Crisis: Tracking the Volatile Security Situation. Available online: https://spire.com/blog/maritime/the-red-sea-crisis-tracking-the-volatile-security-situation/ (accessed on 27 May 2024).

9.   Lee, E.; Mokashi, A.J.; Moon, S.Y.; Kim, G. The Maturity of Automatic Identification Systems (AIS) and Its Implications for Innovation. *J. Mar. Sci. Eng.* **2019**, *7*, 287. [CrossRef]

10.   Perković, M.; Batista, M.; Luin, B. Ship Handling Challenges When Vessels Are Outgrowing Ports. In Proceedings of the Port Management & Navigation Seminar (Trelleborg), Dubai, United Arab Emirates, 12–13 December 2023; p. 64. Available online: https://www.researchgate.net/publication/376456166_Ship_Handling_Challenges_When_Vessels_are_Outgrowing_Ports (accessed on 27 May 2024).

11.   Perković, M.; Gucma, L.; Bilewski, M.; Muczynski, B.; Dimc, F.; Luin, B.; Vidmar, P.; Lorenčič, V.; Batista, M. Laser-Based Aid Systems for Berthing and Docking. *J. Mar. Sci. Eng.* **2020**, *8*, 346. [CrossRef]

12.   Androjna, A.; Perković, M. GNSS Vulnerabilities vs. Cyber Challenges in Maritime Navigation. In CHENS 2024, Brdo pri Kranju-Slovenija; Chiefs of European Navies; 2024; p. 17. Available online: https://www.researchgate.net/publication/380722521_GNSS_Vulnerabilities_vs_Cyber_Challenges_in_Maritime_Navigation (accessed on 27 May 2024).

13.   Liang, M.; Su, J.; Liu, R.W.; Lam, J.S.L. AISClean: AIS data-driven vessel trajectory reconstruction under uncertain conditions. *Ocean. Eng.* **2024**, *306*, 117987. [CrossRef]

14.   Zhou, Y.; Dong, Z.; Bao, X. A Ship Trajectory Prediction Method Based on an Optuna–BILSTM Model. *Appl. Sci.* **2024**, *14*, 3719. [CrossRef]

15. Li, Y.; Song, G.; Yip, T.-L.; Yeo, G.-T. Fuzzy Logic-Based Decision-Making Method for Ultra-Large Ship Berthing Using Pilotage Data. *J. Mar. Sci. Eng.* **2024**, *12*, 717. [CrossRef]
16. Gucma, L.; Naus, K.; Perkovič, M.; Specht, C. Applied Maritime Engineering and Transportation Problems 2022. *Appl. Sci.* **2024**, *14*, 3913. [CrossRef]

Journal of
*Marine Science
and Engineering*

A Molecular Dynamics Approach to Identify the Marine Traffic Complexity in a Waterway

Zihao Liu *, Zhaolin Wu, Zhongyi Zheng and Xianda Yu

Navigation College, Dalian Maritime University, Dalian 116026, China
* Correspondence: zihaoliu@dlmu.edu.cn

Abstract: With the rapid development of the shipping industry in recent years, the increasing volume of ship traffic makes marine traffic much busier and more crowded, especially in the waterway off the coast. This leads to the increment of the complexity level of marine traffic and poses more threats to marine traffic safety. In order to study marine traffic safety under the conditions of increasing complexity, this article proposed a marine traffic complexity model based on the method in molecular dynamics. The model converted ship traffic to a particle system and identified the traffic complexity by analyzing the radial distribution of dynamic and spatial parameters of ships in a Euclid plane. The effectiveness of the proposed model had been validated by the case studies in the waters of Bohai Strait with real AIS (Automatic Identification System) data and simulated data. The results show that the proposed model can evaluate the marine traffic complexity more sufficiently and accurately. The proposed model is helpful for marine surveillance operators to monitor and organize marine traffic under complex situations so as to improve marine traffic safety.

Keywords: marine traffic; traffic complexity; radial distribution function; marine traffic safety; marine surveillance

Citation: Liu, Z.; Wu, Z.; Zheng, Z.; Yu, X. A Molecular Dynamics Approach to Identify the Marine Traffic Complexity in a Waterway. *J. Mar. Sci. Eng.* **2022**, *10*, 1678. https://doi.org/10.3390/jmse10111678

Academic Editors: Marko Perkovic, Lucjan Gucma, Sebastian Feuerstack and Apostolos Papanikolaou

Received: 25 September 2022
Accepted: 4 November 2022
Published: 7 November 2022

Publisher's Note: MDPI stays neutral with regard to jurisdictional claims in published maps and institutional affiliations.

1. Introduction

With the rapid development of the shipping industry, the volume of ship traffic has increased significantly in recent years [1]. The increasing volume of ship traffic makes marine traffic much busier and more crowded, especially in the waterway off the coast, where ship activity is more frequent. Although the increasing ship traffic volume can contribute to the development of the economy, it will also pose more threats to marine traffic safety. On the one hand, the increasing ship traffic volume will lead to an increase in traffic density in the waterway, making the ship traffic more crowded and thus increasing the possibility of collision accidents. On the other hand, the growth of ship traffic also leads to the complexity of ship traffic. The complex ship traffic will make it more difficult for ships to avoid collision [2]. It will also make it more difficult for marine surveillance operators to monitor and organize the ship traffic in the waterway. In dealing with the complex traffic situation, marine surveillance operators may face the cognization pressure only by their subjective judgment. The efficiency of marine traffic supervision will be damaged and further affect marine traffic safety. Therefore, how to identify the complexity of marine traffic sufficiently and accurately under the increasingly complex traffic situation is helpful to alleviate the risk of collision between ships and improve the efficiency and effects of marine surveillance operators in monitoring and organizing the ship traffic in the waterway.

The rapid growth of maritime traffic in the past decade has led to an increase in marine traffic complexity, which also facilitates the development of navigational equipment. Automatic Identification System (AIS) is one of the advanced navigational equipment; it is an advanced telecommunication and information system, which can broad-cast the information of a ship to other surrounding ships and shore stations by VHF [3]. AIS data contains plenty of ship information related to real-time sailing. With the information, a lot of information related to marine traffic situations can be extracted and analyzed. In recent years, in order to study marine traffic safety under the conditions of increasing complexity, scholars in the marine traffic field have proposed a lot of new models or methods with plenty of data. Silveira et al. [4] utilize AIS data to analyze the marine traffic pattern in a dense water area and determine the collision candidate ships by calculating the future distance between ships so as to evaluate the ship collision risk. Wu et al. [5] depict the global ship density and traffic density map with a bulk of AIS data in different resolutions, which can represent the marine traffic safety level to some extent. Yu et al. [6] propose a novel method to identify the near-miss collision risk when multiple ships encounter it. The method is established based on ship motion behavior and can evaluate the multi-ship near-miss collision risk from temporal, spatial, and geographical perspectives. The method can also improve the level of risk assessment by identifying the different levels of near-miss collision risk of different ships and also provide some reference and help to put forward the measures to reduce the risk. Zhang et al. [7] propose a new method to identify collision risk based on a convolutional neural network. The convolutional neural network can analyze and recognize the image built by AIS data to rapidly identify the collision risk under encounter. Bakdi et al. [8] propose an adaptive ship domain to identify the collision risk and grounding risk under complex traffic situation. The model can also be used to identify the marine traffic risk considering maneuvering limitations. The method can improve the accuracy of real-time marine traffic risk identification under large-scale monitoring. Wen et al. [9] propose a marine traffic complexity model according to air traffic control. The model can assess the complexity level in the water area by evaluating the density and the collision risk of ship traffic, respectively. The model is established based on ship relative motion parameters and can assess the overall complexity level of ship traffic by interpolation technique. Rong et al. [10] conduct spatial correlation analysis of ship clusters based on the characteristics of marine traffic. The authors use AIS data to describe the collision hotspots along the Portuguese coast and analyze the correlation between the collision hotspot and the characteristics of marine traffic. The method can be helpful in improving marine traffic safety and reducing the collision risk in the water area. Zhang et al. [11] propose a two-stage black spot identification model, which can detect more risk in the water area. The model has a higher detection rate for marine accidents. The authors use the model to depict the black spot in the Jiangsu section of the Yangtze River by historical data. The model is helpful in optimizing the search and rescue resources and improving the safety management level. Liu et al. [12] propose a novel framework for real-time regional collision risk prediction based on a recurrent neural network approach. After identifying the regional collision risk, the optimized RNN method is used to predict the regional collision risk of a specific water area in a short time. The model is useful for collision risk prediction for the water area under complex traffic situations. Liu et al. [13] propose an improved danger sector model to identify the collision risk of encountering ships. The model considers the course alteration maneuver by taking ship maneuverability limitation into consideration. The model is helpful for calculating the collision risk between ships in complex traffic situations. Zhen et al. [14] propose a novel regional collision risk assessment method, which considers aggregation density under multi-ship encounter situations. The model can more intuitively and effectively quantify the temporal and spatial distribution of regional collision risk under complex traffic situations and improve the efficiency of traffic management. Yu et al. [15] propose an integrated multi-criteria framework for assessing the ship collision risk under different scenarios dynamically. The framework can identify collision parameters and candidates according to ship offices' experience, which is useful for analyzing the ship traffic under complex situations. Merrick et al. [16] analyze the traffic density of the proposed ferry service expansion in San Francisco Bay. They establish a simulation model to estimate the number of vessel interactions and the increment caused by expansion plans. Utilizing the model, the geographic profile of vessel interaction frequency is presented. Altan et al. [17] analyze the marine traffic in the Strait of Istanbul. The ship attributes are tracked by a grid-based analysis method. Through the analysis, they summarize some conclusions about vessel distribution, draught, speed exceed, and the influence on traffic patterns. Ramin et al. [18] use AIS data to research the complex marine traffic in Port Klang and the Straits of Malacca. The method applied was time-series models and associative models. Utilizing the methods, the density of ships in Port Klang and the Straits of Malacca can be predicted and forecasted. Kang et al. [19] use AIS data to estimate the ship traffic fundamental diagrams in the Strait of Singapore, which plays a crucial role in international freight transportation. The ship traffic fundamental diagrams investigate the ship traffic's speed–density relationship and can be used to estimate the theoretical strait capacity. van Westrenen et al. [20] utilize AIS data to analyze the traffic on the North Sea. The near misses are detected by using ship-state information. Through the research, they conclude that the near misses are not spread evenly over the sea but are concentrated in a number of specific locations. Du et al. [21] propose a new method to improve near-miss detection by analyzing behavior characteristics during the encounter process in the Northern Baltic Sea. The ship attributes, perceived risk, traffic complexity, and traffic rule are included in evaluating ship behaviors. Moreover, the risk levels of detected near misses are quantified. Endrina et al. [22] conduct a risk analysis for RoPax vessels in the Strait of Gibraltar. The first two steps of the IMO Formal Safety Assessment are used to present the results with the accidents statistics covering 11 years. In addition, a high-level model risk for collisions was built through an Event Tree, and the individual and social risks were calculated.

It can be found that most of the recent studies on complex marine traffic are based on a large number of ship data. On the one hand, it benefits from the rapid development of computer and information technology, which enables these data to be stored, analyzed, and calculated. On the other hand, it benefits from the availability of a large number of AIS data. The utilization of these large amounts of real AIS data can improve the accuracy of the model, improve the accuracy of the results, and make the results obtained by the model more realistic. The recent studies also make it possible to study complex marine traffic from temporal and spatial perspectives. Previously, we proposed a ship density model based on the radial distribution function [23]. The model can calculate the ship density and traffic density in the specified water area and can evaluate the complexity of marine traffic to some extent. The model can help marine surveillance better understand marine traffic in complex situations and improve their monitoring efficiency. However, the model only considers the ship positions and the distance between them and can only quantify the complexity of marine traffic from a position perspective, which makes the accuracy of the complexity results limited in some cases. Therefore, in this article, a new marine traffic complexity model is supposed to be proposed in order to evaluate the complexity level of marine traffic in a waterway. This model not only considers the spatial distribution characteristic of ship traffic but also incorporates the ship motion parameters in it, which can identify the complexity level more sufficiently and accurately in a waterway and assist marine surveillance operators to better acknowledge, monitor, and organize the marine traffic under complex situations. The reminders of the article are arranged as follows. In Section 2, the marine traffic complexity model was established based on the radial distribution function, which considers the complexity of ship motion and ship position, respectively. In Section 3, the proposed model was validated by some experimental case studies in Bohai Strait waters with simulated data and real AIS data. In Section 4, the effectiveness of the proposed model was discussed, and the advantage of the proposed model was analyzed compared with the previous model. Moreover, the limitations of the proposed model were presented. In Section 5, the conclusion was drawn, and some future studies about this model were presented.

2. The Marine Traffic Complexity Model

2.1. Radial Distribution Function

In statistical mechanics and molecular dynamics, radial distribution function (RDF) is used to examine the interaction and bonding state between particles. RDF is defined to quantify the variation of density as a function of distance from a specified particle in the system [24]. For example, in a molecule system, the RDF of the specified atom can reveal the densities of the surrounding atoms at different distances in a three-dimension space. As shown in Equation (1), $g_i(r)$, the radial distribution of the specified atom i, can be expressed as the number of the surrounding atoms in a spherical shell with the thickness of Δr at a distance of r [25].

$$g_i(r) = \frac{N_i(r, \Delta r)}{\rho V(r, \Delta r)} \tag{1}$$

where $N(r, \Delta r)$ refers to the number of atoms in the spherical shell with the thickness of Δr at a distance of r from atom i. Δr is usually obtained by dividing the radius of the spherical space R_c by the set number of bins N_g, ρ refers to the density of atoms in the molecule; it can be obtained by dividing the number of atoms in the molecule N by the volume of the molecule space V. $V(r, \Delta r) = 4\pi r^2 \Delta r$ refers to the volume of the spherical shell with the thickness of Δr at a distance of r from atom i. An example radial distribution of a particle system is shown in Figure 1.

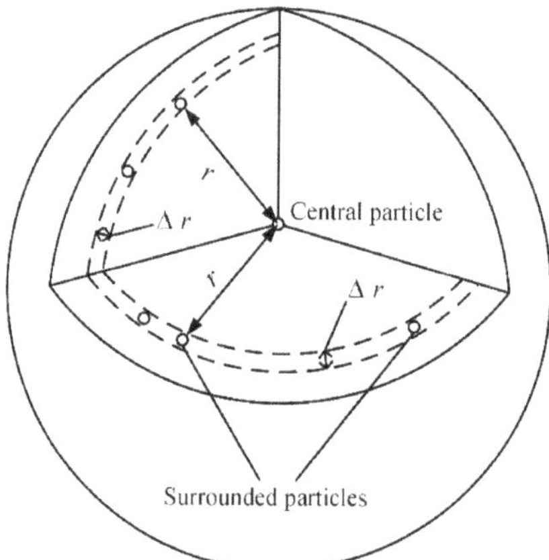

Figure 1. An example radial distribution of a particle system.

Apart from expressing the radial distribution surrounding a specified particle, RDF can also quantify the distribution probability of all particles in a particle system by accumulating all densities of all surrounding particles, as shown in Equation (2).

$$g(r) = \sum_{i}^{N} \frac{N_i(r, \Delta r)}{N\rho V(r, \Delta r)} \tag{2}$$

where $g(r)$ refers to the radial distribution of all particles at a distance of r, N refers to the number of atoms in the molecule.

The RDF is capable of describing the distribution probabilities of all particles inside it at different distances. In the graph of RDF, the height of peaks can express the distribution probability positively, namely the local density at different distances. In addition, the integral of the RDF curve can represent the overall distribution probability of the particles in a system (L), as shown in Equation (3), which is called the coordination number in molecular dynamics. It refers to the number of particles directly linked to the specified particle in a particle system [26]. In other words, it can determine the overall density of the particle system. Take the metal atoms as an example; they are tightly stacked and result in a relatively higher or even highest coordination number [27].

$$L = \int_0^r g(r) = \int_0^r \sum_i^N \frac{N_i(r, \Delta r)}{N\rho V(r, \Delta r)} dr \tag{3}$$

In addition, RDF can also be used to assess the complexity of particle distribution [28]. In the RDF graph, a sharp peak means the particles distribute concentratedly at a specified distance, and a smooth curve means the particles may distribute at any distance. Therefore, the complexity from the perspective of the position of particles can be assessed by observing the shape of the RDF curve, more specifically, by quantifying the discreteness of the RDF curve. Generally, the particles in gas distribute most disorderedly because of its characteristics [27]; the RDF curve of gas-particle systems is always smooth, even close to a straight line. However, the ideal crystal particle system exhibits many sharp peaks, which indicates that the particles inside distribute at some fixed distances [29].

Utilizing the above-mentioned characteristics of RDF, the marine traffic complexity model in this article was established in the following sections.

2.2. The Waterway Traffic Complexity Model

This article aimed to propose a model to identify the marine traffic complexity in a waterway by RDF utilizing its characteristics. To achieve this, the ship traffic in a waterway should be considered as a particle system at first, where the ships were converted to ship particles, and the whole ship traffic was converted to a ship traffic particle system. The converted ship particles and ship traffic particle system are shown in Figure 2. In order to reflect the marine traffic complexity in a waterway sufficiently, the model was built in ship motion perspective and ship position perspective, respectively.

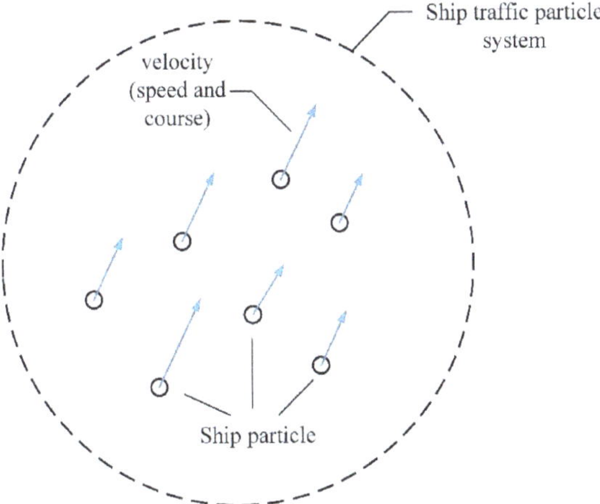

Figure 2. The ship molecules and ship traffic particle system.

2.2.1. Traffic Complexity on Ship Motion

It is known that speed and course are two crucial dynamic parameters for a ship. In an electronic chart or other similar displayed facilities, the speed and course of a ship are shown together by a single line. It is because the line represents a velocity vector that has both size and direction. Considering this characteristic, this article is to build speed and course complexity together with a two-dimensional RDF model in a velocity plane. The speed and course are two crucial ship motion parameters that have a great impact on marine traffic complexity. If ships sailed at different velocities, the complex traffic situation might be formed easier, and the action to be taken to avoid collision may be more difficult. In order to model marine traffic complexity by speed and course, the velocity plane should be built first. As shown in Figure 3, the velocity plane is in a two-dimensional coordinate axis.

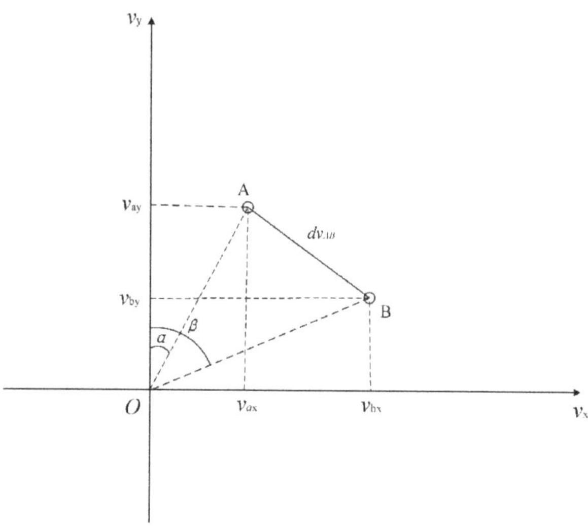

Figure 3. The velocity plane of ship traffic particle system.

It should be noticed that this velocity plane, although exhibited as a cartesian coordinate system, was not used to represent the position of a ship in the waterway but the positions of ships on a phantom plane which is the quantification of velocity in longitude and latitude directions. In fact, if we arranged a ship in this velocity plane with its speed vector, the y-axis and x-axis refer to the projection of ship velocity in a latitude direction and longitude direction, respectively. In Figure 3, (v_{ax}, v_{ay}) and (v_{bx}, v_{by}) refer to the coordinates of Ship A and Ship B in the velocity plane, α and β are the courses of Ship A and Ship B.

In order to illustrate the velocity plane better, two ships are simulated in the plane, which is Ship A and Ship B, respectively, with velocity $\vec{a}$ and $\vec{b}$. Points A and B are the centers of the two ships, which are considered as the positions of the two ships in the velocity plane but not the positions in reality. The position of the ship in the velocity plane indicates its velocity in a cartesian coordinate form. Through the coordinates, we can obtain the speed and course of a ship. Angle α and β are the courses of two ships, and they are represented by arrow lines with direction. In addition, the length of the arrow line represents the magnitude of speed, which are a and b for Ship A and Ship B, respectively.

According to the law of sines and cosines, the velocity $\vec{a}$ and $\vec{b}$ can be decomposed to (v_{ax}, v_{ay}) and (v_{bx}, v_{by}) by using the speed and course parameters of Ship A and Ship B, as expressed by Equations (4)–(7).

$$v_{ay} = \left|\vec{a}\right| \times \cos \alpha \tag{4}$$

$$v_{ax} = \left|\vec{a}\right| \times \sin \alpha \tag{5}$$

$$v_{by} = \left|\vec{b}\right| \times \cos \beta \tag{6}$$

$$v_{bx} = \left|\vec{b}\right| \times \sin \beta \tag{7}$$

where v_{ax} and v_{ay} refer to the speed of Ship A in longitude and latitude direction, v_{bx} and v_{by} refer to the speed of Ship B in longitude and latitude directions, respectively.

In Figure 3, apart from the parameters mentioned above, there is another crucial parameter in the velocity plane, which is the connection line AB. The magnitude of line AB dv_{AB} refers to the distance between two ships in the velocity plane but not the distance in reality. In other words, it can be considered as the quantification of the difference between the velocities of Ship A and Ship B. In this article, the distance dv_{AB} is named velocity distance, and it can be calculated by Equation (8).

$$dv_{AB} = \sqrt{\left(v_{ay} - v_{by}\right)^2 + \left(v_{ax} - v_{bx}\right)^2} \tag{8}$$

As distance can represent the difference between objects on spatial distribution, the distance dv_{AB} in velocity plane calculated by Equation (8) can help us distinguish the difference between two ships on speed and course and their ship motion in the velocity plane. Through the velocity distance dv_{AB}, RDF can be used to quantify the divergence of ships on their dynamics motion parameters.

RDF can represent the spatial distribution probability around the given central particle. By accumulating all radial distribution of the particles in the system, the overall spatial distribution probability, namely the occurrence probabilities of the particles at different locations, can be obtained. Utilizing this characteristic, the radial distribution function can be used to calculate the radial distribution of all ship molecules in the velocity plane so as to the occurrence possibilities of ship molecules on different magnitudes of velocity distance. Equation (6) was used to express the radial distribution of all ship molecules in the ship traffic particle system in the velocity plane. The ship traffic system can be arranged in a waterway or a specified water area.

$$g(r) = \sum_{i=1}^{N} \frac{N_i(dv, \Delta dv)}{\lambda N \rho S(dv, \Delta dv)} \tag{9}$$

where $N_i(dv, \Delta dv)$ refers to the number of ship particles in the circular ring with the width of Δdv at a distance of r from ship particle i, Δdv can be obtained by dividing the radius of the circular space R_{dv} by the set number of bins N_g, ρ refers to the density of the ship traffic particle system in the velocity plane, which can be obtained by dividing the number of ship particles in the ship traffic particle system N by the area of the particle space S, $S(dv, \Delta dv)$ refers to the area of a circular ring with the width of Δdv at the distance of dv in the velocity plane, and λ is an adjustment coefficient.

$N_i(dv, \Delta dv)$ can be obtained by the Algorithm 1.

Algorithm 1: Calculating $N_i(dv, \Delta dv)$

Input: $v_{ix}, v_{jx}, v_{iy}, v_{jy}, N_g, R_{dv}$
Output: $N_i(dv, \Delta dv)$
1: $\Delta dv = R_{dv}/N_g$
2: FOR each ship particle j DO
3:    $RV_{ij} = \sqrt{\left(v_{jx} - v_{ix}\right)^2 + \left(v_{jy} - v_{iy}\right)^2}$
4:    IF $RV_{ij} < R_{dv}$ THEN
5:      $Index = Ceil\left(RV_{ij}/\Delta dv\right)$
6:      $N_i(Index, \Delta dv) = N_i(Index, \Delta dv) + 1$
7:    ELSE
8:      CONTINUE next ship particle
9: $N_i(dv, \Delta dv) = \sum N_i(Index, \Delta dv)$

Δdv can be obtained by dividing the radius of the studied area R_{dv} by the number of bins N_g, indicating that the R_{dv} will be separated into N_g bins. For each ship particle, the velocity distance between it and another ship particle RV_{ij} can be calculated by the coordinates of the two ships in the velocity plane, and the ship particles within the scope of the area should be reserved. For each ship particle reserved, the bin index can be obtained by dividing exactly the velocity distance RV_{ij} by Δdv, and then the ship particle can be accumulated in the corresponding bin. After dealing with all ship particles, $N_i(dv, \Delta dv)$ can be obtained by accumulating the ship particles in all bins.

According to Equation (9), the radial distribution graph of the ship traffic particle system can be depicted. An example radial distribution graph of ship velocity is shown in Figure 4. Similar to other radial distribution graphs, it can reveal the distribution probabilities of ship particles in the velocity space, which can indicate the difference between ships on motion patterns. In order to quantify the overall difference between ships on motion pattern, the characteristic of RDF mentioned in Section 2.1 was used, which can be used to assess the complexity of particle distribution. For a radial distribution graph, the complexity of particle distribution can be assessed by observing the shape of the RDF curve, more specifically, by quantifying the discreteness of the RDF curve. However, it should be noticed that the complexity of particle distribution in the velocity plane does not indicate the complexity of ship position distribution but indicate the complexity of ship velocity distribution. If a radial distribution curve is smooth, it means the difference between ship motions may distribute at any velocity distance. In such a situation, the traffic complexity on ship motion is high. If the curve exhibited plenty of sharp peaks, it means the difference between ship motions appears at several fixed velocity distances. Under this situation, the traffic complexity on ship motion is low. Therefore, in order to quantify the traffic complexity on ship motion by RDF curve, a phantom straight line was assumed in the radial distribution graph, as shown in Figure 4, which represents the most complex situation, because the difference between velocities may be any values. It is a special radial distribution graph, which is parallel to the x-axis, with the expression $g(r) = p$. We can use it to represent the complexity level of any other radial distribution graph. The phantom straight line can be calculated by averaging all values of velocity distance. Then, the variation between the RDF curve with this phantom line was calculated so as to represent the traffic complexity on ship motion. The expression is shown in Equation (10).

$$VD = \frac{1}{N}\sum\sum\nolimits_{i=1}^{N}\left(\frac{2N_i(dv, \Delta dv)}{\lambda N \rho S(dv, \Delta dv)} - p\right)^2 \tag{10}$$

where VD can express the magnitude of velocity difference, p refers to the value of the phantom line, which is parallel to x axis, and it can be calculated by the mean value of the velocity distance distribution.

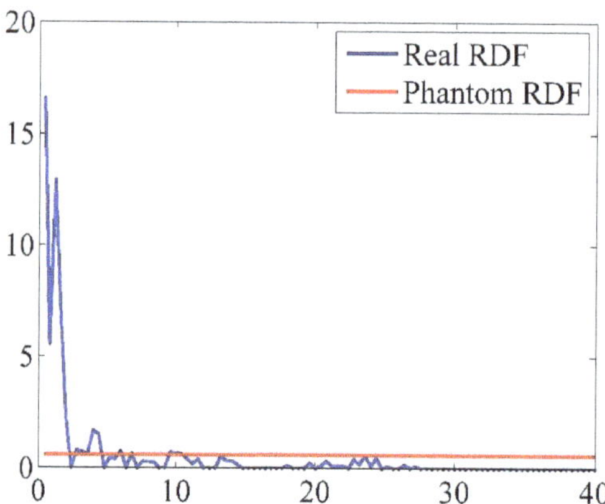

Figure 4. An example radial distribution graph of ship velocity.

After obtaining VD, the negative exponential function applied in [30] was used to map the relationship between it and the traffic complexity on ship motion C_{sm}. The negative exponential function was expressed as Equation (11).

$$C_{sm} = -a_{sm}e^{b_{sm} \times VD} \tag{11}$$

In order to identify the traffic complexity on ship motion by Equation (11), the parameters a_{sm} and b_{sm} should be determined at first. For determining the two parameters, two extreme scenarios were assumed. One of the extreme scenarios is that the magnitude of velocity difference is extremely low and corresponds to an extremely small value for the traffic complexity on ship motion between 0 and 1, which is assumed as 0.05 in this article. Under such a situation, ships in the waterway sail with the same course and speed. The value of VD of such a situation can be calculated when the waterway is up to its traffic capacity and all ships sailed with the same speed and course, which can be obtained from historical statistical data. Another extreme scenario is that the magnitude of the velocity difference is extremely high, and the RDF curve coincides exactly with the phantom straight line. Under this situation, the value of VD is 0, and the corresponding C_{sm} is 1. Utilizing the variables in the above-mentioned two scenarios, the parameters a_{sm} and b_{sm} can be calculated, and the complexity of ship motion can be computed. The complexity of ship motion is a value between 0 and 1 and can identify the complexity level of ship speed and course and indicate the disorder of ship dynamic attributes.

2.2.2. The Complexity on Ship Position

Apart from considering traffic complexity on ship motion, the traffic complexity on ship position should also be considered when identifying marine traffic complexity in a waterway. As modeled in [23], RDF can be used to assess the ship traffic density.

As mentioned in Section 2.1, RDF is defined to quantify the variation of density as a function of distance from a specified particle in the system. By RDF, we can assess the local density at a different distance from the specified particle, namely the distribution probabilities of particles at different distances. In addition, RDF can also express the distribution probability of all particles in a particle system by accumulating all densities of all surrounding particles. In order to quantify the particle density, either local or global, the RDF can be integrated with a certain distance because the integral of RDF can represent the overall distribution probability of the particles in a system, namely the coordination number in molecular dynamics. Therefore, in order to assess the ship traffic density, ship particles should be arranged in a real water plane. For a specified ship particle, we can obtain the distribution probabilities of its surrounding ship particles by RDF and assess the ship density by integrating the RDF curve as expressed by Equations (12) and (13).

$$g_{sp}(d) = \int_0^R \sum_{i=1}^N \frac{N_i(d, \Delta d)}{\lambda N \rho S(d, \Delta d)} \tag{12}$$

$$SD = \int_0^R \sum_{i=1}^N \frac{N_i(d, \Delta d)}{\lambda N \rho S(d, \Delta d)} \tag{13}$$

The value of SD can represent the ship traffic density in a waterway.
The Algorithm 2 gives the computing process of $N_i(d, \Delta d)$.

Algorithm 2 Calculating $N_i(d, \Delta d)$

Input: $long_i$, $long_j$, lat_i, lat_j, N_g, R_d
Output: $N_i(d, \Delta d)$
1: $\Delta d = R_d / N_g$
2: FOR each ship particle j DO
3: $\quad R_{ij} = \sqrt{\left(long_j - long_i \right)^2 + \left(lat_j - lat_i \right)^2}$
4: $\quad$ IF $R_{ij} < R_d$ THEN
5: $\qquad Index = Ceil\left(R_{ij} / \Delta d \right)$
6: $\qquad N_i(Index, \Delta d) = N_i(Index, \Delta d) + 1$
7: $\quad$ ELSE
8: $\qquad$ CONTINUE next ship particle
9: $N_i(d, \Delta d) = \sum N_i(Index, \Delta d)$

The principle of Algorithm 2 is the same as Algorithm 1 in calculating $N_i(dv, \Delta dv)$. The only difference is that the variable in Algorithm 2 corresponds to the real water plane, which is a Euclid plane based on the ship's real position.

The negative exponential function applied in [30] was used to convert the value of SD to the traffic complexity on ship position. The expression of the negative exponential function is shown as Equation (14).

$$C_{sp} = -a_{sp}e^{b_{sp} \times SD} \tag{14}$$

In order to map the relationship between the value of SD and the traffic complexity on ship position by Equation (14), the parameters a_{sp} and b_{sp} should be determined at first. For determining the two parameters, two extreme scenarios were assumed. One extreme scenario is that the waterway is full of ships, and the value of SD under such a situation corresponds to a very large traffic complexity value between 0 and 1, which is assumed as 0.95 in this article. This situation can be determined by calculating traffic capacity in the waterway according to [31]. Another extreme scenario is that the waterway is empty, and the value of SD under such a situation corresponds to zero. Utilizing the variables in the above-mentioned two scenarios, the parameters a_{sp} and b_{sp} can be calculated, and the complexity on ship position can be computed. The complexity on ship position is a value between 0 and 1 and can identify the complexity level on ship position distribution and can indicate the busyness, congestion, and compactness of ships in the waterway.

After obtaining the complexity on ship motion and ship position, the final marine traffic complexity can be calculated by synthesizing these two complexity indexes. The method and expression for calculating collision risk, which is also an index in evaluating marine traffic, was applied for this synthesizing process [32–34], as expressed in Equation (15).

$$comp = \sqrt{\omega_{sm}C_{sm}^2 + \omega_{sp}C_{sp}^2} \tag{15}$$

where ω_{sm} and ω_{sp} are the weight coefficients for the two complexities. The sum of ω_{sm} and ω_{sp} is 1 and can be preset by the traffic situation and water type.

The overall flow chart of the proposed model of marine traffic complexity is shown in Figure 5. The synthesized marine traffic complexity is also a value between 0 and 1, which can represent the complexity level of the ship traffic and can assist marine surveillance operators to better acknowledge, monitor, and organize the marine traffic under complex situations.

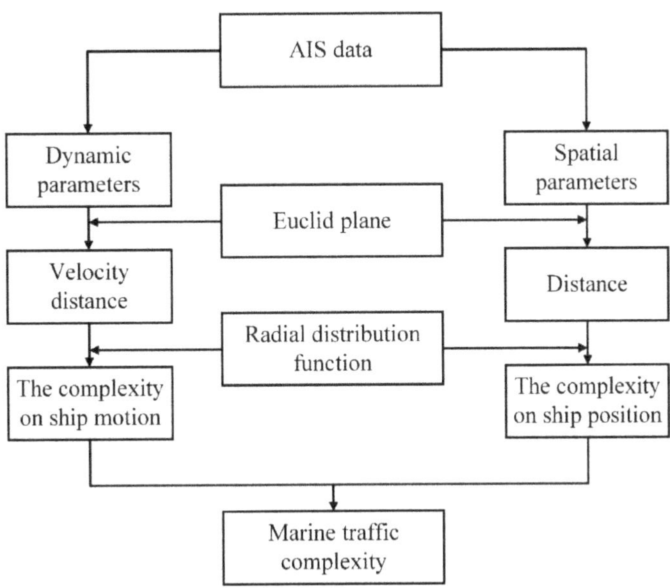

Figure 5. The flow chart of the proposed model.

3. Case Studies

Some experimental case studies were carried out to validate the proposed marine traffic complexity model. AIS data in Bohai Strait was utilized. Bohai Strait is located between the Bohai Sea and the Northern Yellow Sea in China. The traffic conditions for Bohai Strait provide the model validation with prerequisites. This is because the marine traffic in Bohai Strait is busy and crowded [35], and the traffic volume and traffic density are relatively big, which meansship encounters form much more easily and thus increase the complexity level.

In order to use AIS data for experiments, the AIS source data was first decoded and stored in a database by time sequence. After cleaning and sorting the data, the required information is extracted from the database by Structured Query Languages. In addition, to ensure the information extracted from AIS data can represent the instantaneous scenario accurately, the interpolation process is also conducted.

3.1. Case Studies with Simulated Data

Firstly, some simulated data were used to validate the proposed model. In this case study, the data was simulated on the position, motion, and size parameters of ships. The ships were arranged in the water area where the center is 38.6° N, 120.85° E with a radius of 6 nmile, and this is the location where Laotieshan Precautionary Area locates in reality.

Utilizing the simulated data, three scenarios were formed. The ships in each of these three scenarios were sailing at different velocities, namely, different speeds and courses, but their positions were exactly the same. The three scenarios are shown in Figure 6.

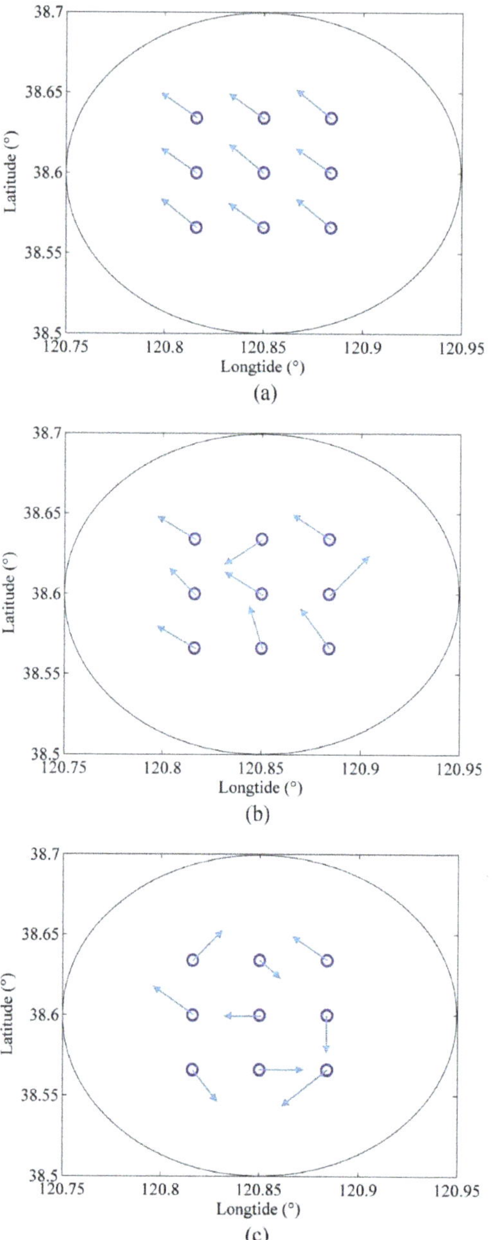

Figure 6. The three scenarios for the first simulated case study (**a**) Scenario 1; (**b**) Scenario 2; (**c**) Scenario 3.

Then, the simulated data of ships was inputted into the proposed model to identify the traffic complexity level of each scenario with related algorithms. The results obtained are shown in Table 1, together with Figure 7.

Table 1. The results of the first simulated case study.

Scenario	Scenario 1	Scenario 2	Scenario 3
Complexity on ship motion	0.1397	0.3037	0.4639
Complexity on ship position	0.2431	0.2431	0.2431
Marine traffic complexity	0.1983	0.2751	0.3703

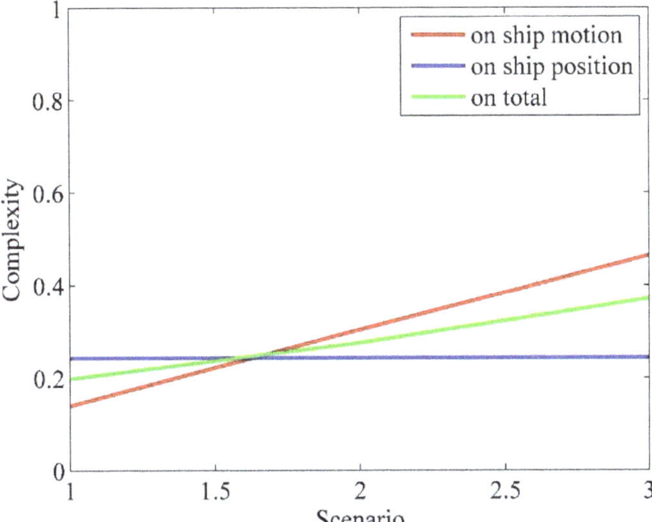

Figure 7. The results of the first simulated case study.

The results show that the complexity on ship position for each scenario is the same, which is 0.2431, but the complexity on ship motion increases gradually from 0.1397 to 0.4639. The increasing complexity on ship motion makes each scenario's final marine traffic complexity different, which also increases gradually, from 0.1983 to 0.3703.

In actuality, it can be observed from Figure 6 that the spatial distribution of ship position is exactly the same for each scenario. However, the velocities of ships are not the same. In the first scenario, the velocities for each ship are close, with speeds close to 10 kn, and the courses are close to 300°. However, in the second and third scenarios, the velocities vary obviously, especially in the third scenario. Therefore, the complexity on ship motion, which considers ship speed and course in this article, of these three scenarios should increase gradually, and the complexity on ship position was supposed to be the same, which is confirmed by the results obtained above.

Furtherly, three other scenarios were formed. The ships in each of these three scenarios were sailing at the same velocity, namely the different speeds and courses, but their positions were different. The three scenarios are shown in Figure 8.

Then, the simulated data of ships was inputted the proposed model to calculate the traffic complexity level of each scenario. The results obtained are shown in Table 2, together with Figure 9.

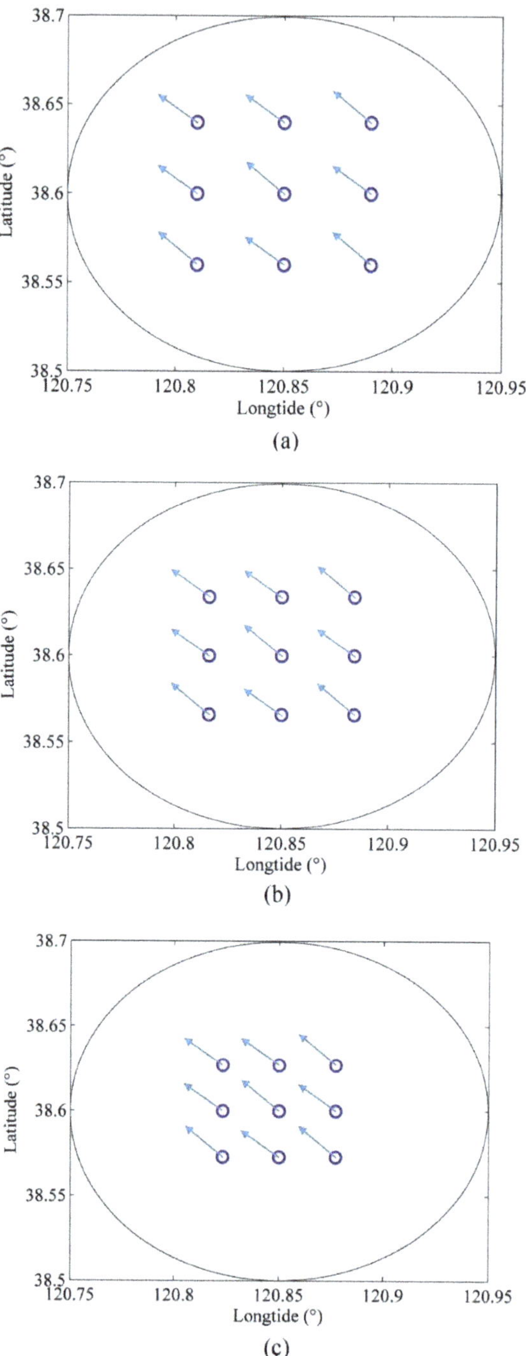

Figure 8. The three scenarios for the second simulated case study (**a**) Scenario 1; (**b**) Scenario 2; (**c**) Scenario 3.

Table 2. The results of the second simulated case study.

Scenario	Scenario 1	Scenario 2	Scenario 3
Complexity on ship motion	0.1397	0.1397	0.1397
Complexity on ship position	0.1496	0.2431	0.5091
Marine traffic complexity	0.1447	0.1983	0.3733

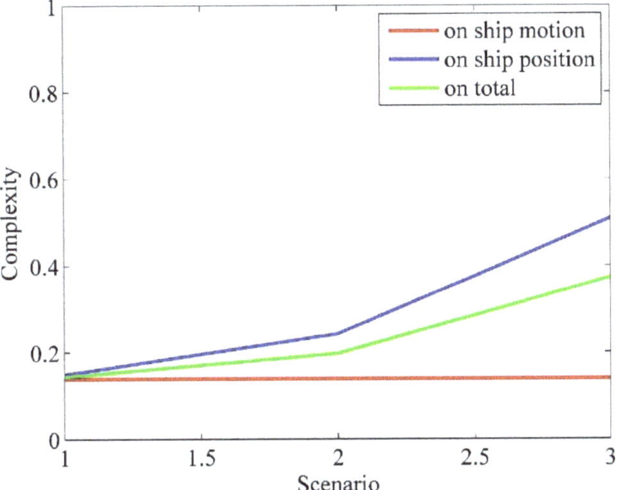

Figure 9. The results of the second simulated case study.

It can be found that the complexity on ship motion for each scenario is the same, which is 0.1397, but the complexity on ship position increases gradually from 0.1496 to 0.5091. The increasing complexity on ship position makes the final marine traffic complexity for each scenario different, which also increases gradually, from 0.1447 to 0.3732.

Actually, it can be observed from Figure 8 that the spatial distribution of ship position is different for each scenario. In the first scenario, the spatial distribution is relatively sparse. In the second and third scenarios, the spatial distribution becomes compact, especially in the third scenario. However, the velocities of ships are exactly the same for each scenario. Therefore, the complexity on ship position, which considers ship position in this article, of these three scenarios should increase gradually, and the complexity on ship motion was supposed to be the same, which is also confirmed by the results obtained above.

3.2. Case Studies with Real AIS Data

In order to further validate the proposed model, real AIS data was used to carry out some other case studies. The used AIS data was from the ships sailed in the Bohai Strait area. The first case study with real AIS data was to validate the proposed model from a temporal perspective. This case study was carried out in the Laotieshan Precautionary Area (Laotieshan PA), located at the northwest entrance and exit of the Laotieshan Traffic Separation Scheme (Laotieshan TSS). The reason why Laotieshan PA was chosen is the obvious difference between the speed and course, as well as position, of the ships that sailed within it. The traffic in Precautionary Area consists of some different traffic flows, as shown in Figure 10, which makes marine traffic relatively complex in it.

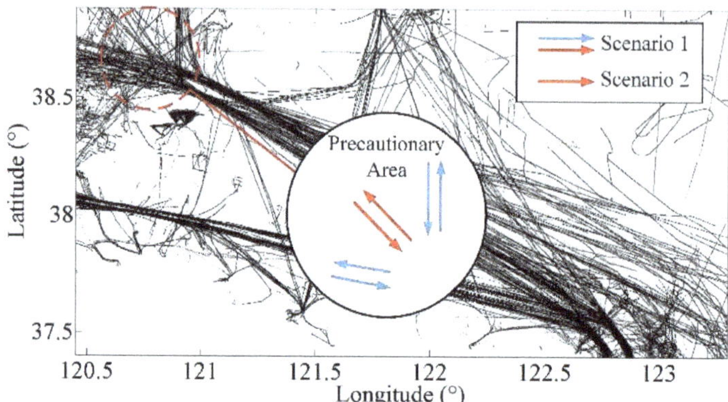

Figure 10. The location of Laotieshan PA and its main traffic flows.

From Figure 10, it can be observed that there are six main traffic flows in Laotieshan PA. Two of them are extended from the direction of Laotieshan TSS. Of the rest four traffic flows, two of them are in the north–south direction, and the other two flows are close to the west–east direction. For this studied area, we divided the traffic into two situations. The first situation is that all traffic flows in PA were considered, which were exhibited as red and blue lines in Figure 10. The second situation is that only the traffic flow which nearly extends the direction of TSS was considered, which is exhibited as red lines in Figure 10.

For the two situations, the complexity on ship motion was calculated based on the proposed model. The calculation was conducted for a one-time node for each hour through a day in sequence. The results are shown in Figure 11.

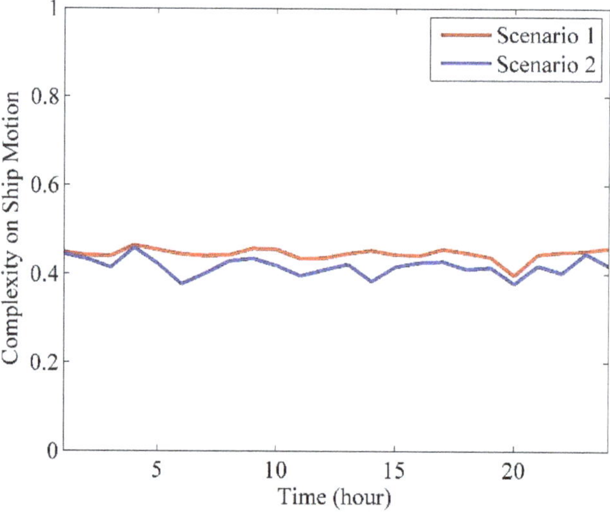

Figure 11. The comparison between the complexity on ship motion for the two situations.

The results show that the complexity on ship motion in the second situation is less than that of the first situation, where the average values are 0.4451 and 0.4169, respectively. The results are consistent with the actual situation. This is because compared with the second situation, there are more different traffic flows considering in the first situation, which makes the speed and course more varied and leads to the increase of complexity level.

Therefore, it is proved that the proposed model can identify the marine traffic complexity on ship motion effectively.

In addition, the complexity on ship position for the studied PA was also calculated based on the proposed model. The calculation was conducted for a one-time node for each hour throughout the day. The results are shown in Table 3.

Table 3. The complexity on ship position for the studied PA.

Time	C_{sp}
0–1	0.1855
1–2	0.0973
2–3	0.0932
3–4	0.0926
4–5	0.0585
5–6	0.0656
6–7	0.0748
7–8	0.0608
8–9	0.1106
9–10	0.1190
10–11	0.1174
11–12	0.1053
12–13	0.1589
13–14	0.1282
14–15	0.2052
15–16	0.1630
16–17	0.0943
17–18	0.1375
18–19	0.1277
19–20	0.0965
20–21	0.1382
21–22	0.1339
22–23	0.1498
23–24	0.1143

In order to validate that the results are effective in identifying the complexity on ship position. A traditional index was used to make a comparison, which is ship density. According to [5], ship density in a region at time t is the number of ships per unit area in this region at this time, which can be calculated by Equation (16).

$$SD = \frac{N}{Time \times S_{Area}} \tag{16}$$

where SD refers to ship density, N refers to the number of ships, *Time* refers to the time interval and S_{Area} refers to the area of the studied water. For the time nodes in Table 3, ship densities (SD) were calculated, and the results are shown in Table 4.

Then, a Pearson correlation analysis was carried out. The analysis result is shown in Table 5.

It can be found that the p-value was less than 0.01 with a correlation coefficient of 0.692, which indicates that the complexity on ship position and ship density are significantly correlated for all time nodes. As ship density is an index that can reflect the busyness and congestion of marine traffic, we have reason to believe the proposed model can identify the complexity related to a ship position distribution effectively.

Furthermore, another case study was carried out to validate the proposed model with real AIS data, which was to validate the proposed model from a spatial perspective. In this case study, ten different water areas of the same size were selected. The locations of these ten water areas are shown in Figure 12.

Table 4. The ship density for the studied PA.

Time	SD
0–1	0.0796
1–2	0.0707
2–3	0.0707
3–4	0.0619
4–5	0.0442
5–6	0.0619
6–7	0.0442
7–8	0.0442
8–9	0.0707
9–10	0.0973
10–11	0.0796
11–12	0.0619
12–13	0.0973
13–14	0.0707
14–15	0.1326
15–16	0.0707
16–17	0.0973
17–18	0.0707
18–19	0.0707
19–20	0.0796
20–21	0.0884
21–22	0.0619
22–23	0.0796
23–24	0.0884

Unit of SD: $pcs/\left(s \times nmile^2\right)$

Table 5. Pearson correlation analysis between C_{sp} and SD for 24 time nodes.

p Value	Correlation Coefficient
<0.01	0.692

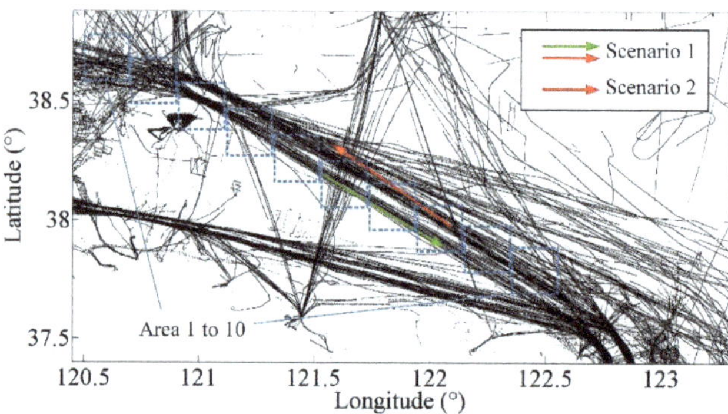

Figure 12. The locations of these ten water areas.

It can be found that the ten water areas are along with the direction of the main traffic flows in Bohai Strait. Firstly, to validate the proposed model can effectively identify the complexity level on ship motion, we also divided the traffic in these areas into two situations. The first situation is that all ships were considered, which is exhibited as the red and green line in Figure 12. The second situation is that only ships sailed to the northwest were considered, exhibited as a red line in Figure 12. For the two situations, the complexity on ship motion was calculated based on the proposed model. The results are shown in Figure 13.

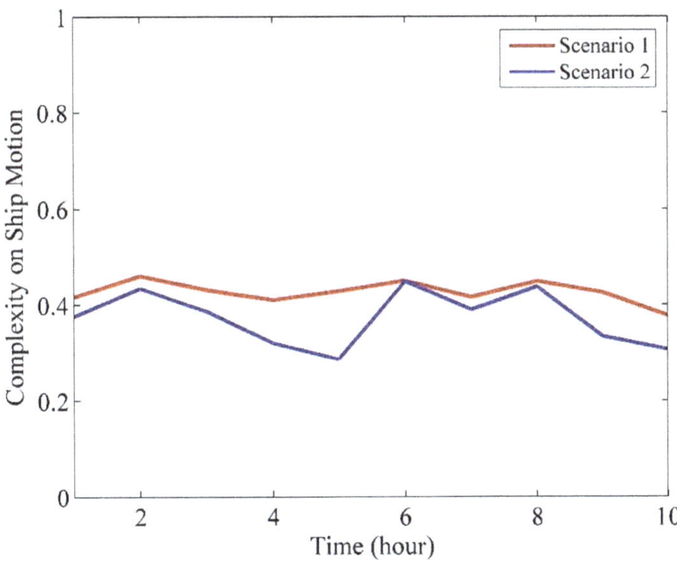

Figure 13. The complexity on ship motion for the spatial case study.

The results show that the complexity on ship motion in the second situation is less than that of the first situation, where the average values are 0.4266 and 0.3725, respectively. The results are consistent with the actual situation because compared with the first situation, only ships heading west were considered, which makes the speed and course less varied and leads to the decrease of complexity level. Therefore, it is proved that the proposed model can identify the marine traffic complexity on ship motion effectively again.

For the complexity on ship position, we also use the index of ship density to evaluate it. The complexity and ship density were calculated, respectively, for the ten studied water areas, and the results are shown in Tables 6 and 7.

Table 6. The complexity on ship position for the ten studied water areas.

Area	C_{sp}
1	0.0855
2	0.1051
3	0.1897
4	0.0927
5	0.1054
6	0.1247
7	0.0902
8	0.0563
9	0.0769
10	0.1310

A Pearson correlation analysis was carried out again to examine the correlation between the two sets of values. The analysis result is shown in Table 8.

Table 7. The ship density for the ten studied water areas.

Area	*SD*
1	0.0354
2	0.0884
3	0.1238
4	0.0973
5	0.0442
6	0.1149
7	0.0973
8	0.0531
9	0.0531
10	0.0973

Table 8. Pearson correlation analysis between C_{sp} and *SD* for the ten studied water areas.

p-Value	Correlation Coefficient
<0.05	0.705

Unit of SD: $pcs/\left(s \times nmile^2\right)$

It can be found that the p-value was less than 0.05 with a correlation coefficient of 0.705, which indicates that the complexity on ship position and ship density are strongly correlated for the ten studied water areas. As mentioned above, since ship density is an index that can reflect the busyness and congestion of marine traffic, we also have a reason to believe the proposed model can identify the complexity related to ship position distribution effectively.

4. Discussion

In this article, a model which can identify the marine traffic complexity was proposed. The proposed model was built with radial distribution function in molecular dynamics. The proposed model consists of two parts, which are the complexity on ship motion and the complexity on ship position, respectively. In identifying the complexity on ship motion, the speed and course of the ship were considered by constructing a two-dimension velocity plane. The velocity distance in the velocity plane was treated as the distance in the radial distribution function model. In identifying the complexity on ship position, the position of the ship was considered by constructing a traditional two-dimensional Euclidean plane. The final marine traffic complexity can be obtained by merging these two complexities. The proposed model can evaluate marine traffic complexity from an objective perspective utilizing the ship attributes extracted from AIS data. It can assist maritime surveillance operators in acknowledging the marine traffic situation in the jurisdiction water area in a more objective way, especially under complex traffic scenarios, where they may face a cognization difficult only by their subjective judgment. It can also facilitate their services for ships and traffic and thus contribute to the enhancement of marine traffic safety.

For validating the proposed model, a series of experimental case studies were carried out. At first, in Section 3.1, two simulated case studies were conducted. As the proposed model consists of two parts, which are the complexity on ship motion and ship position, we validate the two complexities, respectively, by controlling the variables in each simulated case study. In the first simulated case study, we made the ship positions fixed and changed the ship velocities gradually by changing the ship speeds and courses from an ordered state to a disordered state. The complexity on ship motion part in the proposed model can effectively react to this change. In the Second simulated case study, we fixed the ship velocities but changed the ship position gradually to make them more and more compact. The complexity on ship position part in the proposed model can effectively react to this change. The results can prove the effectiveness of the proposed model to some extent. Furthermore, in Section 3.2, two case studies with real AIS data were carried out to validate the proposed model. The proposed model was proved in temporal and spatial perspectives, respectively. For the validation from a temporal perspective, Laotieshan PA was chosen as the studied water area. Two situations of marine traffic in Laotieshan PA were divided, one considers all traffic and the other only considers limited traffic flows. For the two situations, 24 time nodes were selected from each hour through the day, and we calculated the complexity on ship motion for these time nodes. The results show that the complexity of the situation of limited traffic is less than that of the situation of all traffic, which conforms to reality. For the complexity on ship position, we also calculated the results for all time nodes. To evaluate its effectiveness, the ship density index, which can reflect the busyness and congestion of the ship was adopted. A Pearson correlation analysis between the ship density index and the complexity on ship position was conducted. The result shows a significant correlation between them, thus verifying the effectiveness of the results to some extent. In addition, to validate the effectiveness of the proposed model more sufficiently, a case study from a spatial perspective was carried out. Ten water areas were treated as studied areas. For the traffic in the studied area, we also divided it into two situations; one considers all traffic, and the other only considers nearly one-way traffic. Then, the complexity on ship motion was calculated. The results obtained by the proposed model can still react to the change in the situation, which can exhibit that the value of a limited traffic situation is less than that of a full traffic situation. Moreover, we identified the complexity on ship position for the ten studied areas and used ship density as an evaluating index again. The Pearson correlation analysis result shows that there exists a strong correlation between them, thus proving the effectiveness of the results again.

RDF was used to model the ship density and traffic density in [23]. As mentioned in that article, the model can reveal the traffic complexity to some extent. However, the RDF model in [23] can only evaluate the complexity from a spatial distribution perspective; the ship position is the only considered factor. However, ship motion is also important for determining traffic complexity. If the ship speeds and courses were varied and disordered, the encounter between ships may form easier, and thus increase the difficulties for collision avoidance. Under such complex traffic situations, collision accidents are more likely to happen. Therefore, in order to sufficiently reflect the traffic complexity level, only considering ship positions is not enough. In order to prove this, we carried out two sets of case studies in Section 3.1. It can be observed from Figures 6, 7 and Table 1 that when ship positions were fixed, only evaluating the complexity on ship position cannot distinguish the difference between the three scenarios. For distinguishing the complexity levels between them, the complexity on ship motion is supposed to be calculated at the same time. In order to prove the advantage of the proposed model compared to the previous model further, the case study in Section 3.1 was expanded here. The compactness of the ships in Figure 6 is moderate. Here, we changed the ship position distribution to a more compact state, as shown in Figure 14. The complexity results calculated based on the proposed model are shown in Table 9.

Table 9. The results of the expanded case study.

Scenario	Scenario 1	Scenario 2	Scenario 3
Complexity on ship motion	0.1397	0.3037	0.4639
Complexity on ship position	0.5091	0.5091	0.5091
Marine traffic complexity	0.3733	0.4192	0.4870

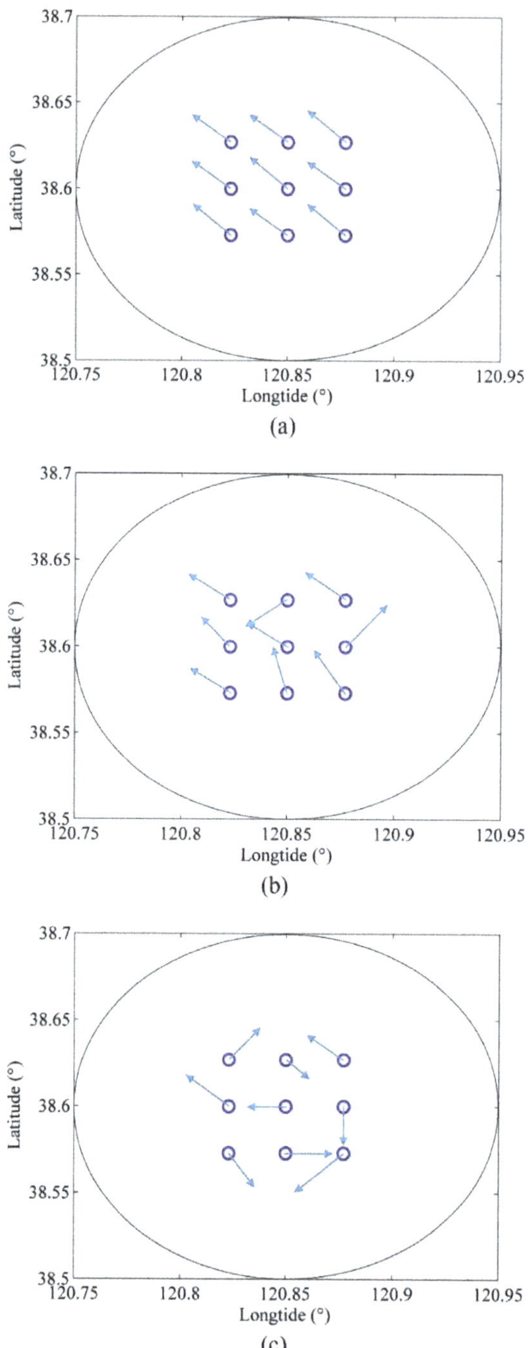

Figure 14. The scenarios for the expanded case study (**a**) Scenario 1; (**b**) Scenario 2; (**c**) Scenario 3.

It can be found that the complexity on ship position in Table 9 can react to the change of compactness, which becomes higher compared to the results in Table 2. However, for the three scenarios in Figure 14, it can be observed that the ship motions are obviously different. However, the complexity on ship position cannot distinguish them, which can only be performed by calculating the complexity on ship motion in the proposed model at the same time. Therefore, the results illustrate the advantage of the proposed model in fully evaluating marine traffic complexity again compared to the previous model.

In addition, the ship density index was used to evaluate the complexity on ship position in Section 3.2, but it should be noted that the ship density is not able to fully reflect the complexity too. Take the scenarios (a) in Figures 6 and 14 as an example, it can be found that the ships become more compact, but their density remains the same, which is both $9/\left(\pi \times 6 \; nmile^2\right)$ for 1 s time period. However, the proposed model can react to this change in ship compactness, increasing from 0.2431 to 0.5091. Therefore, the proposed model also has the advantage of identifying complexity compared with the traditional ship density index.

However, there are also some limitations to the proposed model in this article. Firstly, in modeling dynamic traffic complexity, the speed and course are considered because they are the most important dynamic parameters for ships. However, apart from speed and course, there are some other parameters that may influence traffic complexity, such as rate of turn, encountering angle, encountering type, and changing rate of speed. For example, encountering type includes head-on, crossing, and overtaking; different encountering type may cause different traffic complexity according to a different type of waterway. The relationship between traffic complexity and the factors mentioned above needs to be studied further in the future. Additionally, the proposed model calculates marine traffic complexity by using AIS data. Although AIS is mandatory on most of the ships, there are still some small ships that do not fit AIS, such as fishing ships. Therefore, for the waterway off the coast, where fishing ships may appear, as AIS data is not capable of covering such ships, the results may not be accurate enough to represent the actual traffic complexity in the waterway. To solve this problem, some other data should be used in modeling marine traffic complexity, such as Radar data or traffic observation data. Moreover, the newest AIS data was needed to further validate the model under the current traffic situation. In addition, in synthesizing the two complexities in a final marine traffic complexity, the expression and method in [32–34] were adopted. Considering that the studied area is relatively open, the two coefficients are taken as the same value in case studies. However, for different kinds of water, the two coefficients may be changed, or maritime surveillance can determine the coefficient according to their own purpose. The relationship between the coefficients and the kinds of water is supposed to be studied in the future.

5. Conclusions

Herein, a marine traffic complexity model considering the motion and density of the ship was proposed, which can identify the complexity level of marine traffic in a waterway. To establish the proposed model, the radial distribution function in molecular dynamics was used. Firstly, ship traffic was converted to a ship particle system where ships were converted to ship particles. Then, the complexity on ship motion and ship position were modeled, respectively, by analyzing the radial distribution of ships' dynamic and spatial parameters in a Euclid plane. Finally, the two complexities were merged to obtain the marine traffic complexity by an analytical method. In order to validate the proposed model, some experimental case studies were conducted in Bohai Strait waters with real AIS data and simulated data. The results show that the proposed model can effectively and accurately identify the marine traffic complexity in a waterway, react to the change of variable, and has the advantage compared with the previous model and the traditional density index in identifying complexity level. In addition, by utilizing the proposed model, marine surveillance operators can better acknowledge, monitor, and organize the traffic under complex situations, so as to improve marine traffic safety.

The proposed model also has some limitations which should be improved in future research. Firstly, apart from speed and course, some other dynamic parameters which may influence the traffic complexity, such as rate of turn, encountering angle, encountering type, and changing rate of speed, were not considered. The relationship between traffic complexity and the factors mentioned above needs to be studied further in the future. Secondly, as AIS data is not available for some small ships, such as fishing ships, the results may not be accurate enough to represent the traffic complexity, including such ships in the waterway. Therefore, some other types of data should be used in modeling marine traffic complexity. Thirdly, the merging of the two complexities in this article is under simple consideration; the relationship between the coefficients and the water types should be explored further in future research.

Author Contributions: Conceptualization, Z.L.; methodology, Z.L.; software, Z.L.; validation, Z.L., X.Y.; formal analysis, Z.L.; investigation, Z.L.; resources, Z.W. and Z.Z.; data curation, Z.L.; writing—original draft preparation, Z.L.; writing—review and editing, Z.L., Z.W., and Z.Z.; visualization, Z.L.; supervision, Z.W. and Z.Z.; project administration, Z.L.; funding acquisition, Z.L. All authors have read and agreed to the published version of the manuscript.

Funding: This research was funded by "the Fundamental Research Funds for the Central Universities" (Grant. 3132022134) and the Talent Research Start-up Funds of Dalian Maritime University (Grant. 02500128).

Institutional Review Board Statement: Not applicable.

Informed Consent Statement: Not applicable.

Data Availability Statement: Not applicable.

Acknowledgments: We would like to express our gratitude to the editors and reviewers whose valuable comments and suggestions will make improvements to the quality of this paper.

Conflicts of Interest: The authors declare no conflict of interest.

References

1. Weng, J.; Liao, S.; Wu, B.; Yang, D. Exploring effects of ship traffic characteristics and environmental conditions on ship collision frequency. *Marit. Policy Manag.* **2020**, *47*, 523–543. [CrossRef]
2. Zhen, R.; Shi, Z.; Liu, J.; Shao, Z. A novel arena-based regional collision risk assessment method of multi-ship encounter situation in complex waters. *Ocean Eng.* **2022**, *246*, 110531. [CrossRef]
3. Tu, E.; Zhang, G.; Rachmawati, L.; Rajabally, E.; Huang, G. Exploiting AIS Data for Intelligent Maritime Navigation: A Comprehensive Survey From Data to Methodology. *IEEE Trans. Intell. Transp. Syst.* **2018**, *19*, 1559–1582. [CrossRef]
4. Silveira, P.A.M.; Teixeira, A.P.; Guedes Soares, C. Use of AIS data to characterize marine traffic patterns and ship collision risk off the coast of Portugal. *J. Navig.* **2013**, *66*, 879–898. [CrossRef]
5. Wu, L.; Xu, Y.; Wang, Q.; Wang, F.; Xu, J. Mapping Global Shipping Density from AIS Data. *J. Navig.* **2017**, *70*, 67–81. [CrossRef]
6. Yu, H.; Fang, Z.; Murray, A.T.; Peng, G. A direction-constrained space-time prism-based approach for quantifying possible multi-ship collision risks. *IEEE Trans. Intell. Transp. Syst.* **2019**, *22*, 131–141. [CrossRef]
7. Zhang, W.; Feng, X.; Goerlandt, F.; Liu, Q. Towards a Convolutional Neural Network model for classifying regional ship collision risk levels for waterway risk analysis. *Reliab. Eng. Syst. Saf.* **2020**, *204*, 107127. [CrossRef]
8. Bakdi, A.; Glad, I.K.; Vanem, E.; Engelhardtsen, Ø. AIS-based multiple vessel collision and grounding risk identification based on adaptive safety domain. *J. Mar. Sci. Eng.* **2020**, *8*, 5. [CrossRef]
9. Wen, Y.; Huang, Y.; Zhou, C.; Yang, J.; Xiao, C.; Wu, X. Modelling of marine traffic flow complexity. *Ocean Eng.* **2015**, *104*, 500–510. [CrossRef]
10. Rong, H.; Teixeira, A.; Guedes Soares, C. Data mining approach to shipping route characterization and anomaly detection based on AIS data. *Ocean Eng.* **2020**, *198*, 106936. [CrossRef]
11. Zhang, J.; Wan, C.; He, A.; Zhang, D.; Guedes Soares, C. A two-stage black-spot identification model for inland waterway transportation. *Reliab. Eng. Sys. Saf.* **2021**, *213*, 107677. [CrossRef]
12. Liu, D.; Wang, X.; Cai, Y.; Liu, Z.; Liu, Z. A Novel Framework of Real-Time Regional Collision Risk Prediction Based on the RNN Approach. *J. Mar. Sci. Eng.* **2020**, *8*, 224. [CrossRef]
13. Liu, Z.; Wu, Z.; Zheng, Z. An Improved Danger Sector Model for Identifying the Collision Risk of Encountering Ships. *J. Mar. Sci. Eng.* **2020**, *8*, 609. [CrossRef]
14. Zhen, R.; Shi, Z.; Shao, Z.; Liu, J. A novel regional collision risk assessment method considering aggregation density under multi-ship encounter situations. *J. Navig.* **2022**, *75*, 76–94. [CrossRef]
15. Yu, Q.; Teixeira, A.P.; Liu, K.; Guedes Soares, C. Framework and application of multi-criteria ship collision risk assessment. *Ocean Eng.* **2022**, *250*, 111006. [CrossRef]

16. Merrick, J.R.W.; van Dorp, J.R.; Blackford, J.P.; Shaw, G.L.; Harrald, J.; Mazzuchi, T.A. A traffic density analysis of proposed ferry service expansion in San Francisco Bay using a maritime simulation model. *Reliab. Eng. Syst. Saf.* **2003**, *81*, 119–132. [CrossRef]
17. Altan, Y.C.; Otay, E.N. Maritime Traffic Analysis of the Strait of Istanbul based on AIS data. *J. Navig.* **2017**, *70*, 1367–1382. [CrossRef]
18. Ramin, A.; Mustaffa, M.; Ahmad, S. Prediction of Marine Traffic Density Using Different Time Series Model From AIS data of Port Klang and Straits of Malacca. *Trans. Marit. Sci.* **2020**, *9*, 217–223. [CrossRef]
19. Kang, L.; Meng, Q.; Liu, Q. Fundamental diagram of ship traffic in the Singapore Strait. *Ocean Eng.* **2018**, *147*, 340–354. [CrossRef]
20. van Westrenen, F.; Ellerbroek, J. The effect of traffic complexity on the development of near misses on the North Sea. *IEEE Trans. Syst. Man Cyber. Syst.* **2015**, *47*, 432–440. [CrossRef]
21. Du, L.; Valdez Banda, O.A.; Goerlandt, F.; Kujala, P.; Zhang, W. Improving Near Miss Detection in Maritime Traffic in the Northern Baltic Sea from AIS Data. *J. Mar. Sci. Eng.* **2021**, *9*, 180. [CrossRef]
22. Endrina, N.; Rasero, J.C.; Konovessis, D. Risk analysis for RoPax vessels: A case of study for the Strait of Gibraltar. *Ocean Eng.* **2018**, *151*, 141–151. [CrossRef]
23. Liu, Z.; Wu, Z.; Zheng, Z. Modelling ship density using a molecular dynamics approach. *J. Navig.* **2020**, *73*, 628–645. [CrossRef]
24. Widom, B. *Statistical Mechanics A Concise Introduction for Chemists*; Cambridge University Press: Cambridge, UK, 2002; p. 102.
25. McQuarrie, D.A. *Statistical Mechanics*; HARPER & ROW: New York, NY, USA, 1976; pp. 254–260.
26. IUPAC. *Compendium of Chemical Terminology: IUPAC Recommendations*; Blackwell Scientific Publications: Hoboken, NJ, USA, 1987; p. 335.
27. Bauer, R.C.; Birk, J.P.; Marks, P. *Introduction to Chemistry: A Conceptual Approach*; McGraw-Hill Inc.: New York, NY, USA, 2009; pp. 403–405.
28. Ma, X.; He, J.; Xie, X.; Zhu, J.; Ma, B. Study of C-S-H gel and C-A-S-H gel based on molecular dynamics simulation. *Concrete* **2019**, *351*, 118–122. (In Chinese)
29. Chandler, D. *Introduction to Modern Statistical Mechanics*; Oxford University Press: Oxford, UK, 1987; pp. 199–200.
30. Zhen, R.; Riveiro, M.; Jin, Y. A novel analytic framework of real-time multi-vessel collision risk assessment for maritime traffic surveillance. *Ocean Eng.* **2017**, *145*, 492–501. [CrossRef]
31. Fujii, Y.; Tanaka, K. Traffic capacity. *J. Navig.* **1971**, *24*, 543–552. [CrossRef]
32. Kearon, J. Computer program for collision avoidance and track keeping. In Proceedings of the International Conference on Mathematics Aspects of Marine Traffic, London, UK, 1 September 1977.
33. Lisowski, J. Determining the optimal ship trajectory in collision situation. In Proceedings of the IX International Scientific and Technical Conference on Marine Traffic Engineering, Szczecin, Poland, 19–22 October 2001.
34. Szlapczynski, R. A unified measure of collision risk derived from the concept of a ship domain. *J. Navig.* **2006**, *59*, 477–490. [CrossRef]
35. Liu, J.; Han, X. Survey and Analysis of Vessel Traffic Flow in the Bohai Strait. *Ship Ocean Eng.* **2008**, *37*, 95–98. (In Chinese)

Journal of
Marine Science and Engineering

Article

Decentralized Documentation of Maritime Traffic Incidents to Support Conflict Resolution

Dennis Jankowski [1,*], Julius Möller [2], Hilko Wiards [1] and Axel Hahn [1,2]

[1] German Aerospace Center (DLR), Institute of Systems Engineering for Future Mobility, Escherweg 2, 26121 Oldenburg, Germany

[2] Department of Computing Science, Carl von Ossietzky University Oldenburg, Ammerländer Heerstraße 114-118, 26129 Oldenburg, Germany

* Correspondence: dennis.jankowski@dlr.de

Abstract: For the investigation of major traffic accidents, larger vessels are obliged to install a voyage data recorder (VDR). However, not every vessel is equipped with a VDR, and the readout is often a manual process that is costly. In addition, not only ship-related information can be relevant for reconstructing traffic accidents, but also information from other entities such as meteorological services or port operators. Moreover, another major challenge is that entities tend to trust only their records, and not those of others as these could be manipulated in favor of the particular recording entity (e.g., to disguise any damage caused). This paper presents an approach to documenting arbitrary data from different entities in a trustworthy, decentralized, and tamper-proof manner to support the conflict resolution process. For this purpose, all involved entities in a traffic situation can contribute to the documentation by persisting their available data. Since maritime stakeholders are equipped with various sensors, a diverse and meaningful data foundation can be aggregated. The data is then signed by a mutually agreed upon timestamping authority (TSA). In this way, everyone can cryptographically verify whether the data has been subsequently changed. This approach was successfully applied in practice by documenting a vessel's mooring maneuver.

Keywords: decentralized documentation; traffic incidents; conflict resolution; timestamping authority; maritime data

Citation: Jankowski, D.; Möller, J.; Wiards, H.; Hahn, A. Decentralized Documentation of Maritime Traffic Incidents to Support Conflict Resolution. *J. Mar. Sci. Eng.* **2022**, *10*, 2011. https://doi.org/10.3390/jmse10122011

Academic Editors: Sebastian Feuerstack, Marko Perkovic and Lucjan Gucma

Received: 15 November 2022
Accepted: 14 December 2022
Published: 16 December 2022

Publisher's Note: MDPI stays neutral with regard to jurisdictional claims in published maps and institutional affiliations.

1. Introduction

In the 1960s, the first regulations to make flight data recorders (black boxes) mandatory were adopted in several countries. Back then, primitive analog systems had been used to record primary flight data such as altitude, airspeed, heading, and acceleration [1]. Several decades later, more convenient digital flight data recorders are used that can precisely record over 100 variables simultaneously [2]. Compared to the aviation industry, continuous recording of data for marine casualty investigations was introduced several years later, in 1997, by IMO Resolution A.861(20), which was also included in the International Convention for the Safety of Life at Sea (SOLAS). According to SOLAS chapter V [3], passenger ships and ships of 3000 gross tonnages and upwards, which are constructed after a specific date, shall be fitted with a voyage data recorder (VDR) to assist in casualty investigations. Since then, VDRs have been used to analyze shipping accidents in official investigations; for example, in the investigation of the grounding and partial sinking of the Costa Concordia in 2012 [4].

However, VDRs are tightly secured devices, and their disassembly, and especially data extraction and analysis can be a complex and time-consuming manual process [4]. Also, according to the IMO casualty investigation code, investigations are regulated by the national authorities, which often only initiate official investigations for serious accidents that lead to the total loss of a vessel, loss of human lives, or significant environmental pollution (e.g., as regulated in the Maritime Safety Investigation Act of Germany or the

Merchant Shipping (Accident Reporting and Investigation) Regulations of the UK). This leads to several problems in investigating less severe accidents or critical traffic situations, especially from the perspective of data acquisition. Data from the vessel's sensor systems or surveillance data is one of the essential puzzle pieces for tracing the chain of events that lead to such a situation. As today's vessels often have sizes of 300 m in length or larger, even smaller accidents can lead to major financial losses for the involved participants in an accident. Additionally, law enforcement can become a problem as ships often sail under the flag of micro-states, and international enforcement of the law is extremely costly. In the best case, data acquisition must happen immediately after an incident. Still, in busy port areas, damages to port infrastructure, e.g., quay walls, may not always be recognized directly after the incident, and it may not be clear how the damage occurred. In these cases, there often exists multiple data sources that could contain data that is related to such an event. Nevertheless, if no official authority is coordinating the process of data acquisition and analysis, trust issues may arise between the involved parties regarding the authenticity of the data.

From this, it can be seen that there is a need to collect data in potentially critical maritime traffic situations that may lead to accidents while maintaining trust between all different involved parties in the case that the recorded data may be used in a later analysis of such a situation. Apart from that, the provision of a guarantee between the traffic participants that the data has been recorded at the point of the event, and was not modified, is also important to maintain trust. Therefore, in this paper, we propose a new method for recording vessel traffic data from multiple parties, with the intention of using this data in an incident investigation while guaranteeing non-repudiation of the recordings.

In Section 2, we initially provide relevant background information on current standards for Maritime Accident Investigations, Data Protection, and existing procedures for the secure storage of distributed data. On this basis, we subsequently derive requirements for a system for the documentation of critical traffic situations. Section 3 takes a closer look at the related work. We present existing techniques that address tamper-proof recording of data. In addition, blockchain-based approaches and concepts from the automotive and maritime sectors will also be considered. In Section 4, our approach to decentralized tamper-proof documentation of maritime traffic situations is presented in detail. Afterwards, in Section 5, the approach is tested and evaluated on a mooring maneuver in practice. Finally, Section 6 summarizes the obtained results and provides insight into limitations and future work.

2. Background

To derive requirements for a system for the documentation of maritime incidents, considering the above-mentioned challenges, it is first analyzed how accident investigations are currently carried out. Then concerns are discussed that may arise for different parties regarding a self-organized solution for investigating more minor incidents. Considering these approaches, state-of-the-art methods to securely document incident data are analyzed. Finally, we derive requirements for the conceptualization of a system for the documentation of these incidents.

2.1. Maritime Accident Investigations

Accidents in the maritime industry have been present since the invention of shipping. Countries typically have their own regulations and processes on when and how accident investigations should be carried out, adapting codes and regulations from the IMO (cf. [5]). Results from the analyses of accidents and incidents are then published in reports. As a basis for relevant accident data, we analyzed accident investigation reports from the authorities of Germany [6] and the United Kingdom [7]. The following information is normally included in investigation reports dealing with accidents that are related to traffic (this does not include accidents on board a single ship):

- Data related to the involved vessels (name, type, flag state, size, special features, etc.).
- Planned routes of the involved vessels.

- Type, date, and place of the accident.
- A detailed description of the sequence of events that led to the accident, often including textual descriptions of personnel behavior, navigation decisions, vessel movements, pictures of the involved ships or damage caused, and other information that is specific to the type of accident.
- If available—data dumps from the involved vessels.
- Meta-information on how the investigation was carried out.
- Analysis of the causes of the accident.
- Summary and recommendations.

Especially the description of the sequence of events that led to the accident is a critical part of a report, as it is the main basis for analyzing the causes of the accident. It often includes eyewitness testimony or may be based on data recorded on board the vessel or other maritime surveillance equipment (e.g., VTS centers).

Hence, in the discussed scenario of a smaller incident that is not being investigated by a state authority, the involved parties should aim at preserving data that can represent navigation decisions and vessel movements at the time of the accident as a minimum objective reference for its cause. Other information (such as the data related to the vessels, meta-information on the investigation, and further analysis) can still be derived at a later stage of the incident analysis.

2.2. Data Protection and Sovereignty

In the case of an accident investigation, authorities have the executive power to confiscate VDRs or request data from external surveillance systems. However, this often happens a certain amount of time after the event. For more minor incidents without an official investigation, the procedure of data recording and preservation must be organized differently. Also, at the time of the incident itself, it may not be apparent to all parties that an incident is currently happening. Therefore, it is essential that data in potentially critical situations is continuously recorded and available for later analysis. In the case of the quay wall, it can also occur that the damage is only noticed in a general inspection or by other incoming vessels, and it is not even clear which parties were involved in the incident. In these cases, it is not an option of the port owner to force all ship owners to share their VDR data; thus, other means are needed for more efficient data sharing. On the other hand, it might also not be a possibility for all the involved parties to continuously exchange data with each other just in case something happens. Depending on the shared data, this may also violate data protection regulations (such as the GDPR in the EU, cf. [8]), for example, if they include personal data. Another aspect to consider is the value of the data to a company since, for example, business secrets could be revealed if data about a ship is constantly shared (preferred routes, etc.). In general, the idea of being able to meaningfully control and govern one's own data is a recent trend and is generally referred to as data sovereignty [9]. These factors lead to the conclusion that the recorded incident data cannot be stored centrally in a cloud environment (or similar) but must remain with the involved parties until a conflict arises and needs to be resolved. Even then, the participants should have full decision-making power over their data.

Finally, keeping local records of data with proof that it has not been modified and was recorded at the time of an incident only makes sense if it can be ensured that the recorded data evidence will stand up in court. Only in this way can a serious conflict resolution be performed between the involved parties and other relevant stakeholders (such as insurance companies). Similar to an analog paper document, a digital datum acquires its validity through a digital signature. In the setup of a Public-Key Infrastructure, users possess a public- and a private-key, which can be used to encrypt or sign data. The public key is also typically signed by a certification authority (CA) in the form of a certificate, thus binding it to the physical identity of a user. In order for the user to be able to use the key pair to sign its data in a trustworthy way, a certification authority must be used to certify the public key and apply special verification procedures (in accordance with applicable law) to ensure that the user's identity is correct. Furthermore, key-holders must take measures to prevent their private key from being exposed to other unauthorized parties in such a way that it can be assumed that data was actually signed by the key-holder and the key was not compromised [10].

2.3. Approaches to Immutable and Decentralized Data Storage

Establishing trust between different parties that have no direct trust relationship but aim to exchange information with each other is not a new problem in communications engineering. Typical setups for the exchange of digital information start with the authentication of the involved parties with the help of an identity provider. Therefore, a relying party can identify other participants by trusting the identity provider [11]. However, authentication of a party does not imply full trust in the actions of that party. In the maritime industry, it may be easy to identify a vessel or infrastructure with the help of its registration data at the IMO via an MMSI or even digitally via Identity Providers that connect vessel identities to digital certificates such as the Maritime Connectivity Platform (MCP) [12]. However, when it comes to trusting the correctness of a data asset that was recorded by a specific entity, additional frameworks need to be introduced. As discussed in Section 1, for accident investigation data, it is crucial to ensure that the data were recorded at the time of the accident and not subsequently modified. In similar situations, where the different participants do not trust a common central entity that could guarantee the correctness of the exchanged data, blockchain technology has often been applied.

A blockchain is a specific type of distributed database that only allows the addition of new information if a consensus in the network of involved participants is reached. Furthermore, data is only being added to the blockchain in the form of atomic transactions, which are organized in a chained data structure referencing its respective predecessor. After a transaction is issued, it is impossible to modify it, which is referred to as immutability [13].

However, the inflationary usage of blockchain technology in all possible use cases has been questioned more and more recently [14,15]. Also, there are other less complex solutions to the problem of providing proof that data existed at a certain point in time and was not modified. For instance, a possibility of this kind in regard to timestamps is represented by so-called timestamping authorities [16,17]. A TSA is a service that establishes that some data existed at the specified time based on reliable time sources and cryptographic signatures. Depending on the application, such a TSA can be operated in-house or externally, either as a commercial service or as a service for third parties.

The biggest problem in using these services is that all parties must trust the selected TSA [18]. Suppose one party can obtain the private key or manipulate the TSA's internal system time. In that case, the entire chain of evidence is invalidated since the data can then be signed with different timestamps. To circumvent this, an approach based on "distributed trust" is presented in [19] by randomly selecting subsets from a set of TSAs via a pseudorandom number generator (PRNG). However, even in this case, users of TSAs must always rely on the availability and proper functioning of the TSA they are using to sign (similar to trusting a certificate authority in a PKI).

2.4. Requirements for the Secure Documentation of Critical Maritime Traffic Situations

From the above considerations, the following requirements were derived for the development of a system to securely document critical maritime traffic situations for usage in smaller incident investigations and insurance cases:

(R1) The system must be able to record arbitrary sources of data.
(R2) The system must provide evidence that the data was recorded during the incident and was not modified.
(R3) The system must support decentralized usage such that every involved party can protect the sovereignty of their data.

(R4) It must be possible for a participant to dynamically join a network with other participants to agree on the common data recording of a potentially critical situation (e.g., berthing of a ship or an evasive maneuver).

(R5) If not all the involved participants are equipped with the system, a fallback mechanism must exist such that data is still being recorded by the remaining parties with a minimal loss of trust.

3. Related Work

In the past decade, the field of data security has seen a strong increase in interest with a particular focus on methods to prevent data tampering. This chapter presents some of the most promising approaches to tamper-proof data recording. First, we examine blockchain-based approaches, which are often associated with cryptocurrencies, but their possible applications extend far beyond that domain. Furthermore, we explore other techniques for tamper-proof data recording based on cryptographic techniques, and finally close with approaches from the automotive and maritime domains.

3.1. Blockchain-Based Approaches

Various works have already considered the use of blockchain-based approaches for documenting events. These can be roughly divided into approaches that use public blockchains (e.g., Bitcoin, Ethereum) and those that use permissioned blockchains. The former are particularly attractive for their transparency, as everybody can view the blockchain data, while permissioned blockchains are usually used in private settings.

Approaches based on the public Bitcoin blockchain were pursued in [20,21]. Due to the transaction costs associated with public blockchains, these approaches are based on writing summary hashes or anchor points of KSIs (Keyless Signatures Infrastructure) [22] to the blockchain at regular intervals. The data itself is not published. In this way, it is possible to verify whether data has been tampered with by inspecting the summary hashes. The drawback of these approaches is that it is not possible to directly verify the data that was written to the blockchain. Instead, the blockchain user needs to trust the KSI infrastructure that was used to write the summary hashes to the blockchain. Furthermore, it is also apparent that the benefit of the Bitcoin blockchain lies primarily in the permanent publication of the data, as the timestamps themselves are not trustworthy (see [23,24]) and require additional external time verification such as KSI [22].

To achieve better scalability and a higher data rate, logging infrastructures based on permissioned blockchain were presented in [25,26] and achieved data rates of 100–3500 transactions per second. In [27], an approach is described where cars serve as live witnesses to situations and decisions as participants in a blockchain. Based on vehicle-to-vehicle and vehicle-to-infrastructure communication, a shared truth is formed in this way. However, trust requires that as many independent parties as possible are part of the permissioned blockchain network. Depending on the use case with alternating or short-term stakeholders, this could result in a large overhead and is the reason why permissioned blockchain networks are usually based on a network of companies and organizations [27].

3.2. Tamper-Proof Data Recording

In addition to the approaches presented based on blockchains, there are other ways to check data records based on cryptographic properties, to a certain extent, for subsequent changes. For example, the documents can be electronically signed, stored in a data structure, and subsequently checked for validity. In [28], a log server is presented for this purpose, which provides a structure for tamper-evident data logging for a larger number of clients based on Merkle trees by feeding back smaller commitments to the clients. In [29], an approach based on hash-chains is presented. This approach enables logs to be protected on compromised machines without the need to publish anchors of the log by linking each entry with the previous one based on one-way hash functions. Therefore, a modification of the data would render the complete chain compromised. Several approaches based on the trusted hardware features of newer processors are shown in [30,31]. These works use the Trusted Platform Module (TPM) to ensure a tamper-proof log. The TPM creates a chain of trust from the bootloader to the operating system and the log system. The work in [31] uses TPM 2.0 to secure the log even between power cycles.

3.3. Further Approaches

In the automotive sector, the Event Data Recorder (EDR) is currently used to record incident sensor data [32]. Triggered by a crash, these record low-bandwidth data including car and engine speeds, brake status, and accelerations [33]. The data is stored in non-volatile memory within the control unit. These systems are permanently integrated into cars and operate on proprietary interfaces [34], but are themselves probably not adequately protected against manipulation of the collected sensor values, and measures against subsequent overwriting are currently not described [35].

Some works deal with fishery logbooks, i.e., logbooks in which the catches of fishermen can be documented. However, there is no focus on the manipulation of the data, as these are concerned with the architecture of the data exchange [36] or the analysis of these data [37].

A low-threshold approach to analyzing maritime sailing behavior and traffic situations is to look at AIS data. This data is publicly receivable and can be purchased afterwards from service providers but lacks a fine-grained resolution. Vessel Traffic Services [38] use this data (along with other sensors such as radar, cameras, etc.) to provide live assistance and instructions [39]. Since this data is sent publicly at regular intervals, the data received in this way can be used as additional anchor points for a possible conflict resolution.

In summary, the maritime domain still lacks a method for trusted, automatic data recording of multiple sensors without compromising data sovereignty and privacy. The blockchain-based approaches shown are promising but still miss some practicality regarding transaction speed, scaling, and compartmentalization. Some of these shortcomings are solved by using classical approaches based on cryptographic signatures, but these lack the distributed characteristics of the blockchain. As a result, we are not aware of any approach that sufficiently addresses the properties of non-repudiation, decentralization, and performance simultaneously.

4. Concept

As already described in the requirements from Section 2.4, it is crucial for trustworthy documentation that it can be verified that the recorded data has not been changed intentionally or unintentionally. This assurance can be provided by timestamping authorities which is fast and efficient compared to alternative approaches such as blockchain technologies for the certification of data [13,40]. Therefore, this paper presents a TSA-based concept to document critical traffic incidents in a decentralized scenario.

The certification process of TSAs is analogous to the certification processes of a PKI. However, in this case, an authority is used to certify the existence of a date at a specific time instead of the identity of an entity (c.f. Figure 1).

For this purpose, the requestor sends its data that should be certified in hashed form to the TSA. This is done to reduce the payload and to avoid exposing the actual data to the TSA. The TSA adds the current timestamp to the transmitted data hash and hashes the data again. The TSA then sends the signed document back to the requester. The requester needs to persist the original data and the TSA's digital signature. If the requester wants to prove that their data has existed at a certain point in time and has not been changed, they can verify this cryptographically in two steps (c.f. Figure 2).

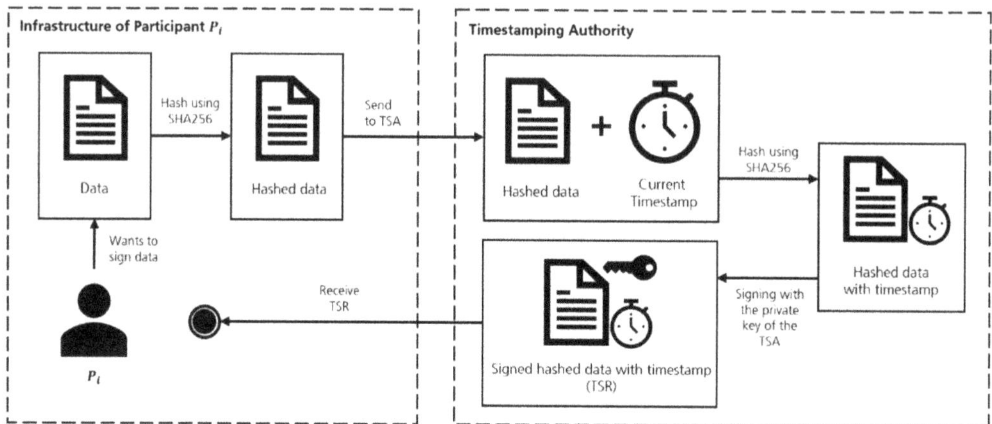

Figure 1. Timestamping process between a requester and a TSA [41].

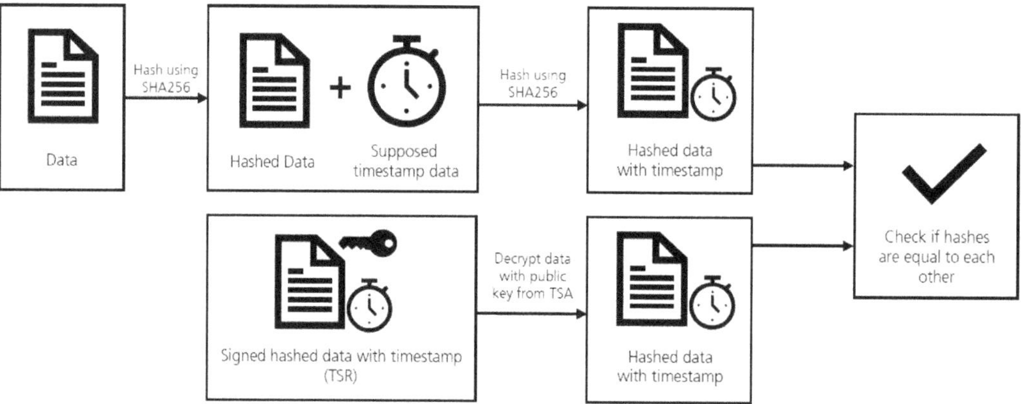

Figure 2. Verification process of timestamped data [41].

In the first step, a hash is created based on the data using the same hash method that was applied by the requester during the timestamping process. Subsequently, the exact timestamp at which the TSA has signed the data is added to this hash. This document is hashed again with the hash procedure used by the certifying TSA. In the second step, the cryptographic signature of the document is decrypted using the TSA's public key. The hash that was derived in this process is then compared with the hash value from step 1. If the two values match, the verification was successful, and it can be assumed that the data already existed at the specified time. If the two hash values differ, there can be many reasons for this. For example, it is possible that the data was changed or that it was signed at a different time. Regardless of the reason, the verification failed in this case.

In this way, it is possible to verify whether data was available at a certain point in time by utilizing a TSA. However, the main issue with using an authority to establish trust is that all the trust depends on the authority itself. Thus, the signature of a TSA has only some value if the parties, to whom something should be proven, also trust this single TSA. Consequently, if an architecture is built based on only one authority, this directly leads to a problem if only one of the involved parties does not trust this central TSA. Especially in the global maritime domain, in which many different stakeholders from various fields and nations meet, an agreement on a worldwide central authority is not feasible. The utilized architecture must therefore ideally support the use of arbitrary TSAs, so that the parties involved can individually agree on a single or a set of trusted TSAs in the specific situation. In this way, trust is maximized as each participant can communicate their trusted TSAs instead of choosing from a predefined set of authorities. Since the participants in a traffic event usually do not know each other beforehand, the agreement on the set of jointly trusted TSAs must happen before a critical traffic situation.

Figure 3 shows the proposed architecture that enables the involved stakeholders to document their data in a trustworthy and decentralized scenario without using a central instance for coordinating the agreement on a TSA.

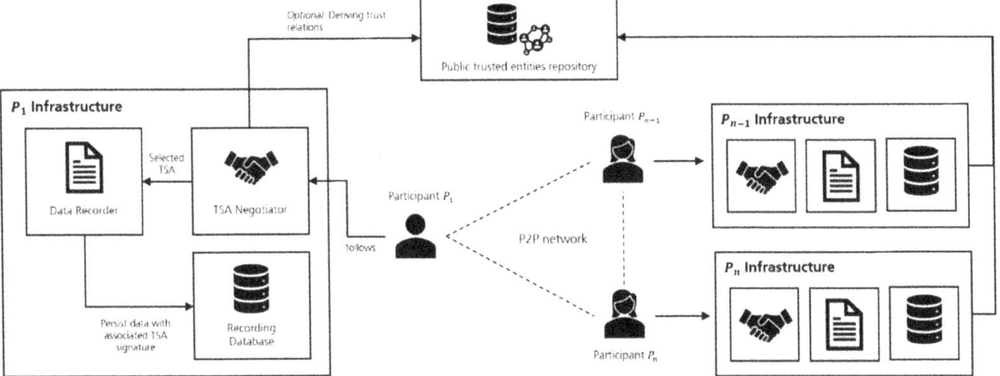

Figure 3. Approach for Decentralized Documentation of Maritime Traffic Incidents.

The entire architecture consists of five components:

- P2P-Network—Basis for communication between the involved participants of an incident that should be recorded.
- TSA Negotiator—Responsible for the derivation of a common set of trusted TSAs between all participants.
- Data Recorder—Coordinates the recording and signing of data streams by a trusted TSA derived from the TSA Negotiator.
- Recording Database—Persists the original data with the associated signature from the used TSA locally.
- Public Trusted Entities Repository—Can optionally be used by participants for publishing which other entities (e.g., organizations, or companies) and TSAs they trust. The repository is centrally managed and supports the participants in deriving a larger set of trusted TSAs based on their trusted relationships.

Each participant deploys the *TSA Negotiator*, the *Data Recorder,* and *the Recording Database* on its local infrastructure so that it has complete control over these components. This reduces the amount of data that leaves the infrastructure of the participant, which in turn protects the participants' sovereignty and minimizes the system's required bandwidth. In the following, the overall process of a to-be-documented situation is described in more detail:

(1) Initiate the documentation: The entire documentation process is initiated by activating a predefined trigger. This trigger can be chosen according to the use case. For instance, the trigger can be the entry of a ship into a certain geographical area (such as the port) or an under- or exceeded speed. Thus, the trigger defines whether an incident is a document worthy incident or not. The definition of a trigger is the responsibility of the participant with the primary interest in documenting a critical incident, such as the port operator, when a vessel enters the harbor.

(2) Establish a communication channel: The basis for communication between the participants is a P2P-Network that is set up dynamically between the participants in

the event of a potential incident. By using a P2P-Network, a central communication channel can be avoided so that the participants can communicate with each other in a completely decentralized way. The network is used to coordinate the communication between the participants to agree on a common set of trusted TSAs that can then be used to sign the record. Since the negotiation already takes place during the initiation of a critical traffic situation, the involved parties must have already established a P2P network between themselves.

(3) Derive a common trusted set of TSAs: After establishing a communication channel, the participants need to find a common trusted set of timestamping authorities with which they will sign their data recordings. For negotiation, each participant sends its own trusted TSAs to every other participant over the P2P network so that each participant knows the trusted TSAs from every other participant. The derivation of the trusted TSAs is performed by a deterministic negotiation protocol of the TSA Negotiator.

(4) Record the data: The Data Recorder is responsible for recording and signing the data that should be documented. Therefore, the Data Recorder periodically divides the data stream from a participant into discrete chunks and sends the hashed chunks to the TSA derived by the TSA Negotiator. Afterwards, the TSA's original data and signature are stored locally in the Recording Database.

In the following, we will discuss in detail how the *P2P Network*, the *TSA Negotiator*, the *Data Recorder,* and the *Public Trusted Entities Repository* work.

4.1. P2P Network

To agree on a common set of TSAs, a communication channel is needed through which all participants can share their trusted TSAs. Since there might not be a central instance for the communication that is trusted by all, a P2P network is used allowing the participants to interact directly with each other without an intermediary. As already described, the P2P network is established dynamically when a critical traffic incident occurs between the involved parties as soon as the defined trigger is released. To be able to set up such a network dynamically, all participants' IPs must be known by everyone. Depending on the application, this may not be the case, so the IP-addresses must first be exchanged with each other. In practice, this can be done in every conceivable way. In the following, we have outlined an exemplary possibility in more detail.

Let P be the set of all participants, so one of the participants $a \in P$ publishes its IP address publicly so that the other participants from the set $P \setminus a$ can obtain it. Subsequently, each participant from $P \setminus a$ sends its own IP-address to participant a. In this way, participant a knows the IP addresses of all participants, enabling him to forward the addresses of all participants to everyone. In this way, each participant receives the IP-addresses of all other participants, which means that each participant is now able to communicate with one another.

Usually, the initiating participant should be responsible for publishing the primary IP address. The IP address can be transmitted via various channels, via a website or AIS messages. It is also conceivable to publish the address as a service in a public Maritime Service Registry (MSR) of the MCP. The service registry acts as a directory where services of maritime stakeholders can be found by entering various parameters such as keywords or geographical regions. For example, a port operator could publish its IP address as a service in the Maritime Service Registry. Vessels wishing to enter the port would then have the opportunity to search the MSR for a corresponding service, in order to find the stored IP-address that is required for the initial contact. After all, the parties have already contacted the port operator; therefore, the latter can set up the P2P network as described above. However, in some cases, it might not be possible for every participant to directly expose an IP-port for communication (e.g., due to some ship-related IT security regulations). Also, as IP connectivity is not always available, it is conceivable to use other technologies (such as VDES) as a backup solution for communication. Another possible solution is an architecture, as proposed by IEC 63173-2 [42], where a service at the shore-side exposes a publicly available interface via IP and the communication to the ship ("last mile") is realized with a more secure communication channel.

4.2. TSA Negotiator

As mentioned in Section 4, the main task of the TSA Negotiator is to ensure that the participants can agree on a common set of timestamping authorities when entering a potentially critical situation. This is realized with a negotiation protocol as is shown in Figure 4. The negotiation protocol is a deterministic protocol so that each participant can run the protocol locally after gathering all information and the results of all participants will be consistent with each other. Therefore, a central instance that coordinates the negotiation can also be avoided here.

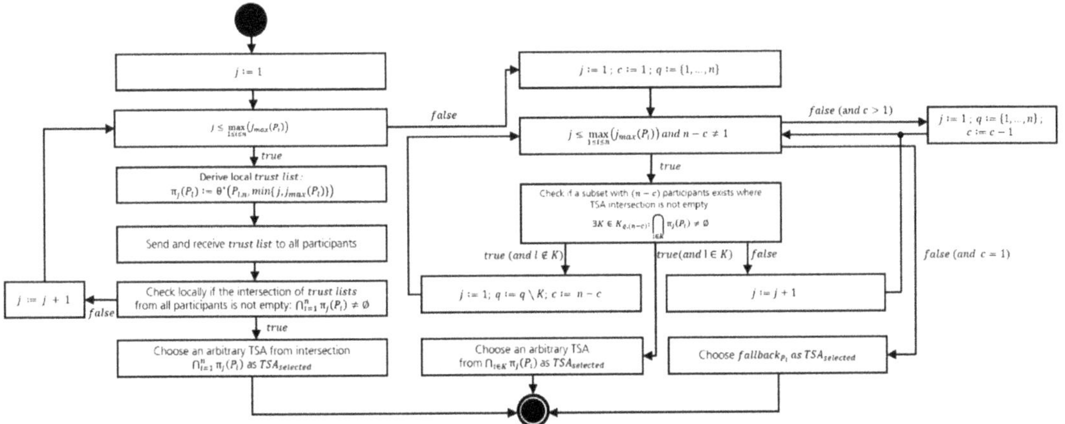

Figure 4. Negotiation Protocol for the agreement on a trusted TSA between different participants.

In the following, we refer to the participants as P_i with $i = \{1, \ldots, n\}$. Furthermore, the local view of a participant is denoted by using P_l for self-references. Initially, each P_i provides a list of trusted TSAs $\theta(P_i)$ and a list of trusted organizations $\phi(P_i)$.

It is assumed that each participant P_i has an internal prioritization of TSAs and wants to maximize the trust by choosing the TSA ranked highest in its local priority list. To ensure that the TSAs with the highest trust from all of the participants are found, the potential TSAs are exchanged in several iterations such as j (a higher number of iterations, j means less trust as the lists of possible TSAs are extended with lower prioritized TSAs by each participant). We call a set of TSAs that is trusted by a participant P_i the *trust list* of P_i. Theoretically, any participant can arbitrarily choose trust lists in every iteration. However, we propose the following method for determining the trust list of a participant P_l in iteration j $\left(\pi_j(P_l)\right)$ of the protocol:

$$\pi_j(P_l) := \theta^*(P_l, \min\{j, j_{max}(P_l)\})$$

with

$$\theta^*(P_i, n) := \theta\left(\left\{P_k | P_k \in \cup_{i=1}^{n} \phi^{(l-1)}(P_i)\right\}\right)$$

With P being a set of participants, $\theta(P)$ is defined as $\theta(P) := \cup_{p \in P}\theta(p)$ and $\phi^l(P_i)$ is the repeated application of ϕ to get the trusted organizations of P_i with ϕ^1, the union of all trusted organizations of each $P_k \in \phi^1$ with ϕ^2, etc. Note that $\phi^0 = \emptyset$.

For example, in the first iteration $j = 1$, P_l will only add the TSAs it trusts directly to its trust list $\pi_1(P_l)$. For $j = 2$, more TSAs will be added by also including the trusted TSAs of organizations or companies that the participant P_l trusts (given by ϕ^1, e.g., by using the public trusted entities repository, see Section 4.4). Subsequently, this can be continued by also adding the TSAs of the trusted organizations of the organizations trusted by $P_l (j = 3)$. These transitive relations can thus be continued for the number of iterations in the protocol. In this way, in each iteration, the probability that a commonly trusted TSA is found is increased, as $\pi_j(P_i) \subseteq \pi_{j+1}(P_i)$. At the same time, the individual participant's trust in the added TSAs decreases. As this process will not carry on indefinitely, each participant P_i defines locally in how many iterations $j_{max}(P_i)$ the own trust list is extended, so that the number of total iterations in the protocol is defined as: $\max\limits_{1<i<n}(j_{max}(P_i))$.

Now for each iteration, each participant derives its trust list for the j-th iteration $\pi_j(P_l)$ and broadcasts it to all participants over the P2P network. In this way, each participant receives $n - 1$ trust lists in total. Then, it is checked locally (per iteration) if there is an intersection of all trust lists:

$$\bigcap_{i=1}^{n} \pi_j(P_i) \overset{!}{\neq} \varnothing$$

If an intersection is found, any TSA from the identified intersection can be used by the participants as $TSA_{selected}$ and utilized to sign the recorded data. If after the last iteration (i.e., $j = \max\limits_{1<i<n}(j_{max}(P_i))$) still no common TSA can be found, it is not possible to identify a TSA that all participants trust. Therefore, the second stage of the protocol is entered and is aimed at finding the largest possible subsets of participants who can agree on a set of TSAs in such a way that:

$$\exists K \in K_{q,k} : \bigcap_{i \in K} \pi_j(P_i) \neq \varnothing$$

Here, we define $K_{q,k} := \{M \subseteq P(q) | |M| = k\}$ with q being an index set for the considered participants in a single round of the second stage of the protocol. Hence, $K_{q,k}$ is the set of all subsets of q with size k. We introduce c as a pruning factor for the participant sets, such that the number of currently considered participants for finding a common TSA is $(n - c)$. Initially, we set $c = 1$ and $q = \{1, \ldots, n\}$, so that all subsets are considered that are missing only a single participant. For each set K in $K_{q,\,(n-c)}$ it is then checked whether a common set of TSAs exists for which $\bigcap_{i \in K} \pi_j(P_i) \neq \varnothing$ is true (similar to the first stage of the protocol). Generally, the search for a common TSA set proceeds identically to the search with all participants so that for each pruning factor, the trust lists are also successively extended in j iterations. If a set is found, then an arbitrary TSA from the set $\bigcap_{i \in K} \pi_j(P_i)$ is selected as $TSA_{selected}$ for signing the data for the participants from K. Since the trusted TSAs of the individual participants do not change during the negotiation, the trust lists do not need to be transmitted again over the P2P network. If a common TSA cannot be found for any K after j iterations, c is increased by 1 so that in the next iteration, all $K \subseteq q$ with one participant less are taken into account.

However, since a valid K is a proper subset of the index set of participants q, even after finding a TSA there will be a set of participants $q \backslash K$ for which no TSA has been found yet. Given that the set $q \backslash K$ refers to at least 2 participants if $c > 1$, it should still be aimed to create a common trust basis between these participants. This is done in the same way as for the first subset with $j := 1$; $q := q \backslash K$; $c := n - c$. This process is repeated until $n - c = 1$, so that no more subsets with at least 2 participants can be formed. If this case occurs, a common TSA cannot be found based on the trust lists of the respective participants under any circumstances. Here, each of these participants P_i should use self-selected TSA ($fallback_{P_i}$) as $TSA_{selected}$ to sign the records and to establish at least a minimum of trust and security in the documentation.

Note that the search for $\bigcap_{i \in K} \pi_j(P_i)$ for a subset K of q assumes that it is better to find a common TSA at least among the largest possible subset of participants, instead of finding no TSA at all. In this way, at least the largest possible trust is established between the participants. Also, if at any step there are multiple subsets of q of equal size for which $\bigcap_{i \in K} \pi_j(P_i) \neq \varnothing$ holds, a choice must be made for one subset K deterministically since the participants in the two sets might overlap. To deterministically select one of the sets, they are hashed and then sorted alphabetically. The alphabetically first set is then defined as the selected set K.

4.3. Data Recorder

After each participant has found a $TSA_{selected}$, the local documentation of the incident can begin. The Data Recorder is responsible for managing the data's recording, signing and persistence. Figure 5 shows the functionality of the Data Recorder.

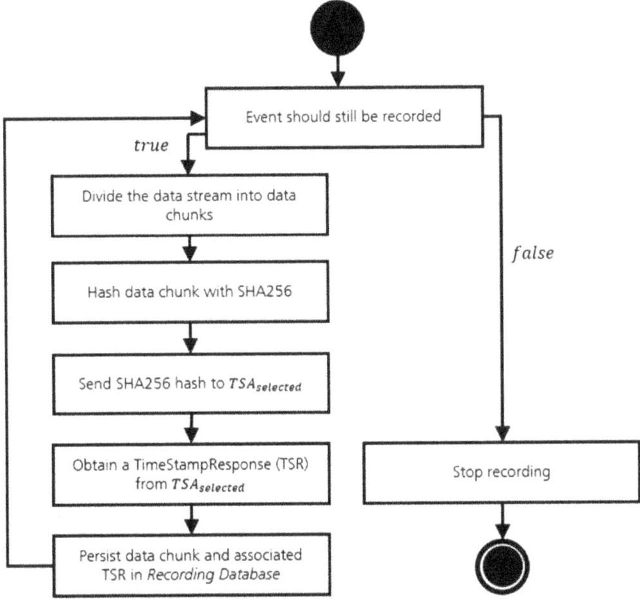

Figure 5. Procedure for local data recording and signing.

Initially, participants decide for themselves whether they want to continue the current recording of the event. Every participant can choose when to end the recording independently from the other participants since the end of an incident can be interpreted differently. As the recorded data is usually live data, the data must first be split into regular chunks so that the data can be signed by the $TSA_{selected}$ at all. Chunks can be formed in different ways (fixed size, fixed time span, etc.). Depending on the use case and individual preferences of the participants, a different way may be favored. Therefore, the participants can also decide by themselves under which conditions they subdivide their data. The chunk is then hashed using SHA-256 and transmitted to the $TSA_{selected}$. Analogous to the signing process in Figure 1, the hashed chunk is hashed again by the $TSA_{selected}$ and signed with the private key of the $TSA_{selected}$. Afterwards, the resulting time stamp response (TSR) is sent back to the appropriate participants so they can persist the TSR together with the original data chunk locally on its own infrastructure. This process is repeated until the participant decides that the hazardous situation has ended, and proof of the process is not required anymore.

4.4. Public Trusted Entities Repository

The Public Trusted Entities Repository is a component that can be used *optionally* by participants for publishing publicly which companies, organizations and TSAs they trust. A participant's trusted entities and TSAs are expressed as a separate set of entities and TSAs. Thus, for example, $\phi(P_A) = \{P_B, P_C, \ldots\}$ represents the trusted entities and $\phi(P_A) = \{P_B, P_C, \ldots\}$ the trusted TSAs of participant P_A. Using these two sets, it becomes possible for participants, analogous to the Web of Trust, to derive transitive trust relations by adding the TSAs of trusted entities to their own trusted TSAs or to look at which other entities their trusted entities rely on in order to add their trusted TSAs to their own *trust list* (c.f. Section 4.2).

Example: If participant *A* trusts organization *B*, then participant *A* can look up in the Public Trusted Entities Repository whether organization *B* has published which TSAs and organizations it trusts. If participant *A* finds an entry for organization *B*, it can view its TSAs and add them to its own set of trusted TSAs.

In this way, the process of finding trusted TSAs is simplified by allowing the participants to derive them via their relationships instead of having to check each TSA individually. At the same time, the repository should increase the number of trusted TSAs per participant to maximize the probability of a common intersection between participants. Nevertheless, the use of the repository is only optional, and participants are free to derive their list of trusted TSAs according to their own preferences.

5. Application and Evaluation

In the following, we present a proof-of-concept prototype to show the applicability of the architectural framework and the negotiation algorithm that was introduced in Section 4. For this purpose, we first describe how the approach was prototypically implemented. Then, an exemplary use-case is presented in which the approach is practically applied under real conditions in a maritime testbed infrastructure. Finally, the results of the application are measured and discussed with respect to the system's performance.

5.1. Implementation

In the following, we introduce the technologies we used for the proof-of-concept implementation of the presented approach. The prototypical implementation will serve as the basis for the following application and evaluation. The full implementation has been published and can be used for research purposes under the CC BY 4.0 license (DOI: 10.5281/zenodo.7323786).

General: The prototype is mainly implemented in Java. The TSA Negotiator and the Data Recorder were implemented analogously to the outlined processes in Figures 4 and 5. For the handling of any cryptographic operations and security-related protocols, such as the local creation of a timestamping request, the library Bouncy Castle was used. Bouncy Castle is a collection of open-source cryptographic programming interfaces [43]. It provides many methods for handling, e.g., X.509 certificates, key pairs, and timestamping authorities.

Timestamping Authority: For the certification of the data, an open timestamping authority of the "German Research Network (DFN)" is used. For research purposes, the TSA can be used free of charge. The exchange of data complies with RFC 3161. The TSA used can easily be replaced by any other RFC 3161-compliant TSA via the adjustment of a single parameter in the implementation [44].

Data Recording Database: For the persistence of the recorded data and the TSA signatures, a MongoDB instance is utilized. MongoDB is a No-SQL database that is well suited for storing semi-structured and unstructured data. Therefore, the database is ideally situated to persist the original records data, which can occur in a wide variety of different file formats, along with the associated signature of the TSA [45].

Communication: For communication between the participants, a simple peer-2-peer was implemented using the client-server framework Netty. The entire TSA negotiation is performed over the P2P network. The data that needs to be transmitted is exchanged with each other via TCP/IP.

Deployment: For a user-friendly deployment on multiple machines, the implementation was fully containerized using Docker [46]. For this purpose, two Docker containers were created containing all components, methods, and interfaces needed.

5.2. Evaluation Scenario

During mooring maneuvers, collisions between vessels and the quay walls occur regularly. Often, minor damages remain undetected at first and are only noticed by the port operator after some time. A subsequent clarification of which vessel caused the damage is no longer possible so that the port remains on the incurred costs. To prevent this, port operators are interested in documenting the berthing maneuvers of vessels in a trustworthy way. To evaluate the presented approach for the decentralized documentation of maritime incidents, we applied our system to a real-world problem of port operators and assessed it with respect to its functionality and performance.

For this purpose, the system was embedded in the *SmartKai* testbed in Cuxhaven, Germany (c.f. Figure 6a). The *SmartKai* testbed was set up for the development of a port-side assistance system to support pilots and captains during the berthing of vessels. This is realized by measuring the distance of a vessel to the quay wall using LiDAR sensors and making it available to the stakeholders in real-time. In contrast to AIS, the measurements are not limited to a single reference point but cover the entire contour of a vessel and are available to the stakeholders at a much higher frequency (5 Hz). In total, the *SmartKai* testbed provides 8 LiDAR sensors, radar, and AIS data over the entire port area, and other environmental data such as tidal, weather, and visibility information. Due to the high number of available sensors and the realistic environmental conditions such as the used hardware and network, the testbed is ideally suited to evaluate the presented approach properly. The functionality of the presented approach is demonstrated by a mooring maneuver of the vessel *Steubenhoeft* at the quay wall in the harbor of Cuxhaven. The *Steubenhoeft* is a dredger that is 40 m long and 10 m wide (c.f. Figure 6b). In principle, the proposed approach could also be used to document any other traffic situation.

(a)

(b)

Figure 6. (a) LiDAR sensor at the quay wall in Cuxhaven, Germany, to measure the distance between the quay wall and vessels (part of the SmartKai infrastructure) (b) Dredger vessel *Steubenhoeft* with which the mooring maneuver that has to be documented was performed.

Within the evaluation mooring scenario, a total of three participants take part (c.f. Figure 7):

(1) The Port Operator who operates the port is interested in ensuring that the quay wall of the port is not damaged and has installed 8 LiDAR sensors at the quay wall with an update frequency of 5 Hz. The sensors' measurements are synchronized so that they can be collected and signed together. Furthermore, the Port Operator trusts TSA1 and TSA2 but does not rely on any other external entity.

(2) The Vessel Traffic Service Operator knows the Port Operator and supports the berthing documentation by providing valuable environmental data about the port area. The VTS operator collects AIS, visibility, tidal, and wind data. The AIS data is received several times per second from surrounding vessels at irregular intervals. The installed wind sensors update with a frequency of 10 Hz, and the visibility sensor every minute. In contrast, the tidal information is retrieved only once per hour. The VTS Operator mainly trusts TSA3. If it is not possible to agree directly on the first priority, the VTS Operator also trusts the TSAs of its trusted entity—the BSH.

(3) The Berthing Vessel is the vessel that wants to moor in the port. The Berthing Vessel stores its received AIS messages and own GPS positions. The GPS data is encoded in the NMEA0183 format and is updated with 1 Hz. The vessel relies primarily on TSA4 and TSA5. However, the vessel is willing to expand its list in two additional iterations: Once by the TSAs of its trusted entities—the DLR and the IALA and another time by the TSAs of the trusted entities of the DLR and IALA.

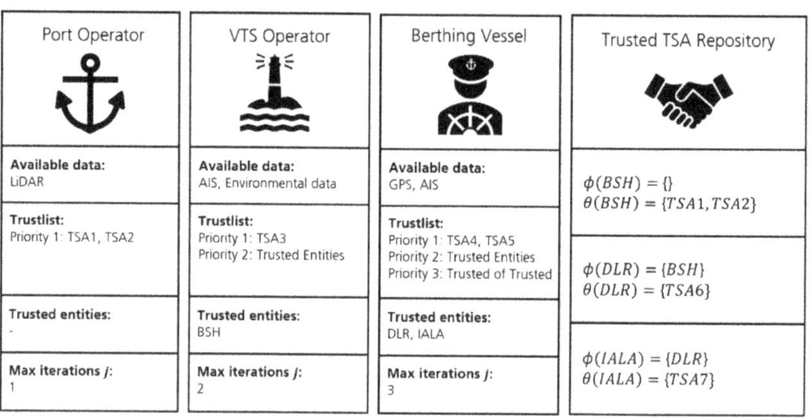

Port Operator	VTS Operator	Berthing Vessel	Trusted TSA Repository
Available data: LiDAR	**Available data:** AIS, Environmental data	**Available data:** GPS, AIS	$\phi(BSH) = \{\}$ $\theta(BSH) = \{TSA1, TSA2\}$
Trustlist: Priority 1: TSA1, TSA2	**Trustlist:** Priority 1: TSA3 Priority 2: Trusted Entities	**Trustlist:** Priority 1: TSA4, TSA5 Priority 2: Trusted Entities Priority 3: Trusted of Trusted	$\phi(DLR) = \{BSH\}$ $\theta(DLR) = \{TSA6\}$
Trusted entities: -	**Trusted entities:** BSH	**Trusted entities:** DLR, IALA	$\phi(IALA) = \{DLR\}$ $\theta(IALA) = \{TSA7\}$
Max iterations *J*: 1	**Max iterations *J*:** 2	**Max iterations *J*:** 3	

Figure 7. Initial situation of the evaluation scenario.

The hardware and software of the three participants is deployed on three separate machines. The Port Operator and VTS Operator instances are located at the quay and are connected to each other via a local network. These machines have an Intel Core i7-8700T processor with 16GB DDR4 RAM. The machine of the Berthing Vessel is located on board. It is based on an Intel Core i7-1185G7 processor with 16GB DDR4 RAM. All machines have an LTE connection.

The Public Trusted Entities Repository is available for all the participants. In our implementation, any entity (such as an organization, an authority, or a company) is allowed to publish their trusted organizations ϕ and trusted TSAs θ. However, in reality, it is primarily organizations, authorities, and institutes that publish trust lists, as they have the resources to verify the trustworthiness of individual TSAs. In addition, these also serve as an anchor of trust so that many participants could also trust their trust lists. In the specific test scenario, the repository has three entries from BSH, DLR, and IALA (c.f. Figure 7). The parties are the public authority "Federal Maritime and Hydrographic Agency of Germany (BSH)," the intergovernmental organization "International Association of Marine Aids to Navigation and Lighthouse Authorities (IALA)," and the research institute "German Aerospace Center (DLR)." Note that the entries in the Trusted TSA Repository are only an example and do not necessarily correspond to reality.

5.3. Application and Results

To demonstrate the presented approach, the evaluation scenario was applied to the four phases presented in Section 4. In addition, it is described how the generated documentation of the mooring maneuver can be used in case of a conflict.

(1) Initiate the documentation: Since the Port Operator wants to protect its port infrastructure, they are also responsible for defining when a traffic situation becomes critical and worthy of documentation. As already described in Section 4, for this purpose, a trigger needs to be defined that determines when a critical traffic event begins and thus the documentation is initiated. In the case of a mooring maneuver, a geographical region in front of the quay wall was utilized (c.f. Figure 8).

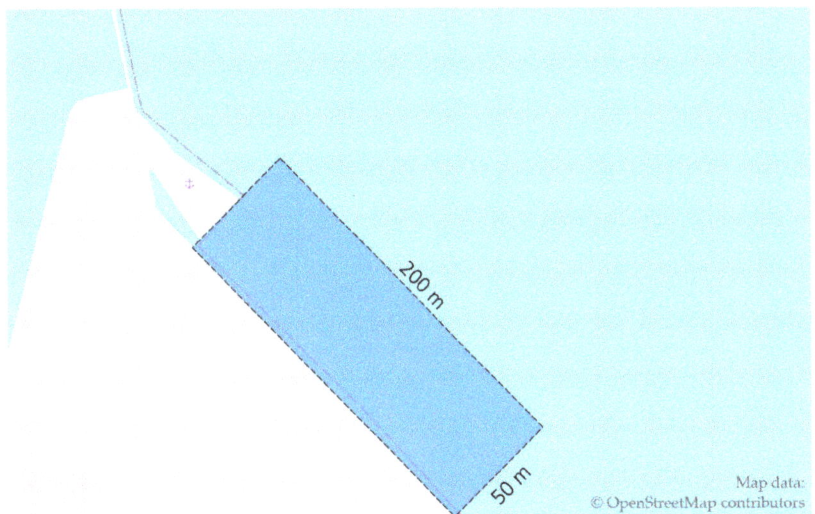

Figure 8. Geographical region in front of the quay wall where a documentation process is initiated when a vessel enters (Cuxhaven, Germany).

As a trigger, we defined a 50 m × 200 m box in front of the Amerikahafen in Cuxhaven. If a Berthing Vessel enters this region, the documentation's initialization trigger is released. All vessels that request to berth in the port are informed that they have to participate in the documentation process when entering the port. Additionally, the IP-address of the port is also published at which a vessel has to report as soon as it enters the critical region.

(2) Establish a communication channel: For establishing a communication channel between all participants, the Berthing Vessel is obligated to contact the IP of the Port Operator by using the prototype. Since the Port Operator and the VTS Operator know each other and the VTS Operator always supports the berthing documentation, the two participants are permanently connected via a P2P network. In this way, the Port Operator is connected to the VTS Operator and the Berthing Vessel. To enable the Port Operator and the VTS Operator to communicate with each other as well, the Port Operator broadcasts each other's IP addresses to all participants. In this way, each participant knows the IP of every other participant and can thus establish a P2P connection.

(3) Derive a common trusted set of TSAs: Once a communication channel is established, participants use it to negotiate a mutually trusted set of TSAs. For this purpose, the participants share their preferred familiar TSAs in j iterations to all other participants via the P2P network. The negotiation proceeds according to the Negotiation Protocol (c.f. Figure 4), as outlined in detail in Table 1.

Table 1. Familiar TSAs from each participant for each negotiation iteration j and common TSA intersection according to the Negotiation Protocol.

Iteration j	Port Operator	VTS Operator	Berthing Vessel	TSA Inters.
1	TSA 1, 2	TSA 3	TSA 4, 5	∅
2	TSA 1, 2	TSA 1, 2, 3	TSA 4, 5, 6, 7	∅
3	TSA 1, 2	TSA 1, 2, 3	TSA 1, 2, 4, 5, 6, 7	TSA 1, 2

- Iteration $j = 1$: According to the evaluation scenario (c.f. Figure 7), in the first iteration, each participant shares their first priority of trusted TSAs. Therefore, the Port Operator broadcasts TSA1 and TSA2, the VTS Operator broadcasts TSA3, and the Berthing

Vessel TSA4 and TSA5. After receiving messages from all other participants, the participant can determine the intersection of the trusted TSAs locally. In iteration $j = 1$, the intersection is an empty set, so another iteration is initiated along the negotiation protocol.

- Iteration $j = 2$: In the second iteration, participants are requested to expand their list of trusted TSAs. Therefore, the VTS Operator and the Berthing Vessel include the TSAs of their trusted entities. The VTS Operator trusts the BSH. Based on private relationships or the use of the Public Trusted Entities Repository, the VTS Operator knows which TSAs are trusted by the BSH. In this case, the BSH trusts TSA1 and TSA2, so they are included into the VTS Operator's TSA list. The Berthing Vessel relies on the DLR and the IALA. The DLR trusts TSA6 and the IALA trusts TSA7. Therefore, both TSAs are also included in the list of the Berthing Vessel. The Port Operator follows a very strict strategy and trusts only his own two TSAs. There is no extension of the list at all. Afterwards, every participant broadcasted their trust list and it is checked again whether a common intersection exists between the participants. After iteration $j = 2$, the intersection at TSA is still empty, so the next iteration is initiated.

- Iteration $j = 3$: In the third iteration, only the Berthing Vessel is further expanding its list of trusted TSAs. For this purpose, the Berthing Vessel relies on the trusted TSAs of the trusted entities from the DLR and IALA. Therefore, TSA1 and TSA2 are also added to the trust list of the Berthing Vessel. The VTS and Port Operators are not expanding their trust list. Again, the intersection of all trust lists is determined locally. In the iteration, the participants find a common basis with TSA1 and TSA2.

This fulfills the condition $\cap_{i=1}^{n} \pi_j(P_i) \neq \varnothing$, so the negotiation protocol terminates at this point. The participants can use either TSA1 or TSA2 to sign their data. If no common TSA has been found at this point, it would not be possible to find a common TSA for all participants, since all participants have already extended their lists to the maximum ($j_{max} = 3$). In this case, an attempt would be made to find a common TSA for as large a subset of participants as possible (c.f. Figure 4, left part).

Record the data: After agreeing on a common set of TSAs, no further communication takes place over the P2P network. Instead, each participant records their data analogously to the process of Figure 5 by chunking the data in 10-s intervals and hashing it with SHA256. Each participant then sends the hashes directly to one of the determined TSAs. In our application, the Port Operator and Berthing Vessel used the TSA1 and the VTS Operator utilized TSA2. Finally, the data chunks of the data stream are persisted in the local recording database together with the signature of the TSA.

Only in case of an actual conflict, e.g., the Berthing Vessel damages the quay wall, the signed records are retrieved from the databases as proof of the actual course of events. In this case, the Port Operator and the VTS Operator would disclose their data to show that the Berthing Vessel damaged the quay wall. In addition, the participants can verify, with the process described in Figure 2, that the data already existed during the berthing maneuver. Since the Berthing Vessel also trusts the signing TSA, it cannot deny that the data has been manipulated afterwards by the Port or VTS Operators. Of course, the Berthing Vessel can also provide its own signed data in order to exonerate itself. In general, a large and heterogeneous data foundation helps to reconstruct the real course of a berthing maneuver. In this way, the data of several participants can be compared with each other in order to check whether they provide a coherent overall picture. For example, the vessel's track recorded by the LiDAR sensor could be compared with the AIS recordings of the vessel itself. Even in the case of real-time manipulation during a critical traffic event, data recordings that do not fit into the overall picture could be detected in this way so that they can still be identified as being manipulated.

Since it is not always foreseeable when a critical traffic situation will occur, it is also essential to evaluate the required time from the initiation of the documentation process to the final signed data. Therefore, as part of the practical application of the approach, we collected data for the duration of creating the communication channel, deriving the common TSA basis, and signing and storing a data chunk (c.f. Table 2). In addition, we also ran the scenario 1000 times simulatively to see if the durations also matched the values from the field test. The simulation was executed on a computer with an Intel Core i9-9900K processor and 32 GB of RAM.

Table 2. Time required by the approach in the evaluation scenario depending on the respective phase.

Phase	Duration in SmartKai Testbed	Average Duration in Simulation
Establishing communication channel	636 ms	30 ms
Derive a common trusted set of TSAs	313 ms	100 ms
Signing and storing of a data chunk	118 ms	106 ms
Total	1067 ms	236 ms

As can be seen, the presented approach requires a total of 1067 milliseconds under real conditions from initiating the communication channel to storing the first signed chunk in the database. This time is composed of the three phases of establishing the communication channel with 636 ms, the TSA negotiation of the participants with 313 ms, the signing of the data with a TSA, and subsequent storage in the local database with a duration of 118 ms. Even faster results were obtained within our 1000 simulative runs. On average, the total duration was 236 ms. The creation of the communication channel with 30 ms was significantly faster than in the testbed. This can be explained by the faster and more stable internet connection in the simulation environment. Furthermore, the derivation of a TSA set with 100 ms and signing and saving of the chunks with 106 ms was also faster than in the testbed.

5.4. Discussion

The functionality of the approach for a decentralized documentation of critical traffic situations was demonstrated based on a realistic evaluation scenario within a maritime SmartKai testbed. Furthermore, the presented approach fulfills the requirements of Section 2.4:

(R1) Firstly, if a participant wishes to document its data recording, it must be sent to the identified TSA in hashed form. This makes the documentation process completely independent of the format and content of the underlying data. The only prerequisite is that it must be possible to hash the documented data. This is no issue for any data stored in files, and even streamed data can be hashed by breaking it down into individual packets.

(R2) Secondly, as proof that the data has not been manipulated after recording, a hash of the data is sent to a TSA approved by all participants. This TSA provides the hash with a timestamp and signs it with its digital signature. In this way, it can be cryptographically verified at any time whether the data has been manipulated after it has been recorded.

(R3) The presented approach puts a strong focus on protecting the sovereignty of individual data providers. Each participant is able to record and persist his data independently. As long as there is no conflict, the data is also not shared with the other participants. Even in the event of a conflict, the data provider has the option not to share the data. Additionally, only a hash of the original data is transmitted to the TSA itself, so even the TSA cannot draw any conclusions about the actual data.

(R4) Furthermore, anyone can join the P2P network used for communication as long as the participant knows the IPs of the other participants. The best way to distribute the IPs among the participants depends on the use case for which the documentation has to be created.

(R5) Lastly, if not all participants are equipped with the system for documentation, it is still possible for the remaining participants to document a situation among themselves in a tamper-proof manner. If no participant is equipped with the system or it is not possible to reach an agreement during TSA negotiation, the fallback TSA mechanism takes effect (c.f. Section 4.2). In this case, the participant documents his data with a TSA selected by himself in order to establish at least a basic trust in the documentation.

In order for the approach to be used for documenting critical traffic situations, it is important that the entire process from establishing contact to documenting remains within a reasonable time frame. Unlike in other domains such as the automotive domain, critical traffic events in the maritime domain can often be determined several minutes in advance, so the required time of 1068 ms should not have a negative impact in most cases.

In addition, the developed approach was also compared with similar existing approaches to discuss the strengths and weaknesses of each approach (cf. Table 3).

Table 3. Comparison of different approaches for the reconstruction of maritime situations (+: fulfilled; o: partially fulfilled; -: not fulfilled).

Approach	Variety of Data	Tamper-Proof	Data Sovereignty	Complexity	Connectivity
Voyage Data Recorder	o	+	+	-	+
Logbook	-	-	+	+	+
Blockchain-based approach	+	+	+	-	-
TSA-based approach	+	+	+	o	-

We compared the existing methods to reconstruct traffic events, such as VDR and logbook, the blockchain-based approach and the TSA-based approach presented in this paper. The evaluation criteria include the possibility of documenting any data (Variety of data), the tamper-proofness of the documentation (Tamper-proof), the local persistence of the data (Data Sovereignty), the complexity of the overall system and the readout process (Complexity), and the dependence on an Internet connection (Connectivity).

The Voyage Data Recorder can record various information in an automated way but is mainly limited to information about its own ship. However, due to its robustness and certification, the data is kept very secure, making it difficult to manipulate it [47]. In addition, the data is stored directly on the ship, so it does not need to be migrated to an external infrastructure. However, as mentioned above, retrieving the data turns out to be time-consuming, which is why the approach is not used in practice to clarify minor incidents [4]. A connection to the Internet is not required.

The most important events on board are documented in a logbook [3]. In general, this is a manual process that is still carried out by hand. Accordingly, a logbook does not record detailed data. In addition, it is only expected that the logbook will be filled out truthfully [48]. Therefore, it can be comparatively easily manipulated. However, on the positive side, the data records are only stored on the vessel itself, and the effort for the documentation and readout is low. The logbook also does not require an internet connection.

The blockchain and TSA-based approaches differ from the classic approaches. Both approaches are, in principle, data-independent so that any data can be documented. By storing a hash on a blockchain or signing the hash with a TSA, subsequent changes to the data can be easily identified. In both cases, the data is communicated externally, but since the data is only transmitted as a hash, no conclusions can be drawn about the original data (cf. Section 4). The complexity of the blockchain solution depends very much on how the blockchain itself is designed. Accordingly, the effort required to enter and verify a hash can be very high. However, since each TSA follows a standardized process and the data is cryptographically signed, the integrity of the data can be verified by examining the signature in a computationally efficient way and without a network of multiple blockchain nodes. Therefore, the overall complexity of the approach can be considered lower than the blockchain-based solution. However, both approaches require an internet connection in order to document the data in a tamper-proof way. In the near future, it is expected that ships will also have a permanent connection via satellites [49–51]. In regions close to land, most ships already have an LTE connection. Therefore, this criterion should not be a major limitation in the future.

Overall, the existing approaches for documenting traffic situations focus on data from the own vessel. Other external data sources are not part of the documentation. Nevertheless, the data from other sources could contribute to a better understanding of the cause of critical traffic situations. Therefore, establishing such a system in the maritime domain would make sense. In principle, the blockchain-based and TSA-based approaches are suitable for tamper-proof documentation. However, the complexity of a TSA-based approach will usually be lower, so it should be preferred.

6. Summary and Conclusions

In this paper, we have presented an approach for the creation of a decentralized and trustworthy documentation of maritime traffic situations. Our research literature reveals that the clarification of minor traffic incidents presents a particular difficulty, since the VDR on board a vessel is often only read by the authorities in the case of major incidents. A forgery-proof data basis is generally not available for clarification of smaller incidents. Furthermore, apart from ship-related data, e.g., the VDR, and a lot of contextual information such as weather information, is recorded by other entities. This contextual information can be very important for a proper documentation of an incident but is currently not addressed by existing approaches.

Therefore, we have taken up this challenge and by first analyzing which methods already exist to persist data in a tamper-proof way. Besides central authorities that perform the documentation and blockchain-based approaches, timestamping authorities that have the task of adding a timestamp to data in a cryptographically verifiable way can also be utilized. Compared to the other approaches, TSAs have the advantage of being able to certify the integrity of data records using standardized and computationally efficient operations. However, as the maritime domain is an international industry with many different participants from a large number of countries, it is unrealistic to assume that all participants will agree on a single central TSA. For the agreement on a TSA, the Negotiation Protocol was introduced, which ensures that a TSA with the maximum trust will be found among the participants. For communication, a P2P network is established when a document worthy situation is initiated. The presented approach also allows storing data records on the documenting party's infrastructure. In this way, the participants retain complete control over their own data. Only in the case of a conflict resolution, the participant needs to share his certified data with the other parties.

To demonstrate the functionality of the approach, we integrated our implementation into the infrastructure of the SmartKai Testbed in Cuxhaven. The approach was successfully tested based on a berthing maneuver involving three different parties. It was shown that even in real deployment environments, the parties could communicate over the dynamically created P2P network to agree along the Negotiation Protocol on a common TSA, to certify their own data using the authority and persist it in a local database.

Nevertheless, it should be noted that the presented approach also has some limitations. For example, a continuous internet connection or additional effort to establish non-IP-based communication to a gateway is required to agree on a TSA and to send the data recordings to the TSA for certification. Especially when the approach should be used on the open sea, where ships usually have limited or no internet connection, this could cause problems. Still, by using alternative communication channels such as the VHF Data Exchange System (VDES), or by caching the required information beforehand, it would be possible to exchange the required information even without a stable internet connection. A further limitation is that only parties that already have the required software installed on their infrastructure can participate in the documentation process. Parties that do not have the software or hardware can not be included in the documentation process. Lastly, the presented approach only grants that the data already existed on the date issued by the TSA. However, the data could have been manipulated before it was signed by the provider. Therefore, it is extremely important to look at data from multiple participants when resolving conflicts and see if they match.

In future work, we envision being able to ensure that a participant's data records have not been manipulated before the time of certification. In addition, we would like to investigate whether alternative communication channels can be used for the TSA negotiation, so that the parties can agree on a common basis of trust even if they do not have a stable internet connection in the initiation of a critical traffic incident. Furthermore, a legal classification of the approach would be interesting in order to be able to assess whether the decentralized documentation may also be used as evidence according to the current legal situation.

Author Contributions: Conceptualization, D.J., J.M., H.W. and A.H.; data curation, D.J., J.M. and H.W.; investigation, D.J., J.M. and H.W.; methodology, D.J., J.M. and H.W.; project administration, D.J.; software, D.J., J.M. and H.W.; supervision, A.H.; validation, D.J. and H.W.; visualization, D.J. and H.W.; writing—original draft, D.J., J.M. and H.W.; writing—review and editing, D.J., J.M., H.W. and A.H. All authors have read and agreed to the published version of the manuscript.

Funding: This work has been funded by the German Federal Ministry of Transport and Digital Infrastructure (BMVI) within the funding guideline "Innovative Hafentechnologien" (IHATEC) under the project SmartKai with the funding code 19H19008E and the FuturePorts project of the German Aerospace Center (DLR).

Institutional Review Board Statement: Not applicable.

Informed Consent Statement: Not applicable.

Data Availability Statement: Not applicable.

Conflicts of Interest: The authors declare no conflict of interest. The funders had no role in the design of the study; in the collection, analyses, or interpretation of data; in the writing of the manuscript; nor in the decision to publish the results.

References

1. Morcom, A.R. Flight data recording systems: A brief survey of the past developments, current status and future trends in flight recording for accident investigation and operational purposes. *Aircr. Eng. Aerosp. Technol.* **1970**, *42*, 12–16. [CrossRef]
2. Yoder, T.A. Development of Aircraft Fuel Burn Modeling Techniques with Applications to Global Emissions Modeling and Assessment of the Benefits of Reduced Vertical Separation Minimums. Master's Thesis, Massachusetts Institute of Technology, Cambridge, MA, USA, 2007. Available online: https://dspace.mit.edu/handle/1721.1/39713 (accessed on 28 July 2022).
3. International Maritime Organization (IMO). SOLAS Chapter V—Safety of Navigation. Available online: https://www.imorules.com/SOLAS_REGV.html (accessed on 5 October 2022).
4. Cantelli-Forti, A. Forensic Analysis of Industrial Critical Systems: The Costa Concordia's Voyage Data Recorder Case. In Proceedings of the 2018 IEEE International Conference on Smart Computing (SMARTCOMP), Taormina, Italy, 18–20 June 2018; pp. 458–463. [CrossRef]
5. IMO. Casualties. 2019. Available online: https://www.imo.org/en/OurWork/MSAS/Pages/Casualties.aspx (accessed on 31 July 2022).
6. Bundesstelle für Seeunfalluntersuchung. Bundesstelle für Seeunfalluntersuchung—Untersuchungsberichte. 2022. Available online: https://www.bsu-bund.de/DE/Publikationen/Unfallberichte/Unfallberichte_node.html (accessed on 31 July 2022).
7. UK Government. Marine Accident Investigation Branch Reports. *GOV.UK.* 2022. Available online: https://www.gov.uk/maib-reports (accessed on 31 July 2022).
8. Voigt, P.; Von dem Bussche, A. *The EU General Data Protection Regulation (GDPR)*; Springer International Publishing: Cham, Switzerland, 2017. [CrossRef]
9. Hummel, P.; Braun, M.; Tretter, M.; Dabrock, P. Data sovereignty: A review. *Big Data Soc.* **2021**, *8*, 2053951720982012. [CrossRef]
10. Mason, S. *Electronic Signatures in Law*; University of London Press: London, UK, 2016.
11. Stallings, W.; Brown, L. *Computer Security: Principles and Practice*, 5th ed.; Global ed.; Pearson: New York, NY, USA, 2018. Available online: https://elibrary.pearson.de/book/99.150005/9781292220635 (accessed on 16 August 2022).
12. Weinert, B.; Park, J.H.; Christensen, T.; Hahn, A. A Common Maritime Infrastructure for Communication and Information Exchange. In Proceedings of the 19th IALA Conference 2018, Incheon, Republic of Korea, 27 May–2 June 2018.
13. Sheth, H.; Dattani, J. Overview of blockchain technology. *Asian J. Converg. Technol.* **2019**, *5*, 1–4. [CrossRef]
14. Koens, T.; Poll, E. What blockchain alternative do you need? In *Data Privacy Management, Cryptocurrencies and Blockchain Technology*; Springer: Berlin/Heidelberg, Germany, 2018; pp. 113–129.
15. Andolfatto, D. Blockchain: What it is, what it does, and why you probably don't need one. *Fed. Reserve Bank Louis Rev.* **2018**, *100*, 87–95. [CrossRef]

16. *ANSI X9.95-2016*; Trusted Time Stamp Management and Security. American National Standards Institute (ANSI): Annapolis, MD, USA, 2016.
17. Zuccherato, R.; Cain, P.; Adams, D.C.; Pinkas, D. Internet X.509 Public Key Infrastructure Time-Stamp Protocol (TSP); No. 3161; RFC Ed. August 2001. Available online: https://www.rfc-editor.org/info/rfc3161 (accessed on 5 November 2022).
18. Buldas, A.; Laud, P.; Lipmaa, H.; Villemson, J. Time-stamping with binary linking schemes. In *Advances in Cryptology—CRYPTO '98*; Krawczyk, H., Ed.; Springer: Berlin/Heidelberg, Germany, 1998; Volume 1462, pp. 486–501. [CrossRef]
19. Haber, S.; Stornetta, W.S. How to time-stamp a digital document. *J. Cryptol.* **1991**, *3*, 99–111. [CrossRef]
20. Moller, J.; Jankowski, D.; Lamm, A.; Hahn, A. Data Management Architecture for Service-Oriented Maritime Testbeds. *IEEE Open J. Intell. Transp. Syst.* **2022**, *3*, 631–649. [CrossRef]
21. Nofer, M.; Gomber, P.; Hinz, O.; Schiereck, D. Blockchain. *Bus. Inf. Syst. Eng.* **2017**, *59*, 183–187. [CrossRef]
22. Jamthagen, C.; Hell, M. Blockchain-Based Publishing Layer for the Keyless Signing Infrastructure. In Proceedings of the 2016 Intl IEEE Conferences on Ubiquitous Intelligence & Computing, Advanced and Trusted Computing, Scalable Computing and Communications, Cloud and Big Data Computing, Internet of People, and Smart World Congress (UIC/ATC/ScalCom/CBDCom/IoP/SmartWorld), Toulouse, France, 18 July 2016; IEEE: New York, NY, USA, 2016; pp. 374–381. [CrossRef]
23. Cucurull, J.; Puiggalí, J. Distributed Immutabilization of Secure Logs. In *Security and Trust Management*; Barthe, G., Markatos, E., Samarati, P., Eds.; Springer International Publishing: Cham, Switzerland, 2016; Volume 9871, pp. 122–137. [CrossRef]
24. Buldas, A.; Kroonmaa, A.; Laanoja, R. Keyless Signatures' Infrastructure: How to Build Global Distributed Hash-Trees. In *Secure IT Systems*; Riis Nielson, H., Gollmann, D., Eds.; Springer: Berlin/Heidelberg, Germany, 2013; Volume 8208, pp. 313–320. [CrossRef]
25. Gervais, A.; Ritzdorf, H.; Karame, G.O.; Capkun, S. Tampering with the Delivery of Blocks and Transactions in Bitcoin. In Proceedings of the 22nd ACM SIGSAC Conference on Computer and Communications Security, Denver, CO, USA, 12 October 2015; pp. 692–705. [CrossRef]
26. Szalachowski, P. (Short Paper) Towards More Reliable Bitcoin Timestamps. In Proceedings of the 2018 Crypto Valley Conference on Blockchain Technology (CVCBT), Zug, Switzerland, 20 June 2018; pp. 101–104. [CrossRef]
27. Putz, B.; Menges, F.; Pernul, G. A secure and auditable logging infrastructure based on a permissioned blockchain. *Comput. Secur.* **2019**, *87*, 101602. [CrossRef]
28. Shekhtman, L.; Waisbard, E. EngraveChain: A Blockchain-Based Tamper-Proof Distributed Log System. *Futur. Internet* **2021**, *13*, 143. [CrossRef]
29. Wust, K.; Gervais, A. Do you Need a Blockchain? In Proceedings of the 2018 Crypto Valley Conference on Blockchain Technology (CVCBT), Zug, Switzerland, 2 June 2018; pp. 45–54. [CrossRef]
30. Crosby, S.A.; Wallach, D.S. Efficient Data Structures for Tamper-Evident Logging. In Proceedings of the 18th Conference on USENIX Security Symposium, Montreal, QC, Canada, 14–18 August 2009; pp. 317–334.
31. Schneier, B.; Kelsey, J. Cryptographic Support for Secure Logs on Untrusted Machines. In Proceedings of the 7th USENIX Security Symposium (USENIX Security 98), San Antonio, TX, USA, 26 January 1998. Available online: https://www.usenix.org/conference/7th-usenix-security-symposium/cryptographic-support-secure-logs-untrusted-machines (accessed on 5 November 2022).
32. Levin, D.; Douceur, J.R.; Lorch, J.R.; Moscibroda, T. TrInc: Small Trusted Hardware for Large Distributed Systems. In *NSDI*; Springer: Berlin/Heidelberg, Germany, 2009.
33. Sinha, A.; Jia, L.; England, P.; Lorch, J.R. Continuous Tamper-Proof Logging Using TPM 2.0. In *Trust and Trustworthy Computing*; Holz, T., Ioannidis, S., Eds.; Springer International Publishing: Cham, Switzerland, 2014; Volume 8564, pp. 19–36. [CrossRef]
34. MVEDR-EC-Motor Vehicle Event Data Recorder Brake and Electronic Control Working Group. IEEE 1616-2021-Standard for Motor Vehicle Event Data Recorder (MVEDR). IEEE. Available online: https://standards.ieee.org/ieee/1616/10329/ (accessed on 5 November 2022).
35. Yao, Y.; Atkins, E. The Smart Black Box: A Value-Driven High-Bandwidth Automotive Event Data Recorder. *IEEE Trans. Intell. Transp. Syst.* **2020**, *22*, 1484–1496. [CrossRef]
36. Singleton, N.; Daily, J.; Manes, G. Automobile Event Data Recorder Forensics. In *Advances in Digital Forensics IV*; Ray, I., Shenoi, S., Eds.; Springer: Boston, MA, USA, 2008; Volume 285, pp. 261–272. [CrossRef]
37. Vinzenz, N.; Eggendorfer, T. Forensic Investigations in Vehicle Data Stores. In Proceedings of the Third Central European Cybersecurity Conference, Munich, Germany, 14–15 November 2019; pp. 1–6. [CrossRef]
38. Wahab, A.; Waseso, B.; Pranoto, H. Synchronization of Catch Fish Data in Fisheries e-Logbook with a Vessel Monitoring System. *Int. J. Adv. Technol. Mech. Mechatron. Mater.* **2021**, *2*, 46–54. [CrossRef]
39. Mion, M.; Piras, C.; Fortibuoni, T.; Celić, I.; Franceschini, G.; Giovanardi, O.; Belardinelli, A.; Martinelli, M.; Raicevich, S. Collection and validation of self-sampled e-logbook data in a Mediterranean demersal trawl fishery. *Reg. Stud. Mar. Sci.* **2015**, *2*, 76–86. [CrossRef]
40. Young, W. What Are Vessel Traffic Services, and What Can They Really Do? *Navigation* **1994**, *41*, 31–56. [CrossRef]
41. Chang, S. Development and analysis of AIS applications as an efficient tool for vessel traffic service. In Proceedings of the Oceans '04 MTS/IEEE Techno-Ocean '04 (IEEE Cat. No.04CH37600), Kobe, Japan, 9–12 November 2004; Volume 4, pp. 2249–2253. [CrossRef]

42. Hepp, T.; Schoenhals, A.; Gondek, C.; Gipp, B. OriginStamp: A blockchain-backed system for decentralized trusted timestamping. *It-Inf. Technol.* **2018**, *60*, 273–281. [CrossRef]
43. Detho, W. Developing a system for securely time-stamping and visualizing the changes made to online news content. *arXiv* **2018**, arXiv:1802.07285. [CrossRef]
44. *IEC TC80*; IEC 63173-2:2022: Maritime Navigation and Radiocommunication Equipment and Systems-Data Interfaces-Part 2: Secure Communication between Ship and Shore (SECOM). IEC: Geneva, Switzerland, 2022. Available online: https://webstore.iec.ch/publication/64543 (accessed on 5 November 2022).
45. Legion of the Bouncy Castle Inc. Bouncy Castle–Documentation. Available online: https://web.archive.org/web/20221012164800/https://www.bouncycastle.org/documentation.html (accessed on 15 November 2022).
46. Deutsches Forschungsnetz (DFN). DFN Timestamping Authority. Available online: https://web.archive.org/web/20220000000000*/https://www.pki.dfn.de/faqpki/faq-zeitstempel (accessed on 15 November 2022).
47. MongoDB, Inc. MongoDB Documentation. Available online: https://web.archive.org/web/20221115034209/https://www.mongodb.com/docs/manual/ (accessed on 15 November 2022).
48. Docker, Inc. Docker Documentation. Available online: https://web.archive.org/web/20221104022709/https://docs.docker.com/ (accessed on 15 November 2022).
49. Piccinelli, M.; Gubian, P. Modern ships Voyage Data Recorders: A forensics perspective on the Costa Concordia shipwreck. *Digit. Investig.* **2013**, *10*, S41–S49. [CrossRef]
50. Schotte, M. Expert Records: Nautical Logbooks from Columbus to Cook. *Inf. Cult.* **2013**, *48*, 281–322. [CrossRef]
51. Wei, T.; Feng, W.; Chen, Y.; Wang, C.-X.; Ge, N.; Lu, J. Hybrid Satellite-Terrestrial Communication Networks for the Maritime Internet of Things: Key Technologies, Opportunities, and Challenges. *IEEE Internet Things J.* **2021**, *8*, 8910–8934. [CrossRef]

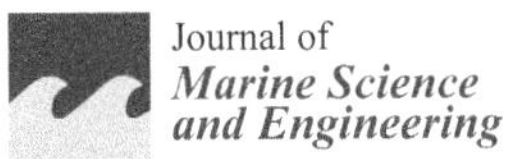

Journal of
*Marine Science
and Engineering*

Article

Identifying the Most Probable Human Errors Influencing Maritime Safety

Xiaofei Ma [1,2], Guoyou Shi [1,2,*], Weifeng Li [1,2] and Jiahui Shi [1,2]

[1]   Navigation College, Dalian Maritime University, Dalian 116026, China
[2]   Key Laboratory of Navigation Safety Guarantee of Liaoning Province, Dalian 116026, China
*   Correspondence: sgydmu@163.com

Abstract: In the traditional and extended shipboard operation human reliability analysis (SOHRA) model, the error-producing condition (EPC) is critical. The weight and proportion of each EPC in one specific task are often determined by the experts' judgments, including most of the modified versions. Due to this subjectivity, the result and recommended safety measures may not be as accurate as they should be. This study attempts to narrow the gap by proposing a novel approach, a combination of SOHRA, entropy weight method, and the TOPSIS model. The entropy weight and TOPSIS method are employed to decide the weight of each EPC based on the foundation of the SOHRA model. A cargo-loading operation from a container ship is analyzed to verify this model. The results suggest that the entropy-weighted TOPSIS method can effectively determine the weights of EPCs, and the eight most probable human errors are identified.

Keywords: human error; loading operation; ship safety; shipboard operation human reliability analysis; entropy-weighted TOPSIS method

Citation: Ma, X.; Shi, G.; Li, W.; Shi, J. Identifying the Most Probable Human Errors Influencing Maritime Safety. *J. Mar. Sci. Eng.* **2023**, *11*, 14. https://doi.org/10.3390/jmse11010014

Academic Editors: Marko Perkovic, Lucjan Gucma and Sebastian Feuerstack

Received: 13 November 2022
Revised: 9 December 2022
Accepted: 16 December 2022
Published: 22 December 2022

1. Introduction

Human error is a hot research topic in the aviation, nuclear energy, healthy service, railway, and maritime fields. This is because many related accidents or incidents have connections with human errors [1–3]. For example, studies [4–6] show that around 80% of maritime accidents are caused by human error, or at least have a connection with human error. Furthermore, human error can cause significant accidents, thus inducing substantial economic losses, environmental pollution, or even human life losses.

As the central part of human factor research, the study of human error can be traced back to the 1930s [7,8]. However, human error, as a trouble-free identified factor, is easy to analyze qualitatively but difficult to study quantitatively. Early on, most of the research did not originate from the maritime field but from aviation [9], nuclear energy [10], railway [11,12], factory [13,14], medical care [15,16], other shoreside safety management [17], etc. Some methods have already been modified to study maritime safety. Meanwhile, the scanty human error data hampered the relevant research [18], especially in the maritime domain. Therefore, some alternative methods, such as human error probability (HEP) [1,19] and human reliability analysis (HRA) [20–22], were developed to obtain human error information more accurately and even predict human error occurrence.

1.1. Human-Error-Related Risk Assessment

Human-error-related risk assessment aims to assess the risk of human error, in other words, to obtain the HEP. According to the values of HEP, the author can assess the human error risks of a specific task. Therefore, HEP is the critical factor utilized in the risk assessment of human error [23].

One frequently used method is the Human Error Assessment and Reduction Technique (HEART), proposed by Williams [24]. Two related factors, human error probability (HEP)

and general error probability (GEP), were introduced to assess the risks of human error. The experts' judgments are an essential step in this method. Over time, the model has been modified many times, and its application has been broadened to other fields, e.g., medical care, the chemical industry, road traffic, maritime traffic, etc. Akyuz et al. [25] presented a modified model, incorporating the Analytic Hierarchy Process (AHP) and HEART method, to explore human errors in the tank-cleaning process onboard chemical tankers. The experts' judgments determined the GEPs and EPCs. Although the AHP method is used to weigh the proportion of each EPC, creating a judgment matrix using experts' judgments is still a crucial step in the AHP method. One year later, they [26] proposed another version by combining HEART and interval type-2 fuzzy sets (IT2FSs). The IT2FSs are used to cope with the linguistic variables judged by the experts. Wang et al. [27] utilized an H-HEART-F approach to assessing the HEP of the task. The Z-numbers and decision-making trial and evaluation laboratory (DEMATEL) method are used to address the experts' judgments and the interdependence of each EPC. They are only used to deal with the fuzziness of experts' judgments; the subjectivity of the research method still exists.

Another common model is the Successive Likelihood Index Method (SLIM) developed by Embrey et al. [28]. It is used to assess the HEP so that mitigation measures can be adopted to minimize human error, especially when scanty human error information is available. Islam et al. [29] studied human failure concerning the maintenance procedures of the ship's main engine through the SLIM model. The experts needed to rate the PSFs of each step and weigh those PSFs to obtain an SLI for each sub-task of the process. Akyuz [2] proposed a fuzzy-based SLIM model to explore the human error quantification in abandoned ship operations. Experts' judgments decided the weights of PSFs in the operations. The fuzzy sets were used to process the subjectivity of the experts' judgments, which could mitigate the subjectivity of the process. Erdem et al. [1] presented a modified SLIM model: IT2FS-SLIM. The IT2FS is used to cope with the subjectivity of the experts' judgments so that the sensitivity of the judgments can be mitigated. Islam et al. [23] developed a monograph to assess human error likelihood for maritime operations through the SLIM model; a series of experts' judgments are applied to rate and weight the PSFs according to the steps of the tasks.

Limited by the scanty information on human error, most research has relied on the experts' judgments, including the rating process and weight determination of related factors. The experts' judgments have significant influence on the sensitivity of the results. Some attempts have been applied to mitigate the subjectivity, but, presently, no alternative methods have been found to replace the experts' judgments. Consequently, subjectivity still exists in this research domain.

1.2. Human Reliability Analysis

It is considered that minimizing human error increases human reliability. Therefore, articles focusing on the HRA are considered human-error-related studies. These articles have made an extensive exploration of human error and reduction strategies.

Hollnagel [30] proposed the Cognitive Reliability and Error Analysis Method (CREAM) model to quantify human error retrospectively and prospectively. Because of its ability for quantification, numerous modified versions have been developed to deal with human error in specific operations. Ung [18] used fault tree analysis, fuzzy Bayesian network, and the CREAM method to study human failure in an oil tanker collision situation. The experts' judgments were used to decide the weights and quantitative effects caused by the ambient factors. The factors with the higher occurrence rate were identified. Akyuz [31] modified the CREAM model to analyze human errors when operating the inerting gas operation in an LPG tanker. The evaluations of common performance conditions (CPCs) were performed by the experts, and the relevant judgments were assigned for each primary process. The sub-tasks with higher HEP were identified and the risk mitigation measures were given. By integrating the CREAM model, the Bayesian network, and evidential reasoning, Yang et al. [32] provided a hybrid strategy. The nine CPCs' interaction was considered in the method, which is viewed as its main evolution. Zhou et al. [33], Xi et al. [3], Shirali et al. [34], and Wu et al. [35] also provide improved approaches to studying HRA quantitatively; experts evaluate all the CPCs in these methods.

In addition, Li et al. [36] proposed an Association Rule Bayesian Networks (ARBN) model to study the external factors influencing human error by analyzing ship collision reports. Two separate Bayesian networks, environment–human BN and ship–human BN, were built to evaluate the human error probability better. The experts' judgment was still an unavoidable step for this method.

Swain et al. [37] proposed the Technique for Human Error Rate Prediction (THERP) model around 30 years ago for nuclear power plant applications. It was utilized to predict human error by calculating the HEP values. Now, the model has been successfully adopted in the maritime field. Zhang et al. [38] presented a modified model, THERP-BN, to evaluate the HEP of emergency operations on an autonomous ship. Based on the experts' judgments, the fuzzy number synthesis method was used to obtain the experts' scores so that the HEP could be calculated. The results provided a good reference for constructing a shore control center. Its main improvement was to make complicated things clear and easy to analyze.

There are more models and methods to explore human errors, such as Human Entropy (HENT) [39], BN-HRA [40], Railway Action Reliability Assessment (RARA) [11], Controller Action Reliability Assessment (CARA) [14], Nuclear Action Reliability Assessment (NARA) [15], and A Technique for Human Error Analysis (ATHENA) [41]. Among the existing models, including modified, revised, and hybrid approaches, only marine maintenance and operations human reliability analysis (MMOHRA), SOHRA, HEART, NARA, CARA, and RARA involve EPC calculation. Few of them could obtain the weight of each EPC without the experts' judgment. This article will attempt to narrow this gap. Of the six identified methods, MMOHRA and SOHRA were explicitly developed for the maritime domain; the other four models were not. The MMOHRA model was proposed based on the framework of the SOHRA method. It is more specific than the SOHRA model because it was created exclusively for marine operations and maintenance. Considering that this research will not use marine maintenance operations for verification, the SOHRA model is preferred for this article.

2. Methodology

This study intends to apply the SOHRA model as the foundation for human error probability evaluation. The entropy weight and Technique of Preference by Similarity to Ideal Solution (TOPSIS) method are combined to obtain the weights of EPCs, in other words, to provide the values of the A_i in Equation (1).

2.1. SOHRA Model

In the maritime domain, quantifying human error is demanding since scanty human error data affect its process. Therefore, adopting empirical techniques such as SOHRA to quantify human error probability is practical. SOHRA, a model modified from HEART, is utilized to study human reliability for shipboard activities. The two main characteristics of SOHRA are generic error probability (GEP) and m-EPC. The GEP derived from the HEART model includes nine different values. Each value corresponds to a generic task type (GTT) (from A to M); GTT and GEP can be seen in Akyuz et al. [42]. The m-EPCs [6], derived from EPCs in the HEART model, are special to maritime operations. They are critical factors, internal or external, which could affect people's performance onboard. Furthermore, because its values are obtained based on numerous maritime accident reports, it is applicable for all shipboard operations, including deck work and engine room work. However, for more accurate calculation purposes, improvement of the existing m-EPCs may be required, such as the mmo-EPCs proposed by Kandemir et al. [4].

Here, the EPC values were collected from six different models: MMOHRA, SOHRA, HEART, NARA, CARA, and RARA. Their values can be found in Kandemir et al. [4]. The six models' EPCs may have different names, such as m-EPCs or mmo-EPCs. Whether they are referred to as EPCs, m-EPCs, or mmo-EPCs, we use EPC here for standardization.

The purpose of the SOHRA model is to calculate the HEP value for specific operations, figure out the steps with higher HEP values, and thus give measures or recommendations to manage human errors. Therefore, the HEP values are calculated by Equation (1) as per the SOHRA model.

$$HEP_j = GEP_j \times \left\{ \prod_i [(mEPC_i - 1)A_i + 1] \right\}, i = 1, 2, \cdots, n; j = 1, 2, \cdots, 9 \tag{1}$$

where i represents the number of m-EPCs in each step of the task, j is the number of GEPs, and A_i is the weight of each m-EPC.

Several steps should be implemented to perform this calculation:

Firstly, the task should be identified, and the steps or sub-tasks should be determined as per the hierarchical task analysis (HTA).

Secondly, based on the steps or sub-tasks, a set of scenarios are defined to match the GTT and m-EPC parameters, which include internal and external conditions.

Thirdly, by applying the majority rule, the experts' judgments can help assign the sub-tasks with appropriate GTT and suitable EPCs. Then, the GEP values can be obtained according to the identified GTT, and the EPC values can be obtained from Kandemir et al. [4]. The EPC value "NA" in NARA, CARA, and RARA models is deemed zero for better calculation. Study [6] shows that the values of EPCs positively correlates with human error. A larger value implies a higher probability of human error. If the EPC's value is less than 1, it indicates the EPC has no connection with human error. Therefore, it is reasonable to replace "NA" with zero.

Fourthly, the entropy weight method and TOPSIS approach are used to determine the values of A_i; details are listed in Section 2.2.

Fifthly, the HEP value can be calculated through Equation (1).

The last step is the recommendation of safety barriers to minimize human error, according to the calculated HEP values.

2.2. Entropy Weight Method and TOPSIS Model

The entropy weight method, first developed by Shannon [43], is utilized to obtain the weights of the targets based on the index variability. It has a certain accuracy as an objective method to determine the weight, compared with subjective methods such as AHP. It replaces the expert weight and reduces subjectivity. Furthermore, the weights determined by this method can be modified because of its high adaptability.

Hwang et al. [44] proposed the TOPSIS model to cope with the multi-criteria decision making (MCDM) problem. The blending of the two approaches could minimize the subjectivity of data weighting and is thus suitable for systematic risk and safety assessment [45]. The blended approach based on the two methods can be achieved by implementing the following steps:

Step 1: Based on Table 1, the decision matrix can be obtained.

$$D = \begin{bmatrix} d_{11} & \cdots & d_{1n} \\ \vdots & \ddots & \vdots \\ d_{m1} & \cdots & d_{mn} \end{bmatrix} \tag{2}$$

where d_{mn} is the value of the n-th EPC against the m-th approach. m represents the number of approaches; n means the total number of EPCs in Table 1. The calculation will be performed when $m = 3$ (the first three approaches) and $m = 6$ (all approaches). Table 2 shows the simple notation of EPCs and approaches.

Table 1. Task description of a cargo-loading operation.

Sub-Task	Description of Task
1. Human Safety	
1.1	Make sure that all crew on deck use PPE
1.2	Use safety belt when working/climbing on containers
1.3	Keep clear from the container passage area
1.4	Be aware of the risks of mislaid equipment on operated container
1.5	Be aware of the risks of lashing operations on bays at which cargo operations are occurring
2. Ship/Cargo Security	
2.1	Check if any oil is dropped off from gantry to the deck
2.2	Check all cellguides against any damage during operation
2.3	Check the seal of loading containers
2.4	Check if the top cover of OT container is damaged and not preventing another container being put on them
2.5	Check the IMO signs/labels of dangerous cargoes
2.6	Check the tightness of all straps on flatrack containers if any
2.7	Make sure that on/off switch is kept off before connection of reefer plug
2.8	Check the temperature setting degree and ventilation and humidity settings (%) of reefer container
2.9	Consider the height while loading HC/OT containers in hold
2.10	Inform C/O in case flatrack container is overhighed or overgauged than declared
2.11	Check if loading containers in balance on port/starboard side of the ship
2.12	Inform the office-charterer-agent if the ship heels more than 5° during loading
2.13	Inform C/O if hook spreaders are used for cargo operations
2.14	Check if the leakage containers onboard
2.15	Inform C/O in case damaged container is observed
2.16	Prepare interchange report for damaged containers
2.17	Inform the charterer and management office if the container is heavily damaged
2.18	Check the ship's ropes during operation frequently
2.19	Keep the drafts under strict control during cargo operation

Table 2. Simple notation of EPCs and Approaches.

EPCs	Series No.	EPCs	Series No.	EPCs	Series No.	Approaches	Series No.
EPC1	S1	EPC14	S14	EPC27	S27	MMOHRA	T1
EPC2	S2	EPC15	S15	EPC28	S28	SOHRA	T2
EPC3	S3	EPC16	S16	EPC29	S29	HEART	T3
EPC4	S4	EPC17	S17	EPC30	S30	NARA	T4
EPC5	S5	EPC18	S18	EPC31	S31	CARA	T5
EPC6	S6	EPC19	S19	EPC32	S32	RARA	T6
EPC7	S7	EPC20	S20	EPC33	S33	-	-
EPC8	S8	EPC21	S21	EPC34	S34	-	-
EPC9	S9	EPC22	S22	EPC35	S35	-	-
EPC10	S10	EPC23	S23	EPC36	S36	-	-
EPC11	S11	EPC24	S24	EPC37	S37	-	-
EPC12	S12	EPC25	S25	EPC38	S38	-	-
EPC13	S13	EPC26	S26	-	-	-	-

Step 2: the values in the decision matrix are normalized with the following equations to avoid the dimension differences of various factors:

$$d'_{ij} = \frac{d_{ij} - \overline{d_i}}{s_j}, i = 1, 2, \cdots, m; j = 1, 2, \cdots, n \tag{3}$$

$$\overline{d_i} = \frac{1}{n} \sum_{i=1}^{n} d_{ij} \tag{4}$$

$$s_j = \sqrt{\frac{1}{n}\sum_{i=1}^{n}\left(d_{ij} - \overline{d}_i\right)^2}, j = 1, 2, \cdots, m \tag{5}$$

Since the values of d'_{ij} should be positive, there is no indication that they are necessarily positive after processing. Therefore, a transformation should be performed to make them positive:

$$d_{ij} = d'_{ij} + b \tag{6}$$

where b is the minimum value that could ensure all the d'_{ij} are positive.

$$f_{ij} = \frac{d_{ij}}{\sum_{j=1}^{n} d_{ij}}, i = 1, 2, \cdots, m; j = 1, 2, \cdots, n \tag{7}$$

where f_{ij} is the decision matrix after standardization.

Step 3: The information entropy can be calculated according to the final decision matrix and entropy theory:

$$E_j = -\frac{1}{\ln n}\sum_{j=1}^{n} f_{ij}\ln f_{ij}, i = 1, 2, \cdots, m \tag{8}$$

If $f_{ij} = 0$, define $f_{ij}\ln f_{ij} = 0$.

Step 4: Then the entropy weight of all the elements can be obtained:

$$w_i = \frac{1 - E_i}{\sum_{i=1}^{m}(1 - E_i)}, i = 1, 2, \cdots, m \tag{9}$$

Step 5: Based on the standardized decision matrix and the entropy weights, the weighted normalization matrix can be calculated using Equation (10):

$$P = \left(p_{ij}\right)_{m \times n} = w_i \times f_{ij} = \begin{bmatrix} w_1 f_{11} & w_1 f_{12} & \cdots & w_1 f_{1n} \\ w_2 f_{21} & w_2 f_{22} & \cdots & w_2 f_{2n} \\ \vdots & \vdots & & \vdots \\ w_m f_{m1} & w_m f_{m2} & \vdots & w_m f_{mn} \end{bmatrix}, i = 1, 2, \cdots, m; j = 1, 2, \cdots, n \tag{10}$$

Step 6: The positive ideal solution Q^+ and the negative ideal solution Q^- are easy to obtain according to the TOPSIS approach:

$$\text{Positive ideal solution}: Q^+ = \left[\left(\max p_{ij}/i\right)\right] = [q_1^+, q_2^+, \cdots, q_m^+] \tag{11}$$

$$\text{Negative ideal solution}: Q^- = \left[\left(\min p_{ij}/i\right)\right] = [q_1^-, q_2^-, \cdots, q_m^-] \tag{12}$$

Step 7: The relative distances between the p_{ij} and the positive ideal solution (PIS) and negative ideal solution (NIS) are as follows:, respectively,

$$R_j^+ = \sqrt{\sum_{i=1}^{m}\left(p_{ij} - q_i^+\right)^2}, j = 1, 2, \cdots, n \tag{13}$$

$$R_j^- = \sqrt{\sum_{i=1}^{m}\left(p_{ij} - q_i^-\right)^2}, j = 1, 2, \cdots, n \tag{14}$$

Step 8: The final results are the relative closeness of the factors to the positive ideal solution and negative ideal solution:

$$A_j = \frac{R_j^-}{\left(R_j^- + R_j^+\right)}, j = 1, 2, \cdots, n \tag{15}$$

The evaluation value A_j represents the degree of correlation between EPCs and HEP. Its range is from 0 to 1. A large value of A_j means a greater relevance of the two elements and vice versa.

3. Case Study and Results

To verify the proposed approach, a cargo loading operation from a container ship was applied to demonstrate its effectiveness. Cargo loading is one of the critical operations onboard, involving a series of tasks including shipboard cooperation and ship-shore co-operation. The potential risks include crew members' safety, shoreside workers' safety, cargo condition, ship equipment, port facilities, environment damage, etc. These risks correlate well with human activities, and thus are suitable for analyzing human failures and developing corresponding safety barriers to minimize the potential risks.

The identified cargo loading process comes from Erdem et al. [1]. The task descriptions are listed in Table 1. The calculations were performed according to the statement in Sections 2.1 and 2.2. Seven masters were invited as experts for this research, and their information is listed in Table 3. The results are given in this section after elaborating on the calculation process. Based on the experts' judgments, the sub-tasks were assigned to their related GTT and corresponding m-EPCs (Table 4). Figure 1 shows the calculation results of A_{s1} and A_{s2}. S_1 represents scenario 1, S_2 represents scenario 2, and R means the reference data, where A_{s1} is the result of the first scenario (only T1, T2, and T3 models are involved) and A_{s2} is the result of the second scenario (T1, T2, T3, T4, T5, and T6 models are involved). The results indicate that significant differences exist between the two scenarios. The final results will tell which scenario is better. Then, the calculated HEPs were obtained separately, based on the two scenarios, and a comparison is illustrated in Table 5. The reference data are from Erdem et al. [1].

Table 3. Experts' profiles.

Expert No.	1	2	3	4	5	6	7
Rank onboard	master	master	master	master	master	master	master
Sea age	15	21	17	29	32	24	35
Ship type	Container	Container	Container	Container	Container	Container	Container

Table 4. Assigned GTT and m-EPCs.

Sub-Tasks	m-EPCs	GTT
1.1	EPC1, EPC17, EPC22, EPC23	G
1.2	EPC12, EPC15, EPC22, EPC26	G
1.3	EPC12, EPC13, EPC15, EPC22	H
1.4	EPC1, EPC9, EPC12, EPC17, EPC21, EPC22	H
1.5	EPC1, EPC2, EPC17, EPC24, EPC25, EPC26	G
2.1	EPC17, EPC22, EPC26	G
2.2	EPC11, EPC13, EPC17	H
2.3	EPC13, EPC15, EPC17	H
2.4	EPC11, EPC15, EPC24, EPC33	E
2.5	EPC11, EPC17, EPC23, EPC32	H
2.6	EPC1, EPC5, EPC17	G
2.7	EPC1, EPC9, EPC14, EPC15	E
2.8	EPC1, EPC4, EPC5, EPC9	E
2.9	EPC2, EPC9, EPC15, EPC32	E
2.10	EPC2, EPC5, EPC13, EPC20	H
2.11	EPC15, EPC17, EPC22	G
2.12	EPC5, EPC9, EPC12, EPC17, EPC24, EPC26, EPC28	H
2.13	EPC1, EPC12, EPC13, EPC20	G
2.14	EPC12, EPC21, EPC24, EPC26	G
2.15	EPC1, EPC11, EPC24, EPC26	M
2.16	EPC1, EPC13, EPC15	H
2.17	EPC2, EPC15, EPC21, EPC22, EPC29	G
2.18	EPC12, EPC13, EPC14, EPC17, EPC24	H
2.19	EPC15, EPC17, EPC21, EPC24, EPC26, EPC32	H

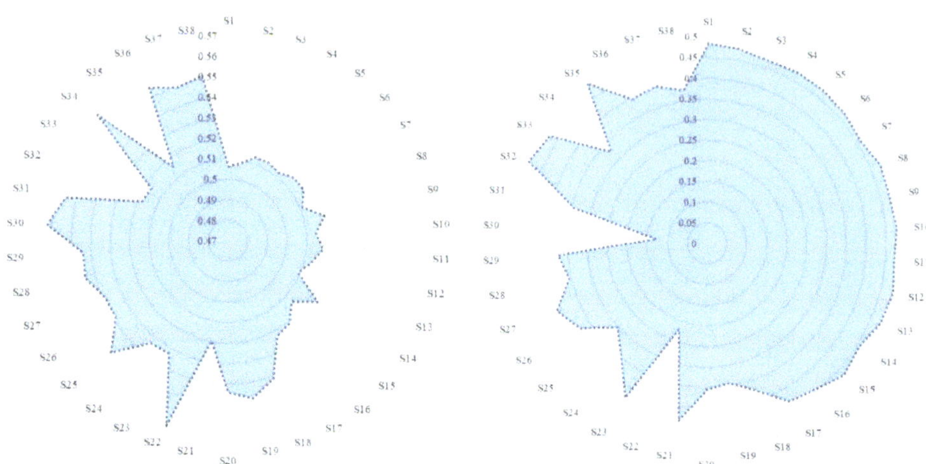

Figure 1. The weights A_{s1} (**left**) and A_{s2} (**right**).

Table 5. HEP comparisons between Scenario 1, Scenario 2, and reference data.

Sub-Task	Scenario 1	Reference Data	Scenario 2
1.1	2.59×10^{-2}	2.59×10^{-2}	2.04×10^{-2}
1.2	4.84×10^{-2}	4.85×10^{-2}	2.88×10^{-2}
1.3	7.50×10^{-3}	7.33×10^{-3}	5.00×10^{-3}
1.4	2.58×10^{-2}	2.90×10^{-2}	1.49×10^{-2}
1.5	1.48×10^{-1}	1.91×10^{-1}	9.49×10^{-2}
2.1	2.30×10^{-3}	2.46×10^{-3}	1.50×10^{-3}
2.2	1.30×10^{-3}	1.29×10^{-3}	1.10×10^{-3}
2.3	1.51×10^{-3}	1.88×10^{-3}	1.22×10^{-3}
2.4	3.41×10^{-3}	3.82×10^{-3}	2.70×10^{-3}
2.5	3.60×10^{-3}	3.58×10^{-3}	2.50×10^{-3}
2.6	2.44×10^{-2}	2.25×10^{-2}	2.03×10^{-2}
2.7	1.26×10^{-2}	1.28×10^{-2}	9.70×10^{-3}
2.8	1.01×10^{-2}	9.90×10^{-3}	8.00×10^{-3}
2.9	1.48×10^{-2}	1.47×10^{-2}	1.15×10^{-2}
2.10	7.40×10^{-3}	7.30×10^{-3}	5.10×10^{-3}
2.11	5.80×10^{-3}	5.94×10^{-3}	4.20×10^{-3}
2.12	1.39×10^{-2}	1.37×10^{-2}	7.20×10^{-3}
2.13	3.88×10^{-2}	3.95×10^{-2}	2.86×10^{-2}
2.14	1.21×10^{-1}	1.14×10^{-1}	8.36×10^{-2}
2.15	4.45×10^{-2}	4.43×10^{-2}	3.20×10^{-2}
2.16	7.30×10^{-3}	7.62×10^{-3}	5.90×10^{-3}
2.17	7.42×10^{-2}	7.45×10^{-2}	4.80×10^{-2}
2.18	8.20×10^{-3}	8.84×10^{-3}	5.80×10^{-3}
2.19	7.10×10^{-3}	7.05×10^{-3}	4.00×10^{-3}

Figure 1 demonstrates the values of A_{s1} and A_{s2}. Figure 2 shows the calculated HEP comparisons between scenario 1 (S1), scenario 2 (S2), and reference data (R), where t1.1 represents sub-task 1.1, and t1.2 means sub-task 1.2, and the remaining labels on the abscissa have a similar basis.

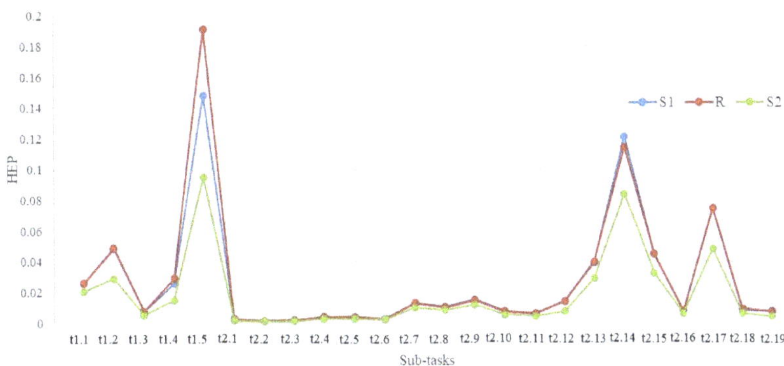

Figure 2. The HEP results of S1, S2, and reference data.

4. Findings and Discussion

The proposed approach can adequately address the operations based on the calculated results and comparison. In light of Figure 2, the results in both scenarios are in good agreement with the reference data, which implies both scenarios are reasonable attempts to calculate the HEP. However, Scenario 1 and the reference data agree better than scenario 2. This result indicates that the weights of EPCs generated by the combination of the MMOHRA model, HEART model, and SOHRA model are more reasonable than that of all six models' combinations. The probable reasons may be as follows:

(1) HEART model is the general foundation of this method; the other five approaches are modified or revised versions of the HEART model.

(2) MMOHRA model and SOHRA model were developed especially for the maritime domain, while the NARA model, CARA model, and RARA model are only for nuclear action reliability assessment, aviation action reliability assessment, and railway action reliability assessment, respectively.

The cargo loading operations of container ships are complex, as listed in Table 1. The two main parts, human safety and ship/cargo security, consist of 24 sub-tasks. Each sub-task involves more than two EPCs, as per the experts' judgments. The eight most frequently involved EPCs are EPC17, EPC15, EPC22, EPC26, EPC24, EPC12, EPC13, and EPC1, as shown in Table 6.

Table 6. The most frequently involved EPCs.

No.	EPC No.	Connotation	No.	EPC No.	Connotation
1	EPC17	Inadequate checking	5	EPC24	Absolute judgment required
2	EPC15	Operator inexperience	6	EPC12	Misperception of risk
3	EPC22	Lack of exercise	7	EPC13	Poor feedback
4	EPC26	Progress tracking lack	8	EPC1	Unfamiliarity

EPC17 (Inadequate checking) is the most frequently appearing EPC. Because the loading operation is a continuous process, all the parameters, such as draft, cargo remaining onboard, and lashing, are dynamically changing, so frequent checking is required during the operation. However, the repetition makes the crew prone to tire, and overlook easy to happen. This situation will induce accidents easily, as the safety barrier is broken at this point. EPC15 (Operator inexperience) stands in the second position, indicating the crew at the operating level does not have enough training or operation experience, as does EPC22 (lack of exercise). Enhanced training programs can remedy this gap. EPC26 (Progress tracking lack) in the loading process is a common non-conformity as per the experts' experiences. During the loading process, duty change, overlook, random errors, fatigue, etc., could suspend progress tracking. EPC24 (Absolute judgment required) is the

required content of competence. EPC12 (Misperception of risk) may occur due to mental stress, fatigue, time shortage, or inexperience. EPC13 (Poor feedback) could happen when unreliable instruments are used or misunderstanding occurs. EPC1 (Unfamiliarity) mostly appears when newly joined crew are on duty or when performing rare tasks.

According to Figure 2, the eight sub-tasks with the highest HEP are shown in Table 7. Sub-task 1.5 with the HEP value 1.48×10^{-1} is located in the first position, and sub-tasks 2.14 and 2.17 are in the second and third positions, respectively. Because they have been analyzed by Erdem et al. [1], to avoid repetition, the remaining five sub-tasks will be discussed in detail.

Table 7. Sub-tasks with the highest HEP.

No.	Sub-Tasks
1	1.5 Be aware of the risks of lashing operations on bays at which cargo operations are occurring
2	2.14 Check if the leakage containers onboard
3	2.17 Inform the charterer and management office if the container is heavily damaged
4	1.2 Use safety belt when working/climbing on containers
5	2.15 Inform C/O in case damaged container is observed
6	2.13 Inform C/O if hook spreaders are used for cargo operations
7	1.1 Make sure that all crew on deck use PPE
8	1.4 Be aware of the risks of mislaid equipment on operated container

Sub-task 1.2 (4.84×10^{-2}) refers to the safety belt while working or climbing on containers. The safety belt is used to protect crew safety. However, modern loading operations emphasize efficiency heavily. The continuous loading process requires endless checking and lashing work but limited time to finish the job. These factors probably increase the mental stress of the duty crew and thus increase the HEP when performing the task. Sub-tasks 2.15 (4.45×10^{-2}) and 2.13 (3.88×10^{-2}) are similar work involving inspection and reporting. Such work may go unnoticed for insufficient inspection rather than for reporting reasons. Table 6 provides the evidence that shows that EPC1 is assigned in both cases. Sub-task 1.1 (2.59×10^{-2}) refers to the proper usage of PPE while working on deck. This task involves safety regulation and safety awareness. It should be common sense that proper PPE should be worn whenever working on deck. Sub-task 1.4 (2.58×10^{-2}) also has a higher HEP value than the remaining sub-tasks, because this kind of risk is not obvious. It takes more time to find the mislaid equipment in an operated container. In contrast, limited time makes it easy to increase the probability of human errors.

The sub-tasks with higher HEP do not indicate that human errors will happen certainly but imply decreased reliability. Meanwhile, the potential risks are increasing. Therefore, a series of error reduction measures are recommended to reduce the chances of the error happening to manage the crew's reliability and strengthen the safety level for this loading operation (Table 8).

Table 8. Recommendations for HEP mitigation.

Sub-Task	EPC	Mitigate Measures
1.5	EPC1	1. Nominate an experienced crew to supervise the lashing operation nearby. 2. Potential dangers should be reminded to the operators
	EPC2	1. Safety meeting to be held before the operation 2. Teamwork is required during performing the task
	EPC17	1. Adequate communication should be maintained 2. Proper instructions should be illustrated before the task
	EPC24	1. Reminders should be made in time in case the situation changes 2. Nominate an experienced crew to help and supervise
	EPC25	1. Proper PPE should be worn before the task 2. The task should be performed according to Chief Mate's instruction
	EPC26	1. Enhance crew situation awareness through adequate training

Table 8. *Cont.*

Sub-Task	EPC	Mitigate Measures
2.14	EPC12	1. Periodical exercises concerning checking the container leakage should be held 2. Any doubt about the container leakage should be reported to Chief Mate
	EPC21	1. Arrange experienced crew while checking to increase the reliability 2. Effective communication should be maintained
	EPC24	1. Necessary training should be performed 2. Checking should be carried out as per instruction
	EPC26	1. Proper records should be kept 2. Procedures should be followed and supervised
2.17	EPC2	1. Prepare reporting templates in case of emergency use 2. Inform the Master earlier in case of emergency
	EPC15	1. Adequate cooperation is required 2. Experienced Master is preferred
	EPC21	1. Incentives should comply with the regulations
	EPC22	1. Frequent training concerning emergency handling should be performed 2. The emergency checklist should be filled up in case any critical steps missing
	EPC29	1. Avoid shouting while communicating 2. More encouragement is suggested during working
1.2	EPC12	1. Arrange a supervisor for this kind of work 2. Adequate reminders should be maintained
	EPC15	1. Safety meeting should be held before commencing work 2. An experienced crew is required to give help and advice
	EPC22	1. Safety checklist should be finished by themselves before working 2. Periodical exercises should be carried out
	EPC26	1. Update the progress in time as per instruction 2. Periodical supervise should be maintained
2.15	EPC1	1. Regular exercises should be held to identify various damaged containers 2. Teamwork is required during performing the task
	EPC11	1. Adequate reminders should be maintained 2. An experienced crew is required to give help and advice
	EPC24	1. Adequate cooperation is required 2. Experienced crew should be assigned to the critical task
	EPC26	1. Update the progress in time as per instruction
2.13	EPC1	1. Periodical checking should be carried out 2. Safety meetings should be held before working
	EPC12	1. Arrange a supervisor for this kind of work 2. Ask for help when the vague danger exists
	EPC13	1. Adequate communication should be kept 2. Reports accordingly as per instruction
	EPC20	1. Demonstration of the task should be exercised 2. Proper training should be held as per the regulation
1.1	EPC1	1. Post the safety instructions in the crew changing room 2. Safety meetings should be held before working
	EPC17	1. The checklist should be filled up before leaving the changing room 2. Periodical supervision should be maintained
	EPC22	1. Periodical exercises should be carried out 2. Demonstration of the task should be exercised
	EPC23	1. Check the equipment before working 2. Proper instruments should be assigned

Table 8. *Cont.*

Sub-Task	EPC	Mitigate Measures
1.4	EPC1	1. Training and exercise should be held to become familiar with the task 2. Teamwork is preferred to reduce one-man error
	EPC9	1. Demonstration of the task should be exercised 2. Experienced crew should be assigned to the critical task
	EPC12	1. Adequate supervision should be kept 2. Strengthen risk awareness through regular safety meetings
	EPC17	1. Adequate communication should be maintained 2. Regular checking should be kept
	EPC21	1. Proper incentives to encourage crew motivation 2. Adequate communication should be kept
	EPC22	1. Periodical exercises should be carried out 2. Safety awareness should be strengthened through demonstration

5. Conclusions

It is impossible to prevent all human errors, but it is possible to minimize their rate of occurrence. Presently, safe and reliable operations onboard ships depend on human reliability. Human error is a dominant factor that can influence human reliability. Therefore, studying human error is significant for controlling human reliability. This article proposes a novel hybrid approach by incorporating the SOHRA model, entropy weight method, and TOPSIS model to calculate human error probability. The entropy weighted TOPSIS approach could effectively reduce subjectivity by replacing the experts' weighting. The weights determined by this method can be modified as well. After comparing with the reference data, it can be concluded that this method is effective. Further, Scenario 1 is better than the other scenario when analyzing their calculated results, which implies the combination of MMOHRA, SOHRA, and HEART models could provide better results concerning the proportions of EPCs. Through this research, an alternative way to obtain the proportions of EPCs is utilized and proved effective.

According to its background, this method could be utilized in various crew operations onboard ships, including deck and engine crew operations. Since differences exist between the deck and engine departments, more specific models could be developed to obtain more accurate HEP values. However, experts' judgments significantly influence this research process. The main reason is that scant human error information could be obtained for the research. At least two methods can be implemented to minimize the subjectivity of experts' decisions. One is the utilization of the Delphi method [46]. The other is to collect enough human error information. Modern technology such as monitoring and recording systems could record human error scenarios, and then, based on the recorded scenarios, the investigation reports of human errors can be obtained, similar to the reports of the Marine Accident Investigation Branch (MAIB, UK) or Australia Transport Safety Bureau (ATSB, Australia). This process could be conducted by companies or government organizations. After adequate human error information is collected, the experts' judgments can be replaced.

Author Contributions: Conceptualization, methodology, software, validation, formal analysis, writing—original draft preparation, writing—review and editing, X.M.; supervision, validation, formal analysis, G.S. and W.L.; formal analysis, writing—review and editing, J.S. All authors have read and agreed to the published version of the manuscript.

Funding: This research was funded by the National Natural Science Foundation of China, grant number 51579025.

Institutional Review Board Statement: Not applicable.

Informed Consent Statement: Not applicable.

Data Availability Statement: Not applicable.

Conflicts of Interest: The authors declare no conflict of interest.

References

1. Erdem, P.; Akyuz, E. An interval type-2 fuzzy SLIM approach to predict human error in maritime transportation. *Ocean Eng.* **2021**, *232*, 109161. [CrossRef]
2. Akyuz, E. Quantitative human error assessment during abandon ship procedures in maritime transportation. *Ocean Eng.* **2016**, *120*, 21–29. [CrossRef]
3. Xi, Y.T.; Yang, Z.L.; Fang, Q.G.; Chen, W.J.; Wang, J. A new hybrid approach to human error probability quantification- applications in maritime operations. *Ocean Eng.* **2017**, *138*, 45–54. [CrossRef]
4. Kandemir, C.; Celik, M. Determining the error producing conditions in marine engineering maintenance and operations through HFACS-MMO. *Reliab. Eng. Syst. Saf.* **2021**, *206*, 107308. [CrossRef]
5. Zhou, Q.J.; Wong, Y.D.; Xu, H.; Thai, V.V.; Loh, H.S.; Yuen, K.F. An enhanced CREAM with stakeholder-graded protocols for tanker shipping safety Ap-plication. *Saf. Sci.* **2017**, *95*, 140–147. [CrossRef]
6. Akyuz, E.; Celik, M.; Cebi, S. A Phase of Comprehensive research to determine marine-specific EPC values in human error assessment and reduction technique. *Saf. Sci.* **2016**, *87*, 63–75. [CrossRef]
7. Read, G.J.M.; Shorrock, S.; Walker, G.H.; Salmon, P.M. Ience: Evolving perspectives on 'Human Error'. *Ergonomics* **2021**, *64*, 1091–1114. [CrossRef]
8. Heinrich, H.W. *Industrial Accident Prevention: A Scientific Approach*; McGraw-Hill: New York, NY, USA, 1931.
9. Kirwan, B.; Gibson, H. CARA: A Human reliability assessment tool for air traffic safety management technical basis and preliminary architecture. In *The Safety of Systems*; Springer: Berlin, Germany, 2007; pp. 197–214.
10. Kirwan, B.; Gibson, H.; Kennedy, R.; Edmunds, J.; Cooksley, G.; Umbers, I. Nuclear Action Reliability Assessment (NARA): A Data-Based HRA Tool. In *Probabilistic Safety Assessment and Management*; Springer: Berlin, Germany, 2004; pp. 1206–1211.
11. Gibson, W.; Mills, A.M.; Smith, S.; Kirwan, B.K. Railway action reliability assessment a railway specific approach to human error quantification. In Proceedings of the Australian System Safety Conference, Adelaide, Australia, 22–24 May 2013; Volume 145, pp. 3–8.
12. Wang, W.; Liu, X.; Qin, Y. A Modified HEART method with FANP for human error assessment in high speed railway dispatching tasks. *Int. J. Ind. Ergon.* **2018**, *67*, 242–258. [CrossRef]
13. Torres, Y.; Nadeau, S.; Landau, K. Classification and quantification of human error in manufacturing. A case study in complex manual assembly. *Appl. Sci.* **2021**, *11*, 749. [CrossRef]
14. Kumar, P.; Gupta, S.; Gunda, Y.R. Estimation of human error rate in underground coal mines through retrospective analysis of mining accident reports and some error reduction strategies. *Saf. Sci.* **2020**, *123*, 104555. [CrossRef]
15. Hsieh, M.; Wang, E.M.; Lee, W.; Li, L.; Hsieh, C.; Tsai, W.; Wang, C.; Huang, J.; Liu, T. Application of HFACS, Fuzzy TOPSIS, and AHP for Identifying Important Hu-Man Error Factors in Emergency Departments in Taiwan. *Int. J. Ind. Ergon.* **2018**, *67*, 171–179. [CrossRef]
16. Sameera, V.; Bindra, A.; Rath, G.P. Human errors and their prevention in healthcare. *J. Anaesthesiol. Clin. Pharmacol.* **2021**, *37*, 328–335. [CrossRef]
17. Hu, W.; Carver, J.C.; Anu, V.; Walia, G.S.; Bradshaw, G.L. Using human error information for error prevention. *Empir. Softw. Eng.* **2018**, *23*, 3768–3800. [CrossRef]
18. Ung, S.-T. Evaluation of human error contribution to oil tanker collision using fault tree analysis and modified fuzzy bayesian network based CREAM. *Ocean Eng.* **2019**, *179*, 159–172. [CrossRef]
19. Zhou, J.L.; Yi, L. A slim integrated with empirical study and network analysis for human error assessment in the railway driving process. *Reliab. Eng. Syst. Saf.* **2020**, *204*, 107148. [CrossRef]
20. Ahn, S.I.; Kurt, R.E. Application of a CREAM based framework to assess human reliability in emergency re-sponse to engine room fires on ships. *Ocean Eng.* **2020**, *216*, 108078. [CrossRef]
21. Zhang, R.; Tan, H.; Afzal, W. A Modified Human Reliability Analysis Method for the Estimation of Human Error Probability in the Offloading Operations at Oil Terminals. *Process Saf. Prog.* **2020**, *40*, 84–92. [CrossRef]
22. Kandemir, C.; Celik, M. A human reliability assessment of marine auxiliary machinery maintenance operations under ship PMS and maintenance 4.0 concepts. *Cogn. Tech. Work* **2020**, *22*, 473–487. [CrossRef]
23. Islam, R.; Yu, H.; Abbassi, R.; Garaniya, V.; Khan, F. Development of a monograph for human error likelihood assessment in marine operations. *Saf. Sci.* **2017**, *91*, 33–39. [CrossRef]
24. Williams, J.C. A data-based method for assessing and reducing human error to improve operational performance. In Proceedings of the IEEE 4th Conference on Human Factor and Power Plants, Monterey, CA, USA, 5–9 June 1988; pp. 436–450. [CrossRef]
25. Akyuz, E.; Celik, M. A methodological extension to human reliability analysis for cargo tank cleaning operation on board chemical tanker ships. *Saf. Sci.* **2015**, *75*, 146–155. [CrossRef]
26. Akyuz, E.; Celik, E. A Modified human reliability analysis for cargo operation in single point mooring (SPM) off-shore units. *Appl. Ocean Res.* **2016**, *58*, 11–20. [CrossRef]
27. Wang, W.; Liu, X.; Liu, S. A hybrid evaluation method for human error probability by using extended DEMATEL with Z-numbers: A case of cargo loading operation. *Int. J. Ind. Ergon.* **2021**, *84*, 103158. [CrossRef]

28. Embrey, D.E.; Humphreys, P.C.; Rosa, E.A.; Kirwan, B.; Rea, K. *SLIM-MAUD: An Approach to Assessing Human Error Probabilities Using Structured Expert Judgement*; United States Nuclear Regulatory Commission: North Bethesda, MD, USA, 1984.
29. Islam, R.A.; Abbassi, R.; Garaniya, V.; Khan, F.I. Determination of human error probabilities for the maintenance operations of marine engines. *J. Ship Prod. Des.* **2016**, *32*, 226–234. [CrossRef]
30. Hollnagel, E. *Cognitive Reliability and Error Analysis Method*; Elsevier: Amsterdam, The Netherlands, 1998. [CrossRef]
31. Akyuz, E. Quantification of human error probability towards the gas inerting process on-board crude oil tankers. *Saf. Sci.* **2015**, *80*, 77–86. [CrossRef]
32. Yang, Z.L.; Abujaafar, K.M.; Qu, Z.; Wang, J.; Nazir, S.; Wan, C. Use of evidential reasoning for eliciting bayesian subjective probabilities in human reliability analysis: A maritime case. *Ocean Eng.* **2019**, *186*, 106095. [CrossRef]
33. Zhou, Q.J.; Wong, Y.D.; Loh, H.S.; Yuen, K.F. A fuzzy and bayesian network CREAM model for human reliability analysis—The case of tanker shipping. *Saf. Sci.* **2018**, *105*, 149–157. [CrossRef]
34. Shirali, G.A.; Hosseinzadeh, T.; Ahamadi Angali, K.; Rostam Niakan Kalhori, S. Modifying a method for human reliability assessment based on cream-Bn: A case study in control room of a petrochemical plant. *MethodsX* **2019**, *6*, 300–315. [CrossRef]
35. Wu, B.; Yan, X.; Wang, Y.; Soares, C.G. An Evidential Reasoning-Based Cream to Human Reliability Analysis in Maritime Accident Process. *Risk Anal.* **2017**, *37*, 1936–1957. [CrossRef]
36. Li, G.; Weng, J.; Hou, Z. Impact analysis of external factors on human errors using the ARBN method based on small-sample ship collision records. *Ocean Eng.* **2021**, *236*, 109533. [CrossRef]
37. Swain, A.D.; Guttmann, H.E. *Handbook of Human Reliability Analysis with Emphasis on Nuclear Power Plant Applications*; Report No. NUREG/CR-1278; United States Nuclear Regulatory Commission: North Bethesda, MD, USA, 1983.
38. Zhang, M.; Zhang, D.; Yao, H.; Zhang, K. A probabilistic model of human error assessment for autonomous cargo ships focusing on human—Autonomy collaboration. *Saf. Sci.* **2020**, *130*, 104838. [CrossRef]
39. El-Ladan, S.B.; Turan, O. Human reliability analysis—Taxonomy and praxes of human entropy boundary conditions for marine and offshore applications. *Reliab. Eng. Syst. Saf.* **2012**, *98*, 43–54. [CrossRef]
40. Abrishami, S.; Khakzad, N.; Hosseini, S.M. A data-based comparison of BN-HRA models in assessing human error proba-bility: An offshore evacuation case study. *Reliab. Eng. Syst. Saf.* **2020**, *202*, 107043. [CrossRef]
41. Cooper, S.E.; Ramey-Smith, A.M.; Wreathall, J.; Parry, G.W. *A Technique for Human Error Analysis (ATHEANA): Technical Basis and Methodology Description*; Nureg/CR-6350; USNRC: North Bethesda, MD, USA, 1996; p. 996.
42. Akyuz, E.; Celik, M.; Akgun, I.; Cicek, K. Prediction of human error probabilities in a critical marine engineering operation on-board chemical tanker ship: The case of ship bunkering. *Saf. Sci.* **2018**, *110*, 102–109. [CrossRef]
43. Shannon, C.E. A mathematical theory of communication. *SIGMOBILE Mob. Comput. Commun. Rev.* **2001**, *5*, 3–55. [CrossRef]
44. Hwang, C.L.; Yoon, K. *Multiple Attribute Decision-Making Methods and Application*; Springer: Berlin, Germany, 1981.
45. Chen, J.; Bian, W.; Wan, Z.; Yang, Z.; Zheng, H.; Wang, P. Identifying factors influencing total-loss marine accidents in the world: Analysis and eva lu-ation based on ship types and sea regions. *Ocean Eng.* **2019**, *191*, 106495. [CrossRef]
46. Duru, O.; Bulut, E.; Yoshida, S. A fuzzy extended DELPHI method for adjustment of statistical time series prediction: An empirical study on dry bulk freight market case. *Expert Syst. Appl.* **2012**, *39*, 840–848. [CrossRef]

Journal of
*Marine Science
and Engineering*

Article

Container Ship Fleet Route Evaluation and Similarity Measurement between Two Shipping Line Ports

Davor Šakan *, Srđan Žuškin *, Igor Rudan and David Brčić

Faculty of Maritime Studies, University of Rijeka, Studentska ulica 2, 51000 Rijeka, Croatia
* Correspondence: davor.sakan@uniri.hr (D.Š.); srdan.zuskin@uniri.hr (S.Ž.)

Abstract: The characterization of ship routes and route similarity measurement based on Automatic Identification System (AIS) data are topics of various scientific interests. Common route research approaches use available AIS identifiers of ship types. However, assessing route and similarity profiles for individual fleets requires collecting data from secondary sources, dedicated software libraries or the creation of specific methods. Using an open-source approach, public AIS and ship data, we evaluate route characteristics for the container ships of a single fleet in a six-month period, calling on two selected ports of the shipping line on the USA East Coast. We evaluate the routes in terms of length, duration and speed, whereas for the similarity measurement we employ the discrete Fréchet distance (DFD). The voyage length, duration and average speed distributions were observed to be moderately positive (0.77), negative (−0.62), and highly positively skewed based on the adjusted Fisher–Pearson coefficient of skewness (1.23). The most similar voyages were from the same ships, with the lowest discrete Fréchet distance similarity value (0.9 NM), whereas 2 different ships had the most dissimilar voyages, with the highest DFD value (14.1 NM). The proposed methodology enables assessment of similarities between individual ships, or between fleets.

Keywords: route similarity; similarity measures; discrete Fréchet distance; fleet analysis; container fleet

Citation: Šakan, D.; Žuškin, S.; Rudan, I.; Brčić, D. Container Ship Fleet Route Evaluation and Similarity Measurement between Two Shipping Line Ports. *J. Mar. Sci. Eng.* **2023**, *11*, 400. https://doi.org/10.3390/jmse11020400

Academic Editors: Marko Perkovic, Lucjan Gucma, Sebastian Feuerstack and Zaili Yang

Received: 27 December 2022
Revised: 30 January 2023
Accepted: 7 February 2023
Published: 11 February 2023

1. Introduction and Background

Route, path, line or way are various terms describing series of interconnected waypoints that a movable entity will traverse or has traversed. For inanimate entities, including land vehicles, aircrafts or ships, routes must be created and adapted to the particulars of the transport mode and voyage objectives. On board ships, the creation of intended routes is still the duty of the navigator(s) and is a part of voyage planning. This is a process, subject to numerous constraints, rules and regulations in which the navigator uses skills, knowledge and experience to create a safe and efficient route for the forthcoming voyage [1]. During the voyage, the vessel movement, traversed positions and other voyage-related data is saved in predetermined intervals by the onboard equipment, and, if necessary, route retrieval is simple on board. For shipping companies, such route data from ships can be useful for various purposes, including safety of navigation, individual voyages, or even more, fleet efficiency metrics based on length of the route, average speed, traversal time or fuel consumption. Using such data for the wider research community, notably in fleet analysis, is not so common since it is rarely public. However, there are other sources of vessel data, such as data broadcasted from the Automatic Identification System (AIS). AIS data is collected and made available by several public and free sources, including the one used in the conducted research, as will be presented in the following chapter. From the sources, such data can be used for various ship route data purposes.

Route extraction in the context of safe route design, based on AIS data and available categories of ships using Kernel Density Estimation with route boundary extraction and centerline extraction, was considered in [2]. In [3], spatial and temporal vessel trajectory clustering in the context of extracting traffic flow information, customary route discovery and anomaly detection was carried out based on AIS data. The trajectories have been investigated in terms of their similarity to Merge Distance. In [4], computer vision techniques were employed for navigation patterns estimation in the context of traffic statistics and a simplified ship maneuvering model, with traffic group collecting based on geometric similarity using AIS data. In [5], the Equivalent Passage Plan method, based on the notion that the vessel in navigation follows a prepared voyage plan, was used as a framework for trajectory simplification and event identification in the context of ship behavior extraction. Further, the Hausdorff distance was used to measure similarity between the simplified and original AIS trajectories. Hausdorff and other similarity methods can be broadly grouped into warping-based methods and shape-based methods. The warping methods include Dynamic Time Warping (DTW), the Longest Common Subsequence (LCSS), and the Edit Distance on Real sequence (EDR). Shape-based methods include the Hausdorff distance, One Way Distance (OWD), and the Fréchet distance [6]. Similarity measures can be used in spatial, temporal and spatiotemporal contexts. For spatial contexts, the measures can be compared topologically and quantitatively, in terms of global and local path similarity, travelled distance, range and shape [7]. After assessing the available various route research interests and similarity methods, it is important to present a brief overview of AIS, the data source on which route analysis and similarity methods are employed in the maritime context.

AIS is a digital short-range automatic data exchange system operating at a very high frequency (VHF) band. Although it was devised as a terrestrial system, AIS signals can be received by satellites as well. Its use is mandatory for all ships under the requirements of the International Convention for the Safety of Life at Sea (SOLAS), determined by the International Maritime Organization (IMO). Although used primary in maritime contexts such as ship identification, collision avoidance or traffic monitoring [8], the available AIS data is a valuable source for various research interests [9]. AIS data is static (e.g., ship identification, basic dimensions, type), dynamic (e.g., position, speed over ground, heading), voyage-related (e.g., navigation status, draft, cargo type) or in form of short safety-related messages. SOLAS-regulated Class A ships send their dynamic data messages in intervals dependent on ship speed and course change. Intervals start from between 2 and 12 s, when the ship is underway, and are 3 min when it is anchored or moored [10]. Further, AIS provides several identifiers for special crafts (e.g., pilot vessels and tugs) and ships such as high-speed crafts (HSC), cargo ships, tankers, passenger ships and ships from the other ships category [11]. Although beneficial for broad ship-type categories, the shortcoming of AIS categorization appears when the research focus is on different categories of ships, such as general cargo ships and container ships. They are both cargo ships; however, they lack separate AIS categories. This makes it more difficult, for example, for evaluation comparing single or multiple fleets, or evaluation of previously stated different cargo ship classes. This was confirmed in the research literature, since the available articles on the subject are rather limited. With the observed gap, we decided to evaluate if the available approaches using AIS default categories in route analysis could be further extended for single or multiple ship fleet analysis.

When assessing ship route research objectives using AIS data [12] and including similarity measures, frequently applied approaches consider general AIS ship categories or created subcategories for route and traffic analysis [13]. In [14], authors examined and delineated maritime routes between ports along the Atlantic coast of the USA using AIS data from 2010 to 2012. Commercial tracks were generated for commercial vessels, with a focus on cargo vessels. The routes were delineated with a 95% boundary of all traffic transiting each route. In [15], authors analyzed ship traffic demand and the spatial–temporal dynamics in Singapore port waters using AIS data. They found that origin-to-destination pairs and navigation routes in the area were stable and found several hotspots with high speeds. In [16], authors evaluated trajectory clustering for marine traffic pattern recognition around the Chinese Zhoushan Islands from January to February 2015. A Density-Based Spatial Clustering of Applications with Noise (DBSCAN) algorithm with improved parameters was proposed. In [17], authors devised a method to transform the ship trajectory into a ship trip semantic object (STSO) with semantic information. Authors used graph theory, and integrated STSO into the nodes and edges of a directed maritime traffic graph for route extraction. The AIS data was data collected from 1 January to 31 December 2017 from crude oil tankers in the Asia-Pacific and Indian Ocean regions. In [18], a ship AIS trajectory clustering method based on the Hausdorff similarity distance and a Hierarchical DBSCAN (HDBSCAN) was used. AIS data on the estuary waters of the Yangtze River in China was used and compared with k-means, spectral clustering and DBSCAN.

As we have presented, there are various route research approaches and interests using AIS data for route and trajectory research. Similarity measurement is used in numerous contexts of AIS data, route and trajectory analysis. Although diverse types of ships were covered in the presented papers, route comparison between other groups of ships besides general AIS ship types is not very well researched. With such observations, we focused our selection on the cargo ship subcategory, which could be suitable for fleet selection and analysis.

Accounting for 13.34% in 2022 of total world fleet by deadweight, container ships are the third largest principal vessel type, behind bulk carriers and oil tankers [19]. Further, the assessment of container ship routes and fleets has several possible advantages over the assessment of other principal vessel types. Vessels are employed in liner services, with companies usually providing public data on their services and vessels. In regular operation, vessels call on the same service ports according to their schedules, thus enabling route comparison in both spatial and temporal domains. Moreover, 4 large carriers (MSC, APM-Maersk, CMA CGM and the COSCO group), accounting for more than 50% of capacity, dominate the market [19], facilitating the collection of significant route data, which can be interpreted for individual fleets or used to make inferences about total container fleet insights.

A broad overview of container ship fleet research topics includes various levels of planning and decision-making, liner shipping networks, scheduling, fleet compositions and deployment, air emissions, optimization and routing, resulting in a considerable body of literature [20–22]. Further, ships and fleet route elements are commonly evaluated in terms of speed and speed optimization [23–25] or trajectory and speed distribution [26]. Speed accuracy is evaluated as well, using the collected speed through water (STW) data of 190 container vessels from Maersk [27]. However, literature on container ship fleet route analysis is scarce. An operational analysis of container ships using AIS and environmental data was carried out [28] based on Hortonworks' big data framework. Trajectories and speeds from several voyages were compared to evaluate voyage energy efficiency for a single 10,000 TEU ship and four 13,000 TEU sisterships. A speed pattern analysis was carried out for 8600, 13,000 and 18,000 TEU ships from several companies, whereas route similarity was not considered. Container ships and fleets were analyzed, using AIS data collected in a database from along the Portuguese coast for top container carriers, services, and vessels [29]. Routes were characterized for individual ships, indicating the sequence of visited ports, respective travel times, the total number of trips and carrying capacity. However, spatial distributions of routes and their similarities were not investigated. In [30], vessel speed, course and path analysis in the Botlek area of the port of Rotterdam was assessed for five container ship categories along with other vessel types. This research was further extended in [31], where a comparison study on AIS data with a characterization of ship traffic behavior was carried out. The data was evaluated for traffic modelling and simulation in restricted waterways. Research areas were a narrow waterway in the port of Rotterdam and the Yangtze wide waterway area near the Su-Tong Bridge. Spatial distributions perpendicular to the water flow and to the channel, sailing speeds, average speeds and course distributions were analyzed. Again, route similarities were not assessed in either of the articles. As can be observed, route similarity measures are usually not considered in either general container or individual fleet analysis.

With that notion in mind, we selected container ships as a subcategory of interest extracted from the total cargo ships category and examined it in terms of route characteristics and similarity.

A fleet of container ships from a single shipping company employed in regular service calling on the ports of Savannah and Charleston on the east coast of the USA was chosen. The selection was chosen due to the similar constrained conditions of the selected ports and possible routes and expected navigational decisions made when planning and executing the voyage.

The remainder of this paper is structured as follows. In Section 2, we present the methods used in the research; we introduce the research area and describe the pre-processing of AIS data, route creation and similarity measurement. In Section 3, we present the research results and findings. Section 4 is the discussion, and we present the conclusions with future research suggestions in Section 5.

2. Methods

To evaluate the differences between the ship routes, several conditions were considered. The ports and area of interest should be located in relationship to one to another to provide modest route choice variability, within a somewhat constrained area. Further, the ports should be called on regularly by major companies of interest to facilitate fleet creation from available collected and processed data. From several candidate pairs, the ports of Savannah and Charleston were chosen. Their respective details and further methodology will be described in the following paragraphs along with a prior general overview, presented in Figure 1.

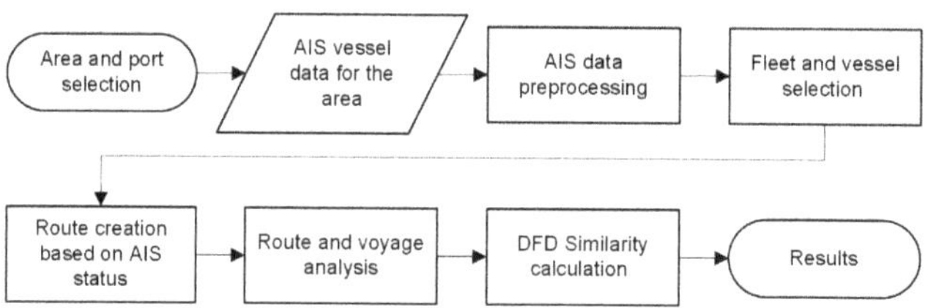

Figure 1. Methodology overview.

The data collected and used is publicly available AIS data from the NOAA AccessAIS portal [32]. The observed period starts on 1 January 2019 and ends on 30 June 2019. The area of interest (AOI) is bounded by 31.791° N, 81.224° W, 33.102° N and 79.453° W, as presented in Figure 2. This is the coastal area between two major USA East Coast ports of interest, Savannah and Charleston. The respective ports, besides handling several other types of cargo, are major hubs for containerized cargo.

During the fiscal year 2019 (ending 30 June 2019), 1848 ships called on the Savannah Garden City Terminal with 4.48 million twenty-foot equivalent units (TEUs) handled. The projected increase in cargo volume for the year 2030 was 48% [33]. In the port of Charleston, the Wando Welch and North Charleston terminals handled 2.93 million TEUs. At all the Charleston cargo terminals, a total of 1696 vessels called in the 2019 fiscal year [34]. Further, vessels from several major container carriers and alliances call on the ports regularly as a part of their services.

According to [35], the distance from Savannah to Charleston is 104 nautical miles (NM). The calculated distance is measured along navigable tracks as the shortest route for safe navigation between the two ports. As we present in the results section, the actual distances for the selected ships are greater due to several reasons, including the size of the ships and the characteristics of the navigational area.

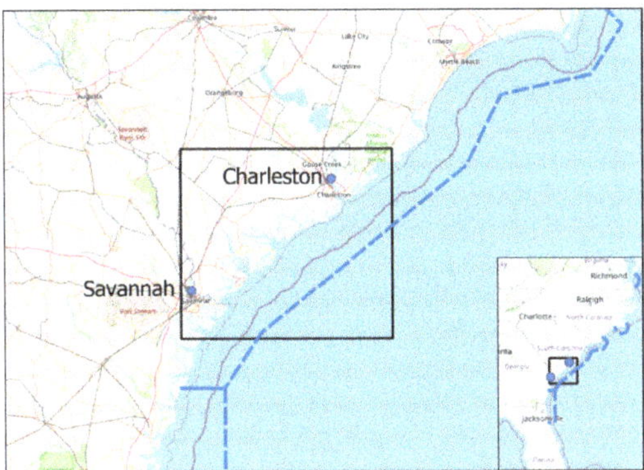

Figure 2. Area of interest with AIS data boundary marked in rectangle with ports of Savannah and Charleston. Right whale Seasonal Management Area (SMA) boundary marked with dashed line.

Besides the importance of the respective ports and the significant vessel traffic they generate, we must emphasize that the coastal area is part of the North Atlantic right whale Seasonal Management Areas (SMA). All vessels with length overall (LOA) of 65 feet (19.81 m) or greater and subject to the jurisdiction of the United States are restricted to speeds of 10 knots or fewer in a continuous 20 NM Seasonal Management Area between 1 November and 30 April annually [36]. The areas were established to prevent collisions of vessels with endangered right whales.

To identify the patterns of interest for route creation, available vessel density data for the AOI was investigated. The data is available as a publicly available annual AIS vessel count dataset for the USA and international waters. The annual data is summarized using 100 m by 100 m geographical grid cells with single transits counted each time a vessel track passes through, starts or stops within the grid cell [37]. A simplified geographical area subset transit density graph can be obtained through the AccessAIS portal as well, as presented in Figure 3.

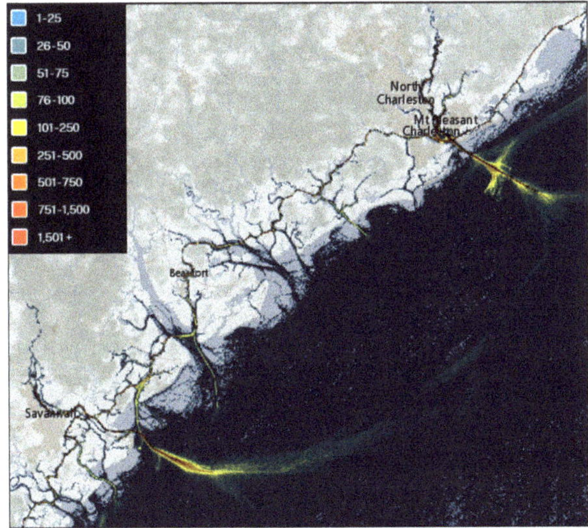

Figure 3. Annual AIS transit vessel counts for the area of interest in the observed period.

We conducted data preparation and preprocessing using the Python programming language; version 3.10.4, the Pandas data analysis library [38], version 1.4.2 [39]; the Shapely package for computational geometry [40] and the NumPy array programming library [41], version 1.21.6. To create routes, we used the MovingPandas library [42], version 0.9.rc3 [43]. The library is based on Pandas, GeoPandas, an extension to Pandas enabling spatial operations on geometric types [44], and HoloViz high-level visualization tools [45]. A trajectory in MovingPandas is defined as a time-ordered series of geometries and can be point or line based.

From the whole dataset, we selected cargo ships by the AIS ship code (70 to 79) used for cargo ships [46]. The cargo ship subset included 724 unique entries. To create a container ship fleet subset, we collected publicly available fleet data about ships and selected ships by their names and identities. The container fleet from the selected company included 22 unique ships with 107,977 cumulative AIS observations and data entries. From the ship identities, a trajectory collection was created for all 22 unique ships. A single trajectory entry in the collection accounts for all ship observations throughout the six-month period. There are several approaches available for the creation and separation of trajectory entries into individual voyages and routes with available MovingPandas functions. They are based on regular time intervals, speed, stopping and observation gaps. Since the AIS data includes the navigational status of the ship—which is 0 for the ship when underway using its engine—the individual voyages were split at entries when the ship status changed from underway to moored or anchored. We must emphasize that the change of a ship's navigational status in AIS must be changed manually on board. If not done in a timely manner, this could lead to false interpretations of a ship's stopping or movement and result in erroneous statuses. However, in the observed dataset, this was not the case, and there were no significant discrepancies between ship status and individual voyage delineation. Further, compared with other available MovingPandas splitting approaches, the preselection by status yielded better results in the splitting up of individual voyages (or trajectories, as implemented in MovingPandas). To reduce the number of unnecessary voyages and split the voyages accordingly, the observational gap value between observations was set to 30 min and the minimum length of the trajectory was set to 200 km (to approximate the distance between ports). With these constraints, the subset was reduced to 13 ships with 23 individual voyages. We created the final subset by choosing the general direction (between 050° and 060°) from individual trajectory start and end points from Savannah to Charleston. The final subset therefore includes 12 voyages (trajectories) from 8 ships. Basic ship particulars are presented in Table 1. The ships in the subset are sorted by overall length and principal dimensions. The abbreviations and acronyms are described as follows: Length Over All (LOA), Breadth extreme (B_{ext}), Draught maximum (D_{max}), Draught per voyage in meters (D_{voy}) and deadweight (DWT) in metric tons. IDs are designated as letters with added numbers for individual voyages.

Table 1. Container ship fleet subset.

Year Built	ID	LOA (m)	B_{ext} (m)	D_{max} (m)	D_{voy} (m)	DWT (t)	GT
2015	G	299.00	48.20	14.80	14.8	113,800	96,253
2015	F1, F2	299.95	48.20	14.80	14.8	112,729	95,263
2010	D	334.00	42.80	15.00	15.0	109,056	89,787
2005	E1, E2, E3	334.07	42.80	14.52	14.5	101,779	91,410
2008	A	347.00	45.20	15.50	15.5	130,700	128,600
2008	C	349.98	42.87	15.00	12.0	120,892	111,249
2008	B	359.99	42.80	15.00	15.0	120,944	111,249
2009	H1, H2	363.61	45.60	15.50	15.5	131,292	131,332

After the creation of individual voyages, we evaluated the route lengths, voyage durations, average speed in knots (kt) and similarity between individual routes. Among numerous available similarity measures, Fréchet distance (FD) or, formally, δ_F is often used [47]. It is described intuitively as a leash length between a person and a dog walking on their respective curves. They can have independent velocities, stop, and start; however, they cannot go backwards. Therefore, δ_F can be described as the shortest possible leash length for the completion of the walk. In a formal way, we can then define a curve as continuous mapping $f : [a, b] \rightarrow V$ where $a, b \in \mathbb{R}$ and $a \leq b$ are in metric space V. Two continuous curves, $f : [a, b] \rightarrow V$ and $g : [a', b'] \rightarrow V$ with their δ_F can be defined as [47]

$$\delta_F(f, g) = \inf_{\alpha, \beta} \max_{t \in [0,1]} d(f(\alpha(t)), g(\beta(t))), \tag{1}$$

where continuous nondecreasing functions $\alpha : [0, 1]$ and $\beta : [0, 1]$ map to $[a, a']$ and $[b, b']$. The continuous curves can be further approximated as discrete polygonal curves. Therefore, an approximative discrete Fréchet distance (DFD) or coupling distance d_{dF} can be calculated. The polygonal curves can then be defined as $P : [0, n] \rightarrow V$ and $Q : [0, n] \rightarrow V$ with corresponding sequences $\sigma(P) = (u_1, \ldots, u_p)$ and $\sigma(Q) = (v_1, \ldots, v_q)$ with coupling L, which is a sequence of distinct pairs of points of curves P and Q, respecting the point order $(u_{a1}, v_{b1}), (u_{a2}, v_{b2}), \ldots, (u_{am}, v_{bm})$. The first and last point pairs must be ordered as $a_1 = 1$, $b_1 = 1$, $a_m = p$, $b_m = q$. For all subsequent points $i = 1, \ldots, q$, we have $a_{i+1} = a_i$ or $a_{i+1} = a_i + 1$, and $b_{i+1} = b_i$ or $b_{i+1} = b_i + 1$. The maximum distance $\| L \|$ between the point pairs for a given coupling L is therefore the coupling distance. We can then define the discrete Fréchet distance as the minimum distance between all possible couplings between P and Q [47].

$$d_{dF}(P, Q) = min\{||L||\} \tag{2}$$

Therefore, using the previous person and dog or frog hopping analogy—since we consider discrete movement—we can describe the pair matching and discrete distance calculation. The person and dog can stay on their current vertices, while the other advances or they both advance to the next vertex on their respective paths. The DFD is then the smallest distance for the whole sequence of jumps to the last vertices.

The advantage of DFD over continuous FD is that it has a reduced computational cost while being a good approximation of the continuous solution. Further, provision of an upper bound on the continuous FD is provided, and the deviation is not greater than the longest edge of the trajectory [48]. Finally, the chosen DFD implementation [49] used in our research is based on the algorithm from [47].

Compared to other similarity measures such as Hausdorff distance, FD (and variants) is more useful for comparing curves, since it considers location and ordering of points of the curve [50]. The chosen DFD implementation is available in the Python similarity measures library, which includes Dynamic DTW, Partial Curve Mapping (PCM), Curve Length (CL) and other similarity measures [51].

The basic implementation of the similarity measures library considers the calculation of similarity between two curves only, so we had to extend the approach as follows. To calculate DFD values, we created 2D arrays of latitude and longitude coordinate pairs from individual ship position entries. The position coordinates, which were defined using the angular World Geodetic System (WGS84), were transformed into local planar-projected Universal Transverse Mercator (UTM) zone 17N system coordinates to improve the accuracy of the implemented DFD approach. Then, a square $n \times n$ mesh grid consisting of individual route arrays with coordinate pairs was created. A vectorized DFD method was then applied to obtain the similarity results, which are presented in the results section.

3. Results

Basic features are summarized in three voyage values: length in nautical miles (NM), duration in hours (h) and average speed in knots (kt). These values, in the form of statistical summary, are presented in Table 2, along with boxplots in Figure 4. The values represent length in nautical miles (NM), voyage duration in hours (h) and average ship speeds in knots (kt).

Table 2. Statistical summary of voyage lengths.

Count (12)	Length (NM)	Duration (h)	Average Speed (kt)
Mean	132.9	14.2	9.5
Std	4.4	1.7	1.0
Min	127.6	10.8	8.1
25%	130.7	13.0	8.8
50%	131.1	14.7	9.2
75%	135.3	15.3	10.0
Max	142.2	16.3	12.0

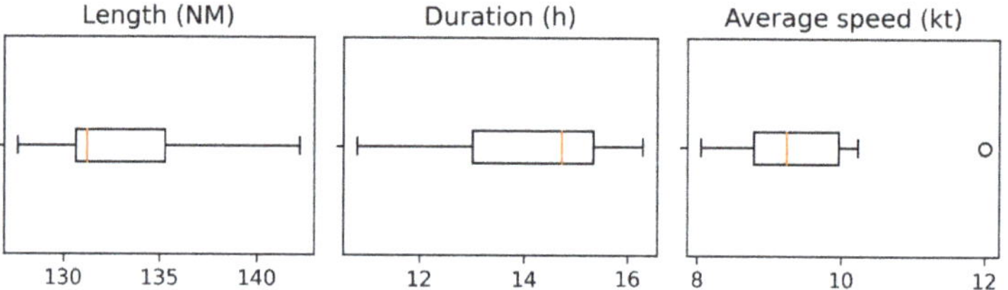

Figure 4. Boxplots for voyage length (NM), duration (h) and average speed (kt).

As is observable, the mean value of voyage length is 132.9 NM with a standard deviation of 4.4 NM, whereas the difference from the shortest to the longest voyage is 14.6 NM. The mean and median length values are close; however, the distribution is moderately positively skewed with a value of 0.77. The skewness value is calculated as an adjusted Fisher–Pearson coefficient of skewness with classification as in [52]. This classification is as follows: normal (-0.5 to $+0.5$), moderately skewed (-1.0 to -0.5 and $+0.5$ to $+1.0$) or highly skewed (<-1.0 or $>+1.0$).

The next value is the distribution for the voyage duration in hours. The difference is 5.2 h between minimum and maximum value, whereas the mean value is 14.2 h. The distribution is moderately negatively skewed (-0.62) as is observable in Figures 4 and 5, in which other distribution characteristics are visible as well.

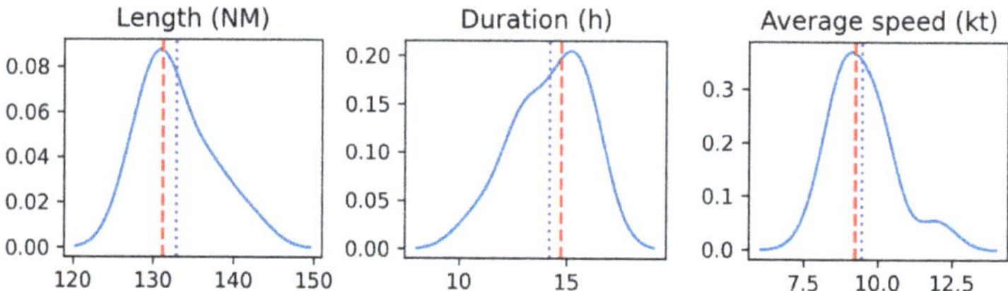

Figure 5. Distribution plots for voyage length, duration and average speed. Dashed vertical line represents median value and dotted line represents mean value.

The last distribution is for the average speed with a difference between minimum and maximum of 3.9 kt. Further, the maximum value can be considered an outlier (outside the interquartile range multiplied by 1.5), which is observable in the boxplot in Figure 4. However, since this not an erroneous value and the sample size is small, we did not remove it from analysis. Further, the distribution is highly positively skewed (1.23), which can be seen in the respective plot in the Figure 5.

Since the number of voyages is small—compared to overall voyages conducted over several years, or total numbers that include other ships and fleets on this service—we did not make further inferences by using methods to test distributions, either for normality or for underlaying distribution determination. This will be addressed further in future research. The actual distribution of individual voyage lengths and speeds can be observed in scatterplots in Figure 6.

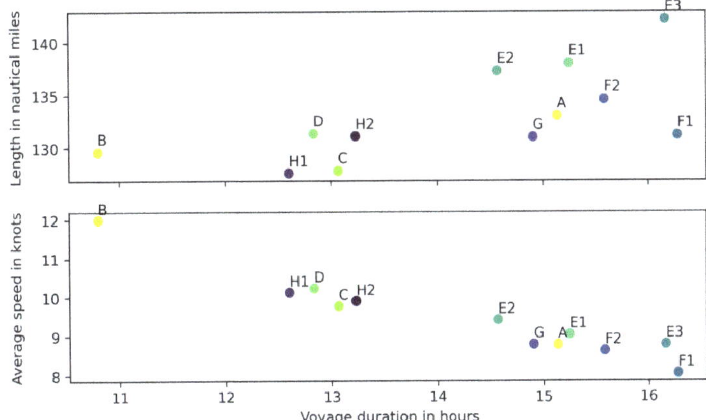

Figure 6. Scatterplot (**top**) of voyage duration (h) and average speed (kt) compared with voyage duration (**bottom**). The letters correspond to ship identification as stated in Table 1.

There are three clusters observable. First, a single entry for ship B, with the shortest voyage length and highest average speed. Second, centered on about 13 h of voyage length, including ships C and D with 2 individual voyages of ship H (H1, H2). The third cluster spreads out between 14 h and 16 h of voyage duration and includes the remaining ships: A, E, F and G.

The characteristics of individual ship routes and their spatial distribution can be observed in Figure 7. Seen beside the visible individual ship routes, the dashed line represents the Seasonal Management Area boundary, and we can see that the ships were sailing outside the area after passing the port approach areas for Savannah and Charleston.

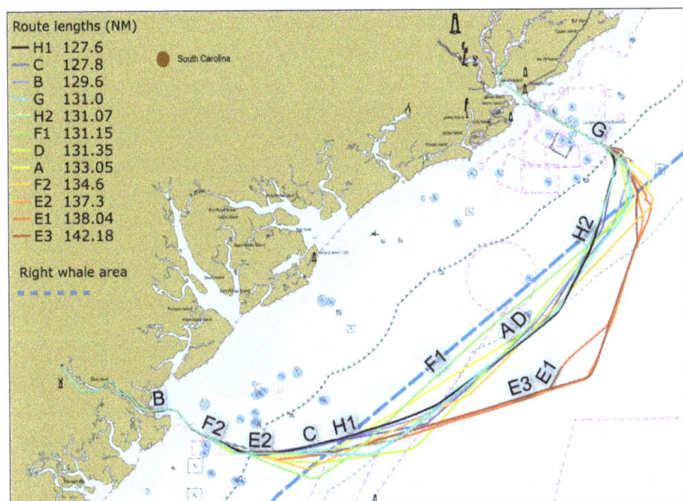

Figure 7. Route lengths for individual ships and voyages (scale 1:600000). Dashed line represents right whale Seasonal Management Area boundary.

Further, we examined the trajectory profiles for the longest and shortest route (H1 and E3) presented in Figure 8. The route length for ship H in voyage H1 (27 January 2019) was 127.6 NM with a passage time of 12.6 h, which was accounted for by an average speed of 10.1 kt. On the other hand, the voyage with the highest average speed of 12 kt, for ship B, was completed in 10.8 h with a total length of 129.6 NM. Ship E, in voyage E3 (14–15 April 2019), completed the respective voyage in 16.2 h with a total length of 142.2 NM with average speed of 8.8 kt. The voyage with the lowest average speed of 8 kt was conducted by ship F in voyage F1, with a total route length of 131.1 NM. The following route comparison could be expressed in terms of highest and lowest speeds, although we present the cases by route lengths only.

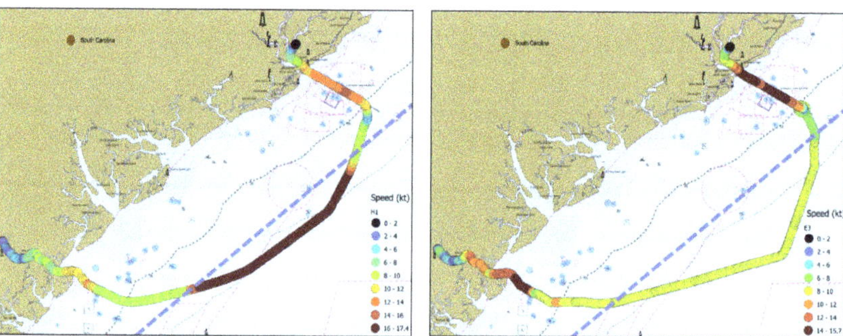

Figure 8. Trajectory profile for shortest (H1–(**left**)) and longest route (E3–(**right**)) with Speed Over Ground (SOG) in knots for individual AIS observations throughout the voyage (scale 1:600000).

Besides the route selection and length, there are differences in speed profiles as well. Ship H1 had a significantly higher speed in the central part of the route between the Savannah and Charleston approaches. For ship H, the speed was over 16 kt to a maximum of 17.4 kt, as was reported in SOG values in AIS data. Ships sailed with required speeds in the SMA and stayed outside the coastal areas in which there are obstructions and fishing facilities which could be dangerous to navigation, as is visible in Figure 9 (left). Further, we can observe differences in the Charleston approach and pilot boarding area (in vicinity of Charleston Entrance; buoys not visible on presented figures). Some of the ships proceeded directly, while two ships (H2, E3) made turns while presumably picking up the pilot.

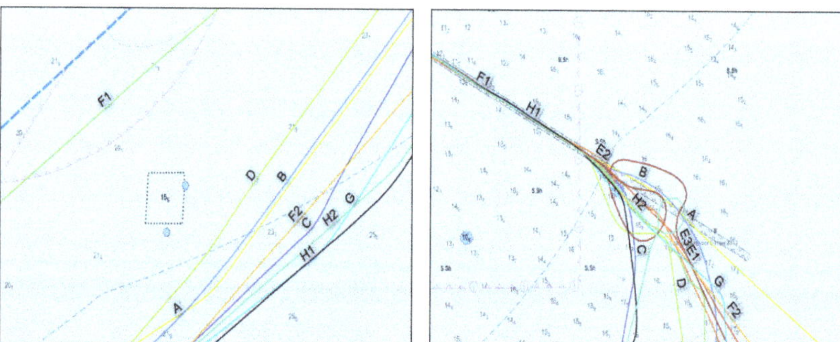

Figure 9. Fish haven obstructions (buoys not shown) with depth of 15.5 m (**left**) and approach area (**right**) for port of Charleston (scale 1:50000).

Finally, we evaluated the routes in terms of similarity. The individual routes are presented in Figure 10 as a reference. We can observe that route parts in constrained areas of navigation such as port approaches or river passages were similar. The differences arose with choices about whether to pass northwards or southwards to avoid several obstructions, which can be observed in previous figures.

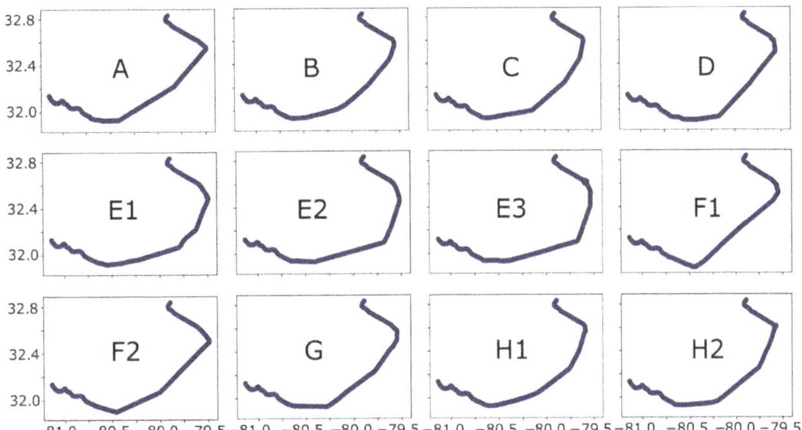

Figure 10. The routes for individual voyages. The x–axis represents longitude, y–axis latitude, for each of the subplots.

The calculated DFD similarity values are presented in Figure 11 as a heatmap. The lower values and darker color intensity indicate higher similarity. We can observe several routes with higher similarity. The calculated values are the distances in nautical miles.

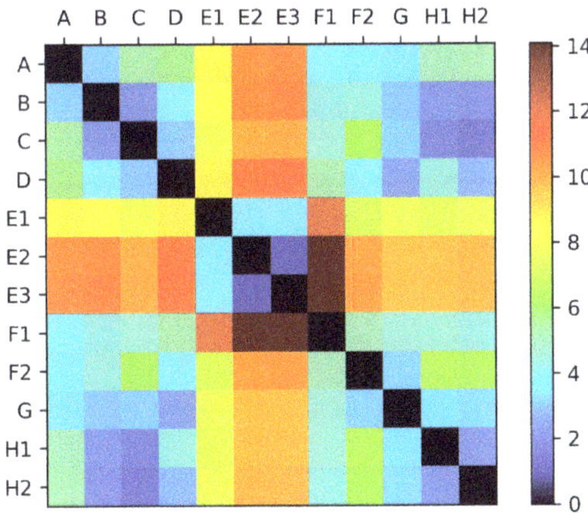

Figure 11. Heatmap plot for discrete Fréchet distance route similarity results (NM). Lower value and darker color intensity indicates higher similarity.

There were similarities among the different voyages from the same ships (E1–E2–E3, F1–F2 and H1–H2). Further, the highest route similarity between different ships was for ship C and ship H, whereas the lowest similarity was between routes from ships E and F. The calculated DFD similarity value complements the interpretation of similarities of visual or geographical routes and the statistical analysis of voyage data.

4. Discussion

The aim of our research was to analyze general route characteristics and evaluate similarity for a container ship fleet of a single company. As the preliminary research literature showed, route and similarity analysis considering fleets is not common. Reasons might include the perception of the availability of data and tools for such analyses. Formatted public AIS data records are available from NOAA, other national providers, and at times as sample or test datasets from private providers. The other challenge is collecting the ship data on which fleet assessment is based. This includes information on ship fleets and categories other than general AIS ship type categories. Nevertheless, the data can be collected from shipping companies and other reliable online sources such as shipping vessel registers and official international or national vessel data providers. However, this is a time-consuming process. Although there are many private providers which can provide aggregated ship data on companies and fleets with detailed ship types, the financial aspect of acquiring data might be worth consideration or even be a barrier for those interested.

For route characterization, we selected the two ports because of their proximity and constraints in the number of possible routes. As expected, the sailed routes taken by ships of different sizes were mostly similar, notably for the same ships. The route differences arose after the port approach areas were passed and the right whale protection area, which all the ships sailed outside, was exited. This was presumably due to the right whale avoidance and speed constraints required during the period of the SMA (from 1 November to 30 April 2019). Further, even after the end of the SMA period, the vessels sailed outside the area. It must be seen in future research if this occurs during the rest of the year. Further, as we presented regarding ships' particulars, the maximum and reported voyage drafts were large, ranging from 14.5 to 15.5 m, with only one vessel (C) reporting a lower voyage draft (12 m) than the maximum (15 m). Therefore, to keep a safe depth below keel, referred to as Under Keel Clearance (UKC), ships must stay in the deeper waters with depths of 20 m or more, which are outside the SMA area. In the shallower area near the coast, there are obstructions and fishing facilities as well, so even in that context it is safer to stay in the deep-water area.

Route characteristics do not differentiate substantially. There are differences in length which we can attribute to collision avoidance, besides route choice by the navigator. However, to confirm that decisively, we should evaluate the surrounding traffic as well. Looking in terms of route length, voyage duration and speed, most of the observations are not very dispersed either from the mean or median values. The distribution shapes reveal moderate to high skewness; however, the number of voyages and routes is too low to give decisive conclusions, and further tests should be conducted regarding the type and shape of the underlaying distribution. Further, assessment of the individual voyages reveals there is some clustering of individual routes, considering both voyage length and duration. The shortest voyage duration was 10.8 h, compared to 16.3 h for the longest voyage. The mean and median values of the average speed are 9.5 kt and 9.2 kt, with only one vessel sailing with an average speed of 12 kt. This along with trajectory analysis reveals differences in the speed profiles of the vessels, at least during parts of their routes. The fastest-average-speed vessel had much more variable speed (H1) throughout the voyage than the vessel with the lowest average speed (E3). To better understand behavior in parts of the voyage, additional trajectory analysis is required, and identification of regularities in behavior, if they exist.

The similarity measurement revealed that the same vessels take more similar routes in different voyages compared to the rest of the vessels. This is observable both in visual shape evaluation (vessels E, F and H), which was applicable in our study, and with DFD similarity measurement. This similarity should be assessed against a baseline route which either is derived as a generalization from numerous routes or is a route devised by a navigator in accordance with established voyage-planning procedures. The DFD pairwise similarity measurement, visual inspection for a small number of routes and methodology, is applicable. However, there is an issue of computational cost for DFD calculation of substantial numbers of routes. We did not generalize the routes due to the small number (several hundred for each route) of AIS data entries—points regularized at approximately one-minute intervals by default in NOAA AIS datasets. The number of entries can be reduced even more with route generalization and the extraction of important route points (e.g., when the vessel alters the course) and omission of the rest. This would reduce computation time when dealing either with substantial numbers of points per voyage or substantial numbers of voyages. Further, other similarity measures should be used as well for comparison. To evaluate usability, in forthcoming research we will assess DFD against other similarities in other shipping areas and with other ships and fleets.

Our methodology builds on established tools and practices for data assessment and similarity measurement, along with our own approach to similarity comparison between the individual ships or entire fleets. Since we conducted the analysis with open-source software based on the Python programming language, other interested researchers can apply the described methodology. Finally, the usage of discrete Fréchet distance to measure similarity of the routes is a valuable addition to route knowledge interpretation. It formalizes the route similarity in a single value; likewise with the route length, which is one of the values with which we compare the routes. The use of DFD values can be applicable in a variety of contexts, such as limiting values for the route selection from a trajectory collection. It can be used for similarity values when comparing planned routes on board ships, besides visual inspection and leg by leg comparison. Further, it can be used as an indication of possible route deviations in the context of the safety of navigation. This can be extended to creating profiles of various ship categories based on different criteria such as fleet, class and sizes.

Finally, we must state that we expected high similarity due the area characteristics, fleet selection, route possibilities and expected ship behavior. Further, this methodology opens new possibilities of knowledge discovery between diverse groups of ships, such as fleets, which are not observable when considering general ship groups.

5. Conclusions

Ship route analysis often considers general route and knowledge extraction in spatial and temporal dimensions. With AIS historical data availability, such a process is simpler than it used to be. However, since the available AIS ship types are limited, analysis for various ship groups has its challenges. With our methodology, we examined route characteristics both in statistical terms and using discrete Fréchet distance for the measurement of similarity for a single container ship fleet. This opens the possibility for the comparison of route creation and execution, not only between similar types of ships, but with fleets as well.

The investigated routes shared similarities in length, duration and average speed; however, there are differences between route legs, where we observed differences in speed profiles. Due to ship size, draft, speed limitations and the avoidance of obstacles in the coastal right whale protection area, ships stayed farther from the coast and outside of the area.

In terms of visual and numerical similarity, we observed that the same ships in different voyages tended to have similar routes. The route similarity between other vessels was also observable, as we expected, due to the relative proximity of, and limited number of potential routes between, the ports. The DFD similarity measure adds value to route interpretation with common values such as route length and duration. We extended the route and similarity analysis from general research on ship categories to fleets, which is not as common. Further, we extended the methods from known software libraries, which were not designed specifically for AIS data analysis, and therefore do not have all the specific functions that could apply when dealing with AIS data. The proposed methodology is based on open-source libraries and data, and can be adapted for further applications. It is our belief that our methodology can improve the reproducibility of similar research approaches and future frameworks.

For future research, we will evaluate longer time periods and datasets to obtain more routes for the comparison of ships in a single fleet and between fleets. We will evaluate methodologies for more distant ports and use other similarity measures. Trajectory analysis will be considered as well, to evaluate the speeds between different route legs.

Author Contributions: Conceptualization, D.Š., S.Ž., I.R. and D.B.; methodology, D.Š. and S.Ž.; software, D.Š.; validation, D.Š., S.Ž., I.R. and D.B.; formal analysis, D.Š.: writing—original draft preparation, D.Š. and S.Ž.; writing—review and editing, I.R. and D.B.; visualization, D.Š.; supervision, S.Ž. and I.R.; project administration, S.Ž. All authors have read and agreed to the published version of the manuscript.

Funding: This research was partially funded by European Union's Horizon Europe under the call HORIZON-CL5-2022-D6-01 (Safe, Resilient Transport and Smart Mobility services for passengers and goods), grant number 101077026, project name SafeNav. Views and opinions expressed are however those of the author(s) only and do not necessarily reflect those of the European Union or Executive Agency (CINEA). Neither the European Union nor the granting authority can be held responsible for them.

Institutional Review Board Statement: Not applicable.

Informed Consent Statement: Not applicable.

Data Availability Statement: Code and data used in the research are openly available in the Zenodo repository [10.5281/zenodo.7573544] under the license CC BY 4.0.

References

1. Šakan, D.; Žuškin, S.; Rudan, I.; Brčić, D. Static Maritime Environment Representation of Electronic Navigational Charts in Global Path Planning. In Proceedings of the 9th International Conference on Maritime Transport, Barcelona, Spain, June 27–28 2022; pp. 1–14. [CrossRef]
2. Lee, J.-S.; Son, W.-J.; Lee, H.-T.; Cho, I.-S. Verification of Novel Maritime Route Extraction Using Kernel Density Estimation Analysis with Automatic Identification System Data. *J. Mar. Sci. Eng.* **2020**, *8*, 375. [CrossRef]
3. Li, H.; Liu, J.; Wu, K.; Yang, Z.; Liu, R.W.; Xiong, N. Spatio-temporal vessel trajectory clustering based on data mapping and density. *IEEE Access* **2018**, *6*, 58939–58954. [CrossRef]
4. Gunnar Aarsæther, K.; Moan, T. Estimating Navigation Patterns from AIS. *J. Navig.* **2009**, *62*, 587–607. [CrossRef]
5. Sánchez-Heres, L.F. Simplification and Event Identification for AIS Trajectories: The Equivalent Passage Plan Method. *J. Navig.* **2019**, *72*, 307–320. [CrossRef]
6. Nie, P.; Chen, Z.; Xia, N.; Huang, Q.; Li, F. Trajectory similarity analysis with the weight of direction and k-neighborhood for ais data. *ISPRS Int. J. Geo.-Inf.* **2021**, *10*, 757. [CrossRef]
7. Ranacher, P.; Tzavella, K. How to compare movement? A review of physical movement similarity measures in geographic information science and beyond. *Cartogr. Geogr. Inf. Sci.* **2014**, *41*, 286–307. [CrossRef]
8. Šakan, D.; Rudan, I.; Žuškin, S.; Brčić, D. Near real-time S-AIS: Recent developments and implementation possibilities for global maritime stakeholders. *Pomor.-Sci. J. Mar. Res.* **2018**, *32*, 211–218. [CrossRef]
9. Tu, E.; Zhang, G.; Rachmawati, L.; Rajabally, E.; Bin Huang, G. Exploiting AIS Data for Intelligent Maritime Navigation: A Comprehensive Survey from Data to Methodology. *IEEE Trans. Intell. Transp. Syst.* **2018**, *19*, 1559–1582. [CrossRef]
10. International Maritime Organization (IMO). Resolution A.1106(29): Revised Guidelines for the Onboard Operational Use of Shipborne Automatic Identification Systems (AIS), London. 2015. Available online: https://edocs.imo.org/Final (accessed on 7 August 2022).
11. ITU-R. *Technical Characteristics for An Automatic Identification System Using Time Division Multiple Access in the VHF Maritime Mobile Frequency Band*; M Series Mobile, radiodetermination, amateur and related satellite services; Recommendation ITU M.1371-5: Geneva, Switzerland, 2014; pp. 1371–1375.
12. Lu, N.; Liang, M.; Yang, L.; Wang, Y.; Xiong, N.; Liu, R.W. Shape-Based Vessel Trajectory Similarity Computing and Clustering: A Brief Review. In Proceedings of the 2020 5th IEEE International Conference on Big Data Analytics, ICBDA 2020, Harbin, China, 8–11 May 2020; pp. 186–192. [CrossRef]
13. Cai, J.; Chen, G.; Lützen, M.; Rytter, N.G.M. A practical AIS-based route library for voyage planning at the pre-fixture stage. *Ocean Eng.* **2021**, *236*, 109478. [CrossRef]
14. Breithaupt, S.A.; Copping, A.; Tagestad, J.; Whiting, J. Maritime Route Delineation using AIS Data from the Atlantic Coast of the US. *J. Navig.* **2017**, *70*, 379–394. [CrossRef]
15. Zhang, L.; Meng, Q.; Fang Fwa, T. Big AIS data based spatial-temporal analyses of ship traffic in Singapore port waters. *Transp. Res. E Logist. Transp. Rev.* **2019**, *129*, 287–304. [CrossRef]
16. Zhao, L.; Shi, G. A trajectory clustering method based on Douglas-Peucker compression and density for marine traffic pattern recognition. *Ocean Eng.* **2019**, *172*, 456–467. [CrossRef]
17. Yan, Z.; Xiao, Y.; Cheng, L.; He, R.; Ruan, X.; Zhou, X.; Li, M.; Bin, R. Exploring AIS data for intelligent maritime routes extraction. *Appl. Ocean Res.* **2020**, *101*, 102271. [CrossRef]

18. Wang, L.; Chen, P.; Chen, L.; Mou, J. Ship AIS Trajectory Clustering: An HDBSCAN-Based Approach. *J. Mar. Sci. Eng.* **2021**, *9*, 566. [CrossRef]

19. United Nations. *UNCTAD Review of Maritime Transport 2022; Review of Maritime Transport/United Nations Conference on Trade and Development, Geneva*; United Nations: Geneva, Switzerland, 2022; ISBN 978-92-1-002147-0.

20. Ksciuk, J.; Kuhlemann, S.; Tierney, K.; Koberstein, A. Uncertainty in Maritime Ship Routing and Scheduling: A Literature Review. *Eur. J. Oper. Res.* **2022**. [CrossRef]

21. Tran, N.K.; Haasis, H.-D. Literature Survey of Network Optimization in Container Liner Shipping. *Flex. Serv. Manuf. J.* **2015**, *27*, 139–179. [CrossRef]

22. Meng, Q.; Wang, S.; Andersson, H.; Thun, K. Containership Routing and Scheduling in Liner Shipping: Overview and Future Research Directions. *Transp. Sci.* **2014**, *48*, 265–280. [CrossRef]

23. Wang, S.; Meng, Q. Sailing Speed Optimization for Container Ships in a Liner Shipping Network. *Transp. Res. Part E Logist. Transp. Rev.* **2012**, *48*, 701–714. [CrossRef]

24. Gao, C.-F.; Hu, Z.-H. Speed Optimization for Container Ship Fleet Deployment Considering Fuel Consumption. *Sustainability* **2021**, *13*, 5242. [CrossRef]

25. Li, X.; Sun, B.; Jin, J.; Ding, J. Speed Optimization of Container Ship Considering Route Segmentation and Weather Data Loading: Turning Point-Time Segmentation Method. *J. Mar. Sci. Eng.* **2022**, *10*, 1835. [CrossRef]

26. Gao, X.; Makino, H.; Furusho, M. Ship Behavior Analysis for Real Operating of Container Ships Using AIS Data. *TransNav Int. J. Mar. Navig. Saf. Sea Transp.* **2016**, *10*, 213–220. [CrossRef]

27. Ikonomakis, A.; Nielsen, U.D.; Holst, K.K.; Dietz, J.; Galeazzi, R. How Good Is the STW Sensor? An Account from a Larger Shipping Company. *J. Mar. Sci. Eng.* **2021**, *9*, 465. [CrossRef]

28. Oh, M.-J.; Roh, M.-I.; Park, S.-W.; Chun, D.-H.; Son, M.-J.; Lee, J.-Y. Operational Analysis of Container Ships by Using Maritime Big Data. *J. Mar. Sci. Eng.* **2021**, *9*, 438. [CrossRef]

29. Botter, R.; Santos, T.; Soares, C.G. Characterizing Container Ship Traffic along the Portuguese Coast Using Big Data. In *Progress in Maritime Technology and Engineering*; CRC Press: Boca Raton, FL, USA, 2018; pp. 93–100.

30. Shu, Y.; Daamen, W.; Ligteringen, H.; Hoogendoorn, S.P. AIS Data Based Vessel Speed, Course and Path Analysis in the Botlek Area in the Port of Rotterdam. In Proceedings of the International Workshop on Next Generation of Nautical Traffic Model, Shanghai, China, 21–24 September 2012.

31. Xiao, F.; Ligteringen, H.; van Gulijk, C.; Ale, B. Comparison Study on AIS Data of Ship Traffic Behavior. *Ocean. Eng.* **2015**, *95*, 84–93. [CrossRef]

32. BOEM; NOAA. MarineCadastre.gov., Automatic Identification System Data 2019 from 2019-01-01 to 2019-06-30. for the Area of Interest [(31.791, −81.224), (33.102, −79.453)]. 2022. Available online: https://marinecadastre.gov/accessais/ (accessed on 17 July 2022).

33. G. F. Department. Georgia Ports Authority Comprehensive Annual Financial Report For the Fiscal Years Ended June 30, 2019 and 2018, Savannah, GA 31402 USA, 2019. Available online: https://gaports.com/wp-content/uploads/2021/10/CAFR-FY19-Final.pdf (accessed on 19 July 2022).

34. Comprehensive Annual Financial Report for Fiscal Year Ended June 30, 2019, Mount Pleasant, South Carolina 29464, USA, 2019. Available online: https://scspa.com/wp-content/uploads/cafr-fy19-final.pdf (accessed on 19 July 2022).

35. National Oceanic National Oceanic and Atmospheric Administration (NOAA). *Distances Between United States Ports 2019*, 13th ed.; NOAA: Washington, DC, USA, 2019; p. 10.

36. Raimondo, G.M.; Spinrad, R.W.; Leboeuf, N.R. *National Oceanic and Atmospheric Administration (NOAA) National Ocean Service Coast Pilot 4 Atlantic Coast: Cape Henry, Virginia to Key West, Florida*, 53rd ed.; NOAA: Washington, DC, USA, 2021. Available online: https://nauticalcharts.noaa.gov/publications/coast-pilot/files/cp4/CPB4_WEB.pdf/ (accessed on 1 August 2022).

37. AIS Vessel Transit Counts 2019 | InPort. Available online: https://www.fisheries.noaa.gov/inport/item/61037 (accessed on 19 July 2022).

38. McKinney, W. Data Structures for Statistical Computing in Python. In Proceedings of the 9th Python in Science Conference, Austin, TX, USA, 28 June 28–3 July 2010; Volume 445, pp. 56–61. [CrossRef]

39. Reback, J.; McKinney, W.; den Bossche, J.V.; Augspurger, T.; Roeschke, M.; Hawkins, S.; Cloud, P.; jbrockmendel; gfyoung; Sinhrks999; et al. *Pandas-Dev/Pandas*; Pandas 1.4.2; Zenodo, 2022. [CrossRef]

40. Gillies, S.; van der Wel, C.; Van den Bossche, J.; Taves, M.W.; Arnott, J.; Ward, B.C. Shapely (2.0). Zenodo, 2022. [CrossRef]

41. Harris, C.R.; Millman, K.J.; Van Der Walt, S.J.; Gommers, R.; Virtanen, P.; Cournapeau, D.; Wieser, E.; Taylor, J.; Berg, S.; Smith, N.J.; et al. Array programming with NumPy. *Nature* **2020**, *585*, 357–362. [CrossRef] [PubMed]

42. Graser, A. MovingPandas: Efficient Structures for Movement Data in Python. *GI_Forum* **2019**, *1*, 54–68. [CrossRef]

43. Graser, A.; Bell, R.; Boiko, G.P.; Parnell, A.; Vladimirov, L.; Theodoropoulos, G.S.; Richter, G.; Elcinto, J.B.; Rodríguez, J.L.C.; Lovelace, R.; et al. *Anitagraser/Movingpandas*; v0.9.rc3; Zenodo, 2022. [CrossRef]

44. Jordahl, K.; den Bossche, J.V.; Fleischmann, M.; McBride, J.; Wasserman, J.; Badaracco, A.G.; Gerard, J.; Snow, A.D.; Tratner, J.; Perry, M.; et al. *Geopandas/Geopandas*; v0.10.2; Zenodo, 2021. [CrossRef]

45. Rudiger, P.; Stevens, J.-L.; Bednar, J.A.; Nijholt, B.; Mease, J.; Andrew; Liquet, M.; B, C.; Randelhoff, A.; Tenner, V.; et al. *Holoviz/Holoviews*; Version 1.14.9; Zenodo, 2022. [CrossRef]

46. MarineCadastre.gov, (NOAA). AIS Vessel Type and Group Codes Used by the Marine Cadastre Project 2018. Available online: https://coast.noaa.gov/data/marinecadastre/ais/VesselTypeCodes2018.pdf/ (accessed on 2 September 2022).
47. Eiter, T.; Mannila, H. *Computing Discrete Fréchet Distance*; Technical Report CD-TR 94/64; Technische Universitat Wien: Wien, Austria, 1994.
48. Tang, B.; Yiu, M.L.; Mouratidis, K.; Zhang, J.; Wang, K. On discovering motifs and frequent patterns in spatial trajectories with discrete Fréchet distance. *Geoinformatica* **2022**, *26*, 29–66. [CrossRef]
49. Jekel, C.F.; Venter, G.; Venter, M.P.; Stander, N.; Haftka, R.T. Similarity measures for identifying material parameters from hysteresis loops using inverse analysis. *Int. J. Mater. Form.* **2019**, *12*, 355–378. [CrossRef]
50. Agarwal, P.K.; Avraham, R.B.; Kaplan, H.; Sharir, M. Computing the discrete Fréchet distance in subquadratic time. *SIAM J. Comput.* **2014**, *43*, 429–449. [CrossRef]
51. Jekel, C.F. Similaritymeasures PyPI. Available online: https://github.com/cjekel/similarity_measures (accessed on 21 July 2022).
52. Piovesana, A.; Senior, G. How Small Is Big: Sample Size and Skewness. *Assessment* **2018**, *25*, 793–800. [CrossRef]

Journal of
*Marine Science
and Engineering*

Article

Multi-Criteria Decision Analysis for Nautical Anchorage Selection

Danijel Pušić * and Zvonimir Lušić

Faculty of Maritime Studies, University of Split, 21000 Split, Croatia; zvonimir.lusic@pfst.hr
* Correspondence: danijel.pusic@pfst.hr

Abstract: Considering that moorings and anchorages for vessels have recently become an important factor in nautical tourism, the selection of their locations is a complex and demanding process. This paper examines numerous criteria from different perspectives to determine the most favourable/optimal locations for nautical anchorages, meeting the conditions and recommendations of professionals from several domains, by applying the methods of multi-criteria analysis. The goal of solving the problem this way is to meet the expectations of future users, spatial planners, possible investors, and concessionaires interested in doing business in these areas, as well as entities that strive to preserve and protect marine and underwater animal life and the environment by preventing their degradation and pollution. However, since there are no precisely defined recommendations for the establishment of nautical anchorages, in the procedures for determining the locations of nautical anchorages, it is possible to use general criteria they must fulfil. The best locations for nautical anchorages may be found, and this research represents a transparent, repeatable, and well-documented approach for methodically solving the problem. This is demonstrated by a comparison of many methods of multi-criteria analysis, utilizing a variety of parameters. On the other side, this calls for proficiency in a wide range of disciplines, including architecture, geodesy, marine safety and transport, architecture, biology, ecology, mathematical programming, operational research, information technology, environmental protection, and others. The best locations for nautical anchorages should be chosen based on the size and number of vessels, available space, depth, distance from the coast, level of protection of the anchorage waters, and many other limiting factors, keeping in mind that the spots which simultaneously satisfy a greater number of significant criteria are preferable. Using multi-criteria analysis methods (AHP (Analytical Hierarchy Process) and TOPSIS (The Technique for Order of Preference by Similarity to Ideal Solution)), evaluating and classifying criteria as well as assigning weight values to selected criteria, this paper investigates the possibility of obtaining the best locations from a group of possible ones. The most important factor when applying multi-criteria analysis methods refer to the following: vessel safety (navigation), hydrometeorological, spatial, economic, and environmental criteria. The main contribution of the paper displays in the proposal to optimize the decision-making process, when determining the optimal locations of nautical anchorages, in accordance with previously defined criteria.

Keywords: nautical anchorage; multi-criteria analysis; criteria; optimization of spatial locations; weight coefficients; concession fields; AHP; TOPSIS

Citation: Pušić, D.; Lušić, Z. Multi-Criteria Decision Analysis for Nautical Anchorage Selection. *J. Mar. Sci. Eng.* **2023**, *11*, 728. https://doi.org/10.3390/jmse11040728

Academic Editors: Marko Perkovic, Lucjan Gucma and Sebastian Feuerstack

Received: 9 March 2023
Revised: 21 March 2023
Accepted: 24 March 2023
Published: 27 March 2023

1. Introduction

The construction of a new nautical anchorages system is a large and long-term investment, and therefore, the determination of new locations is a critical point on the way to the success or failure of the system by exploiting them. One of the main goals when considering the locations for nautical anchorages is to find the most suitable location which can increase the desired conditions defined by the selection criteria.

After finding the nautical anchorages locations, an attempt is made to optimize the number of objectives in determining the suitability of a particular location for a defined system. Such optimization often involves many and sometimes even contradictory factors. Some of the important factors increase the complexity of choosing the right/best among the many possible locations.

The process of selecting the location of nautical anchorages involves a number of perplexing key factors that include maritime, spatial, hydrometeorological, traffic technical, economic, social and other issues, such as sea, coastal, environmental protection, etc. Due to the complexity of the process itself, the simultaneous use of several tools is required for decision support, such as Expert Systems (ES), Geographic Information Systems (GIS), and Multi-Criteria Decision-Making methods (MCDM).

In order to obtain a safe and comfortable stay while anchoring or using certain specific facilities at sea, sailors (both professionals and amateurs) need to take into account many location aspects, especially those considering protection from wind, waves, sea currents, etc.

On the other hand, from the point of view of planning and space utilization, planners and management of local and regional communities, who want to optimize the space at sea, should have in mind a number of factors that enable them to plan the space in the best way, observing a whole series of other factors and especially those who take into account the protection of nature, the sea and coastal areas, the environment, traffic and technical conditions, the current situation on the ground, etc.

Therefore, an attempt is made to compromise on the wishes of future users of nautical anchorages (professional sailors, amateurs and future concessionaires of nautical anchorages, spatial planners, administrations of local communities, and others) by creating a very sophisticated system of interconnections and interdependence, with the aim of selecting the best spaces in the water area predetermined for anchoring vessels while taking advantage of all the benefits that these locations can provide in order to fulfil all expectations or most of them, without disturbing surrounding space, sea, coast, and environment. The former represents the research problem of this work.

The previously defined problem can be solved by applying MCDM methods in order to optimize the selection process of the best locations for setting up nautical anchorages planned to be used as concession fields in the area of Split-Dalmatia County (Croatia).

Given that previous research conducted both in Croatia and around the world shows that there is no unified methodology for selecting the best locations for nautical anchorages, neither for their layout nor from the navigational and safety point of view, for vessels and both their crew and passengers there should be a social and a scientific contribution.

The paper will also present an overview of the factors affecting the anchorage area and prove that the correct selection of the location and the construction of the necessary facilities remarkably affect the safety of people, vessels, and the marine environment. By applying multi-criteria methods and analysis, the best 15 locations of nautical anchorages will be selected from a set of possible 86 available.

The proposal for a systematic solution to the problem posed in this paper was accomplished by applying two methods of multi-criteria analysis 1. AHP (Analytical Hierarchy Process) and 2. TOPSIS (The Technique for Order of Preference by Similarity to Ideal Solution). Using the mentioned methods, several conflicting criteria are taken into account, the weighting values and mutual dependence of which were previously determined on the basis of a survey of future users.

The paper is structured as follows: The motive for the research and the existing problems are described in the introduction, while the second chapter lists materials, methods, previous research, and the most important recommendations for determining the locations of nautical anchorages. The basic steps of the multi-criteria decision-making methods used in this paper are defined in the third chapter, while the fourth chapter analyses the case study. The fifth chapter indicates the most important validation, testing, comparison, and result analysis, while the sixth presents a discussion of the entire procedure.

2. Materials and Methods

2.1. Methodology and Research Plan

Research implementation includes research plan overview and investigation of previous research. The process of collecting and processing data in the field was conducted simultaneously: (a) by gathering the opinions of future users of nautical anchorages through survey research and (b) through the collection, analysis, processing, and storage of data by the author/s at the locations of nautical anchorages for the area of Split-Dalmatia County (in the period from 2018 to 2022). Criteria were selected based on both obtained average scores that the respondents (users) assigned to certain aspects of nautical anchorages and available data on nautical anchorages previously collected (by the author/s). Furthermore, they have defined the goal of each of them as well as the weight values of each criterion and their mutual relations. Both different MCDM methods (AHP, TOPSIS) were applied.

Validation, testing, comparison, and result analysis are shown in the chapter Results. The research methodology in this paper consists of two phases.

In the first phase, the attitudes and opinions of the nautical anchorages users were examined on the basis of a survey questionnaire. The survey questionnaire was made available to visitors of nautical tourism ports and nautical anchorages in Split-Dalmatia County and the entire international community and was published on the pages of the International Association of Maritime Universities [1].

Opinions were collected by forming a questionnaire of future users of nautical anchorages who evaluated five groups of elements (and a total of 18 sub-elements), namely the following: safety (navigational—6 sub-elements); hydrometeorological (with 2 sub-elements); spatial (with 4 sub-elements); economic (with 2 sub-elements) and environmental (ecological) (with 4 sub-elements), which were electronically sent to respondents in the time period from November 2022 to January 2023.

There are a number of factors that need to be considered when identifying suitable sites for anchoring vessels and will often require consultation with a wide range of stakeholders. The most important guidelines and recommendations on factors to consider [2,3].

The survey questionnaire was created and distributed via the web ArcGis survey 123 [4] service. Answers to the questionnaire were received from 74 respondents.

In the second phase, after arranging the data (as a result of survey research) and assessing the significance of certain factors from the perspective of future users of nautical anchorages, the most important criteria were selected and weighted values assigned to each of them.

Then, two different methods of the MCDM method were applied to the input data of 56 bays with 86 locations, i.e., fields of nautical anchorages (variants), for each of which 17 data were known. Essentially, the aim is to select the best locations for nautical anchorages out of the 86 taken into consideration, respecting the criteria that are the most significant and have the most influence. Both methods of the MCDM were implemented using the programming language R (version 4.2.2) [5] and heuristics.

The most important elements on the basis of which the criteria were defined and grouped into five groups are shown in Figure 1. The criteria were derived from various sources, including the Queensland Government's Anchorage Area Design and Management Guideline [3] and The World Association for Waterborne Transport Infrastructure (PIANC) [6]. Later, they were grouped into five categories based on their nature and impact on the selection process.

The parameters for the multi-criteria analysis were obtained from the factor criteria. The most important factors, i.e., criteria, are the following: 1. the surface of the field; 2. the surface of the bay; 3. the percentage share of the field surface in the bay surface; 4. the degree of protection (from wind and waves) of the bay; 5. the distance from the coast; 6. the number of anchorage fields in the same bay; 7. the existence of maritime traffic; 8. official anchorages; 9. the existence of underwater cables and pipelines; 10. the risk of collision; 11. depth; 12. the level of sea changes and the existence of sea currents; 13. proximity to public ports; 14. proximity to existing berths; 15. environmental elements (Environmental

Network Natura 2000); ref. [7] 16. damage from anchoring the vessel to the seabed; and 17. archaeological sites.

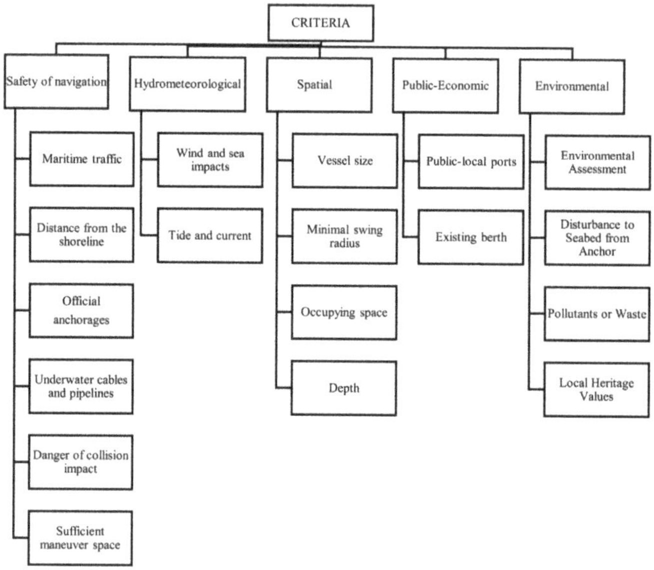

Figure 1. The most important criteria of nautical anchorages [6].

The analysis is limited and carried out according to a limited number of criteria (Figure 1) and certain weight values (based on determining their importance and impact). The selected criteria are unique and coherent, although some are related and interdependent. Some criteria have a greater impact and importance, so they have been singled out. This is the case for, e.g., criteria determined by a border (distance from the coast or depth of the sea, etc.), some of which are imposed by legal regulations, etc.

2.2. Background

The proposed scientific literature for the determination of the best locations for nautical anchorages based on MCDM is relatively small or does not exist at all. There are, however, studies related to methods of MCDM for areas of spatial planning, for example, nautical tourism ports or methods for processing spatially distributed data. Regarding the creation of this paper represents a special challenge to the authors, especially in terms of applying the following: multiple methods of MCDM in parallel over; more numerous input data; using a multi-criteria analysis methodology that combines several different criteria (safety, hydrometeorological, spatial, economic, and ecological); considering the selection of the best locations of nautical anchorages both from the point of view of users and from the point of view of concessionaires, i.e., future investors; contributing to the expansion of scientific methodology related to the subject of research.

In the doctoral dissertation [8], the criteria for assessing the possible impacts of the functioning and operation of the liquefied gas terminal on the marine environment are defined. Using the MCDM, a model that monitors the effect of the selection of individual criteria on the marine environment was created. The pre-elimination criteria were determined in order to reduce, i.e., limit, the observed space from which representative locations previously evaluated by certain criteria were selected. The observed problems were solved with the following methods: GIS, MCDM, and ES, using mathematical models and a regional model of the complete maritime system as a tool.

The research carried out in the doctoral dissertation [9] is related to finding an optimal model with five scenarios and undertaking certain measures for the development of county and local ports in relation to the complementarity of the spatial concept of the port and the city. For the purposes of defining and setting up the model, the most important indicators and measures that affect the level of spatial planning of the port and the city were used. The success of the set model was tested and demonstrated using the port of Rovinj as an example.

The paper [10] presents a conceptual framework for the inclusion of multiple criteria in the assessment of a dry port for developing countries from the perspective of multiple stakeholders. The framework of the work is presented in four steps covered by preliminary research, namely the following: 1. stakeholders are grouped into groups; 2. sub-criteria related to the location of the dry port are listed; 3. individual criteria and sub-criteria are explained; and 4. an MCDM analysis is carried out.

The paper [11] investigates the logistical possibilities of offshore wind farms, specifically the physical characteristics, connections, and appearance of the port to support the phases of installation, operation, and maintenance of offshore wind farm projects. The relative importance of these criteria was determined using the AHP method. The AHP methodology is then applied in a case study as a decision-making tool to enable decision-makers to assess the suitability of a range of ports for offshore wind farms in the UK North Sea.

The paper [12] presents a multi-criteria spatial evaluation (SMCE—Spatial Multi-Decisional Evaluation) intended to identify suitable areas at the regional level for setting up middle-sized coastal fish farms in the Ligurian Sea. The SMCE process follows an integrated approach that can potentially be adapted and applied to any coastal system. The selection of the location is based on the definition of criteria that assess their suitability and how their conditions relate to the entire research area. The results show that SMCE, and especially the procedure, enables the identification of the most suitable areas, which solves the complicated problem of spatial selection of an appropriate location in a simple, fast, and efficient way.

The study [13] is an expert basis for amendments and additions, i.e., the adoption of a new Spatial Plan of the Split-Dalmatia County based on navigational and meteorological features, as well as technical-technological and traffic-navigation features, maritime safety measures, the Natura 2000 Habitats Directive [14], the Register of Strictly Protected Species, technical-technological types of anchorages and the organization of anchorages using expert analysis, with possible future locations defined as nautical anchorages. The cited work also describes the prerequisites that the investor must fulfil in order to obtain the necessary permits and papers to start work.

The research project [15] serves as an expert basis and support for the authorities of the Split-Dalmatia County and the Public Institute for Urbanism in planning the future development and location of ports and anchorages for nautical tourism. The cited project considers the development possibilities of the port of nautical tourism in the Split-Dalmatia County from the point of view of optimal use and environmental protection. All elements of supply and demand of nautical tourism were determined by SWOT (S-Strengths, W-Weaknesses, O-Opportunities, T-Threats) analysis. By defining the advantages, disadvantages, strengths, and weaknesses of Croatian nautical tourism, it is possible to determine its strategy and direction of development. The application of MCDM methods of ports of nautical tourism in the Split-Dalmatia County included a systematic study of proposed micro locations, which resulted in the selection of land and island locations. The next step was the definition of criteria and sub-criteria for their acceptability, and finally, a qualitative and quantitative evaluation of each proposed location was performed.

Report [16] offers resolving a location selection problem by means of an integrated AHP-RAFSI (Ranking of Alternatives through Functional mapping of criterion sub-intervals into a Single Interval) approach. The main goal of this report is to find the optimal EMS (emergency services) as the locations that provide the least response time. The methods used vary and mainly relate to the decision-making variables taken into account. Multi-criteria decision making (MCDM) is used in the case of emergency centre allocation in

Libya. MCDM approach was implemented in two steps. AHP was adopted in the first step to determine the criteria weights, while the results of AHP showed that the response time had the highest weight among other criteria. A ranking of different alternatives was conducted in the second step using RAFSI model to choose the optimal location. Model ranking clearly indicated road-network as the best alternative to locate EMS centres.

The aim of research [17] is to select the best supplier in LISCO in Libya using the Rough AHP method. With increasing awareness of sustainability aspects, both from the academic sector and from the industrial sector, the authors believe that sustainable decision-making techniques for selecting the best suppliers should improve supply chain competencies and help companies maintain a strategically competitive position. The results showed that the Rough AHP method is capable of improving the quality of decision-making by making the process more rational, explicit, and efficient and that in future work this effective method can be generalized to other companies throughout Libya to facilitate the measurement of sustainability performance, consequently providing the company with a robust system while maintaining ecological sustainability.

Research of scientific and professional literature and data that thematically deals with nautical anchorages, criteria, and selection of new locations, especially in Croatia, indicate that there is a modest amount of officially precise collected data on their condition and registered physical traffic in organized nautical anchorages, especially those in Split-Dalmatian counties. In addition, the level of quality of the services provided in them is low, so it is necessary to use more modern analytical decision-making methods to collect accurate and precise data in order to, on the basis of the data thus obtained, use the methods and tools of multi-criteria analysis in the optimal and systematic way of choosing the best locations of nautical anchorages.

Previous research conducted in the world and Croatia on the topic of research problems (determining the best locations for nautical anchorages) showed that there is no single methodology for selecting the best locations for nautical anchorages, as well as their arrangement, especially regarding the safety aspect. Most of the research focuses only on economic factors when choosing anchorages, including other aspects such as tourism, legal, social, sustainable development, protection of the marine environment, seas, and coasts, mostly ignoring safety factors when choosing criteria. The anchorage site selection can be seen in the analysis of the fundamental concept of navigation safety as an important and indispensable part of spatial planning.

Unlike previous research, this work systematically and uniquely takes into account safety (navigation), but also hydrometeorological, spatial, economic, ecological, and other conditions and criteria in the process of selecting the best locations for nautical anchorages. At the same time, the greatest importance is attached to the fundamental concept of safety of navigation and staying at anchorages, as one of the most important factors of the purpose of anchorages and an important element of spatial planning.

2.3. Recommendations for Setting up Nautical Anchorages

In order to determine the optimal solution for the selection of the best locations in the Split-Dalmatia County and the adequate application of MCDM methods, the most important recommendations [18] can be summarized in the list below:

- It is necessary to take into account the existing situation, position, and size of the fields and respect as much as possible the existing boundaries of concession fields, initiatives, conceptual solutions, and proposals if they do not contradict the general principles of maritime safety and protection of the marine environment;
- Anchorages must not interfere with maritime traffic, i.e., both existing and future routes of transit and terminal waterways, and generally need to be in accordance with spatial plans;
- The boundaries of the anchorage must be at a safe distance from the shore, at least twice the width of the largest ship that is expected to sail from a safe isobath, and need to take into account other factors such as possible traffic of other ships, boats and/or

- beaches, bathing areas, and other facilities expected between the anchorage and the shore, where bathers, swimmers, divers and other persons in the sea could endanger the safety of the anchorage;
- The anchorage must be in an area of sufficient depth, whereby the depth of the lowest low water level of living sea minnows must not be less than the expected draft of the ship increased by 1 m;
- The surface of the anchorage may not occupy more than 50% of the bay (exceptionally up to 75%) if it is a bay where there are no other moorings, beaches, infrastructure, or other facilities or activities that require access from the sea or land;
- In the water area, from the nautical anchorage towards the coast and at a distance of up to 150 m from the nautical anchorage in the direction of the open sea, there should be no other artificial installations or structures, including anchorages or measures to direct maritime traffic (regulations on the safety of maritime navigation in internal sea waters and the territorial sea of the Republic of Croatia) [19], and the manner and conditions of supervision and management of maritime traffic;
- Anchorages may not be in the area of official anchorages on the map designated by the port authorities/MMPI (Ministry of the Sea, Transport, and Infrastructure of Croatia) [20] and in places of shelter if they are determined by the ordinance on places of shelter [21];
- Avoid anchoring in the immediate vicinity of submarine cables, submarine installations, and other places where anchoring is prohibited;
- Avoid positioning the anchorage that would be a potential danger of collision, impact, injury, and other risks;
- Anchorages must not limit the manoeuvring space for ships outside the anchorage that need to manoeuvre, and should also provide sufficient manoeuvring space for ships arriving or departing from the anchorage;
- Anchorages must not interfere with the existing moorings of the local population.

3. Methods of Multi-Criteria Analysis When Determining the Optimal Position of Anchorage

3.1. Definition and Description of Multi-Criteria Decision-Making Methods

Multi-criteria analysis plays an important role in the selection of the best variants for finding the best areas (locations) in many areas of spatial planning, optimization of urban and non-urban structures, etc.

According to the manual on multi-criteria analysis (multi-criteria analysis: a manual) published by Communities and Local Government London, the method of MCDM is defined as an approach that explicitly shows all options and their contributions, on the basis of which assistance in the decision-making process is subsequently realized [22,23].

Multi-attribute decision models persist in determining the optimal variant from a set of finite variants $V = \{V_1, V_2, \dots, V_m\}$ which are compared with each other with respect to assigned numerical or non-numerical values belonging to the finite set of criteria $C = \{C_1, C_2, \dots, C_n\}$. Each criterion can aim to reach a maximum or minimum value.

For decision problems with multiple attributes, in which the matrix of consequences contains heterogeneous data, numerical or non-numerical, the homogenization of these data is performed by the normalization process [24], which transforms the matrix of consequences into the matrix $R = (r_{ij})$ $i = 1, m; j = 1, n$, as well as into elements on a certain interval, for example from 0 to 1 $[0, 1]$ or from -1 until $+1$ $[-1, +1]$, etc.

In almost all MCDM problems, there is information about the level/degree of importance of each criterion expressed by the vector $P = \{p_1, p_2, \dots, p_n\}$ that represents the assessment determined by the decision-maker for each criterion.

Any multi-attribute decision problem can be expressed by a matrix A, which is called a consequence matrix (decision matrix) (Table 1), with elements a_{ij} showing the evaluation (consequence) of variant $i, i = 1, 2, \dots, m$ (V_i), by criterion $j, j = 1, 2, \dots, n, (C_j)$.

Table 1. Decision or consequences matrix.

C_i / V_j	C_1	C_2	$\ldots$	C_n
V_1	a_{11}	a_{12}	$\ldots$	a_{1n}
V_2	a_{21}	a_{22}	$\ldots$	a_{2n}
$\ldots$	$\ldots$	$\ldots$	$\ldots$	$\ldots$
V_m	a_{m1}	a_{m1}	$\ldots$	a_{mn}
P	p_1	$\ldots$	p_n	

MCDM methods can be classified into three categories [25], namely:

- Direct methods;
- Indirect methods;
- Methods that use a distance for the construction of hierarchies.

Direct methods build a function defined on a group of variants with real values and select the variants for which the objective function f has the biggest value.

Indirect methods determine a hierarchy on a set of variants based on an algorithm. Methods that use distance choose the variant that is closest to the ideal solution.

In this paper, AHP (direct method) and distance methods (TOPSIS) are used.

3.2. AHP

AHP is an MCDM method originally developed by Prof. Thomas L. Saaty in the 1970s and has been extensively studied and refined ever since, representing one of the most popular analytical techniques and providing a comprehensive and rational framework for structuring and solving multi-criteria decision problem [26].

According to [27], the steps of the AHP method that are to be followed during implementation are described below:

Step 1: Develop a decision hierarchy by decomposing the entire problem into a hierarchy of parameters or criteria;

Step 2: Prioritize among the parameters or criteria of the hierarchy by making a series of judgments based on pairwise comparisons. In this step, the preferences among the criteria are evaluated based on Saaty's scale [27] from 1 to 9, and from 1/9 until 1.

Step 3: Synthesizing the judgment to obtain a set of general priorities for the hierarchy. In this step, the weighted results of the criteria are calculated, which give a relative ranking of the parameters or criteria;

Step 4: Comparing qualitative and quantitative information using informed judgments to derive weights and priorities to check consistency of judgments;

Step 5: Selecting the best alternative based on the available sample data and calculating the final score of each alternative.

In the MCDM method AHP, the decision-making problem is hierarchically structured (Figure 2), given that the decision-making problem is decomposed into subproblems that are analysed independently. At a certain level of the hierarchy, each element (criterion or alternative) is compared with other elements of the same level.

Therefore, based on the matrix of real values (estimates) determined by x_{ij} values, for each criterion (C_j, $j = 1, n$ where n represents the total number of criteria) and each alternative A (A_i, $i = 1, m$, where m represents the total number of alternatives) of the decision-making relationship, the input data are represented by the decision-making matrix D shown mathematically (Equation (1)).

$$C_1 \quad \ldots \quad C_n \qquad D = \begin{matrix} A_1 \\ \ldots \\ A_m \end{matrix} \begin{pmatrix} x_{11} & \ldots & x_{1n} \\ \ldots & \ldots & \ldots \\ x_{m1} & \ldots & x_{mn} \end{pmatrix} \tag{1}$$

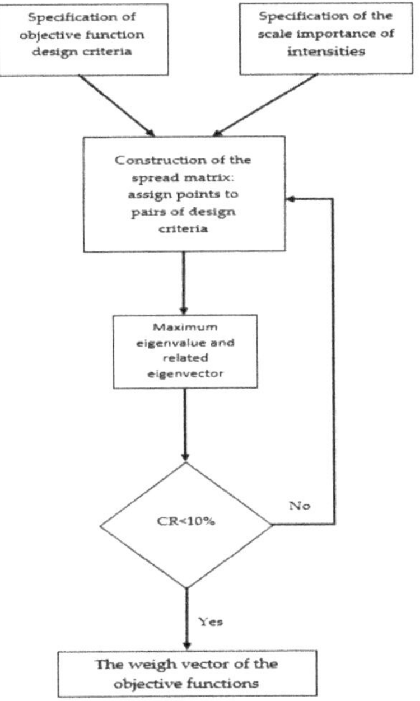

Figure 2. AHP flow diagram [28].

The formulas used during solving the matrices in the AHP steps are given below, as well as the flowchart of the AHP steps [29].

The results of the comparison by pairs of criteria are presented by a square matrix of comparison A of order $n \times n$, where n is the number of observed criteria (alternatives at a later stage). The matrix element a_{ij} of matrix A represents the relative importance of criterion i in relation to criterion j. If $a_{ij} > 1$, criterion i is more important than criterion j, while the reverse is true for $a_{ij} < 1$. If two criteria are of equal importance, then $a_{ij} = 1$.

For consistency, $a_{ij} = 1/aij$ holds for each i, j.

Therefore, $a_{ij} = 1$ holds for every i. Additionally, due to transitivity, $a_{ij} = \alpha_{ik}.\alpha_{kj}$ should hold for every i, j, k. The preferences or relative importance of decision-makers are expressed by Saaty's scale of relative importance, i.e., numbers from 1 to 9, and from 1/9 to 1.

Based on Saatya's scale of relationships between criteria, matrix A was formed. (Equation (2))

$$A = \begin{pmatrix} 1 & a_{12} & \dots & a_{1n} \\ a_{21} & 1 & & a_{2n} \\ \dots & \dots & \dots & \dots \\ a_{n1} & a_{n2} & \dots & 1 \end{pmatrix} = \begin{pmatrix} 1 & a_{12} & \dots & 1/a_{n1} \\ 1/a_{12} & 1 & & a_{2n} \\ \dots & \dots & \dots & \dots \\ 1/a_{n1} & 1/a_{n2} & \dots & 1 \end{pmatrix} \quad (2)$$

whose values along the main diagonal have values of 1.

The vector B (Equation (3)) represents the sums of the elements of the matrix by rows, dimensions n.

$$B = \sum_{i=1}^{n} a_{ij} = b_i, j = 1 \quad (3)$$

Dividing each element of the matrix (of a certain column) with the elements of the vector b, the values of the normalized matrix G are defined by (Equation (4)).

$$G = \begin{pmatrix} g_{11} & g_{12} & \cdots & g_{1n} \\ g_{21} & g_{22} & & g_{2n} \\ \cdots & \cdots & \cdots & \cdots \\ g_{n1} & g_{n2} & \cdots & g_{nn} \end{pmatrix} \tag{4}$$

By calculating the mean values of the elements of the normalized matrix for all columns of the same row, the elements of the vector of weighting coefficients w of dimension n are obtained.

The mean value of all elements of the normalized matrix by column represents a vector of weight coefficients, the sum of which is 1 (Equation (5)).

$$W = \begin{pmatrix} w_1 \\ w_2 \\ \cdots \\ w_n \end{pmatrix} = \sum_{i=1}^{n} w_i = 1, i = 1, n \tag{5}$$

The sum of the products of vectors W and B gives the value λ_{max} which represents the maximum eigenvalue $\lambda \sum_{i=1}^{n} w_i \cdot b_{imax}$.

$A\omega = \lambda_{max}\omega$, forms the matrix of preferences, ω is the eigenvector of order n representing the vector of weight values, while λ_{max} represents the maximum eigenvalue.

The consistency index is calculated according to Equation (6).

$$CI = \frac{\lambda_{max} - n}{n - 1} \tag{6}$$

where n is the number of parameters (criterion).

The consistency ratio is calculated according to Equation (7).

$$CR = \frac{CI}{CRI} \tag{7}$$

where CRI is the consistency index conditioned by the number of criteria.

When the consistency ratio is less than 10%, it is considered that the relationship between the criteria is consistent, and it is passed to the second stage of the AHP method (Table 2).

Table 2. Consistency ratio for the defined number of criteria.

Number of Criteria	1	2	3	4	5	6	7	8	9	10
CRI	0.00	0.00	0.58	0.90	1.12	1.24	1.32	1.41	1.45	1.49

In the process of selecting the best locations in the case study, the AHP method was applied. AHP implies that after the calculation of the consistency among the criteria, the normalization of the elements, the calculation of the mean value of the product of the vectors of normalized values of each variant (86 of them—V_{ij}, $j = 1, k$), the vector of weight values (w_j, $j = 1, k$), and a vector of final values of v_i are defined by (Equation (8)).

$$v_i = \frac{\prod_{j=1}^{k} V_{ij} w_j}{k} \tag{8}$$

A higher value of the variant (v_i) determines a greater influence of the variant in the total rank of the variants (location).

3.3. TOPSIS

The TOPSIS method was developed by Hwang and Yoon in 1981 [30]. It is based on the idea that the chosen alternative should be the one with the shortest Euclidean distance from the ideal solution and the one with the greatest distance from the negative ideal solution. An ideal solution is a hypothetical solution for which all attribute values correspond to the maximum values in the data group that contain satisfactory solutions. A negative ideal solution is a hypothetical solution for which all attribute values correspond to the minimum values in the data group. TOPSIS thus provides a solution that is not only closest to hypothetically the best solution but also farthest from the hypothetical worst one.

The TOPSIS method is based on the idea that the optimal variant must have the minimum distance from the ideal solution.

The method entails defining the objective function $f: V \rightarrow R$, given by the Equation (9).

$$f(V_i) = \frac{\sum_{j=1}^{n} p_j r_{ij}}{\sum_{j=1}^{n} p_{ij}}, \ i = 1, m,$$
(9)

The steps of the TOPSIS method are the following:

Step 1. The normalized matrix $R = (r_{ij})$, $i = 1, \ldots, m, j = 1, \ldots, n$ is built;

Step 2. The weighted normalized matrix $V = (v_{ij})$, $i = 1, \ldots, m, j = 1, \ldots, n$ is built by Equation (10), where:

$$v_{ij} = \frac{p_j r_{ij}}{\sum_{j=1}^{n} p_j}$$
(10)

Step 3. The ideal solutions A, B (Equation (11)) defined as in the Equations (12) and (13) is calculated as following:

$$A = (a_1, a_2, \ldots, a_n)$$
$$B = (b_1, b_2, \ldots, b_n)$$
(11)

where

$$a_j = \begin{cases} maxv_{ij}, if \ C_j \ max \\ \quad 1 \leq i \leq m \\ minv_{ij}, \ if \ C_j \ min \\ \quad 1 \leq i \leq m \end{cases}$$
(12)

$$b_j = \begin{cases} max \ v_{ij}, \ if \ C_j \ min \\ \quad 1 \leq i \leq m \\ min \ v_{ij}, \ if \ C_j \ max \\ \quad 1 \leq i \leq m \end{cases}$$
(13)

Step 4. The distances between solutions Equations (14) and (15) are calculated as follows:

$$S_i = \sqrt{\sum_{j=1}^{n} (v_{ij} - a_j)^2}, i = 1, m$$
(14)

$$T_i = \sqrt{\sum_{j=1}^{n} (v_{ij} - b_j)^2}, i = 1, m$$
(15)

Step 5. The relative proximity from the ideal solution is calculated according to Equation (16):

$$C_i = \frac{T_i}{S_i + T_i}$$
(16)

Step 6. The classification of the set V is performed according to the descending values of C_i obtained in step 5.

4. Case Study—Determination of Nautical Anchorage Locations in Split-Dalmatia County

The Split-Dalmatia County (Figure 3), with its seat in Split, represents the largest Croatian county by area, occupying an area of 14,045 km², of which the land part is 4572 km² (32.5%), and approximately a third is the territorial sea of Croatia with an area of 9473 km² (67.5%).

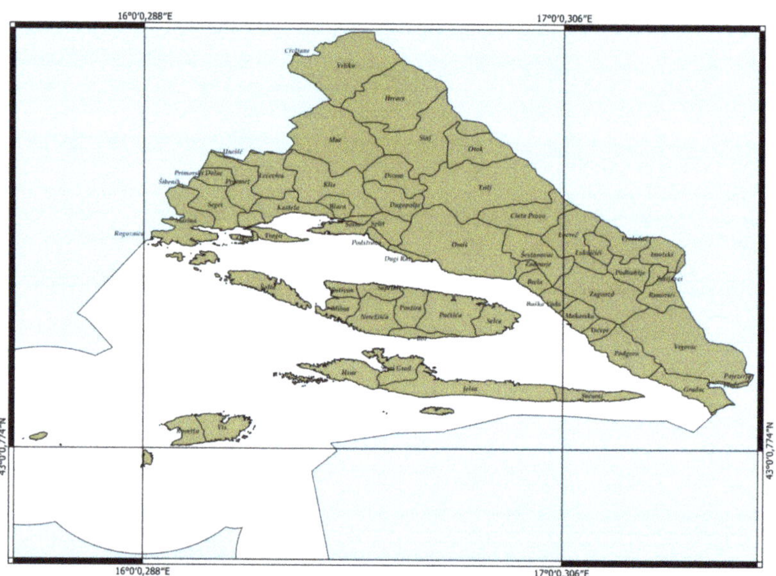

Figure 3. Map of Split-Dalmatia County.

A total of 455,242 inhabitants live in this county (67% in the coastal part, about 7% on the islands, and 26% in the coastal area). The anchorages in the Split-Dalmatia County are located in the central part of the eastern Adriatic coast, which is recognized as one of the most exclusive nautical destinations on the Adriatic [6].

4.1. Dataset

The data used as input in the research were collected in two ways. In the first part of the research, which refers to the understanding of the actual evaluations and the preferences of the characteristics of nautical anchorages from the point of view of "users", the aim is to investigate the real expectations of sailors (professionals and amateurs) regarding the conditions that must be met by nautical anchorages.

Based on the ratings of users (sailors) in the survey research, certain weighting values were assigned to the selected criteria in order to apply MCDM methods based on them, as well as on the basis of data on variants (possible locations of nautical anchorages).

The second phase of the research involves the application of MCDM methods (AHP, TOPSIS) using input data that were obtained as a result of the diligent collection, storage, analysis, and processing of many years of work and experience of the author/s, and refers to a group of 56 bays (Figure 4) with the total of their 86 nautical anchorage fields, each separately described with 20 characteristics (Figure 5).

The spatial arrangement of nautical anchorages in the area of Split-Dalmatia County is shown in the Figure 4.

On the basis of the survey research, the users (sailors) assign the greatest importance to the safety criterion and the least importance to the economic criterion, as shown in Table 3.

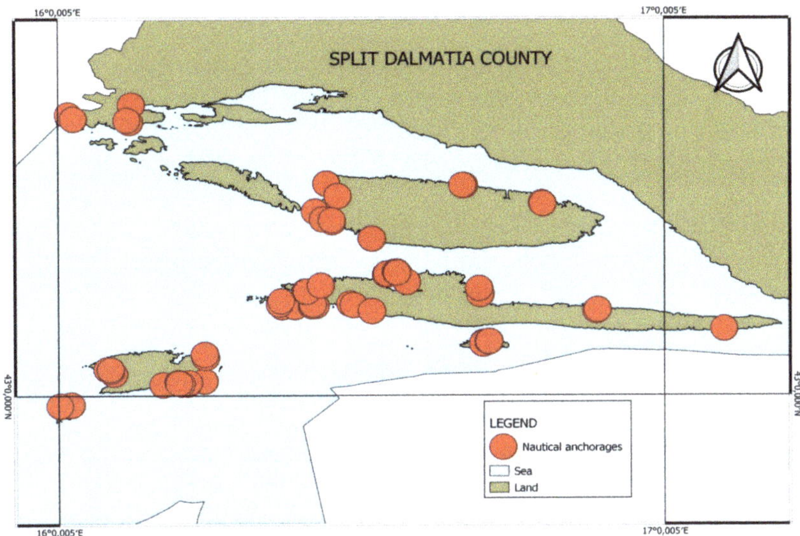

Figure 4. Spatial distribution of nautical anchorages in the area of Split-Dalmatia County.

	1	2	3	4	5	6	7	8	9	10	11	12	13	14	15
AHP	15	34	75	28	32	42	13	41	27	33	17	9	10	16	23
TOPSIS	15	34	42	75	41	37	38	77	79	78	36	28	43	33	23

Figure 5. Excerpt from comparative results of both applied methods of multi-criteria decision making.

Table 3. Survey result—sorted mean grades (from highest to lowest).

The Name of the Criterion	Mean Grade
1. Safety (navigation)—1.1 Underwater installations	4.58
1. Safety (navigational)—1.2 Potential danger	4.51
2. Hydrometeorological criteria—2.1 Protection of the bay	4.41
5. Environmental criteria—5.3 Pollutants or waste	4.27
3. Spatial criteria—3.2 Turning radius	4.26
5. Environmental criteria—5.4 Local heritage	4.15
3. Spatial criteria—3.1 Distance and depth	4.03
1. Safety (navigation)—1.3 Manoeuvre space	3.97
5. Environmental criteria—5.2 Disturbance of the seabed	3.84
5. Environmental criteria—5.1 Impact on the marine environment	3.74
4. Economic criteria—4.2 Public anchorage	3.30
2. Hydrometeorological criteria—2.2 Current and tide	3.26
4. Economic criteria—4.5 Profitability	3.19
4. Economic criteria—4.1 Port area	2.82
4. Economic criteria—4.3 Access to land	2.81
4. Economic criteria—4.4 Traffic and other infrastructure	2.55

For this reason, in the second part of the research, the security elements (available data that were treated as criteria) were assigned the maximum weight values when applying the multi-criteria analysis methods as follows.

The data that have been collected are shown in the Table 3, demonstrating that out of all the important criteria and factors significant for nautical anchorages, users (sailors) emphasize safety factors, namely the following: distance from existing underwater installations from anchorages; protection of the anchorage from wind and waves; the size of the manoeuvring surface, etc., so that in the second phase, when applying the multi-criteria analysis methods, the safety criteria (navigation—protection of the anchorage; the surface of the anchorage field; the surface of the bay; the distance from the coast) will be assigned the highest weight values, as well as the environmental criteria.

The data that have been collected (Table 4) on the locations of nautical anchorages in the area of Split-Dalmatia County are as follows:

- Serial number of the location (No.);
- Location name (Name);
- Island (Island);
- The name of the field in the bay (Field);
- Field surface (Surface F);
- Surface of the bay (Surface B);
- The percentage of the field surface in the bay area (Percentage);
- Protection of the bay (1. Protected; 5: Partially protected; 9: Not protected) (Protection);
- Distance from the coast (Distance);
- Number of anchorage fields in the bay (Number F);
- Existence of maritime traffic (Traffic);
- Existence of an official anchorage (Anchorage);
- Existence of underwater cables and pipelines (Cables);
- Risk of collision (Danger);
- Depth (Depth);
- Tide level and existence of sea currents (Tide);
- Proximity to public ports (Proximity P);
- Proximity to existing berths (Existing B);
- Elements of the environment (Environmental network Natura 2000) (Environment);
- Harm from anchoring a vessel to the seabed (Harmfulness);
- Archaeological sites (Site).

Table 4. Excerpt from collected data on the first eight locations of nautical anchorages.

No	1	2	3	4	5	6	7	8
Name	MILNA Lucice	MILNA Lucice	MILNA Lucice	MILNA Mali bok	MILNA Osibova uvala	MILNA Uvala Slavinjina	NEREZISCA Uvala Blaca	NEREZISCA Uvala Blaca
island	BRAC	BRAC	BRAC	BRAC	BRAC	BRAC	BRAC	BRAC
field	A	B	C	A	A	A	A	B
surface F	2059.39	17,781.05	6744.95	3416.15	5176.56	7756.37	5990.29	2400
surface B	723,767.14	723,767.135	723,767.135	11,800.58	227,733.312	64,415.177	71,424.38	71,424.378
percentage	0.2845377	2.45673631	0.93192267	28.94899	2.27307984	12.0412151	8.386898	3.3601973
openness	5	5	5	9	5	5	5	5
distance	3.9	24.3	24	12.4	2.7	15.2	15.8	0
numberoF	3	3	3	1	1	1	2	2
traffic	1	1	1	1	1	1	1	1
anchorage	1	1	1	1	1	1	1	1
underwaterI	1	1	1	1	1	1	1	1
danger	3	3	3	3	3	3	3	3
depth	5	5	5	5	5	5	5	5
tide	1	1	1	1	1	1	1	1
proximityoP	1	1	1	1	1	1	1	1
existingA	1	1	1	1	1	1	1	1
environment	5	5	5	5	5	5	5	5
harmfulness	5	5	5	5	5	5	5	5
site	1	1	1	1	1	1	1	1

4.2. Description of the Criteria and Initial Settings for the Application of MCDM Methods

The description of the designation and name of the criteria, the unit of measure, and the range of each criterion are defined in Table 5.

Table 5. Label, criterion name, measure unit, and range of input data.

Label	Criterion Name	Measure Unit	Range	
			Min	Max
C1	Field surface (Surface F)	m^2	900	76,654.4
C2	Area of the bay (Surface B)	m^2	11,800.6	723,767
C3	The percentage of the field area in the bay area (percentage)	%	0.28454	39.578
C4	Protection of the bay (Protection)	Whole number: 1—Protected bay; 5—Partially protected; 9—Non protected	1	9
C5	Distance from the coast (Distance);	m	0	81.2
C6	Number of anchorage fields in the bay (Number F)	Whole number	1	4
C7	Existence of maritime traffic (Traffic)	Whole number: If the proximity of the main traffic routes is less than 500 m: 1—No; 5—Yes	1	5
C8	Existence of an official anchorage (Anchorage)	Whole number: If it is in the area of official anchorages: 1—No; 5—Yes	1	5
C9	Existence of underwater cables and pipelines (Cables)	Whole number: The proximity of cables and pipelines is less than 500 m: 1—No; 5—Yes	1	5
C10	Risk of collision (Danger)	Whole number: 1—negligibly small; 2—small; 3—mean; 4—big; 5—very big	1	5
C11	Depth (Depth)	Whole number: 1—Satisfactory; 5—Unsatisfactory	1	5
C12	Tide level and existence of sea currents (tide)	Whole number: 1—small; 3—mean; 5—big	1	5
C13	Proximity to public ports (Proximity P)	Whole number: 1—No; 5—Yes	1	5
C14	Proximity to existing berths (Existing B)	Whole number: 1—No; 5—Yes	1	4
C15	Elements of the environment (Environmental network Natura 2000) (environment)	Whole number: 1—No; 5—Yes	1	5
C16	Harm from anchoring a vessel to the seabed (harmfulness)	Whole number: 1—No; 5—Yes	1	5
C17	Archaeological sites (sites)	Whole number; 1—No; 5—Yes	1	5

Based on the Saaty scale, the relationship between the criteria for the application of the AHP method (Table 6).

The complicated part of the application of the AHP method is the creation of a consistent decision matrix and the establishment of a consistent correlation between the criteria which is, due to the number of criteria (ten), very difficult to implement.

Based on the relationship matrix between the criteria in the AHP method, the value of the maximum eigenvalue (λmax = 11.2671), the consistency ratio (CI = 0.14079), and the consistency index, which is less than 10% (CR = 0.09449), a correct relationship between the criteria is considered established. The consistency index (CR), being less than 10%, also represents a good ratio established between the criteria.

The names of the criteria, the target vector, and the vector of weight coefficients in the TOPSIS method (Table 7).

Table 6. Relationship between criteria according to the Saaty scale (AHP).

Criterion name	Designation	Surface F C1	Surface B C2	Percentage C3	Protection C4	Distance C5	Number F C6	Number F C7	Existing B C8	Environment C9	Harmfulness C10
Surface F	C1	1	1	1	1	2	6	1/2	2	2	5
Surface B	C2	1	1	1	1	1	5	1/2	2	2	4
Percentage	C3	1	1	1	4	2	7	5	3	3	4
Protection	C4	1	1	1/4	1	1	7	2	5	5	5
Distance	C5	1/2	1	1/2	1	1	4	3	2	2	3
Number F	C6	1/6	1/5	1/7	1/7	1/4	1	1/9	1/2	1/2	1
Number F	C7	2	2	1/5	1/2	1/3	9	1	4	4	9
Existing B	C8	1/2	1/2	1/3	1/5	1/2	2	1/4	1	1	2
Environment	C9	1/2	1/2	1/3	1/5	1/2	2	1/4	1	1	3
Harmfulness	C10	1/5	1/4	1/4	1/5	1/3	1	1/9	1/2	1/3	1

Table 7. Elements of the goal vector and their weight values when applying TOPSIS method.

TOPSIS	Criterion Name	Goal Vector	Vector of Weighting Coefficients
Criterion/(name of criteria in the tables)	Label	p_i	w_i
Field surface (Surface F)	C1	max	5
Area of the bay (Surface B)	C2	max	4
The percentage of the field area in the bay area (Percentage)	C3	max	9
Protection of the bay (1. Protected; 5: Partially protected; 9: Not protected) (Protection)	C4	min	13
Distance from the coast (Distance)	C5	max	3.5
Number of anchorage fields in the bay (Number F)	C6	min	1
Existence of maritime traffic (Traffic)	C7	min	1
Existence of an official anchorage (Anchorage)	C8	min	1
Existence of underwater cables and pipelines (Cables)	C9	min	1
Risk of collision (Danger)	C10	min	1
Depth (Depth);	C11	max	1
Tide level and existence of sea currents (Tide)	C12	min	9
Proximity to public ports (Proximity P)	C13	min	1
Proximity to existing berths (Existing B)	C14	min	1
Elements of the environment (Environmental network Natura 2000) (Environment)	C15	min	2
Harm from anchoring a vessel to the seabed (Harmfulness)	C16	min	2
Archaeological sites (Sites)	C17	min	1

5. Result—Validation, Testing, and Comparison Analysis

Obtaining results using two methods of multi-criteria analysis, AHP and TOPSIS, implies [31] the following:

- selection of possible decision options;
- selection of evaluation criteria;
- obtaining performance measures and their fulfilment;
- transformation of data into proportional units (depending on the type of multi-criteria technique method that is applied), which mostly requires entries of decision-makers' preferences;

- determination of criteria and their weighting values, which also largely depends on the preferences of decision-makers;
- ranking or scoring options;
- implementation of sensitivity analysis (weight, performance measures, technique);
- making selection decisions.

In this paper, numerous criteria are analysed in order to determine the most favorable locations of anchorages meeting the conditions prescribed by the recommendations [2,3], while also meeting the expectations of future users, spatial planners, potential investors, and concessionaires who would operate in these areas, as well as entities striving to preserve and protect marine and underwater animal life and the environment, as well as prevent its degradation and pollution. However, since the given recommendations are not precisely defined for the establishment of nautical anchorages, in the procedures for determining the location of nautical anchorages, general and specific criteria are used to satisfy them.

The achieved results indicate that implementation of the methods of multi-criteria analysis enables the selection of the best locations of nautical anchorages in the area of Split-Dalmatia County.

The available criteria (17 in total) for each of the 86 locations represent a decision-making matrix in the implementation of the multi-criteria analysis method. In the AHP method, the relationship between 10 criteria was defined, while in the TOPSIS, all 17 criteria were used, which were assigned goals and weight values.

The list of the sequence of the best locations (per the ordinal numbers of the locations from the input dataset) from the best to the worst (15 of the initial 86) are shown in Figure 5. The same figure indicates the comparative results of two comparisons, shown in order to exhibit the match.

The data (Figure 5) show that, although the order of the 15 best locations obtained by multi-criteria analysis methods is not the same, they differ in a maximum 7 locations.

If the results of the order of the first 15 locations are compared, it is observed that they do not match in a maximum of 7 locations.

The data in Figure 5 show the ordinal numbers of the locations that are in the first 15 best locations obtained by the AHP and TOPSIS method. Shaded (pink) are the ordinal numbers of locations that are in the list of the best 15 obtained both by AHP and based on the TOPSIS method, while those that are not shaded are not recognized as the best by both the AHP and the TOPSIS method. So, locations numbered: 32, 13, 27, 17, 9, 10, 37, 38, 77, 79, 78, 36, and 43 do not appear as the best in both methods (AHP and TOPSIS). Locations marked with numbers 32, 13, 27, 17, 9, 10, and 16 belong to the list of the best 15 locations according to the AHP method, but are not recognized in the list of the best obtained by the TOPSIS method. At the same time, the locations marked with numbers 37, 38, 77, 79, 78, 36, and 43 belong to the list of the best obtained by the TOPSIS method, but AHP method does not recognize them in the list of the best 15.

The sensitivity analysis of the results obtained by the AHP method performed based on the relationship matrix between the criteria. The value of the maximum eigenvalue (λmax = 11.2671), the consistency ratio (CI = 0.14079), while the consistency index is CR = 0.09449 and is less than 10%. The Consistency Index (CR) which, being less than 10%, represents a good ratio established between the criteria.

The reason for this can be found in the difference in the initial settings, the difference in the number, the relationship between the criteria, and the difference in the calculation of the multi-criteria analysis method that are analysed and considered.

Therefore, considering that both applied methods of multi-criteria analysis mostly gave the same or similar results, the research with the proposed methods and the obtained solution represent a strong and effective decision-support tool in the further planning and decision-making process.

6. Discussion

In order to check the robustness, reliability, stability, and accuracy of the obtained solutions, internal and external validation of the results obtained by the TOPSIS method are used.

Internal validation is meant to check the stability of the results by verifying their consistency by comparing the obtained results with data that partially or slightly changes in order to observe the changes that occur. External validation was obtained on the basis of survey research, i.e., by applying a different (alternative) research technique.

As the data on the test (additional/validated) locations were obtained on the basis of survey research (according to other research techniques), and not on the basis of multi-criteria analysis methods, the results are compared by partially changing or adding new data (about five new locations).

As part of the survey, the respondents had the opportunity to (graphically) propose the locations of nautical anchorages (at the site within a group of 86 locations), and to propose both the form and the size of the field of nautical anchorages. The data were cleaned (valueless and unusable ones being rejected), and based on such arranged and organized data, the values of five test locations, i.e., fields, their size and distance from the coast, were generated. As a result of the data obtained in this way, the area of the fields, the share of the percentage of the area of the field in the area of the bay, as well as the distance from the coast, were calculated.

For each of the five proposed test locations (suggested by the surveyed respondents), the input data are identical to the data of the five existing locations from the group of 86 sites, except for the following: 1. field surface; 2. percentage share of the field surface in the bay; and 3. distance from the coast.

The five test locations (Table 8) are marked as follows: 4T, 11T, 42T, 46T, and 74T. Data for the locations with which the order of the obtained results and the values of the elements that change are compared for the locations marked with numbers: 4 and 4T; 11 and 11T; 42 and 42T; 46 and 46T; and 74 and 74T.

Table 8. Location data being validated and location data they are compared with.

No.	4	4T	11	11T	42	42T	46	46T	74	74T
Surface F	3416.2	3278.2	15,056.1	16,588.6	33,300.6	46,613.6	11,217.8	12,947.1	16,066.3	14,430.2
Percentage	29.0	27.8	15.0	16.5	19.3	27.0	4.8	5.6	21.4	19.2
Distance	12.4	5.6	7.4	30.2	15.6	12.6	7.4	29.8	23.2	22.5

Considering that it would be utterly impractical and highly unnecessary to validate the results through both MCDM method used in the empirical research (due to their similarity), the data are validated/tested using the TOPSIS method and include a dataset of 91 locations 5 new and 86 initial locations).

The validation results are presented in Table 9.

Table 9. Ranking and score of testing dataset.

Rank	Score	No.
2 ⌃↑	0.677367199	42T
4	0.608256708	42
18	0.527340126	74
23 ⌄↓	0.512747238	74T
34	0.479203486	4
40 ⌄↓	0.471523883	4T
47 ⌃↑	0.456666886	11T
61	0.438131969	11
89 ⌃↑	0.359936881	46T
90	0.347016121	46

The validation results indicate very small sequence changes. The location was marked with the number 4T, which the survey respondents proposed to have a surface of 3278.2 m^2, with a share of 27.78% in the surface of the bay, a surface of 11,800,584 m^2, and a distance from the coast of 5.6 m, instead of the area of 3416.15 m^2 that has location 4, with a share of 28.9% in the surface of the bay and the distance from the coast of 12.4 m that location 4 had, occupies the 40th position (row 7 of Table 9.) instead of the 34th (row 6 of Table 9.) that was occupied by the location marked with ordinal number 4. Therefore, given that location 4 occupied 34th place, and the location labelled 4T ranks 40th on the list, no significant change has occurred. However, neither the location 4T nor the location marked with serial number 4 belong to the group of the 15 best locations for nautical anchorages. The same happens with the locations marked with the serial numbers 74 and 74T, as well as the locations marked with the serial numbers 11 and 11T and 46 and 46T.

The locations marked with serial numbers 42 and 42T are in the fourth and second positions, respectively, confirming that location 42 (as well as 42T) is an excellent choice for the location of a nautical anchorage. Both are on the list of the top 15 nautical anchorage locations, with the location marked 42T now taking the second position and the location marked 42 the fourth.

This means that although the results achieved at both locations (42 and 42T) are equally good, if it had to be chosen between the data that covered the locations marked 42T and 42, it would be better to decide upon the field surface and distance that were registered at the location marked 42T.

Based on the application of both internal and external validation of the obtained results, we can conclude that the obtained results are very stable and consistent, and that significant changes in the order of the best 15 locations would not change due to insignificant/or minor changes in the input data.

7. Conclusions

The purpose and goal of this paper was to demonstrate the usability of MCDM methods in the process of optimizing the location of nautical anchorages in the Split-Dalmatia County. The MCDM methods include the application of AHP and TOPSIS, based on which the 15 best from the group of 86 possible variants, i.e., locations, are selected in the selection process. The evaluation and assignment of weight values of the criteria were performed based on the ratings given by the respondents (in the first phase) to the most important elements of nautical anchorages. The data were collected through a questionnaire that was created, distributed, and filled in by 74 users (amateur and professional sailors). Thus, the most important criteria from the user's point of view were assigned the highest weight values when applying MCDM methods (second phase), and they mainly relate to the safety of the nautical anchorage. The results showed that both applied MCDM methods, with different initial settings specific to each method (by determining the relationship between the criteria, their weight values, and the objective function of each criterion), gave very similar solutions. The obtained lists of the best MCDM solutions (the most differ in 6 and the least differ in 3) confirm the effectiveness of the MCDM method in selecting the best nautical anchorage locations in Split-Dalmatia County. The obtained consistency ratio among the criteria when applying the multi-criteria AHP method was 9.449231, and is considered acceptable given that they are less than 10% and reflect a good assessment of the criteria and their relationships for the considered case study example. The list of the best 15 locations of nautical anchorages can be used for various deeper and broader research by professional sailors and amateurs, spatial planners, future concessionaires, county offices, scientific and professional staff, and all interested parties, as well as scientific and educational institutions.

As the work takes into account the selection of the best locations of nautical anchorages for smaller vessels and yachts, it is recommended to apply multi-criteria analysis methods and select adequate criteria for larger vessels and/or merchant ships in the future.

Further research may refer to the expansion of multi-criteria decision-making methods by including a larger group of data (other and more numerous locations and criteria, such as vessel size, bottom type, etc.) when selecting the best locations for nautical anchorages for a wider area than the Split-Dalmatia County, for example, the entire area of the Adriatic Sea and beyond. The outcome of this could be a contribution to documenting and enriching the relevant literature in the field of application of multi-criteria decision-making methods in solving more complex problems, and not only in the fields of seafaring and maritime safety, spatial planning, shipping, nautical tourism, and maritime economy.

Author Contributions: Conceptualization, D.P. and Z.L.; methodology, Z.L.; software, D.P.; validation, D.P.; formal analysis, D.P. and Z.L.; investigation, D.P.; resources, D.P.; writing—original draft preparation, D.P.; writing—review and editing, D.P. and Z.L.; visualization, D.P.; supervision, Z.L. All authors have read and agreed to the published version of the manuscript.

Funding: This research received no external funding.

Institutional Review Board Statement: Not applicable.

Informed Consent Statement: Not applicable.

Data Availability Statement: Not applicable.

Conflicts of Interest: The authors declare no conflict of interest.

References

1. International Association of Maritime Universities. Available online: http://www.iamu-edu.org (accessed on 23 January 2023).
2. PIANC Latest Technical Publications. Available online: https://www.pianc.org/publications (accessed on 18 March 2023).
3. Maritime Safety Queensland. *Anchorage Area Design and Management Guideline*; © State of Queensland (Department of Transport and Main Roads): Brisbane, QLD, Australia, 2019.
4. ArcGIS Survey123. Available online: https://survey123.arcgis.com (accessed on 23 January 2023).
5. Ripley, B.D. The R Project in Statistical Computing. *MSOR Connect. Newsl. LTSN Maths Stats OR Netw.* **2022**, *1*, 23–25. [CrossRef]
6. Pušić, D.; Zvonimir, L. Determination of Mooring Areas for Nautical Vessels. In Proceedings of the 20th International Conference on Transport Science, Portorož, 23–24 May 2022; pp. 300–307.
7. European Commission Natura 2000—Environment—European Commission. Available online: https://ec.europa.eu/environment/nature/natura2000/index_en.htm (accessed on 19 January 2023).
8. Jurić, M. Evaluation Model of the Impact of Port Terminals for Liquefied Natural Gases on the Marine Environment. Ph.D. Thesis, University of Rijeka, Rijeka, Croatia, 2019.
9. Schiozzi, D. Models for Development of County and Local Ports with Respect to the Complementary Spatial Concepts Port and the City. Ph.D. Thesis, University of Rijeka, Rijeka, Croatia, 2017.
10. Nguyen, L.C.; Notteboom, T. A Multi-Criteria Approach to Dry Port Location in Developing Economies with Application to Vietnam. *Asian J. Shipp. Logist.* **2016**, *32*, 23–32. [CrossRef]
11. Akbari, N.; Irawan, C.A.; Jones, D.F.; Menachof, D. A Multi-Criteria Port Suitability Assessment for Developments in the Offshore Wind Industry. *Renew. Energy* **2017**, *102*, 118–133. [CrossRef]
12. Dapueto, G.; Massa, F.; Costa, S.; Cimoli, L.; Olivari, E.; Chiantore, M.; Federici, B.; Povero, P. A Spatial Multi-Criteria Evaluation for Site Selection of Offshore Marine Fish Farm in the Ligurian Sea, Italy. *Ocean Coast. Manag.* **2015**, *116*, 64–77. [CrossRef]
13. Račić, N.; Vidan, P.; Lušić, Z.; Slišković, M.; Pušić, D.; Popović, R. *Study of the anchorages of the Split-Dalmatia County I. and II. Phase-base for the Spatial Plan of the Split-Dalmatia County*; Split-Dalmatia County: Split, Croatia, 2019.
14. European Commission. Natura 2000: Sites—Habitats Directive. Available online: https://ec.europa.eu/environment/nature/natura2000/sites_hab/index_en.htm (accessed on 19 January 2023).
15. Kovačić, M.; Domjan, N.; Pušić, D.; Horvat, B.; Bročić, P.; Punda, S. *Action Plan for the Development of Nautical Tourism in the Nautical Tourism in the Split-Dalmatia County*; Hydrographic Institute of the Republic of Croatia: Split, Croatia, 2013.
16. Alossta, A.; Elmansouri, O.; Badi, I. Civil engineering department, Faculty of Engineering, Misurata University, Libya; Mechanical engineering department, Libyan Academy- Misurata Resolving a Location Selection Problem by Means of an Integrated AHP-RAFSI Approach. *Rep. Mech. Eng.* **2021**, *2*, 135–142. [CrossRef]
17. Badi, I.; Abdulshahed, A. Electrical Engineering Department, Misrata University, Misrata, Libya Sustainability Performance Measurement for Libyan Iron and Steel Company Using Rough AHP. *J. Decis. Anal. Int. Comp.* **2021**, *1*, 22–34. [CrossRef]
18. The World Association for Waterborne Transport Infrastructure (PINAC). *PIANC, Report N° 121–2014*; PIANC Secrétariat Général: Bruxelles, Belgium, 2014; ISBN 978-2-87223-210-9.
19. Official Gazette of the Repubilc of Croatia. Available online: https://narodne-novine.nn.hr/clanci/sluzbeni/2013_06_79_1640.html (accessed on 23 January 2023).
20. Ministry of the Sea, Transport and Infrastructure. Available online: https://mmpi.gov.hr/ (accessed on 18 January 2023).

21. Official Gazette of the Repubilc of Croatia. Available online: https://narodne-novine.nn.hr/clanci/sluzbeni/2008_01_3_65.html (accessed on 19 January 2023).
22. De Brucker, K.; Macharis, C.; Verbeke, A. Multi-Criteria Analysis in Transport Project Evaluation: An Institutional Approach. *Eur. Transp. Trasp. Eur.* **2011**, *47*, 3–24.
23. *Multi-Criteria Analysis: A Manual*; Department for Communities and Local Government: London, UK, 2009; ISBN 978-1-4098-1023-0.
24. Neamtu, M. *Jocuri Economice. Dinamică Economică Discretă. Aplicaţii*; Editura Mirton: Timişoara, Romania, 2008; ISBN 978-973-52-0397-9.
25. Ionescu, G.; Cazan, E.; Negrut, A. *Modelarea si Optimizarea Deciziilor Manageriale*; Ed. Dacia: Cluj Napoca, Romania, 1999.
26. Rao, R.V. Comparison of Different MADM Methods for Different Decision Making Situationsof the Manufacturing Environment. In *Decision Making in Manufacturing Environment Using Graph Theory and Fuzzy Multiple Attribute Decision Making Methods: Volume 2*; Rao, R.V., Ed.; Springer Series in Advanced Manufacturing; Springer: London, UK, 2013; pp. 205–242, ISBN 978-1-4471-4375-8.
27. Saaty, T. *The Analytic Hierarchy Process, Planning, Priority Setting, and Resource Allocation*; McGraw-Hill: NewYork, NY, USA, 1980.
28. Vahedi, A.; Meo, S.; Zohoori, A. An AHP-Based Approach for Design Optimization of Flux-Switching Permanent Magnet Generator for Wind Turbine Applications. *Int. Trans. Electr. Energy Syst.* **2015**, *26*, 1318–1338. [CrossRef]
29. Shukla, O.; Upadhayay, L.; Dhamija, A. Multi Criteria Decision Analysis Using AHP Technique to Improve Quality in Service Industry: An Empirical Study. In Proceedings of the International Conference on Industrial Engineering (ICIE-2013), Surat, India, 20–22 November 2013. [CrossRef]
30. Hwang, C.-L.; Yoon, K. *Multiple Attribute Decision Making*; Lecture Notes in Economics and Mathematical Systems; Springer: Berlin/Heidelberg, Germany, 1981.
31. Hajkowicz, S.; Higgins, A. A Comparison of Multiple Criteria Analysis Techniques for Water Resource Management. *Eur. J. Oper. Res.* **2008**, *184*, 255–265. [CrossRef]

Journal of
Marine Science and Engineering

Article

Risk Assessment of Bauxite Maritime Logistics Based on Improved FMECA and Fuzzy Bayesian Network

Jiachen Sun [1,†], Haiyan Wang [1,2,†] and Mengmeng Wang [1,*]

1 School of Transportation and Logistics Engineering, Wuhan University of Technology, Wuhan 430063, China; sunjiachen1969@163.com (J.S.); hywang777@whut.edu.cn (H.W.)
2 National Engineering Research Center for Water Transport Safety, Wuhan University of Technology, Wuhan 430063, China
* Correspondence: mengmeng_stu@163.com
† These authors contributed equally to this work.

Abstract: Because of the many limitations of the traditional failure mode effect and criticality analysis (FMECA), an integrated risk assessment model with improved FMECA, fuzzy Bayesian networks (FBN), and improved evidence reasoning (ER) is proposed. A new risk characterization parameter system is constructed in the model. A fuzzy rule base system based on the confidence structure is constructed by combining fuzzy set theory with expert knowledge, and BN reasoning technology is used to realize the importance ranking of the hazard degree of maritime logistics risk events. The improved ER based on weight distribution and matrix analysis can effectively integrate the results of risk event assessment and realize the hazard evaluation of the maritime logistics system from the overall perspective. The effectiveness and feasibility of the model are verified by carrying out a risk assessment on the maritime logistics of importing bauxite to China. The research results show that the priority of risk events in the maritime logistics of bauxite are "pirates or terrorist attacks" and "workers' riots or strikes" in sequence. In addition, the bauxite maritime logistics system is at a medium- to high-risk level as a whole. The proposed model is expected to provide a systematic risk assessment model and framework for the engineering field.

Keywords: bauxite maritime logistics; risk parameter; FMECA; fuzzy rule-based BN; evidence reasoning (ER); maritime risk

Citation: Sun, J.; Wang, H.; Wang, M. Risk Assessment of Bauxite Maritime Logistics Based on Improved FMECA and Fuzzy Bayesian Network. *J. Mar. Sci. Eng.* **2023**, *11*, 755. https://doi.org/10.3390/jmse11040755

Academic Editors: Marko Perkovic, Lucjan Gucma and Sebastian Feuerstack

Received: 16 March 2023
Revised: 29 March 2023
Accepted: 29 March 2023
Published: 31 March 2023

1. Introduction

As a raw material for many industrial products, the demand for bauxite in China has shown a rising trend in recent years. Due to the shrinking of domestic bauxite resources and the impact of environmental protection policies, China needs to import a large amount of bauxite from abroad every year to meet the demand, and the degree of foreign dependence is continuously increasing. According to Chinese national customs statistics, in 2019 alone, China's bauxite imports reached 101 million tons, accounting for 72.4% of the global bauxite seaborne trade volume. Among them, the imported bauxite mainly comes from countries with rich bauxite resources such as Guinea and Australia.

As one of the key links in the import of bauxite, maritime logistics have the characteristics of long distances, numerous transit nodes, and a complex sea environment. There are many potential unsafe factors. In addition, bauxite has the characteristic of easy fluidization, which further interferes with the robustness and reliability of maritime logistics links. Therefore, it is necessary to evaluate the risks of China's imported bauxite shipping logistics to minimize the impact of potential risk factors in shipping logistics on the stable and continuous imports of China's bauxite, and to ensure the orderly and healthy development of China's aluminum industry.

The failure mode effect and criticality analysis (FMECA) is widely used in the field of risk assessment and reliability analysis. However, the traditional FMECA method has many limitations, which are mainly reflected in the incompleteness of risk characterization parameters, the lack of parameter importance differences, and the limited discrimination of risk priority number (RPN) values [1,2]. In response to the above problems, scholars at home and abroad have carried out a lot of research.

Based on the characteristics of plastic production, Gul, Yucesan, and Gelik [1] proposed an improved FMEA combined with the fuzzy Bayesian network (FBN) to evaluate failures in plastic production. The risk characterization parameters were reconstructed in the model and weighted by the fuzzy best-worst method. Similarly, Wan et al. [3] combined the fuzzy belief rule method with Bayesian networks to establish a new maritime supply chain wind assessment model. In the model, three sub-parameters were introduced to characterize the consequences of the failure event, and then a more complete risk characterization parameter system was constructed, and the parameters were weighted through the analytic hierarchy process (AHP) method. In view of the advantages of the Bayesian network (BN) in dealing with uncertainty, Ma et al. [4] proposed a Bayesian network construction method that combined FMEA and a fault tree analysis (FTA) for the system security assessment. Liu et al. [5] proposed a new risk priority model based on the D number and gray correlation to optimize FMEA and then conduct a risk assessment.

Considering that there is a large amount of uncertain information in the evaluation process, fuzzy set theory has been widely used due to its flexibility, reliability, and strong operability in handling uncertain information. Lee et al. [2] proposed a new fuzzy comprehensive evaluation method using structural importance and fuzzy theory, which effectively dealt with subjective ambiguity under uncertain conditions. In addition, Renjith, Kumar, and Madhavan [6] proposed a fuzzy RPN method to prioritize system failures. This method evaluated risk parameters through fuzzy language variables, and through the IF-THEN rule library to connect the language variables to the fuzzy RPN, which effectively overcomes the shortcomings of the traditional RPN method. Alyami et al. [7] introduced Bayesian networks based on fuzzy rules and developed an advanced failure mode and effect analysis (FMEA) method to assess the criticality of dangerous events in container terminals.

Based on the above research, this paper proposes a systemic risk assessment model combining FBN and improved ER theory based on the improved FMECA (see Figure 1). In this model, a new risk characterization parameter system is constructed, and the parameters are weighted through appropriate methods, while the fuzzy set theory is used to deal with the uncertainty in the expert scoring process. This is accomplished by creating a fuzzy rule-based on the confidence structure, and making full use of Bayesian network reasoning technology, which can effectively obtain an accurate evaluation and clear rating of failure modes. The improved evidence reasoning theory (ER) can effectively integrate the results of fuzzy BN inference to realize the assessment of the system risk.

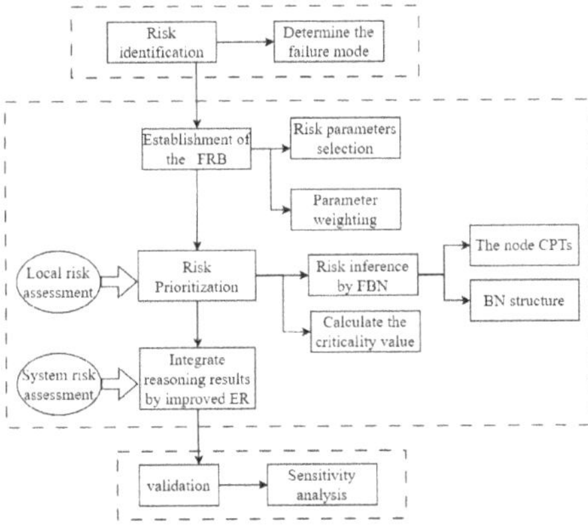

Figure 1. Maritime logistics risk assessment model.

2. Literature Review

Maritime logistics play an increasingly important service sector role to support the fast development of domestic and international trade. Efficiency and effectiveness are the two critical concerns in logistics management [8]. Efficiency means less spending to achieve more output, while effectiveness means reaching the goal in an uncertain situation. Such goal achievement could make the operations leaner and more efficient, but simultaneously make the logistics vulnerable [9]. Therefore, maritime logistics are exposed to various natural and man-made risks [10]. How to make this a safe operation becomes a more critical issue considering various risk factors and attracting more attention from researchers and practitioners. Research addressing container maritime logistics has grown significantly in recent years. There is little research addressing bulk cargo maritime logistics [11], e.g., nickel mineral, bauxite, grain, etc. This study takes references from broader areas in maritime research.

(1)   Maritime Logistics Risk.

Risk is a major influencing factor that could be interpreted as the probability that an event or action may adversely affect the anticipated goal in maritime transportation [12]. Risk identification, analysis, assessment, and mitigation construct the cycle of risk management. Maritime logistics risk management is a decision-making process to minimize the adverse effects of accidental losses based on risk assessment [13]. Compared to traditional risk management, maritime logistics risk management aims to identify and mitigate risks from the perspective of the entire supply chain [14].

Some research focuses on risks related to shipping activities. Siqi Wang et al. [15] identified the risk factors of the dry bulk maritime transportation system from the aspects of people, ships, and cargo, and used the Markov method and the multi-state system method to integrate the probability model of the maritime traffic safety risk state. Jiang, M et al. [16] took the Maritime Silk Road as the research object, and concluded that ship factors are the main factors affecting the occurrence of accidents through an accident data analysis. Øyvind Berle et al. [17] analyzed the failure factors of the dry bulk shipping supply chain from the aspects of supply, capital flow, transportation, communication, internal operation/capacity, and human resources.

There is some research focusing on the risks resulting from the cargo itself. For example, Delyan Shterev [18] studied the safety of liquefiable bulk cargo by sea and pointed out that the relationship between the combined system of ships, cargo, and people can be used to reduce the accidents of liquefiable cargo by sea. Munro and Mohajerani [19] analyzed seven cases of a ship capsizing from the views of cargo fluidization and concluded that the main reason was the excessive moisture of cargo. Ju et al. [11] developed a discrete element method (DEM) to simulate the whole process of liquefaction of fine particle cargo, and identify the key parameters that lead to the liquefaction. Lee [20] analyzed a marine accident caused by nickel ore liquefaction and indicated reducing the risk by changing homogeneous loading into alternate loading through experiments. Daoud, Said, Ennour, and Bouassida [21] combined physical experiments and numerical methods to improve the understanding of cargo liquefaction mechanisms and security in maritime transportation.

(2)   Risk Assessment Methods in Maritime Logistics.

Methods developed and applied in maritime logistics risk management can be divided into qualitative and quantitative. Some research highlights managing the uncertainties in assessing the risks in maritime logistics [3,22,23]. Khan et al. [24] proposed an object-oriented Bayesian network to predict the ship–ice collision probability. Wan et al. [3] established a model combined with a belief rule base and Bayesian network to assess the risk of the maritime supply chain considering the data collected were highly uncertain.

Some literature analyzes and assesses the risk in the maritime supply chain from the perspective of visibility, robustness, and vulnerability. For example, because of the fuzziness and uncertainty of the risk assessment by experts, Emre Akyuz et al. [25] adopted the fuzzy number and bow-tie method to carry out a quantitative comprehensive risk analysis on the liquefaction risk of dry bulk carriers. Liu et al. [26] combined a multi-centrality model and robustness analysis model to analyze the vulnerability of the maritime supply chain, and verified the feasibility of the model through the Maersk shipping line in its Asia–Europe route. Zavitsas, Zis, and Bell [27] analyzed the link between the environment and resilience performance of the maritime supply chain and built a framework to reduce operating costs and risks that may disrupt the maritime supply chain. Vilko et al. [28] identified and assessed the risk of multi-modal maritime supply chains from the perspective of visibility and control. Lam and Bai [10] used the quality function deployment approach to improve the resilience of the maritime supply chain.

Much literature related to risk focuses on the specific risk in the container supply chain, especially after the 911 terrorist attacks in 2001 [13,23,29]. With the development of IT technology, more information could be visible on the official website of a shipping company. Cyber-attack has become one of the risk sources; Polatidis, Pavlidis, and Mouratidis [30] proposed a highly parameterized cyber-attack path discovery method to evaluate the risk of dynamic supply chain maritime risk management. This method is more efficient than the traditional method because it can output the most probable paths instead of all paths.

(3)   Research Gap.

The above discussion indicates that the risks in a maritime logistics context have aroused academic attention, but still are under research, especially in the bulk cargo maritime logistics risk analysis and assessment. More importantly, most literature focuses on the container supply chain [12,13,22,23] or some specific risk in the container supply chain [13,31]. However, some mineral cargo belongs to fluidization cargo that needs to pay more attention to the water percentage in the cargo, such as bauxite, nickel mineral, etc. Some of them need to ventilate and grain. These kinds of maritime logistics are very different from how the container supply chain lies in the fluidization of cargo to become free surface effect or solid–liquid two separate flow layers while in maritime transportation easily. During maritime transportation, the permeability of cargo is low, owing to human error or bad ventilation, resulting in the water percentage in the cargo being up to the limit value or out of the limit, which could further form a free surface effect. If it happened with bad sea conditions or other natural and man-made disasters, it would be easy to make the ship capsize.

Consequently, this paper aims to develop a systemic risk assessment model combining FBN and improved ER theory based on the improved FMECA and improve the safety of bauxite maritime logistics. To start with, the identification of potential failure modes obtained by the failure mode and effects analysis (FMEA), based on the failure relationships embedded in the failure modes, the Bayesian network construction, and the probability parameter defined are discussed.

3. Construction of Risk Assessment Model for Bauxite Maritime Supply Chain Based on FBN

3.1. Determination of Risk Characterization Parameters

3.1.1. Construction of Risk Characterization Parameter System

Risk characterization parameters are important indicators used to characterize and describe the hazards of failure events. A scientific, reasonable, and complete construction of a risk characterization parameter system is conducive to a more comprehensive and accurate grasp of the characteristics of risk factors and improves the accuracy and reliability of the risk assessment results. Traditional FMECA uses the three parameters of occurrence, the severity of consequences, and detection to characterize risk factors [32]. Given the characteristics of maritime logistics, this paper re-introduces three sub-parameters based on the above three parameters to measure and distinguish the severity parameter S after the occurrence of the risk. Based on this, a relatively complete structure is constructed in the form of a risk parameter index hierarchical structure (risk characterization parameter system (see Figure 2)).

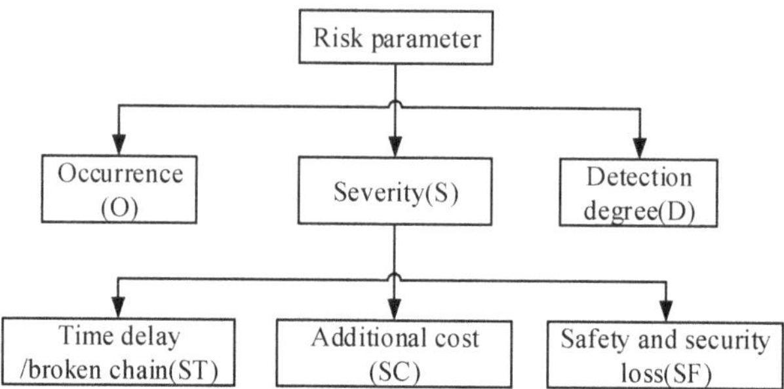

Figure 2. Risk parameter characterization system.

When maritime logistics are affected by an uncertain event, it usually manifests as the normal transportation of ocean logistics being disturbed and the transportation time being delayed. In serious cases, the maritime logistics chain may even be disconnected. For bauxite maritime logistics, it reduces the reliability of maritime logistics services. For time- and temperature-sensitive goods, the consequences of time delays are often more serious than for ordinary goods. Additional costs refer to the increase in a series of additional expenses/costs that are affected by risk factors, such as additional management costs or expenses incurred by risk drivers. Safety and security losses refer to the damage to the physical elements participating in or constituting the maritime logistics after being affected by the failure event, such as personal injury, damage to transported goods, and damage to port infrastructure or ships. To a certain extent, the impact of a risk event on the system can be summarized by the increase in additional costs. This paper aims to measure the consequences' parameters more comprehensively and in detail from different perspectives such as the physical elements of the system, time, and currency to avoid the unclear description caused by the unified overview.

3.1.2. Risk Parameter Weighting Based on AHP-Entropy Weight Method

Given the obvious hierarchical structure relationships and differences in importance between the constructed risk parameter systems, this paper uses appropriate weight coefficients to quantify the relative importance of different parameters.

(1)   AHP

According to the established risk parameter characterization system, the parameters of the same layer are compared in pairs, and a judgment matrix is constructed to express the comparison of the relative importance of the parameters of this layer concerning the parameters of the upper layer. After passing the consistency check, the parameter weight vector is determined by calculating the weight vectors of different experts for the parameters of the same layer, using the arithmetic average method to gather the evaluation results of the experts on the parameters of the same layer to determine the final weight vector of the parameters of the layer. Finally, the weight value of the parameter of the same layer and the weight value of the upper layer parameter to which it belongs are weighted to calculate the comprehensive weight value of the risk parameter in the entire system.

(2)   Entropy method

Entropy is a concept derived from thermodynamics and mainly reflects the degree of chaos in the system. The entropy value in information theory reflects the degree of information carried by an indicator. The smaller the entropy value, the greater the degree of dissociation between the data covered by the indicator, the more information the indicator carries, and the greater the corresponding weight value. When using the entropy method to determine the parameter weight vector, it is necessary to combine expert experience to score these parameters. The score is between 1 and 10. The higher the score, the greater the importance of the risk parameter. The steps for calculating the parameter weight vector using the entropy method are as follows [33].

(i) Construction of evaluation matrix

Assuming that m experts evaluate and score n risk parameters, the original evaluation matrix can be formed $R = (x_{ij})$.

$$R = \begin{pmatrix} x_{11} & \cdots & x_{1n} \\ \vdots & & \vdots \\ x_{m1} & \cdots & x_{mn} \end{pmatrix} \tag{1}$$

Among them, x_{ij} refers to the evaluation value of the jth risk parameter by the ith expert, $i \in (1, 2, \cdots m), j \in (1, 2, \cdots, n)$.

(ii) Initial matrix normalization

Standardizing the evaluation matrix can ensure the correctness and simplicity of the calculation results. The standardization process is as follows:

$$r_{ij} = x_{ij} / \sqrt{\sum_{i=1}^{n} x_{ij}} \tag{2}$$

The evaluation matrix $R^* = (r_{ij})_{m \times n}, r_{ij}(0 \leq r_{ij} \leq 1)$ obtained after standardization is the standard value of the jth risk parameter on the ith expert.

(iii) Calculate the proportion of points

To calculate accurate information entropy, it is necessary to calculate the proportion of points P_{ij} assigned by the ith expert to this parameter under the jth risk parameter.

$$P_{ij} = \frac{r_{ij}}{\sum\limits_{i=1}^{m} r_{ij}} \tag{3}$$

(iv) Calculate information entropy

The formula for calculating the entropy value E_j of the jth index is:

$$E_j = -\frac{1}{\ln m} \sum_{i=1}^{m} P_{ij} \ln P_{ij} \tag{4}$$

The information utility value $d_j = 1 - E_j$ can be obtained according to the value of information entropy.

(v) Calculate the entropy weight of risk parameters

The larger the information utility value, the more important the parameter, and the greater the weight should be when the parameter is weighted. The formula for calculating the entropy weight of risk parameters is:

$$W_j = \frac{d_j}{\sum\limits_{j=1}^{n} d_j} \tag{5}$$

(3) Determine the overall weight

The risk parameter weights obtained by the AHP technique can show the importance that decision-makers place on different parameters. The entropy measure can more objectively reflect the information contained in the risk parameter itself when calculating the weight vector. Hence, by combining these two methods to achieve the purpose of effectively combining the advantages of the two, a more scientific and reasonable parameter weight vector can be obtained. The formula for determining the comprehensive weight of parameters using the AHP-entropy weight method is [32]

$$w_j = \alpha w_j + (1 - \alpha)v_j \tag{6}$$

where v_j and w_j, respectively, represent the parameter weight values obtained according to the AHP and entropy weight method, in which α is the preference coefficients, and $0 \leq \alpha \leq 1$. Considering that the entropy measure can reduce the deviation caused by human factors to a certain extent, and then combined with the opinions of the expert group, this paper defines the preference coefficient as 0.6. From this, the comprehensive weight vector can be obtained $W = (\omega_1, \omega_2, \cdots \omega_n)$. By issuing questionnaires to domain experts (see Table 1 for expert details), after excluding invalid samples, a total of four valid questionnaires were received. Based on the survey results, a judgment matrix can be further obtained. Then, we conducted a consistency check on the judgment matrix, and the low inconsistency ratios (<0.1) of all pairwise comparisons verified the rationality of the results. It turns out that the calculated weights are consistent. Afterwards, the entropy weight method and AHP were combined to obtain the weight of each risk characterization parameter (as shown in Table 2).

Table 1. Experts' knowledge and experience.

Experts	Position	Company	Working Experience
1	A professor, head of port management studies	A university in China	Involved in maritime transport safety management and maritime supply chain management
2	Senior operational managers	A leading port in China	Involved in port safety and operational services
3	A qualified master mariner	A shipping company in China	Involved in bauxite international transportation
4	Senior security officers	A bauxite company in China	Involved in bauxite customs and transport security

Table 2. Risk parameters' weight.

Risk Parameter		Local Weight		Overall Weight
Occurrence likelihood (O)		0.30		0.30
Difficulty of detection (D)		0.25		0.25
Severity of consequences (S)	Time delay/disruption (ST)		0.25	0.11
	Additional cost (SC)	0.45	0.53	0.24
	Safety and security loss (SF)		0.22	0.10

3.2. Determining the Magnitude of Risk Index Based on Fuzzy Logic

3.2.1. Fuzzy Rating Indication

When using the established risk characterization parameter system to assess the hazard of potential risk events in maritime logistics, due to the lack of historical data in the engineering field and the particularity of the risk parameters themselves, in practice, failures are usually evaluated using predetermined scores, thus making a judgment on the event's severity in a certain aspect. The classification of the characteristic parameters of dangerous events is often a vague concept, usually relying on expert experience to give low, relatively low, high, relatively high, and other vague language judgments, and determining the corresponding evaluation level based on the vague language. Therefore, the fuzzy set theory is introduced into the determination of parameter levels that rely on expert ratings to ensure that the quantitative results of risk parameters are more accurate and in line with the actual situation. Experts give an evaluation level to the parameters according to the pre-defined fuzzy rating set and then construct the fuzzy number according to the membership function. Common membership functions include triangles and trapezoids.

According to the relevant literature, the shape of the membership function has a significant impact on the outcomes of fuzzy operations, and triangular and trapezoidal fuzzy numbers are thought to be more efficient [34]. Among them, trapezoidal fuzzy numbers can be turned into crisp values, interval numbers, and triangular fuzzy numbers by adjusting parameters, which can intuitively explain fuzzy variables [35]. Hence, the trapezoidal membership function is more scalable and widely applicable in the decision-making of complex problems [36]. Since the trapezoidal membership function is more in line with the objective evaluation situation, the trapezoidal fuzzy number is used to deal with the linguistic variables of the expert evaluation.

Assuming that real numbers a, b, c, and d ($a < b = c < d$) form the four endpoints of the trapezoidal fuzzy number, then its membership function $\mu(x)$ can be defined as follows:

$$\mu(x) = \begin{cases} 0 & (x \leq a \text{ or } x \geq b) \\ \frac{x-a}{b-a} & (a \leq x \leq b) \\ 1 & (b \leq x \leq c) \\ \frac{d-x}{d-c} & (c \leq x \leq d) \end{cases} \tag{7}$$

Before the expert rating, it is necessary to define the fuzzy rating set with the attribute level and the corresponding membership function. This paper divides the constructed risk characterization parameter variables into five levels. The different level attributes of the corresponding parameters, linguistic variables, and corresponding fuzzy numbers are shown in Tables 3–7. The rating variables of these five parameters are represented by trapezoidal fuzzy numbers. Among them, the probability of the occurrence of failure, the degree of detection, and the variables that characterize the consequences of time delay, additional cost, and safety and security loss corresponding to the rating standard belong to the membership function as shown in Figure 3.

Table 3. Probability of risk O fuzzy rating.

Grade	Linguistic Variables	Definition	Fuzzy Number
O_1	Low	Occurs less than once a year	(0, 0, 1, 2)
O_2	Relatively low	Expected to happen every few months	(0.5, 2, 3, 4.5)
O_3	Medium	Expected at least once a month	(3, 4, 6, 7)
O_4	Relatively high	Expected at least once a week	(5.5, 7, 8, 9.5)
O_5	High	Expected at least once a day	(8, 9, 10, 10)

Table 4. Fuzzy rating of risk detection degree D.

Grade	Linguistic Variables	Definition	Fuzzy Number
D_1	Easy	Can be easily found through risk inspection	(0, 0, 1, 2)
D_2	Relatively easy	Can be detected through regular risk inspection	(0.5, 2, 3, 4.5)
D_3	Medium	Not easily detected by regular risk inspection	(3, 4, 6, 7)
D_4	Relatively hard	May be detected through rigorous risk inspection	(5.5, 7, 8, 9.5)
D_5	Difficult	Unable or difficult to pass rigorous risk inspections	(8, 9, 10, 10)

Table 5. Time delay ST fuzzy rating.

Grade	Linguistic Variables	Definition	Fuzzy Number
ST_1	Short	Delay time less than 6 h	(0, 0, 1, 2)
ST_2	Relatively short	The delay time does not exceed 5% of the planned transportation time	(0.5, 2, 3, 4.5)
ST_3	Medium	The delay time exceeds the planned transportation time by 5–20%	(3, 4, 6, 7)
ST_4	Relatively long	The delay time exceeds the planned transportation time by 20–40%	(5.5, 7, 8, 9.5)
ST_5	Long	The delay time exceeds 40% of the planned transportation time	(8, 9, 10, 10)

Table 6. Extra cost SC fuzzy rating.

Grade	Linguistic Variables	Definition	Fuzzy Number
SC_1	Few	No more than 1% of the total cost	(0, 0, 1, 2)
SC_2	Relatively few	2–5% over the total cost	(0.5, 2, 3, 4.5)
SC_3	Medium	Over 6–20% of the total cost	(3, 4, 6, 7)
SC_4	Relatively many	21–40% over the total cost	(5.5, 7, 8, 9.5)
SC_5	Much	More than 40% of the total cost	(8, 9, 10, 10)

Table 7. Safety and security loss SF fuzzy assessment level.

Grade	Linguistic Variables	Definition	Fuzzy Number
SF_1	Light	The goods, equipment, or system are slightly damaged, but the functions are complete, and the maintenance is convenient and fast; the number of minor injuries does not exceed 2	(0, 0, 1, 2)
SF_2	Relatively light	The equipment or system is slightly damaged, and the maintenance is more convenient; the damage rate of the goods is 1–5%; three people or more have been slightly injured	(0.5, 2, 3, 4.5)
SF_3	Medium	Equipment or system is medium-damaged, and maintenance is not convenient; the proportion of cargo damage reaches 5–10%; 1–2 people are medium-injured	(3, 4, 6, 7)
SF_4	Relatively serious	The equipment or system is seriously damaged and inconvenient to maintain; the damage rate of goods reaches 10–20%; 1–2 people are seriously injured	(5.5, 7, 8, 9.5)
SF_5	Severe	The equipment or system is seriously damaged, and transportation cannot be carried out; the proportion of goods damaged is more than 20%; personnel deaths occur	(8, 9, 10, 10)

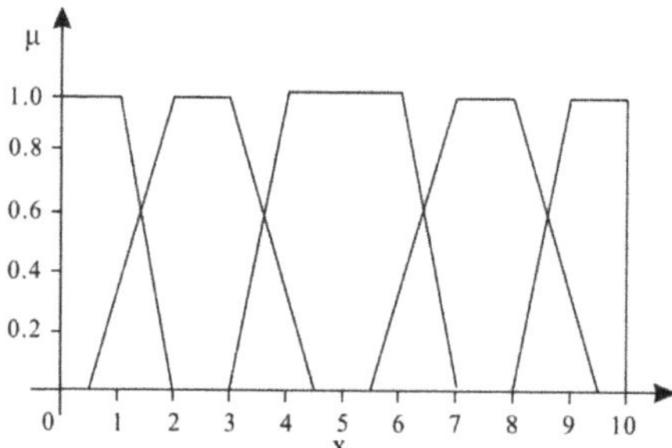

Figure 3. Membership function of a parametric variable.

3.2.2. Fuzzy Rating Result Calculation

After obtaining the evaluation value given by the expert in the form of a trapezoidal fuzzy number (see Table 1 for expert details), the evaluation information among different experts is integrated with the help of an uncertain ordered weighted averaging (UOWA) operator. This method determines the weight of the experts by comparing the degree of difference between the fuzzy number of the evaluation value given by each expert and the average fuzzy number obtained by combining the opinions of different experts. The specific calculation method is as follows.

There is an expert group composed of n experts to evaluate a certain type of failure mode, and convert the evaluation level of the Kth expert to the ith parameter variable into the j trapezoidal fuzzy number as $R_k = \left(a_{ik}^j, b_{ik}^j, c_{ik}^j, d_{ik}^j \right)$, among, $i \in (O, D, ST, SC \text{ and } SF)$. $j = (1, 2, 3, 4, 5)$, $K = (1, 2 \dots n)$. The following uses the UOWA operator to synthesize the fuzzy number of the evaluation opinions of the experts.

(1) Calculate the arithmetic mean of the trapezoidal fuzzy numbers as $\tilde{R}_m = \left(a^j_{im}, b^j_{im}, c^j_{im}, d^j_{im}\right)$, among:

$$a_{im}{}^j = \frac{1}{n}\sum_{k=1}^{n} a^j_{ik} \quad b_{im}{}^j = \frac{1}{n}\sum_{k=1}^{n} b^j_{ik} \quad c_{im}{}^j = \frac{1}{n}\sum_{k=1}^{n} c^j_{ik} \quad d_{im}{}^j = \frac{1}{n}\sum_{k=1}^{n} d^j_{ik} \tag{8}$$

(2) Calculate and measure the distance between R_k and $\tilde{R}_m$.

$$d(R_k, \tilde{R}_m) = \frac{1}{4}\left(\left|a^j_{ik} - a^j_{im}\right| + \left|b^j_{ik} - b^j_{im}\right| + \left|c^j_{ik} - c^j_{im}\right| + \left|d^j_{ik} - d^j_{im}\right|\right) \tag{9}$$

(3) Calculate the similarity between R_k and $\tilde{R}_m$. For trapezoidal fuzzy numbers $R_k = \left(a^j_{ik}, b^j_{ik}, c^j_{ik}, d^j_{ik}\right)$, if $\tilde{R}_m = \left(a^j_{im}, b^j_{im}, c^j_{im}, d^j_{im}\right)$ is their mean value, then

$$s(R_k, \tilde{R}_m) = 1 - \frac{d(R_k, \tilde{R}_m)}{\sum\limits_{k=1}^{n} d(R_k, \tilde{R}_m)} \tag{10}$$

is the similarity between the trapezoidal fuzzy number $R_k = \left(a^j_{ik}, b^j_{ik}, c^j_{ik}, d^j_{ik}\right)$ and the arithmetic mean of the trapezoidal fuzzy array $\tilde{R}_m = \left(a^j_{im}, b^j_{im}, c^j_{im}, d^j_{im}\right)$

(4) Fuzzy number assembly.

The aggregation method of UOWA operators can synthesize the fuzzy number of different expert evaluation values to obtain the final result.

$$R = (a^j_i, b^j_i, c^j_i, d^j_i) = DUOWA(R_1, R_2 \cdots, R_n) = \sum_{k=1}^{n} w_k R_k = \sum_{k=1}^{n} \frac{s(R_k, \tilde{R}_m)}{\sum\limits_{k=1}^{n} s(R_k, \tilde{R}_m)} \times R_k \tag{11}$$

Among them, $\dfrac{s(R_k, \tilde{R}_m)}{\sum\limits_{k=1}^{n} s(R_k, \tilde{R}_m)}$ is defined as the weight coefficient.

After using the UOWA operator to gather expert evaluation opinions, the result obtained is still a trapezoidal fuzzy number. By combining it with the membership function, the fuzzy evaluation results of different parameters under a specific failure mode after comprehensive expert opinions can be obtained. The specific conversion process is shown in Figure 4.

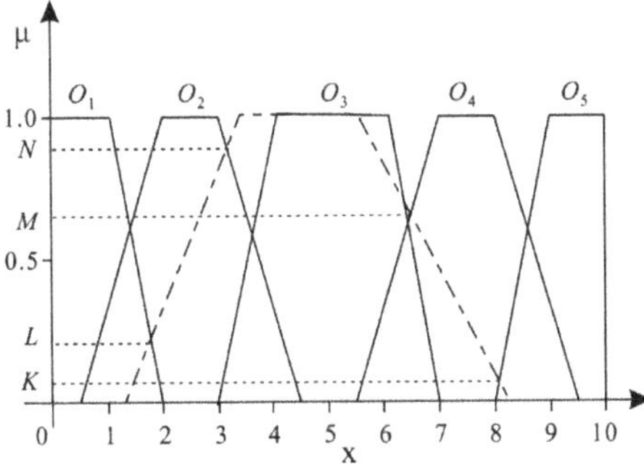

Figure 4. Schematic diagram of level evaluation of failure occurrence.

By combining the fuzzy number of the fuzzy rating evaluated by the expert group with the membership function of each attribute rating, the membership degree corresponding to the highest point $u_{(x)}$ is obtained, and finally, the fuzzy parameter variable set of the corresponding attribute evaluation is obtained. According to this method, the fuzzy set of each risk attribute parameter can be obtained, respectively. As shown in Figure 4, the dotted line represents the trapezoidal fuzzy number obtained after gathering expert opinions. The fuzzy number intersects with the figure O_1, O_2, O_3, O_4, O_5, and the abscissas of the highest intersection point are $L, M, 1, N$ and K, respectively. The fuzzy set for the occurrence probability level O formed by this is:

$$\tilde{O} = \frac{L}{O_1} + \frac{N}{O_2} + \frac{1}{O_3} + \frac{M}{O_4} + \frac{K}{O_5} \tag{12}$$

Which can also be expressed as $\tilde{O} = (L, N, 1, M, K)$, and among them, $L, N, M, K \in [0, 1]$. By normalizing the set of fuzzy membership degrees, the prior probability evaluation value of the risk factor under specific parameters can be obtained.

3.3. Identification of Failure Mode of Bauxite Ocean Maritime Logistics

Bauxite maritime logistics refer to the entire logistics process of goods transported by sea from the port of departure to the port of destination. From the perspective of transportation elements, it can be divided into transportation nodes and transportation routes. Transportation nodes include bauxite export ports, hub trans-shipment ports, and destination ports. The risk of ships at port nodes is mainly due to subjective human factors. Transportation routes refer to specific transportation processes other than non-transport nodes. Ships will suffer risks due to interference from the external environment, cargo status, the ship's operating conditions, and crews during the voyage. To ensure the scientificity, objectivity, and rationality of the selected risk factors, this paper consults with experienced experts in the industry field and consults relevant materials to obtain the main potential risk factors of China's imported bauxite maritime logistics (see Table 8).

Table 8. List of failure modes of bauxite maritime logistics.

Symbol	Failure Mode
FM_1	Worker riots
FM_2	Port congestion
FM_3	Improper operation by the crew
FM_4	Piracy or terrorist attack
FM_5	Terrible sea conditions
FM_6	Bauxite-free surface effect
FM_7	Ship facilities and equipment failure

Among these failure modes, the first one is related to bauxite transit reliability. Guinea has been China's largest source of imported bauxite in recent years. However, the domestic political situation in Guinea is unstable, and worker riots or port strikes have occurred from time to time. For example, in September 2017, riots broke out in the Bokeh bauxite area of Guinea. Protesters exchanged fire with the police, and bauxite production activities were severely hindered. This kind of failure mode can cause injuries or damage to cargo and equipment, delaying ships, and causing severe production standstills and supply disruptions.

The second failure mode is related to the port's loading and unloading capacity and the arrival of ships. Once the port is congested, it will inevitably lead to a delay in the transportation of bauxite, which will lead to an increase in transportation costs; the third failure mode is related to the transport personnel, i.e., the crew themselves. In general, operational errors caused by a lack of safety awareness or lack of emergency response skills, and the crew's mental health problems caused by long-term sea voyages will all have a certain impact on maritime transportation.

The fourth failure mode occurs during the transportation of goods. On China's bauxite import route from Guinea in West Africa to Yantai in Shandong, the Gulf of Guinea and the Strait of Malacca are almost the only way to go. However, pirate attacks often occur in these areas. This kind of situation is bound to bring huge property losses and casualties. The fifth failure mode is related to the natural environment. Bad sea conditions (such as heavy rain, typhoon, tsunami, etc.) will have a huge impact on the normal navigation of the ship. In severe cases, the ship will capsize, and the safety of the crew will be threatened.

The sixth failure mode is related to the cargo itself. Since the bauxite contains a certain amount of water during the actual transportation, it is easy to fluidize during the bumpy transportation at sea to form a free surface effect, thereby reducing the stability of the ship and causing the loss of cargo or the hull of the ship to capsize. The seventh failure mode is caused by the transport ship. The performance of the ship is degraded due to the aging of the ship's operating facilities or equipment or the hidden danger of the ship's hull structure and integrity, which make it easy to cause failures in the process of cargo transportation and to capsize.

3.4. Constructing a Fuzzy Rule Database System Based on a Confidence Structure

In the process of establishing the traditional fuzzy rule base, IF-THEN rules are used to express the association between the attribute of the premise and the attribute of the conclusion. A basic fuzzy rule base consists of a series of simple IF-THEN rules. Although the established fuzzy rule base can express the fuzzy situation, it is difficult to reflect the slight change of the premise attributes in the conclusion part. Moreover, due to insufficient expert experience and evidence, it is difficult to maintain a completely deterministic relationship between the attributes of the premise and the conclusion. Therefore, the fuzzy rule base form based on the confidence structure is adopted, and the confidence degree is used to express the degree of trust in the conclusion part under the given preconditions. R_K : IFA_1^K and $A_2^K, \cdots, A_n^K, THEN\{(\beta_{1K}, D_1), (\beta_{2K}, D_2), \cdots, (\beta_{NK}, D_N)\}, (\sum_{i=1}^{N} \beta_{iK} \leq 1, 0 \leq \beta_{iK} \leq 1), A_j^K\{j \in (1, 2, \cdots, M), K \in (1, 2, \cdots, L)\}$ represents the first category of the jth antecedent attribute in the created Kth fuzzy rule, M represents the total number of antecedent attributes, L represents the total number of rules in the fuzzy rule base, $D_i(i \in \{1, 2, \cdots N\})$ indicates the type of i conclusion, and its corresponding confidence is β_{iK}. Under the condition of entering the a priori A_j^K attribute, if $\sum_{i=1}^{N} \beta_{iK} = 1$, it indicates that the Kth rule created is complete, if $\sum_{i=1}^{N} \beta_{iK} = 0$, it means that the input condition attribute cannot be judged.

According to the characteristics of the constructed FBN, it is necessary to construct a fuzzy rule base based on the confidence structure of the risk parameters C and S, respectively. This paper takes the fuzzy rule base of creating C parameters as an example to introduce the process of creating the rule base. In the process of creating the rule base, the parameters O, S, and D are used as the antecedent attributes, taking the criticality of the failure C as the conclusion attribute to create a fuzzy rule base based on the confidence structure. R_l : IFO_i and S_j and $D_i, THEN\left\{ \left(\beta_1^l, C_1\right), \cdots, \left(\beta_m^l, C_m\right) \right\}$, among them, the variables $i, j, k, m \in \{1, \cdots 5\}$ are used to estimate the criticality of failure modes, and $C_1, C_{1,} \cdots, C_5$, are expressed as "low, relatively low, medium, relatively high, and high".

The determination of the confidence level in the rule base can be based on the accumulated knowledge of past events or subjective experience from domain experts. The former method often requires a large amount of objective data to support, and the knowledge of domain experts is used to reasonably determine all the rules in the rule base. Confidence is often subjective and difficult, especially when faced with a large rule base. Given this situation [7], a proportional method is proposed to ensure the rationalization of the confidence distribution in the rule base. However, the main drawback of this method is that it does not consider the impact of parameter weights. When building the confidence of the rule base, this paper uses the proportional method to determine the confidence distribution based on the importance of each risk parameter. In the process of creating this rule base, all the attribute parameters in the IF part and the THEN part are described by five grade variables. Therefore, the confidence of a certain grade variable in the conclusion attribute parameter can be determined by making the antecedent attributes' variables belong to the same normalized weights of the risk parameters of the grade variables that are summed. Since the antecedent attribute contains three five-valued risk parameters, the constructed fuzzy rule library based on the confidence structure has a total of 125 rules. Due to space limitations, only some of the rules in the rule base are listed in Table 9.

Table 9. Fuzzy rule base based on confidence structure.

Rules	Antecedent Attributes (Input)			Criticality C (Output)				
	O	S	D	Low	Relatively Low	Medium	Relatively High	High
1	Low	Light	Easy	1	0	0	0	0
2	Low	Light	Relatively easy	0.75	0.25	0	0	0
3	Low	Light	Medium	0.75	0	0.25	0	0
4	Low	Light	Relatively difficult	0.75	0	0	0.25	0
5	Low	Light	Difficult	0.75	0	0		0.25
6	Low	Relatively light	Easy	0.55	0.45	0	0	0
...	...	...	...	...	...	...	...	...
123	High	Severe	Medium	0	0	0.25	0	0.75
124	High	Severe	Relatively difficult	0	0	0	0.25	0.75
125	High	Severe	Difficult	0	0	0	0	1

3.5. Bayesian Network Construction

Because the Bayesian network has a good ability to describe the uncertain non-linear relationship between events, it can handle the fuzzy rule base system based on the confidence structure well, and it has efficient reasoning ability. Therefore, this paper will use Bayesian network inference technology to describe and implement the fuzzy rules based on the confidence structure.

Based on the relationship between the characteristic attributes of the failure mode, combined with the constructed risk parameter characterization system, the definition takes the failure occurrence degree O, the detection degree D, and the parameter time delay ST, additional cost SC, and safety and security cost SF that characterize the degree of risk consequence as the root node, taking the consequence severity S as the intermediate node, and the failure mode criticality C as the leaf node to construct the topology model shown in Figure 5.

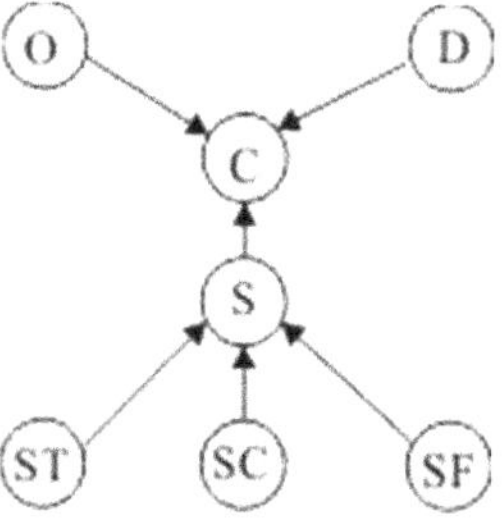

Figure 5. BN topology model.

To make better use of BN for reasoning, the fuzzy rule library based on the confidence structure needs to be transformed into the form of a conditional probability table. Taking rule 2 in Table 10 as an example, the following transformation can be carried out. R_2: IF $O_1, S_1, D_2, THEN\{(75\%, low), (25\%, \text{Relatively low}), (0\%, \text{Medium}), (0\%, \text{Relatively high}),\}(0\%, high)$ That is, given $O_1, S_1,$ and $D_2,$ the probability of child node $C_m (m = 1, 2, \cdots, 5)$

is $(0.75, 0.25, 0, 0, 0)$, or it can be expressed as $P(C_m|O_1, S_1, D_1) = (0.75, 0.25, 0, 0, 0)$. Therefore, the fuzzy rule library of the confidence structure can be transformed into the form of the conditional probability distribution.

Table 10. BN conditional probability distribution.

| C | O_1 | | | | | | | | $\cdots$ | O_5 | | | |
| | S_1 | | | | $\cdots$ | S_5 | | | $\cdots$ | S_5 | | | |
	D_1	D_2	$\cdots$	D_5	$\cdots$	D_1	$\cdots$	D_5	$\cdots$	D_1	D_2	$\cdots$	D_5
C_1	1	0.75	$\cdots$	0.75	$\cdots$	0.55	$\cdots$	0.3	$\cdots$	0.25	0	$\cdots$	0
C_2	0	0.25	$\cdots$	0	$\cdots$	0	$\cdots$	0	$\cdots$	0	0.25	$\cdots$	0
C_3	0	0	$\cdots$	0	$\cdots$	0	$\cdots$	0	$\cdots$	0	0	$\cdots$	0
C4	0	0	$\cdots$	0	$\cdots$	0	$\cdots$	0	$\cdots$	0	0	$\cdots$	0
C_5	0	0	$\cdots$	0.25	$\cdots$	0.45	$\cdots$	0.7	$\cdots$	0.75	0.75	$\cdots$	1

After the rule base is transformed into the created BN to describe the conditional probability distribution of the correlation degree between nodes, the analysis of the criticality of the failure mode is transformed into the calculation of the edge probability of the child node C. According to the method in Section 2, the expert evaluation opinions are represented in the form of trapezoidal fuzzy numbers and processed and transformed. The parameter variables can be assigned to different linguistic variables in the form of membership degrees, and finally discrete fuzzy subset forms, O, D, ST, SC, and SF. In the BN reasoning process, the sum of the probability values of different states of all nodes must meet the condition equal to 1; therefore, the membership degrees of the different states of the root node need to be normalized, and the formula is as follows.

$$P_i = P_i^* / \sum_i^n P_i^* \tag{13}$$

Among them, P_i^* is the confidence value of the state of the ith node parameter before normalization, and n is the total number of node parameter states. After normalization, the prior probability values of the different state variables of the root node parameters can be obtained, from which the probability distribution of the intermediate nodes can be calculated, $P(S_j) = \sum_1^5 \sum_1^5 \sum_1^5 P(S_j|ST_l, SC_m, SF_n) P(ST_l) P(SC_m)(SF_n) j = (1, \cdots 5)$. Next, the edge probability of node NC can be obtained, $P(C_m) = \sum_1^5 \sum_1^5 \sum_1^5 C_m|O_i, S_j, D_k) P(O_i P(S_j) O(D_k)$.

3.6. Use the Utility Function to Sort the Criticality

To determine the criticality level of each failure mode, it is necessary to sort the criticality of each failure mode and clarify the criticality between different failure modes to help managers quickly and accurately make differentiated management strategies. The criticality of each failure mode obtained by using FBN reasoning is given in the form of fuzzy subsets. Therefore, it is necessary to convert the multi-index risk status value of each risk factor into a clear value for sorting. Therefore, we need to use an appropriate utility function vector to quantify the degree of difference between different risk states, to achieve the overall judgment of the criticality level of the failure mode. Using the utility function U to clarify the criticality of the failure mode is as follows.

$$CI = P(C_m) \times U_m \tag{14}$$

This paper takes the utility function vector $U_m = (1, 25, 50, 75, 100)^T$.

3.7. Maritime Logistics System Risk Assessment

The use of FBN can only evaluate the criticality from the failure mode itself, and it cannot effectively evaluate the system risk level. Based on this, the improved ER theory is used to fuse the evaluation results of the hazard degree of each failure mode, and the linguistic variables represented by the confidence structure are used as the input value of the ER to calculate the maritime logistics system risk, and to achieve the purpose of evaluating the maritime logistics system risk from an overall perspective.

ER is an important uncertainty reasoning method, which is widely used in information fusion, expert systems, fault diagnosis, and the military. It can better represent uncertain information and synthesize expert opinions. However, the traditional ER theory has the problems of "conflict of evidence" and "robustness". Therefore, we use the DS synthesis algorithm based on the weight assignment and matrix analysis to overcome the limitations of traditional ER to merge the results of each failure mode inferred by FBN [37]. Assuming that the evaluation results of three risk factors are integrated, the evaluation level is above the five levels, and the identification framework $\theta = \{L, RL, M, RH, H\}$, the evaluation results of the three risk factors are shown in Table 11.

Table 11. Evaluation results of three types of risk factors.

Risk Factor	Identification Framework				
	L	**RL**	**M**	**RH**	**H**
FM_1	a_1	a_2	a_3	a_4	a_5
FM_2	b_1	b_2	b_3	b_4	b_5
FM_3	c_1	c_2	c_3	c_4	c_5

(1) The steps of the DS evidence fusion method based on weight distribution and the matrix analysis are as follows: assuming that the matrix $A = (a_1, a_2, a_3, a_4, a_5)$, $B = (b_1, b_2, b_3, b_4, b_5), C = (c_1, c_2, c_3, c_4, c_5)$, multiply A by B, transpose of the matrix, and obtain the matrix M_1.

$$M_1 = A^T \times B = \begin{pmatrix} a_1 \times b_1 & \cdots & a_1 \times b_5 \\ \vdots & & \vdots \\ a_5 \times b_1 & \cdots & a_5 \times b_5 \end{pmatrix} \tag{15}$$

(2) In the matrix M_1, the sum of the non-main diagonal elements is the degree of conflict between the risk factors A and B.

The column matrix M_1' formed by the main diagonal elements of M_1 should be multiplied with matrix C to obtain matrix M_2.

$$M_2 = M_1' \times C = \begin{pmatrix} a_1 \times b_1 \times c_1 & \cdots & a_1 \times b_5 \times c_5 \\ \vdots & & \vdots \\ a_5 \times b_1 \times c_1 & \cdots & a_5 \times b_5 \times c_5 \end{pmatrix} \tag{16}$$

Then, the conflict degree K of the three types of risk factors is the sum of all non-main diagonal elements of the matrix M_1 and M_2. Following the same steps, a limited number of remaining risk factors can be merged, and the conflict degree K of all risk factors can be obtained.

(3) Use the improved synthetic formula for weight distribution to calculate the following.

$$m(A) = \begin{cases} 0, \ A = \varnothing \\ \sum_{Ai \cap Bj \cap \cdots = A} m_1(A_i) \times m_2(B_j) \times \cdots K \times q(A), A \neq \varnothing \end{cases} \tag{17}$$

where $q(A) = \frac{\sum_{i=1}^{n} m_i(A)}{n}$ represents the average degree of support for all risk factors, and will be allocated to A in the proportion K.

3.8. Sensitivity Analysis

When a new model is proposed and constructed, it needs to go through rigorous testing to verify the reliability and rationality of the proposed model. Especially when it involves a subjective judgment belief structure, the validity of the model is more necessary. A sensitivity analysis is used to study the sensitivity of input to output variables. The input variables can be either parameters or variables. In this research, the confidence parameter corresponding to the root node variable of FBN is used as the input part, and the focus is on the input parameters and the degree to which the change in confidence affects the confidence level of the failure mode. If the constructed FBN is reasonable, the sensitivity analysis should at least satisfy the following two axioms [7].

Axiom 1. *A slight increase/decrease in the prior subjective probability of each input node should result in a relative increase/decrease in the posterior probability value of the output node.*

Axiom 2. *The total magnitude of the influence of the combination probability change from the x attribute (evidence) on the risk priority value should always be greater than the influence from the $x - y (y \in x)$ attribute (sub-evidence) set.*

4. Risk Assessment of Bauxite Maritime Logistics

4.1. Failure Mode Criticality Assessment

To obtain the prior probability of risk events more accurately, this paper uses expert judgment to issue questionnaires to four business managers and researchers who have been engaged in bauxite transportation for a long time. In the questionnaire, these experts need to separately evaluate the identified failure modes. The content of the assessment includes five risk parameters and their fuzzy language variable rating levels. Taking the risk factor "worker riot or strike" as an example, Table 12 shows the evaluation of the failure mode by four experts under the five risk parameters. Using the evaluation result of the UOWA operator comprehensive expert group to obtain the assembled trapezoidal fuzzy number, according to the membership function graph of the risk parameter level, the ordinate of the highest point intersected by the trapezoid corresponding to each parameter level is obtained. After normalization, the prior probability distribution of the failure mode under different risk parameters can be obtained. Similarly, the prior probability distributions of all failure modes concerning each parameter can be obtained.

Table 12. "Worker riots" expert assessment results.

Risk Parameter	Linguistic Variables	Trapezoidal Fuzzy Number	Prior Probability Distribution
O	L, RL, L, L	(0.083, 0.333, 1.333, 2.417)	(0, 0, 0.285, 0.460, 0.255)
D	M, RL, M, RH	(3, 4.175, 5.825, 7)	(0, 0.265, 0.471, 0.265, 0)
ST	H, RH, H, H	(7.583, 8.667, 9.667, 9.917)	(0, 0, 0, 0.426, 0.574)
SC	H, RH, RH, H	(6.75, 8, 9, 9.75)	(0, 0, 0.053, 0.474, 0.474)
SF	H, RH, M, H	(6.3, 7.455, 8.64, 9.29)	(0, 0, 0.154, 0.475, 0.371)

After obtaining the prior probability of the root node of the FBN topology structure under different failure modes, combined with the established fuzzy rule base system, the criticality assessment result of the failure mode can be calculated using Equation (13). The specific calculation can be operated by the software Netica. Figure 6 shows the evaluation result of the risk factor "worker riot or strike" using the software for risk reasoning, and the risk status of the risk factor $P(C) = (17.2\%, 19.4\%, 14.5\%, 27.4\%, 21.4\%)$. That is, the confidence level of the "low" risk status is 17.2%, the "relatively low" confidence is 19.4%, the "medium" confidence is 14.5%, the "relatively high" confidence is 27.4%, and the "high" confidence is 21.4%. In the software, any risk input modification related to the five risk parameters can trigger the change of the node state, which helps to automatically conduct a real-time risk assessment of any target risk factor in the bauxite maritime logistics link.

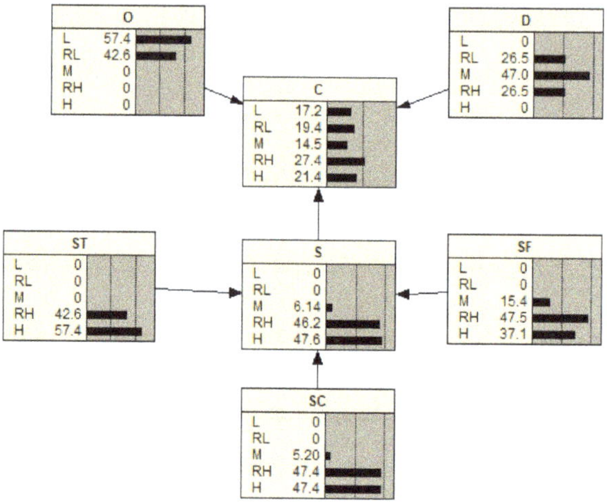

Figure 6. Results of risk assessment of FM1 using Netica software.

The risk status of the risk factors expressed in language variables needs to be further clarified by utility functions to prioritize the risks. The risk assessment value of the risk factor "worker riots" can be expressed as the utility function vector calculation in Section 3.4 $CI_{FM_1} = \sum_{m=1}^{5} p(C_m)U_m = 17.2\% \times 1 + 19.4\% \times 25 + 14.5 \times 50 + 27.4\% \times 75 + 21.4\% \times 100 = 54.222$. Similarly, the values of other risk factors' CI can be obtained, and the results obtained according to the constructed risk assessment model are shown in Table 13. In the bauxite maritime supply chain, the hazard level of risk factors is ranked in order $FM_4 > FM_1 > FM_7 > FM_2 > FM_5 > FM_6 > FM_3$. It can be analyzed from the CI value that "pirate or terrorist attacks" and "worker riots" are the most harmful, and they are the key risk factors affecting the reliability of bauxite maritime logistics links.

Table 13. Failure mode risk assessment result.

Failure Mode	Criticality Assessment Fuzzy Subset					CI Value	Rank
	Low (%)	Relatively Low (%)	Medium (%)	Relatively High (%)	High (%)		
FM_1	17.20	19.40	14.50	27.40	21.40	54.222	2
FM_2	21.60	32.30	24.30	14.30	7.65	38.816	4
FM_3	39.10	34.90	13.40	11.30	1.27	25.561	7
FM_4	14.20	14.60	21.30	32.90	17.00	56.117	1
FM_5	10.70	41.00	40.40	7.96	0	36.527	5
FM_6	21.90	34.90	29.10	12.90	1.27	34.439	6
FM_7	3.48	34.00	45.40	17.20	0	44.135	3

4.2. Risk Assessment of Bauxite Maritime Logistics System

After obtaining the hazard level of each failure mode of maritime logistics, an improved synthetic algorithm based on weight distribution and the matrix analysis is used to fuse the assessment results of individual risk factors to obtain the risk status of the bauxite shipping supply chain system. The fusion results are shown in Figure 7. China's studied imported bauxite shipping supply chain system risk index is described as 18.32% low, 30.18% relatively low, 26.93% medium, 17.71% relatively high, and 6.94% high. The system risk value is 41.419 calculated by using the utility function vector. The confidence that the maritime logistics system is at a medium- to high-risk level reaches 51.58%, indicating that the overall system risk is relatively high.

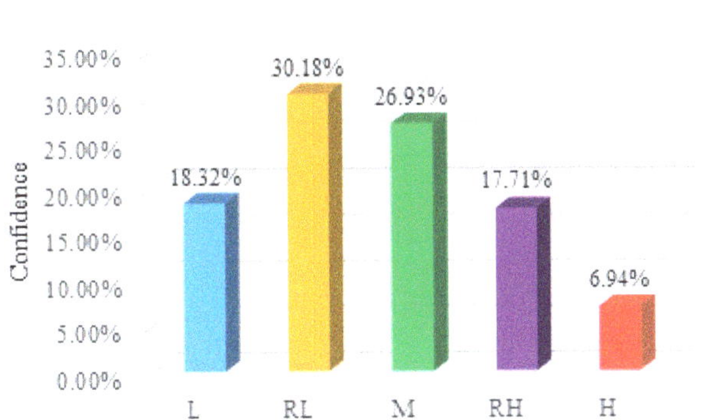

Figure 7. Maritime logistics system risk assessment.

4.3. Sensitivity Analysis Results

According to the axioms introduced in Section 3.8, the sensitivity analysis is carried out to test the validity and reliability of the Bayesian network based on the fuzzy rule base system. The linguistic variables of all risk parameters should be positively correlated with the CI value, that is, when the language variable of each risk parameter slightly increases or decreases, the value of the risk assessment result's CI value should also become higher or lower. Next, the subjective probability of 10% is re-assigned to the different language variables of each parameter and makes the CI value change in an incremental direction. If the constructed model is reasonable, then the CI value should increase accordingly. Taking "worker riots" as an example, in a single risk parameter and various risk parameter combinations, we raise the value of each language variable belonging to "H" by 10%, and the current minimum language of the modified risk parameter. The prior probability of the variable is reduced by 10% to keep the total confidence constant, and then axiom 1 and axiom 2 are tested. The results are shown in Table 14. When the prior probability that the language variable belongs to "H" in a single-risk parameter increases by 10%, the final risk assessment result CI value also increases to varying degrees. For example, when the confidence that the risk parameter "O" is in the "H" state increases by 10%, the CI value increases by 2.97. When the confidence that the risk parameter "D" is in the "H" state increases by 10%, the CI value increases by 1.875, and axiom 1 is verified. In addition, when the risk parameters adopt different numbers of combination types, as the number of combinations increases, the risk assessment CI value also keeps changing in ascending order. For example, when the different combination types of risk parameters are "O", "O D", "O D ST", "O D ST SC", and "O D ST SC SF", the risk assessment values have changed to "2.97", "4.845", "5.47", "5.72", and "6.22", respectively. Therefore, axiom 2 is verified in this model.

Table 14. Sensitivity analysis of different combinations of risk parameters.

Number	Combination	CI Value	Change Value	Serial Number	Combination	CI Value	Change Value
1	Initial	54.222	-	17	O D ST	59.692	5.47
2	O	57.192	2.97	18	O D SC	59.917	5.695
3	D	56.097	1.875	19	O D SF	59.817	5.595
4	ST	54.597	0.375	20	O ST SC	58.317	4.095
5	SC	54.822	0.6	21	O ST SF	58.217	3.995
6	SF	54.722	0.5	22	O SC SF	58.292	4.07
7	O D	59.067	4.845	23	D ST SC	56.972	2.75
8	O ST	57.567	3.345	24	D ST SF	56.972	2.75
9	O SC	57.792	3.57	25	D SC SF	57.197	2.975
10	O SF	57.692	3.47	26	ST SC SF	55.597	1.375
11	D ST	56.472	2.25	27	O D ST SC	59.942	5.72
12	D SC	56.697	2.475	28	O D ST SF	59.942	5.72
13	D SF	56.597	2.375	29	O D SC SF	60.167	5.945
14	ST SC	55.097	0.875	30	O ST SC SF	58.567	4.345
15	ST SF	55.097	0.875	31	D ST SC SF	57.472	3.25
16	SC SF	55.572	1.35	32	O D ST SC SF	60.442	6.22

5. Conclusions

The supply network for bauxite shipping is becoming more intricate in the highly competitive and unstable global bauxite market. China is a large consumer and importer of bauxite. Under the influence of many un-certain risks, it is crucial to ensure the smooth operation of the maritime logistics of imported bauxite. Therefore, it is essential to develop a reliable and adaptable technique to evaluate the risk associated with bauxite maritime logistics. In the face of traditional FMECA methods, there are problems such as incomplete risk-characterization parameters, failure to reflect the difference in parameter importance, and a limited discrimination of RPN values. This research suggests a systemic risk assessment model combining the FBN and improved ER theory based on the improved FMECA. To accomplish the goal of describing the risk factors more thoroughly and accurately, the improved FMECA adds three sub-parameters to the consequence parameter. It also uses the AHP-entropy technique to weigh the risk parameters. Additionally, a fuzzy rule base system based on the confidence structure is created by fusing the fuzzy set theory and expert knowledge. BN reasoning technology is then used to realize the risk inference of complex systems in uncertain environments, and the weighted utility function vector is used to calculate a variety of risk status indicators. The clusters are converted into numerical values, and then the risk factors are sorted. The improved ER theory is used to aggregate individual risk events to realize the overall judgment of the bauxite maritime logistics system risk. The results show that in the bauxite shipping supply chain, "pirate or terrorist attack" is the most important risk factor, followed by "worker riots", "ship facilities and equipment failures", "port congestion", "bad sea conditions", "bauxite free surface effect", and "improper operation by the crew". Therefore, from a controllable point of view, in the process of importing bauxite in China, the protection of ships on the route should be increased, the ships should be regularly maintained and repaired, and the ship operation skills and emergency response capabilities should be improved. At the same time, full attention should be paid to the free liquid surface effect of bauxite, the water content of bauxite should be reduced as much as possible, and effective measures should be taken to stabilize the goods and reduce the bumps during the transportation of the goods. In addition, China's imported bauxite maritime logistics system is at a medium- to high-risk level as a whole. In summary, the main contributions of this paper are as follows:

(1) A systematic risk assessment model is proposed, which can carry out a risk assessment of the system from the local and overall dimensions and can effectively improve the scientificity and accuracy of the risk assessment of the system in an uncertain environment.

(2) The improved FMECA can effectively overcome the limitations of the traditional FMECA method, making it more suitable for the field of risk analysis, and improving the reliability and rationality of the risk assessment.

(3)   The improved ER theory is used to realize the assessment of the overall system risk of maritime logistics, which provides a new perspective for the field of risk assessment.

(4)   The proportional method combined with parameter weights is applied to construct the fuzzy rule base, and to rationalize the confidence distribution in the fuzzy rule base.

However, the bauxite supply chain is faced with various risk challenges. This paper mainly focuses on the sea transportation of bauxite, while risk assessments in other aspects, such as mining, land transportation, and processing need to be further explored in the future.

Author Contributions: Conceptualization, H.W.; methodology, J.S. and H.W.; software, J.S.; validation, H.W., J.S. and M.W.; formal analysis, J.S.; investigation, H.W. and M.W.; data curation, J.S. and M.W.; writing—original draft preparation, J.S.; writing—review and editing, H.W.; visualization, J.S. and M.W.; supervision, H.W.; funding acquisition, H.W. All authors have read and agreed to the published version of the manuscript.

Funding: This research was funded by the National Natural Science Foundation of China, grant number 51909202.

Institutional Review Board Statement: Not applicable.

Informed Consent Statement: Not applicable.

Data Availability Statement: Not applicable.

Acknowledgments: The authors would like to thank the anonymous reviewers for their valuable comments and the editors' help with this article. The authors also appreciate the valuable contribution of the participating experts in the paper.

Conflicts of Interest: The authors declare no conflict of interest.

References

1.   Gul, M.; Yucesan, M.; Celik, E. A manufacturing failure mode and effect analysis based on fuzzy and probabilistic risk analysis. *Appl. Soft Comput.* **2020**, *96*, 106689. [CrossRef]

2.   Lee, Y.-S.; Kim, D.-J.; Kim, J.-O.; Kim, H. New FMECA methodology using structural importance and fuzzy theory. *IEEE Trans. Power Syst.* **2011**, *26*, 2364–2370. [CrossRef]

3.   Wan, C.; Yan, X.; Zhang, D.; Qu, Z.; Yang, Z. An advanced fuzzy Bayesian-based FMEA approach for assessing maritime supply chain risks. *Transp. Res. Part E Logist. Transp. Rev.* **2019**, *125*, 222–240. [CrossRef]

4.   Ma, D.; Zhou, Z.; Jiang, Y.; Ding, W. Constructing Bayesian network by integrating FMEA with FTA. In Proceedings of the 2014 Fourth International Conference on Instrumentation and Meassurement, Computer, Communication and Control, Harbin, China, 18–20 September 2014; IEEE: Piscataway, NJ, USA, 2014; pp. 696–700.

5.   Liu, H.-C.; You, J.-X.; Fan, X.-J.; Lin, Q.-L. Failure mode and effects analysis using D numbers and grey relational projection method. *Expert Syst. Appl.* **2014**, *41*, 4670–4679. [CrossRef]

6.   Renjith, V.; Kumar, P.H.; Madhavan, D. Fuzzy FMECA (failure mode effect and criticality analysis) of LNG storage facility. *J. Loss Prev. Process Ind.* **2018**, *56*, 537–547. [CrossRef]

7.   Alyami, H.; Lee, P.T.-W.; Yang, Z.; Riahi, R.; Bonsall, S.; Wang, J. An advanced risk analysis approach for container port safety evaluation. *Marit. Policy Manag.* **2014**, *41*, 634–650. [CrossRef]

8.   Heckmann, I.; Comes, T.; Nickel, S. A critical review on supply chain risk–Definition, measure and modeling. *Omega* **2015**, *52*, 119–132. [CrossRef]

9.   Wagner, S.M.; Bode, C. An empirical examination of supply chain performance along several dimensions of risk. *J. Bus. Logist.* **2008**, *29*, 307–325. [CrossRef]

10.   Lam, J.S.L.; Bai, X. A quality function deployment approach to improve maritime supply chain resilience. *Transp. Res. Part E Logist. Transp. Rev.* **2016**, *92*, 16–27. [CrossRef]

11.   Ju, L.; Vassalos, D.; Wang, Q.; Wang, Y.; Liu, Y. Numerical investigation of solid bulk cargo liquefaction. *Ocean Eng.* **2018**, *159*, 333–347. [CrossRef]

12.   Nguyen, S.; Wang, H. Prioritizing operational risks in container shipping systems by using cognitive assessment technique. *Marit. Bus. Rev.* **2018**, *3*, 185–206. [CrossRef]

13.   Yang, Y.-C. Risk management of Taiwan's maritime supply chain security. *Saf. Sci.* **2011**, *49*, 382–393. [CrossRef]

14.   Thun, J.-H.; Hoenig, D. An empirical analysis of supply chain risk management in the German automotive industry. *Int. J. Prod. Econ.* **2011**, *131*, 242–249. [CrossRef]

15.   Wang, S.; Yin, J.; Khan, R.U. The multi-state maritime transportation system risk assessment and safety analysis. *Sustainability* **2020**, *12*, 5728. [CrossRef]

16. Jiang, M.; Lu, J.; Yang, Z.; Li, J. Risk analysis of maritime accidents along the main route of the Maritime Silk Road: A Bayesian network approach. *Marit. Policy Manag.* **2020**, *47*, 815–832. [CrossRef]
17. Berle, Ø.; Asbjørnslett, B.E.; Rice, J.B.J.R.E.; Safety, S. Formal vulnerability assessment of a maritime transportation system. *Reliab. Eng. Syst. Saf.* **2011**, *96*, 696–705. [CrossRef]
18. Shterev, D. Safety problems in maritime transport of cargoes which are able to liquefy. *Trans Motauto World* **2021**, *6*, 27–29.
19. Munro, M.C.; Mohajerani, A. Liquefaction incidents of mineral cargoes on board bulk carriers. *Adv. Mater. Sci. Eng.* **2016**, *2016*, 5219474. [CrossRef]
20. Lee, H.L. Nickel ore bulk liquefaction a handymax incident and response. *Ocean Eng.* **2017**, *139*, 65–73. [CrossRef]
21. Daoud, S.; Said, I.; Ennour, S.; Bouassida, M. Numerical analysis of cargo liquefaction mechanism under the swell motion. *Mar. Struct.* **2018**, *57*, 52–71. [CrossRef]
22. Yang, Z.; Bonsall, S.; Wang, J. Facilitating uncertainty treatment in the risk assessment of container supply chains. *J. Mar. Eng. Technol.* **2010**, *9*, 23–36. [CrossRef]
23. Nguyen, S.; Chen, P.S.-L.; Du, Y.; Shi, W. A quantitative risk analysis model with integrated deliberative Delphi platform for container shipping operational risks. *Transp. Res. Part E Logist. Transp. Rev.* **2019**, *129*, 203–227. [CrossRef]
24. Khan, B.; Khan, F.; Veitch, B.; Yang, M. An operational risk analysis tool to analyze marine transportation in Arctic waters. *Reliab. Eng. Syst. Saf.* **2018**, *169*, 485–502. [CrossRef]
25. Akyuz, E.; Arslan, O.; Turan, O. Application of fuzzy logic to fault tree and event tree analysis of the risk for cargo liquefaction on board ship. *Appl. Ocean Res.* **2020**, *101*, 102238. [CrossRef]
26. Liu, H.; Tian, Z.; Huang, A.; Yang, Z. Analysis of vulnerabilities in maritime supply chains. *Reliab. Eng. Syst. Saf.* **2018**, *169*, 475–484. [CrossRef]
27. Zavitsas, K.; Zis, T.; Bell, M.G. The impact of flexible environmental policy on maritime supply chain resilience. *Transp. Policy* **2018**, *72*, 116–128. [CrossRef]
28. Vilko, J.; Ritala, P.; Hallikas, J. Risk management abilities in multimodal maritime supply chains: Visibility and control perspectives. *Accid. Anal. Prev.* **2019**, *123*, 469–481. [CrossRef] [PubMed]
29. Fan, L.; Wilson, W.W.; Dahl, B. Risk analysis in port competition for containerized imports. *Eur. J. Oper. Res.* **2015**, *245*, 743–753. [CrossRef]
30. Polatidis, N.; Pavlidis, M.; Mouratidis, H. Cyber-attack path discovery in a dynamic supply chain maritime risk management system. *Comput. Stand. Interfaces* **2018**, *56*, 74–82. [CrossRef]
31. John, A.; Paraskevadakis, D.; Bury, A.; Yang, Z.; Riahi, R.; Wang, J. An integrated fuzzy risk assessment for seaport operations. *Saf. Sci.* **2014**, *68*, 180–194. [CrossRef]
32. Bowles, J.B.; Peláez, C.E. Fuzzy logic prioritization of failures in a system failure mode, effects and criticality analysis. *Reliab. Eng. Syst. Saf.* **1995**, *50*, 203–213. [CrossRef]
33. Oluah, C.; Akinlabi, E.; Njoku, H.O. Selection of phase change material for improved performance of Trombe wall systems using the entropy weight and TOPSIS methodology. *Energy Build.* **2020**, *217*, 109967. [CrossRef]
34. Nieto-Morote, A.; Ruz-Vila, F. A fuzzy approach to construction project risk assessment. *Int. J. Proj. Manag.* **2011**, *29*, 220–231. [CrossRef]
35. Lyu, H.-M.; Shen, S.-L.; Zhou, A.; Yang, J. Risk assessment of mega-city infrastructures related to land subsidence using improved trapezoidal FAHP. *Sci. Total Environ.* **2020**, *717*, 135310. [CrossRef] [PubMed]
36. Wang, G.; Liu, L.; Shi, P.; Zhang, G.; Liu, J. Flood risk assessment of metro system using improved trapezoidal fuzzy AHP: A case study of Guangzhou. *Remote Sens.* **2021**, *13*, 5154. [CrossRef]
37. Zhao, G.; Chen, A.; Lu, G.; Liu, W. Data fusion algorithm based on fuzzy sets and DS theory of evidence. *Tsinghua Sci. Technol.* **2019**, *25*, 12–19. [CrossRef]

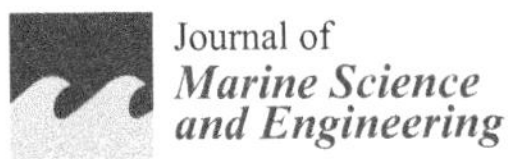

Journal of
*Marine Science
and Engineering*

Article

Detecting Maritime GPS Spoofing Attacks Based on NMEA Sentence Integrity Monitoring

Julian Spravil [1,†], Christian Hemminghaus [1], Merlin von Rechenberg [1,2], Elmar Padilla [1] and Jan Bauer [1,*]

1   Fraunhofer FKIE, Cyber Analysis & Defense, Fraunhoferstraße 20, 53343 Wachtberg, Germany
2   Institute of Computer Science 4, University of Bonn, Friedrich-Hirzebruch-Allee 8, 53115 Bonn, Germany
*   Correspondence: jan.bauer@fkie.fraunhofer.de; Tel.: +49-241-8021465
†   Current address: Fraunhofer IAIS, NetMedia, Schloss Birlinghoven 1, 53757 Sankt Augustin, Germany.

Abstract: Today's maritime transportation relies on global navigation satellite systems (GNSSs) for accurate navigation. The high-precision GNSS receivers on board modern vessels are often considered trustworthy. However, due to technological advances and malicious activities, this assumption is no longer always true. Numerous incidents of tampered GNSS signals have been reported. Furthermore, researchers have demonstrated that manipulations can be carried out even with inexpensive hardware and little expert knowledge, lowering the barrier for malicious attacks with far-reaching consequences. Hence, exclusive trust in GNSS is misplaced, and methods for reliable detection are urgently needed. However, many of the proposed solutions require expensive replacement of existing hardware. In this paper, therefore, we present MAritime Nmea-based Anomaly detection (MANA), a novel low-cost framework for GPS spoofing detection. MANA monitors NMEA-0183 data and advantageously combines several software-based methods. Using simulations supported by real-world experiments that generate an extensive dataset, we investigate our approach and finally evaluate its effectiveness.

Keywords: GPS spoofing; anomaly detection; NMEA-0183; Maritime Cyber Security; GNSS; cyber and electromagnetic activities

Citation: Spravil, J.; Hemminghaus, C.; von Rechenberg, M.; Padilla, E.; Bauer, J. Detecting Maritime GPS Spoofing Attacks Based on NMEA Sentence Integrity Monitoring. *J. Mar. Sci. Eng.* **2023**, *11*, 928. https://doi.org/10.3390/jmse11050928

Academic Editors: Marko Perkovic, Lucjan Gucma and Sebastian Feuerstack

Received: 29 March 2023
Revised: 20 April 2023
Accepted: 24 April 2023
Published: 26 April 2023

1. Introduction

Commercial seagoing vessels, such as container ships, bulk carriers, and tankers, are among the most important means of transportation today. However, they are easy targets for different attacks in the domain of cyber and electromagnetic activities (CEMA), motivated by industrial espionage and economic sabotage to piracy and terrorism. Since the impact of such attacks can pose serious threats not only to the economy but also to humans and the environment, safety and security are of paramount importance. Because the shipping industry as a whole is moreover responsible for the international supply of goods, there is a serious risk of major economic and ecological damage caused by CEMA targeting this industry, which is by no means immune to such attacks [1,2]. Therefore, it must be protected effectively, as recognized in the last two decades by governments and organizations placing Maritime Cyber Security on their agendas.

Navigation in shipping has been fundamentally changed by the advent of civil global navigation satellite systems (GNSSs), the best-known representative of which is the Navigational Satellite Timing and Ranging (NAVSTAR) Global Positioning System (GPS). Today's maritime systems are increasingly computer-aided and heavily depend on the availability of GNSS for accurate positioning, navigation, and timing (PNT), which is usually complemented by highly interconnected sensors within integrated bridge systems (IBSs) on board modern vessels. Despite significant efforts by all GNSS operating stakeholders to upgrade security or deploy new, more secure generations of systems, civil GNSSs currently do not have sufficient security measures as practically demonstrated in the case of GPS [3]. However, cryptographically authenticated GNSS signals have also recently been shown to remain vulnerable to spoofing attacks [4].

The fact that GNSS satellite signals are relatively weak when being received on the Earth's surface makes them intrinsically vulnerable. Thus, an attacker can effectively carry out jamming attacks to prevent a GNSS receiver from processing legitimate signals. Nonetheless, more and more trust is being placed in GNSSs in a risky manner.

While jamming attacks might be easily detected by the crew, since they would cause obvious failures in navigational instruments, so-called GNSS spoofing attacks can remain undetected and, thus, be much more harmful. In a spoofing attack, an adversary generates counterfeit signals, which are difficult to distinguish from legitimate ones and cause receivers to incorrectly calculate position and/or timing. Particularly in maritime off-shore scenarios, attacks on PNT are often more difficult to detect and, thus, perhaps more threatening. Because the majority of the previous work refers to GPS [3,5–8], we also focus, without loss of generality, on GPS and GPS spoofing in this paper. However, other navigation satellite systems, e.g., Galileo, GLONASS, and BeiDou, are all based on the same principle of measuring time differences in signal propagations from satellites. Thus, methods used in this paper can, in general, be transferred to other satellite navigation systems.

Vessels have a long service life. Since maritime systems are strongly embedded, accordingly outdated technologies are used that are usually prone to various emerging types of CEMA attacks. Because particularly spoofing attacks can be very devastating, appropriate countermeasures are urgently needed. Several methods for detecting such attacks exist, as surveyed in [9]. However, the proposed methods cannot be applied to existing receivers without limitations in many cases. They often require expensive and cumbersome hardware upgrades or replacements that are usually evaluated as uneconomical when assessing cybersecurity risks. Taking legacy systems into account, we believe that it is essential to consider efficient software-based approaches for the detection of GPS spoofing, which can be seamlessly retrofitted into those systems.

In this work, therefore, we propose a modular framework for GPS spoofing detection in the maritime sector based on anomaly detection. We advantageously combine different software-based methods and operate solely on network traffic. Thereby, this approach enables cost-effective retrofitting, either by software updates or by additional low-cost off-the-shelf hardware devices. In summary, our contributions are as follows:

- We identify GPS spoofing detection methods from the literature that can operate on data provided by the NMEA-0183 and can be implemented at low cost;
- we propose a MAritime Nmea-based Anomaly detection framework (MANA) that incorporates these methods;
- we generate and provide an extensive dataset including diverse spoofing attacks; and
- we finally evaluate and compare the effectiveness of spoofing detection methods and demonstrate the potential of their combination to compensate for each other's weaknesses.

2. Maritime Systems On Board Vessels

A reliable and accurate position and time estimation is crucial for navigation and the situational picture in IBSs. Hence, for redundancy, the majority of vessels are equipped with two GNSS devices [10] that are supplemented by a variety of additional sensors, networked in a maritime system [11–13]. A typical system architecture is exemplarily shown in Figure 1, and a brief introduction to these systems with a focus on GPS integration is given below.

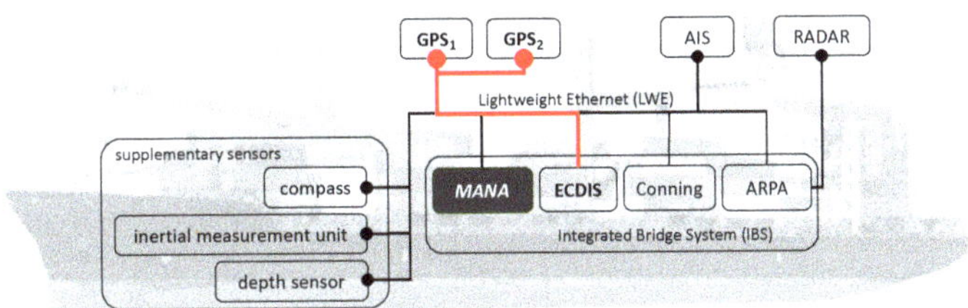

Figure 1. Simplified system architecture of a maritime system. There are multiple GPS receivers on board, along with supplementary sensors, providing navigational data to the IBS via the IP-based Lightweight Ethernet (LWE) [13]. Because the data provided by GPS receivers (red) cannot necessarily be assumed to be trustworthy, the derived PNT information must be verified, which can be achieved using our approach, MANA (gray), implemented as an additional detector component in the IBS.

2.1. GPS Dependency of Nautical Electronics

The PNT data generated by GPS receivers are strongly involved in navigation. In addition, derived measurements, e.g., speed and heading, are used for multiple purposes in maritime systems. The electronic chart display and information system (ECDIS) integrates those measurements into a digital chart to provide a situation picture to navigators to support conning decisions. For collision avoidance, the radio-based automatic identification system (AIS) broadcasts vessels' positions and course information to other maritime entities. Radar makes use of PNT, since automatic radar plotting aid (ARPA) is illustrated relative to the vessel's location and orientation. Autopilots need PNT to calculate necessary course corrections.

Since PNT data are used by many navigational aids, the impact of unreliable and manipulated information can be devastating. Recent incidents show that minimal course deviations may lead to groundings with costly global financial losses [14]. Hence, in the case of CEMA, alternative position estimation and tracking systems are recommended [15].

2.2. NMEA-0183 and Maritime System Networks

Distribution of nautical data on board vessels nowadays relies on Ethernet (IEEE 802.3 product family) as a well-established network standard. The NMEA-0183 standard specified by the National Marine Electronics Association (NMEA) defines the earlier used transmission and encoding of nautical data via 4800-baud serial data bus interfaces. Although the original serial transmission of NMEA-0183 is a legacy technology, the ASCII-based encoding and message format of nautical data via so-called NMEA sentences is still used in modern IP-based protocols. *NMEA over IP* encapsulates sentences in UDP datagrams or TCP streams to distribute nautical information via uni-, multi-, or broadcasts. A more complex IP-based protocol is Lightweight Ethernet (LWE), which is standardized in IEC 61162-450 [16]. In addition to the use of NMEA sentences, LWE defines multicast groups and protocol extensions for the distribution of data files in the maritime system. Nautical devices and sensors, e.g., GNSS receivers, can be integrated into the multicast using additional network hardware, i.e., LWE gateways.

Other protocols, e.g., NMEA 2000 and NMEA OneNet, use more transmission-efficient binary encoding schemes. Although not human-readable, the encoded information is almost equal to that of ASCII-based NMEA sentences. Thus, without loss of generality, we will focus on ASCII-based NMEA in our concept and implementation. With respect to GNSS, NMEA sentences moreover not only contain functional data, e.g., latitude and longitude coordinates, elevation and azimuth angles, and time but also quality information, i.e., carrier-to-noise density (C/N_0), the number of visible satellites, and their IDs.

Overall, the multicast distribution of GPS-related and other sensor data via NMEA sentences in the network enables system-wide monitoring, cf. Figure 1. This is leveraged by our framework, MANA, in order to detect anomalies in PNT streams that are possibly caused by spoofing.

3. GPS Spoofing

GPS signals are an easy attack target. Due to the low signal strength at the Earth's surface, the signals can be effortlessly blocked [17]. In addition, the civil GPS lacks encryption, authentication, or any further security measures to protect the signal integrity. In fact, the data structure, modulation schemes, and spreading codes are publicly available [18]. Altogether, these peculiarities enable jamming and spoofing. In a jamming attack, adversaries try to suppress original GPS signals using artificial interference. As a result, benign signals become unrecognizable to receivers and can no longer be used for PNT. However, receivers are usually aware of whether they are subject to jamming attacks and can react accordingly [17]. An overview of the threat jamming poses to the maritime domain and the main countermeasures techniques is provided in [19]. With spoofing attacks, the situation is fundamentally different. In such attacks, adversaries generate signals that mimic legitimate signals. Often, their goal is to deceive the targeted receiver without being detected so that incorrect PNT estimates are calculated.

3.1. Maritime GPS Spoofing

Besides a variety of cyber and CEMA threats against a vessel in the maritime context (cf. e.g., [20]), GPS spoofing attacks represent a major threat and attract attention from researches, practitioners, and industry. Those attacks are known to be feasible and seriously affecting maritime navigation [3]. Recent reports also highlight the risk posed by GPS jamming and spoofing attacks [15,21]. In 2017, a total of 25 ships reported the wrong position near the port of Novorossiysk, which pointed to the Gelendzhik airport [22]. It is suspected that GPS spoofing caused this incident.

In fact, a maritime environment makes the success of a spoofing attack more likely than in other domains. At sea, there are only a few landmarks to correlate GPS positions with derived information. Without a regular position input, alternative measures, such as dead reckoning, lose their accuracy over time. Therefore, smooth long-term attacks are difficult to detect. Additionally, navigators often unconsciously tend to blindly trust their devices [23]. In the domain of maritime navigation, the risk of attacks on GNSS has also long been noticed by the International Maritime Organization (IMO). The urgent demand for adequate integrity monitoring was first recognized in 2001 within the resolution MSC.915 (22) [24], and further resolutions followed. A compact summary of the evolution of the IMO integrity concepts, such as e-Navigation, can be found in [25]. The need for resilient GPS is also part of these concepts. The Department of Homeland Security provides the Resilient PNT Conformance Framework [26], with four levels of resilience reflecting the users' needs. In this context, our framework can be classified as a level 1 resilient PNT but offers the potential to reach level 2 by adding a complementary PNT source (cf. Section 4.2).

3.2. Attack Model and Scenarios

To spoof GPS receivers, an attacker can employ different techniques. Jafarnia-Jahromi et al. [5] distinguish between three types of attackers with increasing complexity. The attackers' signals range from simple unsynchronized signals to complex, near-authentic signals with even matching angles of arrival. Similarly, we define three attacker types, namely the replay attacker (A_R), the meaconing attacker (A_M), and the simulator attacker (A_S). An overview of the attackers is given in Figure 2.

The simplest attacker is A_R. Equipped with a single antenna, the attacker can generate arbitrary GPS signals or record original ones. These signals are then (re)played at a later point in time. Thus, the signals are usually not synchronized to real GPS signals [5]. Furthermore, except for superimposed signal power, no specific signal takeover strategy is performed. Attacks of this category were already demonstrated two decades ago [27].

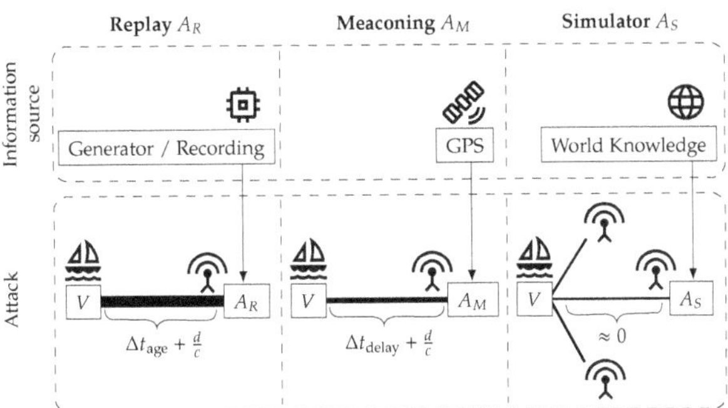

Figure 2. Illustration of the three attacker types (A_R, A_M, and A_S), their information source, and a visualization of the different attack modes on victim V. The signal strength as received by V is reflected by the line strength, ranging from superimposed (left) to normal levels (right). For each mode, the additional pseudo-distance caused by the attacker is shown, where Δt_{age} is the age of the recorded signals, Δt_{delay} is the signal processing delay, d is the actual distance between V and the attacker, and c is the propagation speed of the radio signal.

The meaconing attacker A_M can additionally receive and process genuine GPS signals and react to them with a small delay. The simplest attack this attacker can execute is meaconing, i.e., replaying authentic signals in near-real time. To actively break the lock of a receiver to the authentic satellites, the attacker can execute a jamming attack beforehand. A GPS receiver has a lock when a stable reception from a set of satellites required for PNT is established. Initially, the attacker, therefore, uses signals with a high signal strength to overpower the legitimate signals and to force the receiver to lock onto the spoofed signals. Once the attacker has obtained the lock, i.e., controls all the signals at the receiver, the spoofing power can be reduced.

Meaconing can also be effective against encrypted signals, as the signal content remains unaltered [8]. Advanced forms of meaconing allow spoofing of arbitrary positions by individually delaying each signal [4]. However, a lock on all counterfeit signals is not guaranteed and depends on factors such as the target's speed [28]. As a consequence, the resulting position of the victim is not always the attacker's position and to some extent unpredictable. This issue can be addressed by relaying the authentic signals over the Internet to the attackers' transmitter near the victim, as demonstrated with consumer hardware [29].

Besides real-time signal processing, the simulator attacker A_S can operate multiple antennas to match the original signal alignments. However, such a setup is costly and brings limitations in terms of antenna placement and effective range [5]. The precise position and other motion properties of the victim have to be known, enabling highly accurate signal construction. To manipulate the victim's receiver, the attacker performs a seamless takeover by matching the legitimate signals and carefully increasing the signal strength [7,8,18]. Real-world experiments involving a single-antenna A_S were performed in a lab environment [6] and on board a yacht [3].

4. GPS Spoofing Detection

The first spoofing detection considerations and techniques were already mentioned in the 1990s by MITRE [30]. A theoretical basis was given by Warner and Johnston [17] proposing different techniques, e.g., based on signal strengths or signal time of arrival (TOA) monitoring, to check the integrity of PNT data and to mitigate possible attacks. Following these theoretical considerations, a variety of practical implementations and improvements gradually emerged, beginning with Receiver Autonomous Integrity Monitoring (RAIM) [31]. A comprehensive overview of existing methods and techniques, as well as their complexity and effectiveness, that have been presented over the years can be found in [5,8,9].

The variety of anti-spoofing techniques can be classified according to the layer at which the countermeasure can be applied. For this purpose, we differentiate between system, hardware, firmware, and software layers. An overview of existing approaches for GPS spoofing countermeasures is provided in Table 1. Note that software-based approaches can also be implemented in firmware, but that involves increasing complexity and costs. Because the goal of our solution is to be retrofittable with low effort and cost, we focus on the software-based approaches, which are briefly presented in the following subsection and discussed in terms of their applicability in the maritime context.

Table 1. Classification of known GPS spoofing detection techniques with decreasing complexity of retrofitting per class and their effectiveness against different attack scenarios.

	Spoofing Countermeasure Method	Replay (A_R)	Meaconing (A_M)	Simulator (A_S)
System	Symmetric encryption of full spreading code (e.g., [8,32])	●	◐	◐
	Spread spectrum security code (e.g., [32])	●	◐	◐
	Navigation message authentication (e.g., [32,33])	●	◐	◐
Hardware	L1/L2 power level comparison (e.g., [34,35])	●	●	○
	L1/L2 power level code phase comparison (e.g., [34])	●	●	○
	DOA monitoring (e.g., [36,37])	●	●	◐
Firmware	Synthetic antenna array (e.g., [38])	●	●	◐
	Signal strength monitoring (e.g., [17,39])	●	◐	○
	Doppler monitoring (e.g., [40,41])	●	●	◐
	Code and phase rates consistency check (e.g., [34])	◐	○	○
	TOA monitoring (e.g., [17,42])	◐	◐	○
	PRN code and data bit latency (e.g., [43–45])	●	○	○
	Auxiliary peak tracking APT (e.g., [18])	◐	●	◐
	Signal quality monitoring (e.g., [46–48])	◐	◐	◐
	Distribution analysis of correlator output (e.g., [49,50])	◐	◐	◐
Software	C/N_0 Monitoring (*CNM*) (e.g., [34,39])	●	◐	○
	Physical Cross-Check (*PCC*) (e.g., [51–54])	◐	◐	○
	Clock Drift Monitoring (*CDM*) (e.g., [8,18])	●	◐	○
	Ephemeris Data Validation (*EDV*) (e.g., [34,55])	●	◐	○
	Pairwise Distance Monitoring (*PDM*) (e.g., [7,56])	●	●	◐

Notation: Effective (●), semi-effective (◐), and ineffective (○) GPS spoofing detection regarding individual attacker models defined in Section 3.2, cf. Figure 2.

4.1. Software Controls and Related Work

Software methods are flexible, easy to retrofit, and, thus, also cost-effective. In the literature, there are methods for software-based spoofing detection, such as anomaly detection approaches that monitore the C/N_0 [34,39] or internal clock drifts [8,18]. Other approaches validate ephemeris data [34,55] or implement cross-checks with physical constraints [51,52]. Similar to methods from other categories, the effectiveness of these approaches is diverse, as shown in Table 1. Advanced attackers cannot be detected by most software-based methods. However, they still provide reliable detection for replay and meaconing attackers.

An effective approach is presented by Tippenhauer et al. [7], which requires multiple (at least two) GPS receivers. It is based on the fact that the pairwise distances between multiple receivers in a static constellation are constant with regard to the respective determined position of each receiver (except for GPS inaccuracies). Because a second receiver is available on most commercial vessels [10], this particular approach can be easily implemented in the maritime environment, as already suggested by Zalewski [56]. With high-quality receivers typically used in maritime systems, we expect the inaccuracy of position determination to be reasonably low, which presumably increases the detection probability of this Pairwise Distance Monitoring (*PDM*) approach. Zalewski applied Tippenhauer's approach to the maritime context. Using a mathematical model and simulation, he shows that it is practically not feasible to spoof multiple GPS receivers by a single transmitter so that their relative distance remains. Recently, another approach using multiple receivers was presented in [57]. Instead of considering the effects on position, the authors focus on the time provided by GNSS to protect electrical substations in the energy sector. Similar to our approach, they rely on NMEA sentences, but they are limited to the *PDM* method for detecting spoofing attacks.

In [58], software-based approaches for anomaly detection using GPS spoofing in the context of unmanned aerial vehicles (UAVs) are explored. The authors propose machine-learning algorithms and show that promising detection results can be achieved with different one-class classifiers, which require only non-anomalous data for training. The work represents an interesting approach, the methods of which can, in principle, complement our framework. For their evaluation, they created and shared a dataset of three UAV flight recordings, i.e., one benign, one with GPS jamming, and one with spoofing [59]. However, the purpose-built dataset is restricted to timestamped positions of a single receiver and, thus, inappropriate for the analyses of all the above-mentioned *PDM* methods. In the automotive sector, Lemieszewski [53] recently dealt with the detection of spoofing attacks. The author uses a *PCC* method and correlates the GNSS position estimates with the speedometer of the vehicle, also using NMEA sentences.

A spoofing detection and mitigation approach for the maritime context is proposed in [60]. The authors use RAIM [31] in combination with a *PCC* method based on a motion model of a ship in order to detect offsets to the predefined route, while their mitigation mechanism is based on a genetic algorithm. In addition to route information, their motion model requires other sensor inputs. Our approach, in contrast, entirely relies on NMEA data provided by GPS devices and does not require route information. Nonetheless, the method of Singh et al. could in general be applied in conjunction with our framework.

Similar to our approach, Lee et al. [61] base their spoofing detection on NMEA data but in the context of smartphones. In a physical laboratory environment, the authors first generate spoofing signals that dictate positions on a specific route. These signals are then processed by a stationary GPS receiver, and the effect is investigated. The authors suggest monitoring position, velocity, and time for changes or cross-referencing the results with other sensor information. Another method they describe is to check the relationship between the signal strength and the distance to the corresponding satellite. Lastly, a comparison with other positioning systems based on the residuals of the position calculation is suggested. However, their investigation focuses solely on static scenarios, which limits their general applicability to real-world systems.

In summary, in the field of GNSS spoofing detection, several related works exist that build on software-based methods and, among them, some that obtain their data also from NMEA messages. However, these works often focus on one specific approach. Our work, in contrast, presents a holistic, modular framework that adapts and combines different methods of existing work, offering flexible configuration and tailoring them to the needs of ships in the maritime domain. The later evaluation (cf. Section 7) will show that this combination is necessary to cope with the entire attack space. Before presenting our framework in Section 5, the following paragraph first concludes by briefly discussing complementary approaches from the area of resilient PNT that are orthogonal to our work.

4.2. Complementary Approaches

In addition to the countermeasures presented above that are limited exclusively to the GNSS domain, there are novel approaches for alternative localization technologies. Those technologies can also be used to detect anomalies and potential attacks against GNSS-based positioning. In this context, Oligeri et al. [62] propose a GPS spoofing detection and localization approach that leverages the Public Land Mobile Network (PLMN) infrastructure of terrestrial mobile communication. In [54], the authors even extend their work to the use of WiFi networks and present a crowd-sourced approach. However, this promising work, which leverages the existence of existing land-based infrastructure, is not applicable at sea.

Similarly, in the maritime domain, the R-Mode (Ranging Mode), which is currently under development, is based on so-called signals of opportunity, cf. [63]. These are independent terrestrial signals from existing maritime infrastructures such as AIS, Very high frequency Data Exchange System (VDES), or other maritime radio signals that can be leveraged for ranging. Worldwide coverage of up to 40% of all vessels is predicted [64]. Under good conditions, i.e., in the middle of three R-mode transmitters, a real-time horizontal positioning accuracy of 95% at 12 m could be achieved in the Baltic Sea testbed [65].

Furthermore, Naus et al. [66] show that maritime navigation radar can be used to detect anomalies in position determination. However, the feasibility of radar navigation, in general, depends massively on the availability of characteristic echo marks. The performance at the open sea is therefore questionable. Moreover, terrain navigation is extensively used by underwater vehicles and demonstrated to be feasible for surface vessels in coastal waters to improve GNSS-based navigation [67].

In relation to our approach introduced in this paper, the alternative technologies mentioned represent complementary countermeasures that can be successfully combined. Such a combination, i.e., secured GNSS information correlated with nautical data from additional sensors and augmented by terrain, radar, or PLMNs localization techniques, has great potential not only to mitigate GNSS spoofing attacks but to entirely prevent them.

5. NMEA-Based GPS Spoofing Detection Framework

Building on a versatile selection of multiple existing software approaches presented in the previous section and leveraging the availability of PNT and NMEA data in the network, our framework MANA aims to provide low-cost yet effective integrated integrity checks as a countermeasure against GPS spoofing. Moreover, the framework allows for comprehensive comparison of the selected approaches to find out if a combined solution can compensate for their individual deficiencies. MANA, which is modularly implemented in Python 3, relies on standardized NMEA sentences. Thus, it enables generic, very flexible, and easily retrofittable deployment by adding a detector component at a central position in the maritime network, cf. Figure 1. To the best of our knowledge, no directly related work provides a comparable framework. In the subsequent sections, we will briefly describe MANA's concept and the essential details of individual methods.

5.1. Concept of MANA

The concept of our framework and its data processing workflow are visualized in Figure 3. A stream of NMEA sentences, dispatched by at least two GPS receivers within a network, is taken as input. Alternatively, recorded network trace files can be used as input. For each detection method, a set of fields within NMEA sentences is defined that are relevant to the individual detection approach. These fields are continuously monitored such that every state change triggers the corresponding detection method(s). The relevant required information can be obtained from different and partially redundant NMEA sentences generated by GPS receivers. However, in the end, the set of required sentences can be reduced even to the three GPS-related types $GPGGA, $GPRMC, and $GPGSV. An overview of the contained information, its relationship to the respective detection methods, and alternative NMEA sentences is given in Figure 4.

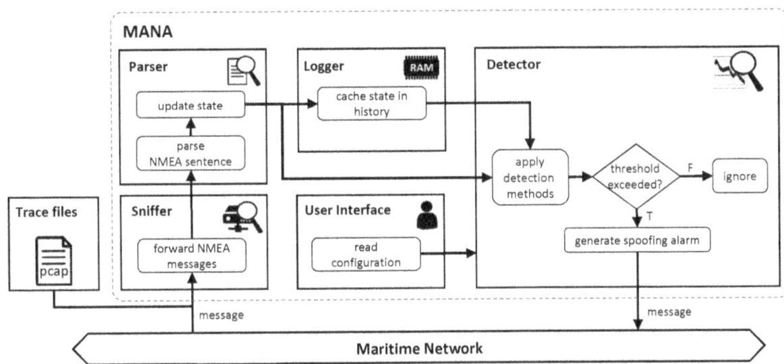

Figure 3. Conceptual overview of the GPS spoofing MANA framework and its components.

The detector generates an alarm if any of the methods indicates potential spoofing. Based on individual methods as basic building blocks, our framework allows the implementation of more complex and sophisticated detection, i.e., by composing the outputs of those methods. This offers great potential to further increase the overall detection capability; however, a detailed investigation is out of the scope of this paper. In order to evaluate if the different methods can compensate for each other's weaknesses, a simple strategy for combining different methods is nonetheless elaborated in Section 7.2.

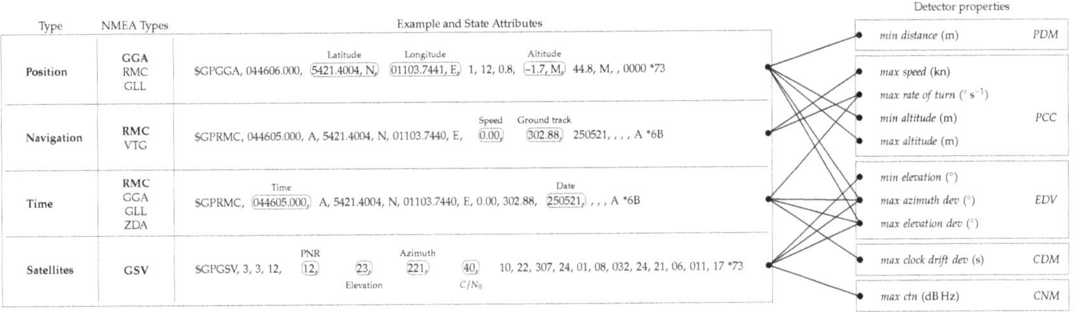

Figure 4. Visualization of the relationship (indicated by the lines) between state attributes extracted from NMEA sentences and the detection methods. The minimal working set of NMEA types is marked in bold. However, the other listed types can also be used as alternatives. Note that the asterisk (*) is the separator symbol for the following checksum in the ASCII notation of NMEA sentences.

5.2. Methods for NMEA-Based Detection

MANA comprises five software-based detection methods from the literature, introduced in Section 4, namely *PCC*, *PDM*, *EDV*, *CDM*, and *CNM* (cf. Table 1 and Figure 4). With the assumption that the data are smooth or predictable under normal conditions but not in case of a spoofing attack, these methods are implemented using thresholds.

Physical Cross-Check (PCC): The *PCC* method is subdivided with respect to the physical features considered for the respective cross-check, i.e., speed over ground (SOG) (PCC_{sog}), rate of turn (ROT) (PCC_{rot}), and height above sea level (PCC_{height}). While SOG and ROT are expected to stay within certain limits, these limits can be exceeded in the event of a perceived spoofing attack. Sudden jumps in position can result in an unreasonably high SOG and ROT if the distance or angular deviation from the course is extreme enough. To restrict the attackers' possibilities, we use maximum thresholds for PCC_{sog} and PCC_{rot} (cf. Figure 4), which must be plausibly selected for the respective vehicle. While the SOG threshold should be set slightly higher than the maximum speed of the vehicle, more tolerance is required for the ROT threshold. The ROT is subjected to a measurement error, which is particularly high when the vessel is slowly or not moving. Therefore, it is useful to consider only ROT measurements when the SOG exceeds a certain threshold, e.g., half of SOG threshold. The altitude is expected to be close to sea level. Thus, static thresholds above and below the sea level are defined for the PCC_{height} method (cf. Figure 4).

Pairwise Distance Monitoring (PDM): PDM monitors the static formation of two receivers and requires position data simultaneously recorded for each receiver, but these are generally not synchronized. Conventional GPS receivers only update positions with a frequency of 1 Hz. Hence, the states of the two receivers have to be aligned in time. For this purpose, one of the receivers is selected as a reference. Then, the other receiver's state is linearly interpolated between its temporally adjacent values, i.e., the time, latitude, and longitude immediately before and after the corresponding reference. Moreover, we use an exponential moving average (with $\alpha = 0.1$) to reduce the effects of small-scale noise in the position sequences. Once the estimates of both receivers are aligned, the measured distance between the geographic position of the receivers can be derived. A spoofing attack is detected if the distance is smaller than the minimum threshold (cf. Figure 4).

Ephemeris Data Validation (EDV): Satellite positions are to some extent predictable. As a data source, we use so-called two-line elements (TLEs). If the TLEs are up-to-date, the predicted satellite positions can be used to validate the estimated ones. However, the decrease of TLEs' accuracies with increasing age [68] and the integer accuracy of NMEA sentences with respect to elevation and azimuth need to be considered. This is checked by the *EDV* method by defining a maximum allowed deviation for both angles. Furthermore, the elevation of all satellites needs to be above a static minimum threshold (cf. Figure 4), ensuring that they are actually visible to the receiver, since implausible constellations indicate potential attacks.

Clock Drift Monitoring (CDM): With regard to *CDM*, a linear clock drift Δt of each device is assumed, which is reasonable for a certain interval in time. The drift of an individual device can be derived by continuously comparing the received GPS time with the local system time. If enough drift measurements are available, a drift function is derived by a linear regression providing the expected clock drift Δt_{exp} for a given point in time. The difference between Δt_{exp} and Δt is then compared with the predefined threshold (cf. Figure 4) and, in cases where deviation of the clock drift exceeds this threshold, a potential spoofing attack is indicated.

C/N_0 *Monitoring (CNM):* This method monitors carrier noise density and derives indications of possible attacks in cases where the noise level exceeds the given threshold (cf. Figure 4). An evaluation of the *CNM* detection method requires real-world data or a simulation that includes a realistic signal propagation model with the associated random processes, particularly C/N_0. Our simulation environment, which is used for the evaluation (in Section 7), does, however, not include such a signal propagation model. Thus, *CNM* is excluded from the later evaluation.

6. Simulation Environment and Dataset

For ease of feasibility and better reproducibility, we use simulations to evaluate the effectiveness of our approach but collected real data in experimental field trials to (i) calibrate simulation models and (ii) appropriately determine thresholds for the detection algorithms, i.e., for the parametrization of MANA. For the latter, it is important to find the right balance between triggering false alarms (false positives) and missing the detection of spoofing attacks (false negatives). For this purpose, we carried out two experiments, a static and a dynamic scenario. In the first experiment (Section 6.1.1), a stationary measurement was conducted to collect static GPS tracks that enable trace-based modeling of GPS errors in our later simulation. In the second experiment (Section 6.1.2), we used two mobile receivers moving in a static formation, i.e., at a fixed distance to each other, and continuously collected all data provided via NMEA sentences. The goal of the second experiment was to record the clock drift and position measurements of the GPS receivers, both for modeling in the simulation and to support the determination of thresholds for clock drift and distance between GPS receivers, required for the *CDM* and *PDM* detection methods (cf. Figure 4).

6.1. Simulation Environment

We use an in-house simulation environment for maritime networks including various sensor information, such as course over ground, heading, compass, and AIS, with a special focus on GPS and its spoofing. The simulation environment is capable of simulating multiple ships, where each ship calculates its current position and time by triangulation based on the algorithm presented in [69]. Physical vessel-related parameters such as velocity and turn rate were set to be appropriate for the commercial seagoing vessels that we consider in this paper. The values used for the evaluation are listed under the category *Ship* in Table 2.

Table 2. Simulator properties and parametrization.

Category	Parameter	Value
	Velocity	20 kn
	Rate of turn	$0.5\,°\mathrm{s}^{-1}$
Ship	Number of GPS receivers	2
	Distance between GPS receivers	4 m
	Distribution	Gaussian
Clock error	μ	0.0012
	σ	0.0076
Clock drift	Drift per second	10.55 µs

For a more realistic simulation, we implemented a trace-based approach to model natural GPS noise using data from the first experiment, which is described in more detail in Section 6.1.1. In addition, random noise is artificially added to the time derived from GPS using a Gaussian distribution and a clock drift according to a second field experiment for which a detailed description follows in Section 6.1.2. According to the results of the experiment, we used a clock drift of ($\approx$10.55 µs/s). Individual samples are further randomly distributed around the regression line according to the measured distribution, for which the parameters can be found in Table 2 under the category *Clock error*.

Furthermore, based on the results of the second experiment, which showed that an insufficient distance (of 2.5 m) between the receiver pair limits reliable detection, the distance was increased in the simulative evaluation. According to [6], a distance between 3 and 5 m can be considered to be feasible for the *PDM* method. Thus, we initially chose a distance of 4 m, cf. Table 2.

6.1.1. Modeling GPS Errors

To include natural GPS noise in the simulation, we chose a trace-based approach for modeling GPS errors. Therefore, we recorded representative GPS tracks files that can be used as trace files by the simulator. Thus, in the first experiment, two GPS receivers were installed in fixed positions, and their outputs, i.e., NMEA sentences, were recorded for multiple hours. The tracks of both receivers were analyzed regarding their spatiotemporal characteristics. As expected and shown in Figure 5a, we observed an averaged error in the position estimation of each receiver of up to several meters induced by natural noise and a varying C/N_0. In addition, a clear temporal correlation of errors was observed for both distance and azimuth.

Especially when processing algorithms, e.g., detection methods (cf. Section 5.2), average the (simulated) GPS error in time, and this temporal correlation can have a crucial impact. Therefore, errors must not be randomly selected from the recorded tracks without considering their temporal correlation. Hence, we utilize recordings of static GPS receivers as a noise source and construct a sequence of offsets (distance and azimuth) to the fixed position in our simulations. For each replication and each receiver, a random index is chosen within the sequence as a starting position. At each time step, this index is incremented by one, and the corresponding offsets are added to the simulated position measurement. In this way, individual error models are obtained for each receiver.

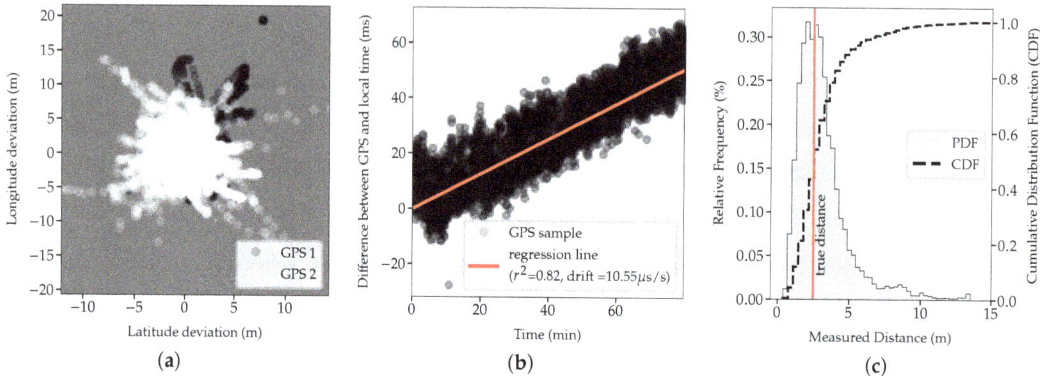

Figure 5. Experimental data from field trials to calibrate simulation models and derive thresholds for detection methods. (**a**) Simultaneous GPS measurements of two static receivers showing a measurement error of up to several meters. (**b**) The time difference between local and GPS time reveals a linear internal clock drift. (**c**) The PDF of measured distances between the position estimates of the two receivers is centered around the true distance.

6.1.2. Modeling Clock Drift

Besides the GPS error, modeling the clock drift is necessary for an evaluation of the different detection methods, particularly the *CDM* detection method. Additionally, the *CDM* detection method requires a threshold for the clock drift to distinguish between benign clock drift measurements and measurements that indicate a spoofing attack. Furthermore, we have to simulate multiple GPS receivers with a sufficiently large distance between them for the *PDM* detection method to work. Therefore, the second experiment was carried out to collect clock drift data and to determine suitable values for the GPS receivers' distance. For the second experiment, the tracks of two mobile GPS receivers were recorded. While moving the receivers, they had a fixed distance to each other of 2.5 m. The log file collected in the experiment provides rich GNSS-related data generated by both receivers, including various NMEA data.

Concerning the internal clock drift, a slight drift was found that is similar for both receivers (roughly 50 ms during 80 min). This drift is constant and, thus, grows linearly over time, as can be seen in Figure 5b, exemplarily shown for the first device. Nevertheless, there is a variance in the sequence of time samples, which can be attributed to the data processing within the receiver and for logging. Using linear regression, the clock drift is determined to be approximately 10.55 µs/s, which is typical for quartz oscillators as integrated into common devices. With the clock drift value and the measured variance, the clock drift can be accordingly modeled for realistic simulation, and a balanced threshold for the *CDM* detection method can be derived.

6.1.3. Calibrating Detection Thresholds

To determine suitable thresholds to parametrize various detection methods of MANA, we have to find a trade-off between false positives and false negatives. Due to the combination of different detection methods in our framework, occasional false negatives of individual methods are expected to be tolerated, because they are complemented by the indications of the other methods. Therefore, we use rather relaxed thresholds overall in

MANA. For the detection of spoofing based on *CDM*, the internal clock drift has to be known for each receiver. To this end, linear regression over a certain time interval could also be used by the detection algorithm in order to "learn" the clock drift. Based on the clock drift and distribution of measurements, the detection algorithm uses a threshold of ± 100 ms to identify possible spoofing while avoiding false positives.

PDM is based on monitoring the measured distance between two GPS receivers and comparing the measurements with the known real distance. Consequently, for the method to work, the distance between the GPS receivers has to be larger than the measuring error (cf. Figure 5a). Even though the second experiment shows that the distance derived from the positions is centered around the true distance (Figure 5c), it can be seen, at the same time, that the estimated distance can become very small due to the measurement errors of both receivers. Thus, it may falsely trigger *PDM*'s minimum threshold. Hence, the distance between the GPS receivers in the simulation has to be significantly larger than 2.5 m for the *PDM* detection method to work.

6.2. MARSIM Dataset

Based on the simulation environment, we generated a dataset, hereinafter referred to as the MARitime SIMulated (MARSIM) dataset. It consists of multiple scenarios (benign and attacked) with a duration of 120 s, provided as network traffic recordings (in packet capture (pcap) format). Each recording contains the NMEA sentences transmitted within the network of a single vessel heading north from a fixed point with constant velocity, cf. Table 2. By filtering background network traffic, the recordings are reduced to two PNT transmitting devices, namely the pair of GPS receivers, which are placed at a fixed distance of 4 m from each other. All information that is necessary for the detectors discussed in this work can be retrieved from these records. The system clock results from the timestamps of the entries and can thus be compared with the GNSS time within the recordings. With a probability of 50 %, spoofing attacks were started after 60 s in a scenario. The attacks are executed by the attackers A_R, A_M, and A_S (cf. Figure 2).

- A_R replays a recording with a certain *age* of a ship that followed a similar route as the victim but shifted eastward by a specified *distance*.
- A_M performs a meaconing attack with a *delay* to its own static position. The attacker is thereby located relative to the victim at a *distance* to the east at the onset of the attack.
- A_S finally constructs signals that will slowly shift the victim's position with a velocity defined by *shift speed* and an azimuth angle of *shift angle*.

All three attackers are equipped with a single sending antenna. Note that for mobile targets, a multi-antenna A_S attack is very complex and extremely difficult to realize in practice. Since most vessels are, moreover, not equipped with more than two GNSS receivers [10], which also cannot acquire the signals' angle of arrival, and the detection of such sophisticated attacks is, in any case, not expected to be feasible using software-based approaches. Hence, we only simulate single-antenna attackers that have actually been demonstrated in practice [3,6].

The mentioned parameters of the individual attackers are expected to have an impact on the detection capabilities of our framework. The parameter space is listed in Table 3. The two-dimensional parameter space has 19 gradations per dimension, resulting in a total of 361 parameter configurations per attacker. A set of 20 benign and 20 spoofed recordings that differs in the noisy position and time measurements are created for each parameter configuration. Hence, in total, our labeled dataset consists of 43,320 scenarios and can be scientifically used for the development and benchmarking of GPS spoofing detection and prevention methods.

Table 3. Parameter space for each attacker model.

Attacker	Parameter		Parameter Space
Replay A_R	*distance*	Physical separation between attacker and victim	$\{0\,\mathrm{m}, 2\,\mathrm{m}, ..., 36\,\mathrm{m}\}$
	age	Age of the replayed recording	$\{0\,\mathrm{ms}, 15\,\mathrm{ms}, ..., 240\,\mathrm{ms}\} \cup \{1\,\mathrm{min}, 1\,\mathrm{d}\}$
Meaconing A_M	*distance*	Physical separation between attacker and victim	$\{0\,\mathrm{m}, 2\,\mathrm{m}, ..., 36\,\mathrm{m}\}$
	delay	Time delay introduced by the attacker	$\{0\,\mathrm{ms}, 15\,\mathrm{ms}, ..., 270\,\mathrm{ms}\}$
Simulator A_S	*shift angle*	Azimuth in which the victim's position is shifted	$\{0°, 10°, ..., 180°\}$
	shift speed	Speed with which the victim's position is shifted	$\{0\,\mathrm{kn}, 4\,\mathrm{kn}, ..., 72\,\mathrm{kn}\}$

7. Performance Evaluation

In this section, we evaluate the effectiveness of MANA to detect GPS spoofing attacks based on the MARSIM dataset (cf. Section 6.2). After describing our methodology and metrics in Section 7.1, we investigate the capability of each method of our framework and evaluate their individual potential in Section 7.2.

7.1. Methodology and Metrics

For our dataset, we propose a binary classification task. Thus, for each file of the dataset, the method needs to decide whether the corresponding scenario is spoofed or not. Furthermore, we group the scenarios by the type of attacker involved, i.e., A_R, A_M, and A_S (cf. Figure 2). The performance is measured using the common metrics *precision* and *recall*, defined as:

$$precision = \frac{true\,positives}{true\,positives + false\,positives}\,, \quad recall = \frac{true\,positives}{true\,positives + false\,negatives}\,.$$

Intuitively, recall represents the actual detection capability. Precision, on the other hand, can be considered as the reliability of a method's indication.

We experimentally adjusted the parameters *max speed* and *max rate of turn* to minimize false detection of spoofing attacks under benign conditions (i.e., relaxed thresholds, cf. Table 4) and consequently achieve high *precision*. CDM's *max clock drift dev* threshold is set based on our observations in Section 6.1.3, whereas for *min distance* and *min speed*, half of the actual distance between receivers and *max speed*, respectively, are used. While this decision may inherently have a negative impact on the *recall*, we expect that a combination of methods can compensate for the potentially lower sensitivity of each method.

Table 4. Configuration of MANA's detection methods.

Method	Parameter	Value
PDM	*min distance*	$2\,\mathrm{m}$
	min speed	$15\,\mathrm{kn}$
PCC	*max speed*	$30\,\mathrm{kn}$
	max rate of turn	$7.5°\,\mathrm{s}^{-1}$
CDM	*max clock drift dev*	$100\,\mathrm{ms}$

7.2. Evaluation Results

The results of our evaluation reveal that the success of spoofing detection methods investigated in our experiments depends, in different ways, on the attackers' parameters. Some methods depend on both attack parameters (PCC_{sog} and PCC_{rot}), others only on a single parameter (CDM), and still others are found to be independent of the selected parameters (PDM), as can be seen in Figure 6. According to the relaxed thresholds mentioned above, the mean *precision* over all displayed methods is 0.92.

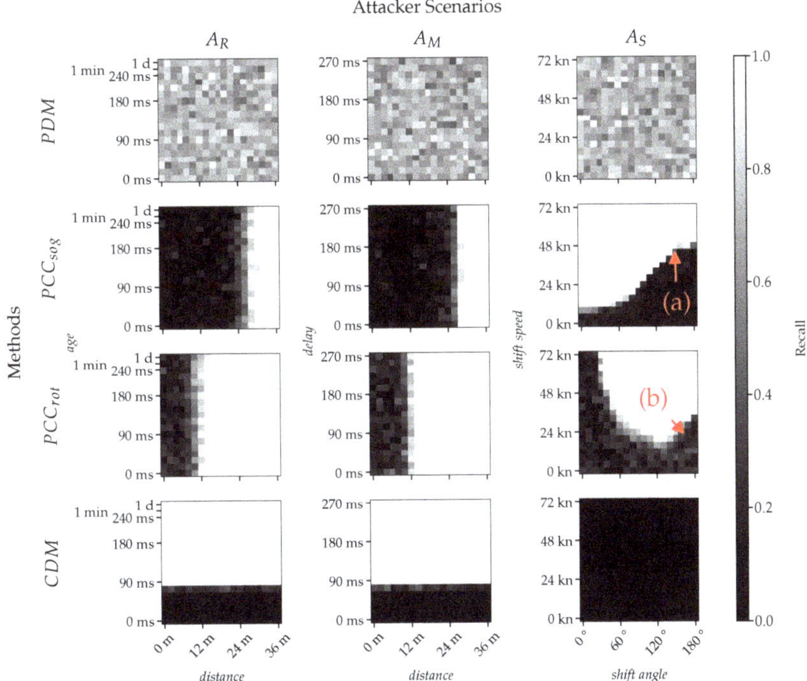

Figure 6. The recall achieved by individual detection methods depends on the type of attacker. Each heatmap has a resolution of 19×19 pixels exploring the parameter space for the given attacker, cf. Table 3. The grayscale of each pixel represents the *recall* calculated over 40 scenarios (20 benign and 20 spoofed) for the given attacker and its parameter constellation. Note that the brighter a pixel, the better the respective method performs. It is observable that the detection capabilities of the methods differ and are often influenced by the attacker parameters either in only one or in both dimensions. Note that both labels (a) and (b) highlight specifics of the evaluation that will be addressed in the textual description (in Section 7.2).

The methods in the focus of this evaluation are *PDM, PCC, EDV,* and *CDM*. In accordance with the classification of known GPS spoofing detection techniques in Table 1, we anticipate *PDM* to perform best and the other methods to perform slightly lower, while we suspect that the effectiveness of *PCC* methods varies considerably.

By checking the fixed distance between receivers, *PDM* (first row in Figure 6) shows the most consistent performance over all scenarios. While there are missed detections (false negatives) for all attacker models, they do not appear to be correlated with the attackers' parameterization. Missed detections are caused by noisy position measurements that keep the calculated distance between receivers above the *min distance* threshold (cf. Table 4). In Section 7.2.2, the impact of the actual distance between receivers on the detection performance of *PDM* will be investigated separately.

In comparison, the rather simple PCC_{sog} method, which detects spoofing attacks based on unrealistically high SOG estimates, gives all three attackers a fairly large margin to remain undetected (black areas in Figure 6). Replay and meaconing (A_R and A_M) are only limited to a similar extent by the choice of the *distance* between the victim and the recorded ship or between victim and attacker. The exceeding of this 26 m limit is almost always detected. The transition between the undetectable (black) and the detectable area (white) is smoother for A_R than for A_M, probably due to the movement of the recorded vessel that is replayed in comparison to the static position of A_M. The relationship between the A_S parameters and the detection capability is more complex. While an increasing *shift speed* in the direction of travel (i.e., *shift angle* of 0°) does not offer much margin for the attacker (roughly 12 kn), the margin significantly increases up to about 46 kn as soon as the *shift angle* passes the mark of approximately 140° (see label *a)* in Figure 6). The victim interprets a *shift angle* of $>140°$ and a *shift speed* of up to 40 kn as traveling in the opposite direction, explaining the observed results. This effect is visualized in Figure 7.

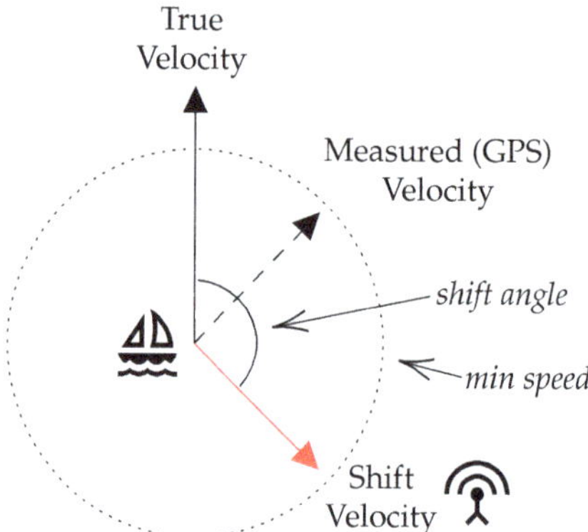

Figure 7. Visualization of the attack performed by A_S and the effect on the *min speed* property of PCC_{rot}. The attacker shifts the victim's position by a constant velocity defined by *shift speed* and *shift angle*. The victim observes a combination (dashed) of the shift velocity (red) and its true velocity (black) based on GPS measurements. Even if the true velocity is above the *min speed* threshold, this combination can fall below, resulting in the effect observed in Figure 6 at labels (a) and (b).

Similar to PCC_{sog}, PCC_{rot} is insensitive to changes in signal *age* or *delay* caused by replay or meaconing attacks (A_R and A_M). However, the attackers are already exposed at a *distance* of roughly 12 m, leading to a significantly increased ROT measurement. The detection of the simulator attacker A_S again depends on both *shift speed* and *shift angle*. A low *shift speed* allows for a large *shift angle* and vice versa. An exception to this rule is marked by label (b) in Figure 6 and has similar reasons as for label (a) (see Figure 7). The area of missing detections is caused by the *min speed* requirement of PCC_{rot} (cf. Section 5.2). If the shift vector composed of *shift speed* and *shift angle* points approximately in the opposite direction of travel, the total speed derived from GPS may fall below that threshold. If the *min speed* threshold is decreased, the caused blind spot eventually disappears. Nonetheless, the threshold needs to be sufficiently high so that the ROT measurement based on noisy GPS positions is reliable and does not cause false positives.

CDM directly monitors the local system clock of the device executing MANA with respect to the time provided by GPS. This results in a limitation of at most 75 ms for the signal *age* of A_R and the *delay* of A_M. A time jump of this magnitude with additional noise (cf. Figure 5b and Section 6.1.2) often exceeds the allowed drift range of 100 ms. Thus, both *age* and *delay* need to stay below this mark to avoid being detected (black area in Figure 6). All other parameters are time-independent and have no influence on *CDM*'s outcome.

The two methods, PCC_{height} and EDV, are not shown in Figure 6, since neither of these proved effective in our simulations. PCC_{height} and EDV detect A_R only if the signal *age* exceeds several hours or minutes, respectively, whereas A_M and A_S were found to be hardly detectable with these methods in our scenarios.

7.2.1. Ensemble of Methods

The performance of an ensemble of all investigated methods can be seen in Figure 8. Here, the binary outputs of the individual methods are combined with a logical OR. Thus, as long as at least one method is triggered, spoofing is detected. Note that this initial approach is rather simple and weights each method equally. The exploration of improved combinatorial methods is part of our future work. For attackers A_R and A_M, a combination of the methods *PDM*, *PCC*$_{rot}$, and *CDM* is sufficient, while A_S requires *PCC*$_{sog}$ in addition to *PCC*$_{rot}$ and *PDM*. Overall, it turns out that a significant area of the parameter space for each attacker is solely covered by the *PDM* method (highlighted in red in Figure 8), which thus significantly contributes to the combined performance.

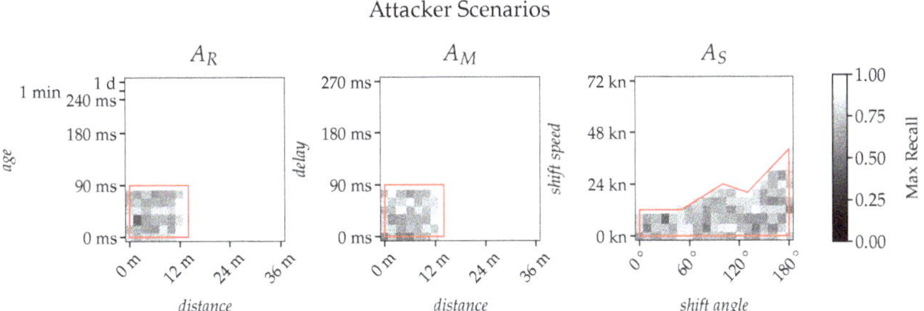

Figure 8. Maximum achieved recall considering all methods from Figure 6 shows the potential effectiveness of combining methods. This combination covers large parts of the attacks represented in the dataset. For each attacker, however, small areas exist (highlighted in red) that are solely covered by the *PDM* method.

7.2.2. Impact of Distance between Receivers on *PDM*

The performance of *PDM* depends on the distance between receiver pairs. To investigate the impact of this distance on the recall, we varied the distance of two receivers between 2 and 10 m with 100 increments in an A_S scenario. For each increment, 100 repetitions are conducted. We keep the *shift angle* and *shift speed* fixed to zero, effectively disabling the attacker. However, the detection capabilities of *PDM* are not related to the parameter choice of the attacker, cf. Figure 6. Hence, the only varying factor is the GPS noise. The results are depicted in Figure 9. As can be seen, the distance significantly impacts the capability of *PDM*. The impact vanishes at a distance of roughly 6 m, which is consistent with the results of Jansen et al. [6].

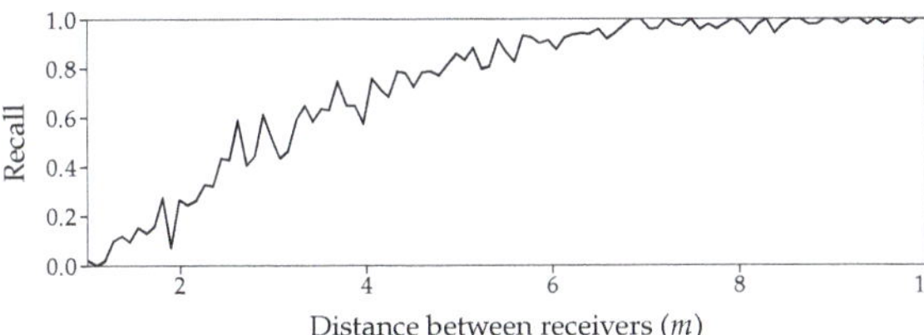

Figure 9. Performance of the *PDM* method in terms of recall as a function of distance between two receivers. Note that the *min distance* threshold is again always set to half the true distance between receivers (x-axis) and that distance measurements are smoothed with a moving average (with $\alpha = 0.1$).

8. Discussion and Outlook

Simulating reality in all its details is generally too complex and not target-oriented. Therefore, our simulator comes with some reasonable abstractions, in the context of which the *CNM* method has also been excluded from the evaluation as explained in Section 5.2. Still, an investigation of *CNM*'s detection capabilities, especially on advanced signal take-over strategies described by Tippenhauer et al. [7], would be desirable and is left for future work. Other random processes are simplified in the simulation for the sake of abstraction. A trace-based approach was used to model the GPS error. Further modeling of random probability distributions was chosen and parametrized based on field experiments (cf. Section 6).

Our simulation equips each attacker with only a single antenna, although advanced simulator attackers A_S can be in possession of multiple antennas (cf. Figure 2). The effectiveness of *PDM* weakens when attackers use multiple antennas. However, this could be simply counteracted by adding more receivers to the victim to further reduce the dimensional freedom of the attacker, as proposed in [7]. As a rule of thumb, *PDM* requires always one more receiver for detection than antennas available to the attacker.

Overall, the methods in our framework MANA showed promising results. While the *PDM* method, in particular, covers the entire attack space, *PCC* methods have the potential to further improve the effectiveness of MANA's detection capabilities in many scenarios. In this context, the ensemble of individual methods plays an important role. Although our simplified approach is sufficient for many scenarios, there is a limitation: All methods are weighted equally, requiring strict thresholds to avoid false positives. Therefore, many insights of relatively weak methods are discarded, decreasing the overall performance. A possible solution to this problem may be to support the ensemble with neural networks to predict a single spoofing indicator. This approach would even allow temporal information to be included in the detection process. Such a network could be trained with our dataset and later refined with short network recordings of the actual vessel in which the detector will be installed, eliminating the need to manually set thresholds.

Nevertheless, it should be noted that there will never be a complete guarantee for the detection of spoofing attacks. Therefore, for the purpose of countering CEMA-based GNSS attacks, it is necessary for practice to develop complementary, GNSS-independent localization systems such as the R-Mode, described in Section 4.2, and to integrate them into the localization process for reliable situational awareness.

In view of the current autonomy trend and the first unmanned vessels now reaching practical operation, the need becomes more urgent. For manned ships, GNSS spoofing can be detected and even mitigated to some degree by using other navigation technologies such as radar or visual aids to verify the vessel's position. If crew members suspect that GNSS signals are being falsified, they can also manually adjust the vessel's course and inform other vessels or port authorities of any navigational problems. In contrast, for unmanned vessels, the effects of spoofing can be much more severe, as these vessels tend to rely almost entirely on GNSS for navigation and communications and generally depend more heavily on automated systems. However, they may not yet have the ability to verify their position and accordingly adjust their course. Without crew members on board, it may also take longer to detect and respond to spoofing attacks. Therefore, the risk of collisions, groundings, or other accidents is much higher.

9. Conclusions

The fact that GPS signals are weak when they reach the Earth's surface makes it easy for adversaries to attack navigation systems, which is particularly problematic in the maritime domain, which is heavily reliant on GNSS. In this context, the paper provides a brief overview of research on detecting and mitigating GPS spoofing attacks at different layers. Since a simple and cost-effective retrofit is crucial for maritime practice, we focused on software-based methods. We proposed MANA, a novel framework comprising a selection of methods that continuously monitors and analyzes information derived from GPS, delivered as NMEA sentences via the maritime network.

Through a maritime simulator that uses real-world data to model realistic behavior, a comprehensive labeled dataset was generated, including legitimate and spoofed samples. This dataset provides not only the basis for our evaluation but can also be used for future benchmarking. In our evaluation, we compared the effectiveness of MANA's individual methods and show that it greatly differs. Pairwise Distance Monitoring (*PDM*) that requires multiple receivers was identified to be the most promising method in our software-based approach and is particularly applicable to maritime systems. We show that *PDM* achieves reliable detection for a wide range of common GPS spoofing attacks while still leaving room for improvements. As our evaluation suggests, alternative methods could support

PDM and thereby compensate for remaining deficiencies in many cases. In our future work, we thus plan to extend our evaluation and investigate the potential of a smart ensemble of individual detection methods of MANA.

Author Contributions: Conceptualization, J.S., C.H. and J.B.; methodology, J.S., C.H. and J.B.; software, J.S. and C.H.; validation, J.S., C.H. and J.B.; investigation, J.S., C.H., M.v.R. and J.B.; writing—original draft preparation, J.S., C.H. and J.B.; writing—review and editing, J.S., M.v.R. and J.B.; visualization, J.S., C.H., M.v.R. and J.B.; supervision, J.B. and E.P.; All authors have read and agreed to the published version of the manuscript.

Funding: This work is part of the project MUM2 (https://www.mum-project.com, accessed on 23 April 2023). It was partially funded by the German Federal Ministry of Economic Affairs and Climate Action (BMWK) within the "Maritime Research Programme" with contract number 03SX543B managed by the Project Management Jülich (PTJ). The authors are responsible for the contents of this publication.

Institutional Review Board Statement: Not applicable.

Informed Consent Statement: Not applicable.

Data Availability Statement: The MARSIM dataset introduced in Section 6.2 (pcap and csv files) and the source code of MANA are available at: https://github.com/fkie-cad/mana, accessed on 23 April 2023. Moreover, a Wireshark dissector for maritime protocols developed by our research group is available as open source at: https://github.com/fkie-cad/maritime-dissector, accessed on 23 April 2023. This dissector can be used to interpret and analyze maritime network traffic, especially the NMEA sentences within the recorded pcap files of MARSIM.

Acknowledgments: The authors thank Konrad Wolsing for supporting the field experiments.

Conflicts of Interest: The authors declare no conflict of interest.

Abbreviations

The following abbreviations are used in this manuscript:

AIS	automatic identification system
APT	auxiliary peak tracking
ARPA	automatic radar plotting aid
C/A	coarse/acquisition code
CDM	Clock Drift Monitoring
CEMA	cyber and electromagnetic activities
C/N_0	carrier-to-noise density
CNM	C/N_0 Monitoring
DOA	direction of arrival
ECDIS	electronic chart display and information system
EDV	Ephemeris Data Validation
GNSS	global navigation satellite system
GPS	Global Positioning System
IEC	International Electrotechnical Commission
IBS	integrated bridge system
IMO	International Maritime Organization
LWE	Lightweight Ethernet
MANA	MAritime Nmea-based Anomaly detection
MARSIM	MARitime SIMulated
MCS	Maritime Cyber Security
NAVSTAR	Navigational Satellite Timing and Ranging
NMEA	National Marine Electronics Association
PCC	Physical Cross-Check
PDM	Pairwise Distance Monitoring
PLMN	Public Land Mobile Network

PNT	positioning, navigation, and timing
PRN	pseudo-random noise
RAIM	Receiver Autonomous Integrity Monitoring
ROT	rate of turn
SOG	speed over ground
TLE	two-line element
TOA	time of arrival
UAV	unmanned aerial vehicle
VDES	Very high frequency Data Exchange System

References

1. Tam, K.; Jones, K. Factors Affecting Cyber Risk in Maritime. In Proceedings of the International Conference on Cyber Situational Awareness, Data Analytics and Assessment (Cyber SA), Oxford, UK, 3–4 June 2019; pp. 1–8. [CrossRef]
2. Androjna, A.; Perkovic, M. Impact of Spoofing of Navigation Systems on Maritime Situational Awareness. *Trans. Marit. Sci.* **2021**, *10*, 361–373. [CrossRef]
3. Bhatti, J.; Humphreys, T.E. Hostile Control of Ships Via False GPS Signals: Demonstration and Detection. *J. Inst. Navig.* **2017**, *64*, 51–66. [CrossRef]
4. Motallebighomi, M.; Sathaye, H.; Singh, M.; Ranganathan, A. Cryptography Is Not Enough: Relay Attacks on Authenticated GNSS Signals. *arXiv* **2022**, arXiv:2204.11641v3. [CrossRef]
5. Jafarnia-Jahromi, A.; Broumandan, A.; Nielsen, J.; Lachapelle, G. GPS Vulnerability to Spoofing Threats and a Review of Antispoofing Techniques. *Int. J. Navig. Obs.* **2012**, *2012*, 127072. [CrossRef]
6. Jansen, K.; Tippenhauer, N.O.; Pöpper, C. Multi-Receiver GPS Spoofing Detection: Error Models and Realization. In Proceedings of the Conference on Computer Security Applications (ACSAC), Los Angeles, CA, USA, 5–8 December 2016; pp. 237–250. [CrossRef]
7. Tippenhauer, N.O.; Pöpper, C.; Rasmussen, K.B.; Capkun, S. On the Requirements for Successful GPS Spoofing Attacks. In Proceedings of the International Conference on Computer and Communications Security (CCS), Chicago, IL, USA, 17–21 October 2011; pp. 75–86. [CrossRef]
8. Psiaki, M.L.; Humphreys, T.E. GNSS Spoofing and Detection. *Proc. IEEE* **2016**, *104*, 1258–1270. [CrossRef]
9. Meng, L.; Yang, L.; Yang, W.; Zhang, L. A Survey of GNSS Spoofing and Anti-Spoofing Technology. *Remote Sens.* **2022**, *14*, 4826. [CrossRef]
10. Januszewski, J. Shipborne satellite navigation systems receivers, exploitation remarks. *Sci. J. Marit. Universiy Szczec.* **2014**, *40*, 67–72.
11. Lund, M.S.; Gulland, J.E.; Hareide, O.S.; Jøsok, Ø.; Weum, K.O.C. Integrity of Integrated Navigation Systems. In Proceedings of the Conference on Communications and Network Security (CNS), Beijing, China, 30 May–1 June 2018; pp. 1–5. [CrossRef]
12. Hemminghaus, C.; Bauer, J.; Wolsing, K. SIGMAR: Ensuring Integrity and Authenticity of Maritime Systems using Digital Signatures. In Proceedings of the International Symposium on Networks, Computers and Communications (ISNCC), Dubai, United Arab Emirates, 31 October–2 November 2021. [CrossRef]
13. Rødseth, Ø.J.; Christensen, M.J.; Lee, K. Design challenges and decisions for a new ship data network. In Proceedings of the International Symposium on International Symposium Information on Ships (ISIS), Berlin, Germany, 27–28 September 2011; pp. 149–168.
14. BBC News. Suez Blockage Is Holding Up $9.6 bn of Goods a Day. 2021. Available online: https://www.bbc.com/news/business-56533250 (accessed on 17 March 2023).
15. Androjna, A.; Brcko, T.; Pavic, I.; Greidanus, H. Assessing Cyber Challenges of Maritime Navigation. *J. Mar. Sci. Eng.* **2020**, *8*, 776. [CrossRef]
16. IEC. *Maritime Navigation and Radiocommunication Equipment and Systems–Digital Interfaces–Part 450: Multiple Talkers and Multiple Listeners–Ethernet Interconnection (IEC 61162-450:2018)*; International Electrotechnical Commission (IEC): Geneva, Switzerland, 2018.
17. Warner, J.S.; Johnston, R.G. *GPS Spoofing Countermeasures*; Technical Report LAUR-03-6163; Vulnerability Assessment Team, Los Alamos National Laboratory: Los Alamos, NM, USA, 2003.
18. Ranganathan, A.; Ólafsdóttir, H.; Capkun, S. SPREE: A Spoofing Resistant GPS Receiver. In Proceedings of the Conference on Mobile Computing and Networking (MobiCom), New York, NY, USA, 3–7 October 2016; pp. 348–360. [CrossRef]
19. Medina, D.; Lass, C.; Marcos, E.P.; Ziebold, R.; Closas, P.; García, J. On GNSS Jamming Threat from the Maritime Navigation Perspective. In Proceedings of the 22th International Conference on Information Fusion (FUSION), Ottawa, ON, Canada, 2–5 July 2019; pp. 1–7. [CrossRef]
20. Caprolu, M.; Pietro, R.D.; Raponi, S.; Sciancalepore, S.; Tedeschi, P. Vessels Cybersecurity: Issues, Challenges, and the Road Ahead. *IEEE Commun. Mag.* **2020**, *58*, 90–96. [CrossRef]
21. Awan, M.S.K.; Al Ghamdi, M.A. Understanding the Vulnerabilities in Digital Components of An Integrated Bridge System (IBS). *J. Mar. Sci. Eng.* **2019**, *7*, 350. [CrossRef]

22. Burgess, M. When a Tanker Vanishes, All the Evidence Points to Russia. 2017. Available online: https://www.wired.co.uk/article/black-sea-ship-hacking-russia (accessed on 17 March 2023).

23. Wu, J.; Thorne-Large, J.; Zhang, P. Safety First: The Risk of Over-Reliance on Technology in Navigation. *J. Transp. Saf. Secur.* **2021**, *14*, 1220–1246. [CrossRef]

24. Maritime Safety Committee (MSC). *Resolution MSC.915(22)–Revised Maritime Policy and Requirements for a Future Global Navigation Satellite System (GNSS)*; MSC 915(22); International Maritime Organization (IMO): London, UK, 2001.

25. Zalewski, P. GNSS Integrity Concepts for Maritime Users. In Proceedings of the European Navigation Conference (ENC), Warsaw, Poland, 9–12 April 2019; pp. 1–10. [CrossRef]

26. Department of Homeland Security. Resilient Positioning, Navigation, and Timing (PNT) Conformance Framework Version 2.0. 2022. Available online: https://www.dhs.gov/sites/default/files/2022-05/22_0531_st_resilient_pnt_conformance_framework_v2.0.pdf (accessed on 20 April 2023).

27. Warner, J.S.; Johnston, R.G. A simple demonstration that the global positioning system (GPS) is vulnerable to spoofing. *J. Secur. Adm.* **2002**, *25*, 19–27.

28. Coulon, M.; Chabory, A.; Garcia-Pena, A.; Vezinet, J.; Macabiau, C.; Estival, P.; Ladoux, P.; Roturier, B. Characterization of Meaconing and its Impact on GNSS Receivers. In Proceedings of the Technical Meeting of The Satellite Division of the Institute of Navigation (ION GNSS+), Online, 21–25 September 2020; pp. 3713–3737. [CrossRef]

29. Lenhart, M.; Spanghero, M.; Papadimitratos, P. DEMO: Relay/replay attacks on GNSS signals. *arXiv* **2022**, arXiv:2202.10897. [CrossRef]

30. Key, E.L. *Techniques to Counter GPS Spoofing. Internal Memorandum*; MITRE Corporation: Mclean, VA, USA, 1995.

31. Brown, R.G. Receiver Autonomous Integrity Monitoring. In *Global Positioning System: Theory and Applications*; Parkinson, B., Spilker, J., Eds.; AIAA Inc.: Washington, DC, USA, 1996; Volume 2, Chapter 5, pp. 143–165.

32. Scott, L. Anti-Spoofing & Authenticated Signal Architectures for Civil Navigation Systems. In Proceedings of the Technical Meeting of the Satellite Division of The Institute of Navigation (ION GPS/GNSS 2003), Portland, OR, USA, 9–12 September 2003; pp. 1543–1552.

33. Kerns, A.J.; Wesson, K.D.; Humphreys, T.E. A blueprint for civil GPS navigation message authentication. In Proceedings of the IEEE/ION Position, Location and Navigation Symposium (PLANS), Monterey, CA, USA, 5–8 May 2014; pp. 262–269. [CrossRef]

34. Wen, H.; Huang, P.Y.R.; Dyer, J.; Archinal, A.; Fagan, J. Countermeasures for GPS signal spoofing. In Proceedings of the Technical Meeting of the Satellite Division of The Institute of Navigation (ION GNSS), Long Beach, CA, USA, 13–16 September 2005; Volume 5, pp. 13–16.

35. Akos, D.M. Who's Afraid of the Spoofer? GPS/GNSS Spoofing Detection via Automatic Gain Control (AGC). *Navigation* **2012**, *59*, 281–290. [CrossRef]

36. Konovaltsev, A.; Cuntz, M.; Haettich, C.; Meurer, M. Autonomous spoofing detection and mitigation in a GNSS receiver with an adaptive antenna array. In Proceedings of the Technical Meeting of The Satellite Division of the Institute of Navigation (ION GNSS+), Nashville, TN, USA, 16–20 September 2013; pp. 2937–2948.

37. McDowell, C.E. GPS Spoofer and Repeater Mitigation System Using Digital Spatial Nulling. U.S. Patent 7,250,903, 31 July 2007.

38. Nielsen, J.; Broumandan, A.; Lachapelle, G. Spoofing Detection and Mitigation with a Moving Handheld Receiver. *GPS World* **2010**, *21*, 27–33.

39. Jafarnia-Jahromi, A.; Broumandan, A.; Nielsen, J.; Lachapelle, G. GPS spoofer countermeasure effectiveness based on signal strength, noise power, and C/N_0 measurements. *Int. J. Satell. Commun. Netw.* **2012**, *30*, 181–191. [CrossRef]

40. Li, J.; Zhu, X.; Ouyang, M.; Shen, D.; Chen, Z.; Dai, Z. GNSS spoofing detection technology based on Doppler frequency shift difference correlation. *Meas. Sci. Technol.* **2022**, *33*, 095109. [CrossRef]

41. Chu, F.; Li, H.; Wen, J.; Lu, M. Statistical Model and Performance Evaluation of a GNSS Spoofing Detection Method based on the Consistency of Doppler and Pseudorange Positioning Results. *J. Navig.* **2019**, *72*, 447–466. [CrossRef]

42. Zeng, Q.; Li, H.; Qian, L. GPS spoofing attack on time synchronization in wireless networks and detection scheme design. In Proceedings of the Military Communications Conference (MILCOM), Orlando, FL, USA, 29 October–1 November 2012; pp. 1–5. [CrossRef]

43. Humphreys, T.E.; Ledvina, B.M.; Psiaki, M.L.; O'Hanlon, B.W.; Kintner, P.M. Assessing the Spoofing Threat: Development of a Portable GPS Civilian Spoofer. In Proceedings of the Technical Meeting of the Satellite Division of The Institute of Navigation (ION GNSS), Savannah, GA, USA, 16–19 September 2008; pp. 2314–2325.

44. Lo, S.C.; Enge, P.K. Authenticating aviation augmentation system broadcasts. In Proceedings of the IEEE/ION Position, Location and Navigation Symposium (PLANS), Indian Wells, CA, USA, 4–6 May 2010; pp. 708–717. [CrossRef]

45. Lo, S.; De Lorenzo, D.; Enge, P.; Akos, D.; Bradley, P. Signal authentication: A secure civil GNSS for today. *Inside GNSS* **2009**, *4*, 30–39.

46. Phelts, R.E. Multicorrelator Techniques for Robust Mitigation of Threats to GPS Signal Quality. Ph.D. Theses, Stanford University, Stanford, CA, USA, 2000.

47. Pini, M.; Fantino, M.; Cavaleri, A.; Ugazio, S.; Presti, L.L. Signal Quality Monitoring Applied to Spoofing Detection. In Proceedings of the Technical Meeting of The Satellite Division of the Institute of Navigation (ION GNSS), Portland, OR, USA, 20–23 September 2011; pp. 1888–1896.

48. Miralles, D.; Bornot, A.; Rouquette, P.; Levigne, N.; Akos, D.M.; Chen, Y.H.; Lo, S.; Walter, T. An Assessment of GPS Spoofing Detection Via Radio Power and Signal Quality Monitoring for Aviation Safety Operations. *IEEE Intell. Transp. Syst. Mag.* **2020**, *12*, 136–146. [CrossRef]

49. White, N.A.; Maybeck, P.S.; DeVilbiss, S.L. Detection of interference/jamming and spoofing in a DGPS-aided inertial system. *IEEE Trans. Aerosp. Electron. Syst.* **1998**, *34*, 1208–1217. [CrossRef]

50. Wei, X.; Aman, M.N.; Sikdar, B. Light-Weight GPS Spoofing Detection for Synchrophasors in Smart Grids. In Proceedings of the International Conference on Power Electronics, Drives and Energy Systems (PEDES), Jaipur, India, 16–19 December 2020; pp. 1–4. [CrossRef]

51. Psiaki, M.L.; O'hanlon, B.W.; Powell, S.P.; Bhatti, J.A.; Wesson, K.D.; Humphreys, T.E. GNSS Spoofing Detection using Two-Antenna Differential Carrier Phase. In Proceedings of the Technical Meeting of the Satellite Division of The Institute of Navigation (ION GNSS+), Tampa, FL, USA, 8–12 September 2014.

52. Dasgupta, S.; Rahman, M.; Islam, M.; Chowdhury, M. A Sensor Fusion-Based GNSS Spoofing Attack Detection Framework for Autonomous Vehicles. *IEEE Trans. Intell. Transp. Syst.* **2022**, *23*, 23559–23572. [CrossRef]

53. Lemieszewski, Ł. Transport safety: GNSS spoofing detection using the single-antenna receiver and the speedometer of a vehicle. *Procedia Comput. Sci.* **2022**, *207*, 3181–3188. [CrossRef]

54. Oligeri, G.; Sciancalepore, S.; Ibrahim, O.A.; Di Pietro, R. GPS spoofing detection via crowd-sourced information for connected vehicles. *Comput. Netw.* **2022**, *216*, 109230. [CrossRef]

55. Nighswander, T.; Ledvina, B.; Diamond, J.; Brumley, R.; Brumley, D. GPS Software Attacks. In Proceedings of the International Conference on Computer and Communications Security (CCS), Raleigh, NC, USA, 16–18 October 2012; pp. 450–461. [CrossRef]

56. Zalewski, P. Real-time GNSS spoofing detection in maritime code receivers. *Sci. J. Marit. Univ. Szczec.* **2014**, *38*, 118–124.

57. Laverty, D.; Kelsey, C.; O'Raw, J. GNSS Time Signal Spoofing Detector for Electrical Substations. In Proceedings of the IEEE Power & Energy Society General Meeting (PESGM), Denver, CO, USA, 17–21 July 2022. [CrossRef]

58. Whelan, J.; Sangarapillai, T.; Minawi, O.; Almehmadi, A.; El-Khatib, K. Novelty-based Intrusion Detection of Sensor Attacks on Unmanned Aerial Vehicles. In Proceedings of the ACM Symposium on QoS and Security for Wireless and Mobile Networks (Q2SWinet), Alicante, Spain, 16–20 November 2020; pp. 23–28. [CrossRef]

59. Whelan, J.; Sangarapillai, T.; Minawi, O.; Almehmadi, A.; El-Khatib, K. UAV Attack Dataset. 2020. Available online: https://ieee-dataport.org/open-access/uav-attack-dataset (accessed on 23 April 2023).

60. Singh, S.; Singh, J.; Singh, S.; Goyal, S.B.; Raboaca, M.S.; Verma, C.; Suciu, G. Detection and Mitigation of GNSS Spoofing Attacks in Maritime Environments Using a Genetic Algorithm. *Mathematics* **2022**, *10*, 4097. [CrossRef]

61. Lee, D.K.; Miralles, D.; Akos, D.; Konovaltsev, A.; Kurz, L.; Lo, S.; Nedelkov, F. Detection of GNSS Spoofing using NMEA Messages. In Proceedings of the European Navigation Conference (ENC), Dresden, Germany, 23–24 November 2020; pp. 1–10. [CrossRef]

62. Oligeri, G.; Sciancalepore, S.; Ibrahim, O.A.; Di Pietro, R. Drive Me Not: GPS Spoofing Detection via Cellular Network: (Architectures, Models, and Experiments). In Proceedings of the Conference on Security and Privacy in Wireless and Mobile Networks (WiSec), Miami, FL, USA, 15–17 May 2019; pp. 12–22. [CrossRef]

63. Johnson, G.; Swaszek, P.; Alberding, J.; Hoppe, M.; Oltmann, J.H. The Feasibility of R-Mode to Meet Resilient PNT Requirements for e-Navigation. In Proceedings of the Technical Meeting of the Satellite Division of The Institute of Navigation (ION GNSS+), Tampa, FL, USA, 8–12 September 2014; pp. 3076–3100.

64. Koch, P.; Gewies, S. Worldwide Availability of Maritime Medium-Frequency Radio Infrastructure for R-Mode-Supported Navigation. *J. Mar. Sci. Eng.* **2020**, *8*, 209. [CrossRef]

65. Grundhöfer, L.; Rizzi, F.G.; Gewies, S.; Hoppe, M.; Bäckstedt, J.; Dziewicki, M.; Galdo, G.D. Positioning with medium frequency R-Mode. *NAVIGATION J. Inst. Navig.* **2021**, *68*, 829–841. [CrossRef]

66. Naus, K.; Waż, M.; Szymak, P.; Gucma, L.; Gucma, M. Assessment of ship position estimation accuracy based on radar navigation mark echoes identified in an Electronic Navigational Chart. *Measurement* **2021**, *169*, 108630. [CrossRef]

67. Hagen, O.K.; Ånonsen, K.B. Using Terrain Navigation to Improve Marine Vessel Navigation Systems. *Mar. Technol. Soc. J.* **2014**, *48*, 45–58. [CrossRef]

68. Kelso, T.S. Frequently Asked Questions: Two-Line Element Set Format. CelesTrak, Satellite Times. 2004. Available online: http://celestrak.com/columns/v04n03/ (accessed on 17 March 2023).

69. Oszczak, B. GNSS positioning algorithms using methods of reference point indicators. *Artif. Satell.* **2014**, *49*, 21–23. [CrossRef]

Journal of
Marine Science and Engineering

MDPI

Article

Evaluating the Vulnerability of YOLOv5 to Adversarial Attacks for Enhanced Cybersecurity in MASS

Changui Lee [1] and Seojeong Lee [2,*]

[1] Korea Conformity Laboratories, Changwon 51395, Republic of Korea; phdculee@gmail.com
[2] Division of Marine System Engineering, Korea Maritime and Ocean University, Busan 49112, Republic of Korea
* Correspondence: sjlee@kmou.ac.kr; Tel.: +82-51-410-4578

Abstract: The development of artificial intelligence (AI) technologies, such as machine learning algorithms, computer vision systems, and sensors, has allowed maritime autonomous surface ships (MASS) to navigate, detect and avoid obstacles, and make real-time decisions based on their environment. Despite the benefits of AI in MASS, its potential security threats must be considered. An adversarial attack is a security threat that involves manipulating the training data of a model to compromise its accuracy and reliability. This study focuses on security threats faced by a deep neural network-based object classification algorithm, particularly you only look once version 5 (YOLOv5), which is a model used for object classification. We performed transfer learning on YOLOv5 and tested various adversarial attack methods. We conducted experiments using four types of adversarial attack methods and parameter changes to determine the attacks that could be detrimental to YOLOv5. Through this study, we aim to raise awareness of the vulnerability of AI algorithms for object detection to adversarial attacks and emphasize the need for efforts to overcome them; these efforts can contribute to safe navigation in MASS.

Keywords: adversarial attack; perturbed image; YOLO; object classification; MASS

Citation: Lee, C.; Lee, S. Evaluating the Vulnerability of YOLOv5 to Adversarial Attacks for Enhanced Cybersecurity in MASS. *J. Mar. Sci. Eng.* **2023**, *11*, 947. https://doi.org/10.3390/jmse11050947

Academic Editors: Marko Perkovic, Lucjan Gucma and Sebastian Feuerstack

Received: 30 March 2023
Revised: 26 April 2023
Accepted: 26 April 2023
Published: 28 April 2023

1. Introduction

The impressive advancements made in object detection and classification algorithms that use deep neural networks (DNNs) have significantly affected numerous industries, including the maritime industry, particularly with the development of maritime autonomous surface ships (MASS). By incorporating artificial intelligence (AI) technology, object detection and classification algorithms can detect obstacles in real time, assist in human decision-making during ship navigation [1], and ultimately enable autonomous navigation. However, despite the advancements in object detection and classification algorithms, these systems present vulnerabilities [2]. Adversarial attacks represent one of the most significant security threats to AI systems. The attacks occur during the training phase of deep neural-network algorithms using a training dataset. An attacker adds small and carefully crafted perturbations to the input data that are difficult to detect by humans [3]. Consequently, the model trained with the perturbed data will have significantly degraded performance in its output.

The introduction of MASS has resulted in a new type of threat to the maritime industry in the form of AI-related threats, such as adversarial attacks. As these attacks can compromise the safe operation of MASS, awareness of, and interest in, these threats are crucial for maritime industry stakeholders. In this study, we aimed to emphasize the vulnerability of object detection and classification algorithms to adversarial attacks and the importance of developing strategies to overcome them. To achieve this, we created perturbed images by modifying the settings of four different types of adversarial attack methods. These images were thereafter used as training data for you only look once version 5 (YOLOv5), an object detection and classification algorithm, to simulate adversarial attacks. By evaluating the accuracy of the model trained on these perturbed images, we confirmed the potential catastrophic impacts of adversarial attacks on object detection and classification algorithms. Ultimately, our goal is to increase awareness of the vulnerability of AI algorithms for object detection to adversarial attacks and develop strategies to mitigate their effects, thereby contributing to safe navigation in MASS.

2. Background

2.1. DNN Algorithms for Object Classification

Object detection and classification are crucial tasks in computer vision that have seen remarkable progress recently due to advancements in deep learning algorithms. Several popular object detection algorithms have emerged, each with unique strengths and limitations [4]. Recent object detection and classification algorithms, such as region-based convolutional neural networks (R-CNN), mask R-CNN [5], you only look once (YOLO) [6], and single shot detector (SSD) [7], have numerous characteristics in common. Accuracy is among the most significant characteristics of any object detection or classification algorithm, and recent algorithms aim for high accuracy in detecting and localizing objects within an image or video stream. Speed and efficiency are crucial features, as many applications require the real-time processing of considerable data. Algorithms such as YOLOv5 and EfficientDet [8] have demonstrated high processing speeds and resource efficiency, without compromising accuracy. Moreover, mask R-CNN and YOLOv5 use instance segmentation, thus providing visualizations of detected objects with bounding boxes and confidence scores. These can be used to manage occluded items and simplify the interpretation of the output.

YOLOv5 is a powerful and versatile object detection algorithm that has garnered significant attention recently, due to its high accuracy, real-time processing speed, and computational efficiency [6,9]. The model is designed to utilize resources efficiently, and hence is suitable for real-time processing on devices with limited computational resources. It can manage object types, sizes, and orientations, and uses transfer learning to improve generalization to new datasets. YOLOv5 can also visualize detected objects with bounding boxes and confidence scores, making it easy to interpret and understand its output. Among the studies using YOLOv5, Nader Al-Qubaydhi et al. proposed a method [10] for detecting unauthorized unmanned aerial vehicles (UAVs) using YOLOv5 and transfer learning. Their study employed transfer learning to adapt the YOLOv5 framework to the Kaggle drone dataset. They fine-tuned the last three YOLOv5 and convolutional layers to match the number of classes in the dataset and introduced data augmentation techniques to enhance the dataset and improve training. The trained model was evaluated by constructing a number-of-iterations-versus-mAP curve at different points, and the results demonstrated high accuracy in detecting unauthorized UAVs.

2.2. AI-Specific Security Threats

AI systems are becoming increasingly popular across industries for their ability to automate decision-making, improve efficiency, and drive innovation. However, because of increased Internet connectivity and usage, these systems have recently become more vulnerable to cybersecurity threats [11,12].

AI-specific security refers to the unique security risks and challenges that AI systems encounter, and the specific security measures that organizations should implement to protect their systems against such threats. The ISO/IEC TR 24028 technical report [12] provides guidelines for managing the security and privacy of AI systems, thus promoting their safe and responsible use. The report highlights several AI-specific security threats, including adversarial attacks, data poisoning, model stealing, and evasion, privacy, integrity, and denial of service attacks.

Adversarial attacks involve intentionally modifying the input data of a machine-learning model to cause the model to make incorrect predictions. These attacks may vary according to the target machine-learning application, including image classification, speech recognition, and autonomous vehicles. Data poisoning involves deliberately inputting misleading or inaccurate data into an AI system with to manipulate its output. Attackers may introduce inaccurate or biased training data into an AI system to influence its decision-making. Data poisoning can result in AI systems that make biased or discriminatory decisions. Model stealing is a technique by which an attacker attempts to acquire an AI model by analyzing its output and using reverse engineering techniques. This aspect can enable the attacker to replicate the model and use it for malicious purposes, such as creating deepfake images or videos. Evasion attacks are techniques used to evade detection by security mechanisms, such as malware or intrusion detection systems. Privacy attacks compromise the privacy of individuals and organizations by exploiting vulnerabilities in AI systems. Attackers may use these vulnerabilities to access sensitive information or to execute espionage and other malicious activities. Integrity attacks compromise data integrity or AI models by modifying or deleting data. These attacks can result in AI systems that produce inaccurate or unreliable results. Denial of service attacks disrupt the availability of AI systems by overwhelming them with requests or other forms of traffic. These attacks can result in significant disruptions to AI systems and financial losses.

Among security threats, adversarial attacks are particularly significant due to their potential to compromise the accuracy and reliability of AI systems [13]. These attacks are intentional modifications of input data designed to compromise the accuracy and reliability of the output of the AI system. The primary objective of an adversarial attack is to add the minimal perturbation to the input data that can result in the desired misclassification. These attacks pose a severe security threat to critical systems such as those used for medical diagnosis or in autonomous vehicles [14,15]. Adversarial attacks can be categorized into different types based on the purpose of the attack and prior knowledge of the attacker. For instance, based on our assumptions on the knowledge of the attacker, these can be classified as white- or black-box attacks. In a white-box attack, the attacker has complete knowledge and access to the model, including architecture, inputs, outputs, and weights [14]. In contrast, in a black-box attack, the attacker only has access to the inputs and outputs of the model and is unaware of the underlying architecture or weights [16].

The basic principle of an adversarial attack is to generate perturbed data by synthesizing specific noise, indistinguishable from a conventional image [4,14], as depicted in Figure 1. When these perturbed data are input into a trained learning model, they appear as a ship to the human eye; however, the deep learning model will classify them as a lighthouse. Therefore, the purpose of an adversarial attack is to lower the accuracy of the model by causing misclassification.

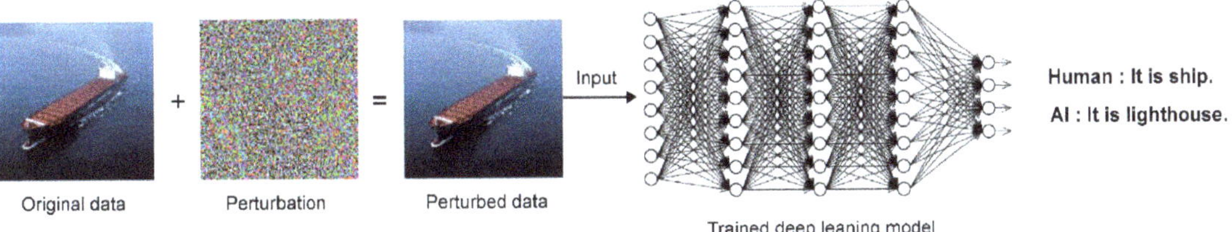

Figure 1. Principle of an adversarial attack.

2.3. Methods for Generating Adversarial Examples

An adversarial example indicates a perturbed data input specifically designed to induce inaccurate classifications by DNN-based algorithms. Adversarial attacks are a critical issue in machine learning security, and numerous methods are available for generating adversarial examples.

Goodfellow et al. proposed the fast gradient sign method (FGSM) [13]. This is a simple and efficient method in which the gradient of the input data loss function and a small amount of noise in the direction of the gradient are summed to generate perturbed inputs that can be easily misclassified by the AI model. Even though the FGSM is easy to implement, it may not produce the most robust adversarial examples.

Kurakin et al. introduced a powerful variant of the FGSM known as the iterative fast gradient sign method (I-FGSM) [17]. This method applies the attack iteratively with a smaller step size, resulting in more robust adversarial examples. However, the increased complexity and number of iterations may make this method more computationally expensive than the FGSM.

Dong et al. proposed the momentum iterative fast gradient sign method (MI-FGSM) [18] as a further extension of the I-FGSM in which a momentum term was added to the perturbation update rule. This aspect can prevent oscillations and accelerate convergence. However, the increased complexity and number of hyperparameters can make the tuning process more challenging compared with the I-FGSM.

Madry et al. proposed the projected gradient descent (PGD) [19] algorithm as a variant of the I-FGSM in which random perturbations are added to the input at each iteration. This modification can prevent the attack from stabilizing around a local minimum, thus improving the generalization of the adversarial examples. However, as in the previous method, the tuning of the PGD requires more computational resources compared with the I-FGSM.

2.3.1. Fast Gradient Sign Method

The FGSM [13] algorithm is a simple and effective method for generating adversarial examples in deep learning models. By adding a minimal perturbation to the input data in the direction of the gradient of the loss function of the input data, the algorithm can cause the model to misclassify the data, even though the perturbed image may appear similar to the original one to a human observer [20]. The algorithm of the FGSM is summarized in Table 1.

Table 1. FGSM algorithm.

Input	An input image x, a target class y, and the size of perturbation ϵ.
Output	An adversarial example x', misclassified by the deep learning model, with a perturbation size satisfying $kx' - k\infty x \le \epsilon$, where k is a scalar and $k\infty$ is its L-infinity norm, and $\epsilon = kx' - k\infty x$.
Algorithm	1. Calculate the gradient of the loss function J for the input image x: gradient = $\nabla x\, J(\theta, x, y)$, where θ represents the parameters of the deep learning model.
	2. Calculate the perturbation by scaling the sign of the gradient with a small ϵ value: perturbation = sign(gradient) * ϵ, where sign() denotes the sign function, and * represents element-wise multiplication.
	3. Add the perturbation to the input image to obtain the adversaria example: $x' = x +$ perturbation.
	4. Clip the pixel values of the adversarial example to ensure that they remain within the valid range: $x' = $ clip$(x', 0, 1)$.
	5. Return the adversarial example x'.

2.3.2. Iterative FGSM

The I-FGSM [17] is an extension of the FGSM that generates adversarial examples by applying multiple iterations of the FGSM at a small step size α. The algorithm clips the perturbation at each iteration to ensure that its L∞ norm does not exceed the specified size ϵ. The I-FGSM can generate more effective adversarial examples than those generated by the FGSM, particularly when combined with other techniques such as momentum or randomization [20,21]. The algorithm of the I-FGSM is presented in Table 2.

Table 2. I-FGSM algorithm.

Input	An input image x, a target class y, the size of perturbation ϵ, the number of iterations T, and the step size α.
Output	An adversarial example x', misclassified by the deep learning model, with a perturbation size satisfying $kx' - k\infty x \leq \epsilon$, where k is a scalar and $k\infty$ is its L-infinity norm, and $\epsilon = kx' - k\infty x$.
Algorithm	1. Initialize the perturbation δ to zero.
	2. For t = 1 to T: a. Calculate the gradient of the loss function J for the input image x: gradient = $\nabla x\, J(\theta, x + \delta, y)$, where θ represents the parameters of the deep learning model. b. Add a scaled version of the sign of the gradient to the perturbation: $\delta \leftarrow \delta + \alpha\, \text{sign(gradient)}$. c. Clip the perturbation δ so that its $L\infty$ norm is at most ϵ: $\delta \leftarrow \text{clip}(\delta, -\epsilon, \epsilon)$. d. Update the adversarial example by adding the perturbation to the input image: $x' \leftarrow x + \delta$. e. Clip the pixel values of the adversarial example to ensure that they remain within the valid range: $x' \leftarrow \text{clip}(x', 0, 1)$.
	3. Return the adversarial example x'.

2.3.3. Momentum Iterative FGSM

The MI-FGSM [18] is an extension of the I-FGSM that generates adversarial examples by adding a momentum term to the update rule. This term prevents oscillations, resulting in a faster convergence to the adversarial example compared to the previous models. The momentum term is computed using the average of the previous gradients; it is scaled by a factor α to control its contribution to the update. The MI-FGSM algorithm can generate more effective adversarial examples than the I-FGSM, particularly when combined with other techniques such as randomization or ensemble methods [20,22]. The algorithm of the MI-FGSM is summarized in Table 3.

Table 3. MI-FGSM algorithm.

Input	An input image x, a target class y, the size of perturbation ϵ, the number of iterations T, and the decay factor μ.
Output	An adversarial example x', misclassified by the deep learning model, with a perturbation size satisfying $kx' - k\infty x \leq \epsilon$, where k is a scalar and $k\infty$ is its L-infinity norm, and $\epsilon = kx' - k\infty x$.
Algorithm	1. Initialize the perturbation δ to zero.
	2. For t = 1 to T: a. Calculate the gradient of the loss function J for the input image x: gradient = $\nabla x\, J(\theta, x + \delta, y)$, where θ represents the parameters of the deep learning model. b. Add a scaled version of the sign of the gradient to the perturbation: $\delta \leftarrow \mu\delta + (\epsilon/T)\, \text{sign(gradient)}$. c. Clip the perturbation δ so that its $L\infty$ norm is at most ϵ: $\delta \leftarrow \text{clip}(\delta, -\epsilon, \epsilon)$. d. Update the adversarial example by adding the perturbation to the input image: $x' \leftarrow x + \delta$. e. Clip the pixel values of the adversarial example to ensure that they remain within the valid range: $x' \leftarrow \text{clip}(x', 0, 1)$.
	3. Return the adversarial example x'.

2.3.4. Projected Gradient Descent

The PGD [19] algorithm is a variant of the I-FGSM that generates adversarial examples by adding random perturbations to the input at each iteration. These perturbations prevent the model from converging to local optima, thus resulting in more diverse adversarial examples. The PGD algorithm clips the perturbation at each iteration to ensure the $L\infty$ norm does not exceed the specified size ϵ. The randomness factor δ controls the magnitude of the random perturbations and can be adjusted to modify the exploration–exploitation balance. The PGD algorithm can generate more robust adversarial examples than those obtained using the I-FGSM, particularly when combined with other techniques such as ensemble or regularization methods [21,22]. The algorithm of the PGD is presented in Table 4.

Table 4. PGD algorithm.

Input	An input image x, a target class y, the size of perturbation ϵ, the number of iterations T, and the step size α.
Output	An adversarial example x', misclassified by the deep learning model, with a perturbation size satisfying $kx' - k\infty x \leq \epsilon$, where k is a scalar and $k\infty$ is its L-infinity norm, and $\epsilon = kx' - k\infty x$.
Algorithm	1. Initialize the perturbation δ to zero. 2. For t = 1 to T: a. Calculate the gradient of the loss function J concerning the input image x: gradient = $\nabla x\, J(\theta, x + \delta, y)$, where θ represents the parameters of the deep learning model. b. Add a scaled version of the gradient to the perturbation: $\delta \leftarrow \delta + \alpha\, \text{sign}(\text{gradient})$. c. Project the perturbation onto the $L\infty$ ball of radius ϵ: $\delta \leftarrow \text{clip}(\delta, -\epsilon, \epsilon)$. d. Update the adversarial example by adding the perturbation to the input image: $x' \leftarrow x + \delta$. e. Clip the pixel values of the adversarial example to ensure that they remain within the valid range: $x' \leftarrow \text{clip}(x', 0, 1)$. 3. Return the adversarial example x'.

3. Materials and Methods

3.1. Problem Setup

In this study, we aimed to simulate adversarial attacks that manipulate the training dataset to degrade the performance of object detection and classification models. A critical aspect of training AI for industrial use is securing suitable datasets. Although datasets for objects such as dogs and cats are common for object detection and classification, it is difficult to obtain datasets suitable for the maritime industry, such as boats, ferries, and buoys. Attackers take advantage of this and distribute perturbed images that are difficult for humans to detect but contain feature information that causes misclassification. At first glance, the perturbed image may appear to be slightly noisy, but this can be mistaken for optical noise. However, the perturbed image contains feature information that causes misclassification and cannot be removed by methods such as blurring, which are used to remove optical noise. Therefore, the system developer (victim) unknowingly uses the perturbed images as part of their training dataset because they appear normal to the developer. This means that an attack can occur without the system developer even noticing it. If the system developer attempts to remove noise from the training dataset, the feature information that causes misclassification will remain, thus compromising the model's performance. Moreover, because of the plateaued performance, the system developer may mistakenly assume that the model's performance is optimal and use it in real-world scenarios where failures to detect objects would cause misclassifications and precipitate serious accidents.

In AI, research into adversarial attack methods is ongoing, e.g., being investigated as a significant threat to using AI for diagnosing disease [3]. In the maritime industry, such as in the use of MASS, research is being conducted on systems that apply AI to detect and classify objects. However, in the maritime industry, the potential risks are not well-known because such adversarial attacks have not yet been experienced. To address this gap, we conducted experiments to determine the adversarial attack methods that could be most detrimental to object detection and classification models.

3.2. Experimental Scenario

In the experiments, we generated perturbed images using various adversarial attack methods and parameter settings and evaluated the accuracy of the models that were trained with these perturbed images. Thereby, we aimed to determine the most effective adversarial attack methods and raise awareness of their potential risks in the maritime industry.

The experimental scenario is divided into four phases, as depicted in Figure 2. In the first phase, the modified Singapore maritime dataset (SMD-Plus) proposed by Kim et al. [9] is pre-processed such that images and annotations are suitable for the YOLOv5 model. In the second phase, an attacker generates the perturbed images. This study assumed that the attacker could generate perturbed images using various methods. Therefore, six pre-trained DNN algorithms and four adversarial attack methods were used to generate the perturbed dataset using Python's PyTorch open-source framework. In the third phase, the system developer collects data for model training and trains the YOLOv5 model [23,24]. During this phase, the attacker deploys a perturbed dataset, and the system developer collects and verifies the data. The slight noise in the perturbed image is considered to be optical noise and passes the verification process. This assumption is fundamental to adversarial attacks [12,13,17–19], and this is the reason for the development of new algorithms to render it more difficult for detection by humans. Therefore, in this scenario, the perturbed image is assumed to pass the verification process without any issues and is included in the training dataset. The adversarial attack occurs during this phase without the explicit intervention of the attacker. In the fourth phase, the trained model is tested on conventional benchmark and perturbed datasets to examine the impact of adversarial attacks [23–27].

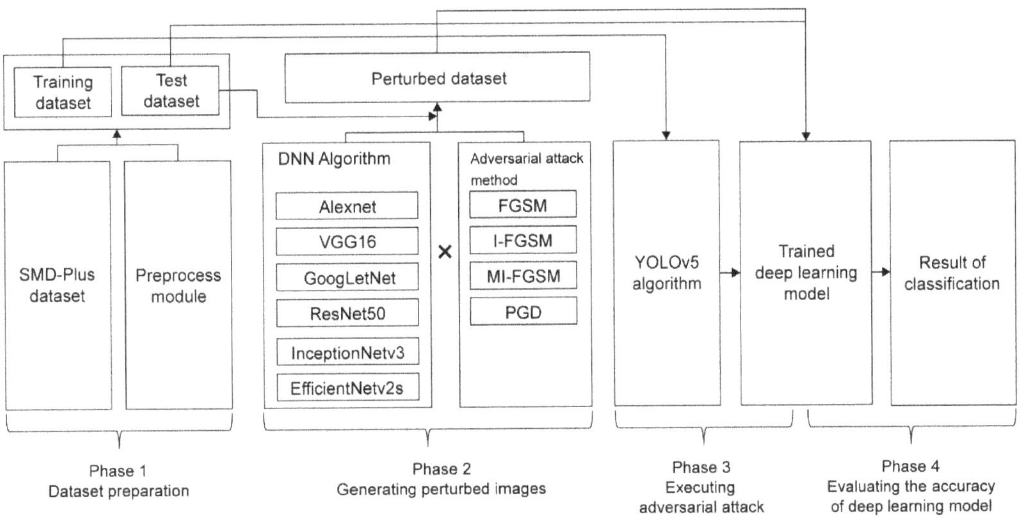

Figure 2. Diagram of the proposed experiment.

4. Results: Experiment on Adversarial Attacks against YOLOv5

4.1. Dataset Preparation

In domain-specific DNN applications, obtaining a proper dataset for training can be challenging, particularly for maritime environments, due to the scarcity of publicly available datasets and the cost of collecting and annotating images. The SMD dataset provides high-quality videos with labelled bounding boxes for ten types of objects in marine environments. The SMD dataset consists of high-definition videos captured at a resolution of 1920×1080 pixels. The dataset is divided into two parts: on-shore videos, consisting of 40 video clips, and on-board videos, consisting of 11 video clips. Additionally, each frame in the video dataset is labeled with bounding boxes and labels. However, this dataset presents some label errors and imprecise bounding boxes; therefore, it is not ideal as a benchmark dataset for object classification. To address this issue, the SMD-Plus dataset [9] was developed to improve the accuracy of bounding box annotations for small maritime objects. Moreover, in the SMD-Plus dataset, visually similar classes were merged to provide more training data for object recognition. Table 5 presents the classification of the training classes in the SMD-Plus dataset.

Table 5. Details of SMD-Plus dataset.

Class	Class Identifier	Number of Objects
Boat	1	14,021
Vessel/Ship	2	125,872
Ferry	3	3431
Kayak	4	3798
Buoy	5	3657
Sailboat	6	1926
Others	7	24,993

Because the SMD-Plus dataset comprises videos and annotations, we split the videos into frame-by-frame images to train YOLOv5. The provided annotations include object classes and the locations of bounding boxes for each video frame. These annotations were converted from the file format developed with the MATLAB ImageLabeler tool into an annotation format suitable for YOLOv5 [9,23]. Notably, 80% of the samples were used as the training dataset; the remaining 20% were used as the test dataset.

4.2. Generating Perturbed Images

Generating perturbed images requires both a DNN algorithm and an adversarial attack method. To assess the effect of adversarial attack methods on deep learning models, we developed perturbed datasets by implementing four different adversarial attack algorithms using PyTorch open-source code and a test dataset as input for six pre-trained models, namely, AlexNet, VGG16, GoogLeNet, ResNet50, InceptionNetv3, and EfficientNetv2s [28]. The objective of this experiment was to investigate the combinations of DNN algorithms, adversarial attack methods, and changes in the ϵ value that are most detrimental to YOLOv5 by generating perturbed images for adversarial attacks. The DNN algorithm, adversarial attack method, and hyperparameters (including the ϵ value) are determined by the attackers at the moment they generate the perturbed image. As presented in Table 6, we considered ϵ as the independent variable varying from 0.01 to 0.3; the other variables were kept constant as control variables. We generated 120 perturbed datasets, one for each combination of the pre-trained model and epsilon value.

Table 6. Parameters of the methods used for the generation of adversarial examples.

	FGSM	I-FGSM	MI-FGSM	PGD
Hyperparameters	$\epsilon = 0.01, 0.05, 0.1,$ 0.2, 0.3	$\epsilon = 0.01, 0.05, 0.1,$ 0.2, 0.3 $T = 20$ $\alpha = 0.01$	$\epsilon = 0.01, 0.05, 0.1,$ 0.2, 0.3 $T = 20$ $\mu = 0.001$	$\epsilon = 0.01, 0.05, 0.1,$ 0.2, 0.3 $T = 20$ $\alpha = 0.01$

We used an AMD Ryzen 9 5950X processor with 64 GB of main memory and an NVIDIA GeForce RTX 3080 Ti to generate the perturbed images. The runtime of the different methods is reported in Table 7. The execution time was not affected by the ϵ value.

Table 7. Execution time of pre-trained DNN algorithms for the generation of perturbed image.

	Execution Time (s)			
	FGSM	I-FGSM	MI-FGSM	PGD
AlexNet	19	43	22	31
VGG16	27	291	132	190
GoogLeNet	21	132	97	128
ResNet50	21	187	102	137
InceptionNetv3	25	207	145	199
EfficientNetv2s	30	327	231	312

Figures 3 and 4 depict the images generated using the FGSM and PGD with AlexNet with different ϵ values. When the ϵ value is small, changes made to the image are not easily distinguishable. However, as the ϵ value increases, changes become more noticeable. Nevertheless, it is challenging to distinguish whether this is caused by an adversarial attack or simple optical noise.

The results of the changes in the ϵ value affect not only the addition of noise that can be discerned by the human eye but also the performance degradation of the targeted model. In the following section, we simulate the performance degradation of the targeted model according to the ϵ value.

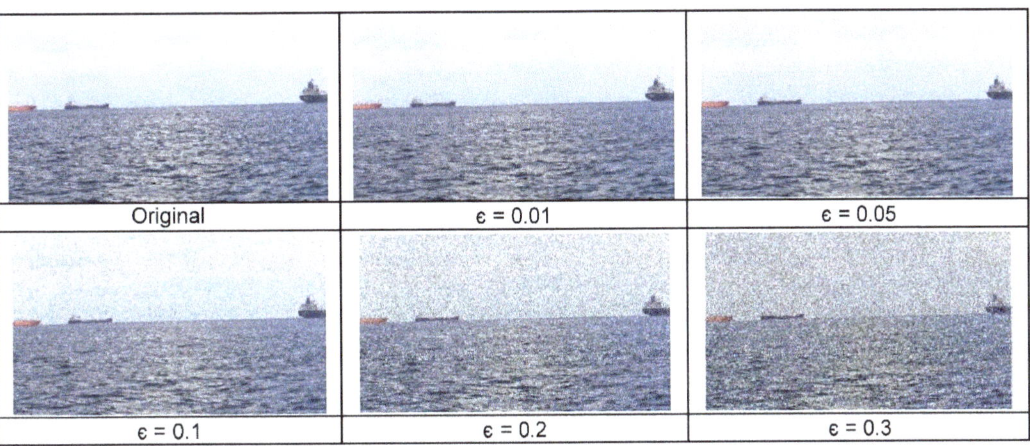

Figure 3. Perturbed images using the FGSM method with AlexNet.

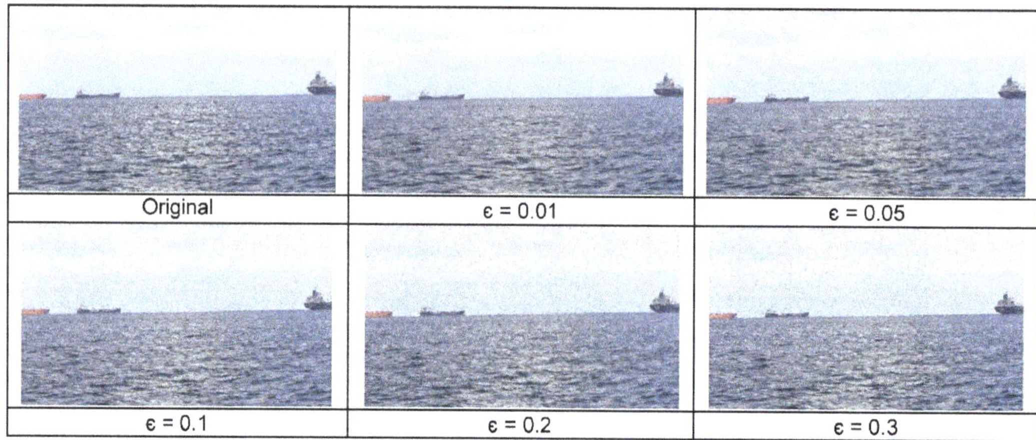

Figure 4. Perturbed images obtained using the PGD method with AlexNet.

4.3. Simulating an Adversarial Attack on a Deep Learning Model

We employed transfer learning using the YOLOv5s model as the base model to improve the training speed and accuracy of the object detection and classification model. The YOLOv5 model has four structures, YOLOv5s, YOLOv5m, YOLOv5l, and YOLOv5x, according to speed and accuracy. We assumed that YOLOv5s would be more vulnerable to adversarial attacks because of its low number of neural network layers and lower model accuracy. Transfer learning is a common learning method that is used to improve the performance of a model on relatively small datasets. The method uses a pre-trained model trained on a large amount of data as the base model. We used the fine-tuning method, which involves re-training the entire neural network, and adjusted the last convolutional layers of YOLOv5 to match the number of classes in the dataset, increasing them from 3 to 8. We set the training parameters to 100 epochs, learning rate of 0.12, and a batch size of 16 [10,19,22].

To evaluate the robustness of the YOLOv5-based deep learning model, we performed a comprehensive analysis by inputting a single original test dataset and the 120 perturbed datasets generated using various adversarial attack methods. By measuring the change in accuracy under these different scenarios, we aimed to assess the ability of the model to resist adversarial attacks and its overall performance in detecting objects in images.

We used the original test dataset to perform object classification. The YOLOv5s model trained using the SMD-Plus dataset achieved an accuracy of 0.896. The experiment also tested the accuracy of the model for each of the six different pre-trained DNN algorithms using the four adversarial attack methods and five ε values. The results are summarized in Tables 8–13.

Table 8. Accuracy of transfer learned model of YOLOv5s for different ε values using AlexNet.

	FGSM	I-FGSM	MI-FGSM	PGD
ε = 0.01	0.873	0.861	0.841	0.831
ε = 0.05	0.810	0.791	0.837	0.776
ε = 0.1	0.612	0.740	0.768	0.681
ε = 0.2	0.417	0.681	0.633	0.631
ε = 0.3	0.132	0.671	0.491	0.614

Table 9. Accuracy of transfer learned model of YOLOv5s for different ε values using VGG16.

	FGSM	I-FGSM	MI-FGSM	PGD
ε = 0.01	0.831	0.822	0.811	0.818
ε = 0.05	0.618	0.751	0.801	0.716
ε = 0.1	0.437	0.693	0.715	0.679
ε = 0.2	0.105	0.651	0.631	0.678
ε = 0.3	0.063	0.643	0.551	0.653

Table 10. Accuracy of transfer learned model of YOLOv5s for different ε values using GoogLeNet.

	FGSM	I-FGSM	MI-FGSM	PGD
ε = 0.01	0.827	0.858	0.841	0.829
ε = 0.05	0.766	0.765	0.818	0.771
ε = 0.1	0.551	0.750	0.761	0.728
ε = 0.2	0.266	0.721	0.731	0.731
ε = 0.3	0.115	0.711	0.565	0.710

Table 11. Accuracy of transfer learned model of YOLOv5s for different ε values using ResNet50.

	FGSM	I-FGSM	MI-FGSM	PGD
ε = 0.01	0.841	0.849	0.833	0.838
ε = 0.05	0.633	0.788	0.761	0.712
ε = 0.1	0.410	0.711	0.763	0.653
ε = 0.2	0.160	0.667	0.622	0.614
ε = 0.3	0.061	0.651	0.531	0.609

Table 12. Accuracy of transfer learned model of YOLOv5s for different ε values using InceptionNetv3.

	FGSM	I-FGSM	MI-FGSM	PGD
ε = 0.01	0.827	0.832	0.832	0.819
ε = 0.05	0.568	0.736	0.776	0.718
ε = 0.1	0.355	0.649	0.737	0.608
ε = 0.2	0.037	0.615	0.619	0.600
ε = 0.3	0.011	0.601	0.456	0.587

Table 13. Accuracy of YOLOv5s for different ε values using EfficientNetv2s.

	FGSM	I-FGSM	MI-FGSM	PGD
ε = 0.01	0.830	0.841	0.845	0.809
ε = 0.05	0.568	0.776	0.767	0.711
ε = 0.1	0.437	0.677	0.691	0.638
ε = 0.2	0.055	0.663	0.611	0.627
ε = 0.3	0.009	0.661	0.431	0.614

5. Discussion

The experiment performed in this study clarifies the vulnerability of object classification algorithms, specifically those using deep neural networks, to adversarial attacks. The results demonstrate that all algorithms and adversarial attack methods result in a significant decrease in accuracy when the ε value exceeds 0.2. This outcome highlights the significance of selecting a suitable ε value to develop effective defense strategies against these attacks.

Although the FGSM has the advantages of a higher success rate and faster generation time for perturbed images compared with other methods, the resulting image may contain a large amount of noise. Additionally, our results indicated that AlexNet generates perturbed images significantly faster than the other DNN algorithms, making it an ideal choice when reducing generation time is crucial. This may be because AlexNet has a simpler layer configuration than other DNN algorithms.

A crucial aspect of adversarial attacks is adding perturbations that are not easily detectable by humans. Consequently, the FGSM may not be an effective adversarial attack because it adds high-level noise to the image, thus making it more probable for humans to detect the attack. On the contrary, the PGD method consistently demonstrated a high success rate for attacks across all algorithms. Due to the FGSM adding noise of epsilon only once to the original image, humans can easily detect the noise. However, PGD gradually adds noise several times. Furthermore, perturbations generated with an ϵ value up to 0.1 were not easily detectable by humans for all DNN algorithms due to the difficulty in distinguishing them from optical noise. Figure 5 depicts the accuracy of each DNN algorithm, and the time required to generate perturbed images using the PGD method with an ϵ value of 0.1. As the ϵ value increases, it becomes easier for humans to detect the noise.

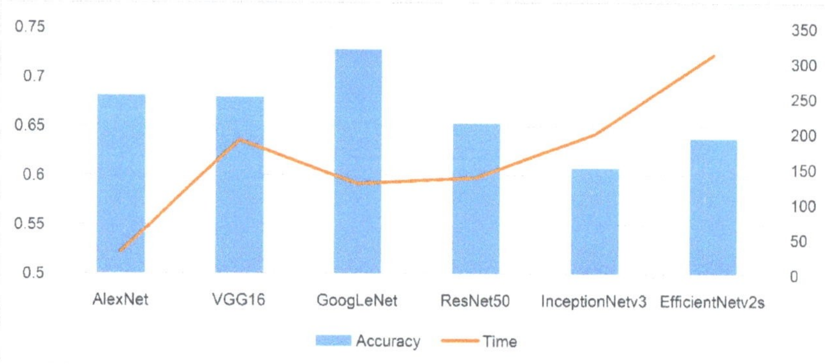

Figure 5. Accuracy of each DNN algorithm and time required to generate perturbed images using the PGD method for $\epsilon = 0.1$.

Based on our results, different approaches should be recommended depending on the priority of the defense resource. For instance, if generation time is a critical factor, it is more appropriate to use PGD with AlexNet and an ϵ value of 0.1. In contrast, if the success rate represents a high priority, PGD with InceptionNetv3 and an ϵ value of 0.1 is a more suitable strategy. Finally, if a compromise is necessary, the use of PGD with ResNet50 and an ϵ value of 0.1 is recommended. However, these recommendations should not be generalized and applied indiscriminately. In fact, the performance of object classification algorithms using YOLOv5 with transfer learning may vary for distinct datasets and adversarial attack methods; thus, it is necessary to consider the specific context when designing defense mechanisms against adversarial attacks. The experiment developed in this study provides a substantial foundation for further research and development in this critical area of cybersecurity.

6. Conclusions

6.1. Contributions

AI technologies are essential for enabling the operation of MASS. Object detection and classification algorithms are critical for improving navigation and collision avoidance and in optimizing the performance and efficiency of the vessels. However, the vulnerability of AI systems to adversarial attacks is a significant concern, as these attacks can compromise the accuracy and reliability of these systems and have real-world consequences. Despite the lack of experience with adversarial attacks in the maritime industry, the potential risks of such attacks are significant. Therefore, the SMD-Plus dataset, which includes classes such as ferries, boats, and buoys that MASS may encounter during actual operation, was used to generate adversarial images in various ways. We thereafter performed adversarial attacks on YOLOv5 with transfer learning, an object detection and classification algorithm. The experimental results demonstrated that the time required for generating perturbed images varied depending on the DNN algorithm and adversarial attack method. Moreover, we found that changes in the ϵ value can affect the vulnerability of the system to adversarial attacks. Experimentally, we determined the adversarial attack methods that are most harmful to object detection and classification models.

In AI, the risk of adversarial attacks has long been recognized, and studies on adversarial attacks and mitigation methods have been ongoing. However, the optimization of these studies for the specific characteristics of each industry and the awareness of their necessity is important. Recently, studies on these attacks have also been conducted in the medical field. In the maritime industry, stakeholders are developing and studying systems that apply AI models to realize MASS. However, they have not fully recognized the risks of adversarial attacks or experienced them. Nevertheless, to ensure the safe operation of MASS, recognizing the risks of adversarial attacks and developing measures for their mitigation are crucial. By presenting a case of adversarial attacks using maritime datasets, this study has contributed to raising awareness among stakeholders on cybersecurity in AI. Moreover, our findings will be used to investigate experimental defense technologies to mitigate vulnerability to adversarial attacks, ultimately contributing to the enhancement of cybersecurity in MASS. The development of effective defense strategies could further improve the security and safety of autonomous ships, rendering them a more reliable transportation mode.

6.2. Limitations

Even though this study does not consider technical advancements in adversarial attack methods, its aim is to provide information about security threats to object detection and classification algorithms through adversarial attack methods. Therefore, we simplified our experiments to help gain empathy for the risks of adversarial attacks. For this, we used known adversarial attack algorithms and limited hyperparameters and assumed that the attacks could not be detected when the perturbed images were acquired and validated. Therefore, the results of this experiment cannot be generalized. In the future, we intend to extend our research on the vulnerability of AI systems considering different object detection and classification algorithms in addition to YOLOv5 and various hyperparameters. Furthermore, we will continue to research methods to identify and mitigate vulnerabilities that could pose even more critical threats to the maritime industry.

Author Contributions: Conceptualization, C.L. and S.L.; methodology, C.L.; software, C.L.; validation, C.L. and S.L.; formal analysis, C.L.; investigation, C.L. and S.L.; resources, C.L.; data curation, C.L.; writing—original draft preparation, C.L.; writing—review and editing, S.L.; visualization, C.L. All authors have read and agreed to the published version of the manuscript.

Funding: This research was supported by Korea Institute of Marine Science & Technology Promotion (KIMST) funded by the Ministry of Oceans and Fisheries, Korea (RS-2023-00256086).

Institutional Review Board Statement: Not applicable.

Informed Consent Statement: Not applicable.

Data Availability Statement: The source code presented in this study is openly available in https://github.com/ChanguiLee/YOLOv5toAdversarialAttacks (accessed on 14 April 2023).

Conflicts of Interest: The authors declare no conflict of interest.

References

1. Al-Shatti, A.; Khaksar, W. Artificial intelligence in autonomous maritime navigation: A comprehensive review. *J. Navig.* **2021**, *74*, 756–788.
2. Tomic, T.; Peneva, J. Maritime autonomous surface ships: A review of recent developments and challenges. *J. Navig.* **2020**, *73*, 827–843.

3. Apostolidis, K.D.; Papakostas, G.A. A survey on adversarial deep learning robustness in medical image analysis. *Electronics* **2021**, *10*, 2132. [CrossRef]

4. Girshick, R.; Donahue, J.; Darrell, T.; Malik, J. Rich feature hierarchies for accurate object detection and semantic segmentation. In Proceedings of the IEEE Conference on Computer Vision and Pattern Recognition, Columbus, OH, USA, 23–28 June 2014; pp. 580–587. [CrossRef]

5. He, K.; Gkioxari, G.; Dollár, P.; Girshick, R. Mask R-CNN. In Proceedings of the IEEE International Conference on Computer Vision, Venice, Italy, 22–29 October 2017; pp. 2980–2988.

6. Redmon, J.; Divvala, S.; Girshick, R.; Farhadi, A. You only look once: Unified, real-time object detection. In Proceedings of the IEEE Conference on Computer Vision and Pattern Recognition, Las Vegas, NV, USA, 27–30 June 2016; pp. 779–788. [CrossRef]

7. Liu, W.; Anguelov, D.; Erhan, D.; Szegedy, C.; Reed, S.; Fu, C.Y.; Berg, A.C. SSD: Single shot multibox detector. In *European Conference on Computer Vision*; Springer: Cham, Switzerland, 2016; pp. 21–37.

8. Tan, M.; Pang, R.; Le, Q.V. EfficientDet: Scalable and Efficient Object Detection. In Proceedings of the 2020 IEEE/CVF Conference on Computer Vision and Pattern Recognition (CVPR), Seattle, WA, USA, 13–19 June 2020; pp. 10778–10787. [CrossRef]

9. Kim, J.-H.; Kim, N.; Park, Y.W.; Won, C.S. Object detection and classification based on YOLO-V5 with improved maritime dataset. *J. Mar. Sci. Eng.* **2022**, *10*, 377. [CrossRef]

10. Al-Qubaydhi, N.; Abdulrahman, A.; Turki, A.; Abdulrahman, S.; Naif, A.; Bandar, A.; Munif, A.; Abdul, R.; Abdelaziz, A.; Aziz, A. Detection of unauthorized unmanned aerial vehicles using YOLOv5 and transfer learning. *Electronics* **2022**, *11*, 2669. [CrossRef]

11. Maimunah, A.; Rosadi, R. A review of artificial intelligence application in maritime transportation. *J. Mar. Sci. Eng.* **2019**, *7*, 445.

12. *ISO/IEC TR 24028:2020*; Information Technology—Artificial Intelligence—Overview of Trustworthiness in Artificial Intelligence. ISO: London, UK, 2020.

13. Goodfellow, I.J.; Shlens, J.; Szegedy, C. Explaining and harnessing adversarial examples. *arXiv* **2015**, arXiv:1412.6572.

14. Biggio, B.; Roli, F. Wild patterns: Ten years after the rise of adversarial machine learning. *Commun. Secur.* **2018**, *84*, 317–331. [CrossRef]

15. Athalye, A.; Engstrom, L.; Ilyas, A.; Kwok, K. Synthesizing robust adversarial examples. In Proceedings of the 35th International Conference on Machine Learning, Stockholm, Sweden, 10–15 July 2018.

16. Kurakin, A.; Goodfellow, I.; Bengioet, S. Adversarial machine learning at scale. In Proceedings of the 33rd Conference on Neural Information Processing Systems (NeurIPS), Vancouver, BC, Canada, 8–14 December 2019.

17. Kurakin, A.; Goodfellow, I.; Bengio, S. Adversarial examples in the physical world. *arXiv* **2016**, arXiv:1607.02533.

18. Dong, Y.; Liao, F.; Pang, T.; Su, H.; Zhu, J.; Hu, X.; Li, J. Boosting adversarial attacks with momentum. In Proceedings of the IEEE Conference on Computer Vision and Pattern Recognition, Salt Lake City, UT, USA, 18–23 June 2018; pp. 9185–9193. [CrossRef]

19. Madry, A.; Makelov, A.; Schmidt, L.; Tsipras, D.; Vladu, A. Towards deep learning models resistant to adversarial attacks. *arXiv* **2018**, arXiv:1706.06083.

20. Tan, H.; Wang, L.; Zhang, H.; Zhang, J.; Shafiq, M.; Gu, Z. Adversarial attack and defense strategies of speaker recognition systems: A survey. *Electronics* **2022**, *11*, 2183. [CrossRef]

21. Alotaibi, A.; Rassam, M.A. Adversarial machine learning attacks against intrusion detection systems: A survey on strategies and defense. *Future Internet* **2023**, *15*, 62. [CrossRef]

22. Tramèr, F.; Kurakin, A.; Papernot, N.; Boneh, D.; McDaniel, P. Ensemble adversarial training: Attacks and defenses. *arXiv* **2017**, arXiv:1705.07204.

23. Krizhevsky, A.; Sutskever, I.; Hinton, G.E. ImageNet classification with deep convolutional neural networks. In *Advances in Neural Information Processing Systems*; Morgan Kaufmann Publishers, Inc.: Burlington, MA, USA, 2012; pp. 1097–1105.

24. Simonyan, K.; Zisserman, A. Very deep convolutional networks for large-scale image recognition. *arXiv* **2014**, arXiv:1409.1556.

25. Szegedy, C.; Liu, W.; Jia, Y.; Sermanet, P.; Reed, S.; Anguelov, D.; Erhan, D.; Vanhoucke, V.; Rabinovich, A. Going deeper with convolutions. In Proceedings of the IEEE Conference on Computer Vision and Pattern Recognition, Boston, MA, USA, 7–12 June 2015; pp. 1–9.

26. He, K.; Zhang, X.; Ren, S.; Sun, J. Deep residual learning for image recognition. In Proceedings of the IEEE Conference on Computer Vision and Pattern Recognition, Las Vegas, NV, USA, 27–30 June 2016; pp. 770–778.

27. Szegedy, C.; Vanhoucke, V.; Ioffe, S.; Shlens, J.; Wojna, Z. Rethinking the inception architecture for computer vision. In Proceedings of the IEEE Conference on Computer Vision and Pattern Recognition, Las Vegas, NV, USA, 27–30 June 2016; pp. 2818–2826.

28. Tan, M.; Le, Q.V. EfficientNet: Rethinking model scaling for convolutional neural networks. In Proceedings of the International Conference on Machine Learning, Long Beach, CA, USA, 10–15 June 2019; pp. 6105–6114.

Journal of
*Marine Science
and Engineering*

Article

A Network Model for Identifying Key Causal Factors of Ship Collision

Jianzhou Liu, Huaiwei Zhu, Chaoxu Yang and Tian Chai *

College of Navigation, Jimei University, Xiamen 361021, China; 202112861003@jmu.edu.cn (J.L.);
zhw1645@163.com (H.Z.); y1810652154@163.com (C.Y.)
* Correspondence: chaitian@jmu.edu.cn

Abstract: In the analysis of the causes of ship collisions, the identification of key causal factors can help maritime authorities to provide targeted safety management solutions, which is of great significance to the prevention of ship collisions. In order to identify the key causal factors leading to ship collisions, we first construct a network model of ship collisions, in which the nodes represent the causal factors, and the edges represent the interrelationship between the causal factors. Second, based on the constructed network model, we propose a successive safety analysis method. This method can quantify the importance of each causal factor, and the quantified results allow us to identify the key causal factors of ship collisions. Finally, we verify the validity of the model using numerical cases.

Keywords: complex networks; cascading failures; causes of accidents; ship collisions

1. Introduction

Maritime transport is an important part of global trade, which accounts for 90% of the total global trade [1]. However, the maritime environment is complex and volatile, and it is extremely challenging to ensure the safe navigation of ships and avoid maritime traffic accidents. According to a related study [2], among the many maritime traffic accidents, ship collisions have the highest incidence rate. Meanwhile, the occurrence of ship collision accidents often causes serious consequences, such as casualties, property damage, and environmental pollution [3]. Therefore, how to prevent ship collision accidents has been the focus of attention of maritime departments.

Usually, accidents or unsafe events are caused by a number of uncertainties [4]. These uncertainties together form a complex system, and anomalies in multiple factors within the system may induce an accident to occur. In maritime traffic systems, ship collisions occur when multiple factors interact to induce the formation of causal chains that lead to accidents. In each causal chain, there are often some key factors, and the abnormality of the key factors will lead to the abnormality of a series of factors. Therefore, finding the key factors and managing them can reduce the chain reaction of the causal chain and thus reduce the probability of accidents.

In practical ship safety management, the identification of critical factors is very important. Due to the limited human and material resources on board, the crew cannot apply the same level of protection to all causes of accidents. Therefore, the identification of critical factors can help the crew to prioritize the needs of the causes of accidents. This significantly improves the efficiency of ship safety management. Based on this fact, we propose a network model to identify the critical causes of ship collisions. We analyze the model from the perspective of protection, quantify each cause of accident through network efficiency, and finally identify the critical causes of accidents through the quantified results.

The rest of the paper is structurally organized as follows: Section 2 describes the work related to ship collision accident studies. Section 3 describes the basic concept of cascading failures. Section 4 presents the construction method and identification method of the model.

Citation: Liu, J.; Zhu, H.; Yang, C.; Chai, T. A Network Model for Identifying Key Causal Factors of Ship Collision. *J. Mar. Sci. Eng.* **2023**, *11*, 982. https://doi.org/10.3390/jmse11050982

Academic Editors: Marko Perkovic, Lucjan Gucma and Sebastian Feuerstack

Received: 14 April 2023
Revised: 24 April 2023
Accepted: 3 May 2023
Published: 5 May 2023

Section 5 uses numerical cases to demonstrate the validity of the model. Section 6 presents the conclusions and future work.

2. Literature Review

At present, many scholars have conducted studies on ship collision accidents. According to their research directions, these studies can be divided into ship collision accident causation analysis and ship collision accident risk assessment.

The purpose of ship collision causation analysis is to explore the root causes of ship collisions and to reduce the probability of accidents by strengthening the protection against the root causes of accidents. For example, Zhang [5] used hierarchical analysis to establish a ship–bridge collision risk evaluation model, and this model can provide decision-making suggestions for bridge siting through the qualitative and quantitative assessment of the risk level of ship–bridge collisions. Ugurlu [6] used fault trees and multiple correspondence analyses to quantitatively and qualitatively analyze 513 ship collisions, and the final results of the study showed that 94.7% of the collisions were caused by human errors. Afenyo et al. [7] applied Bayesian networks to the collision scenario between ships and icebergs in order to study the root causes of ship–iceberg collisions in the Arctic shipping route, and through sensitivity analysis, they concluded that mechanical equipment, miscommunication, and communication equipment failure are the main causes of ship–iceberg collisions. Chai et al. [8] collected 300 reports of ship collisions, extracted the causal factors and causal chains from them, constructed a ship collision causal network with the causal factors as nodes, and analyzed the dynamic changes in the nodes of the network by setting different thresholds to finally determine the root causes of ship collisions.

Ship collision risk assessment aims to assess the collision risk of ships navigating in different water environments, and the results of their studies are usually the probability of ship collisions in different environments. For example, Zhen et al. [9] proposed a real-time multivessel collision assessment framework to evaluate the collision risk of ships in complex waters, in which the clustering method of spatial density was first used to identify ship encounter scenarios; then, the ship collision risk assessment function was constructed using the distance-closest point of approach (DCPA), the time-closest point of approach (TCPA) and relative bearing (RB), and finally, the collision risk of ships in complex waters was evaluated based on the magnitude of the function value. Wu et al. [10] proposed a fuzzy logic method for ship–bridge collision risk assessment, which takes ship characteristics, natural environment, and other factors as variables and fuzzifies these variables; after fuzzifying the input variables, IF–THEN rules are established and used for fuzzy reasoning to derive the bridge collision risk and determine the ship–bridge collision risk level. Huang et al. [11] used the speed method to evaluate the overlap probability of two ships' positions in the future by setting different reachable speeds and finally used the overlap probability to evaluate the collision risk of ships. Chen et al. [12] proposed a collision risk assessment method based on speed barriers, and they evaluated the collision risk between multiple ships from the perspective of speed and verified the effectiveness of this method by using the data of the automatic ship identification system.

There is no doubt that the above research studies have a positive impact on the prevention of ship collisions. However, with the rapid increase in the level of ship automation, and the potential causes of accidents have become more complex. In the process of analyzing the causes of accidents, it is necessary to consider the correlation between the causes of accidents as much as possible. However, the number of causes can be hundreds or thousands, and it will be a challenge to analyze the interrelationship of these causes effectively. Fortunately, complex networks can provide a good method for this challenge.

Complex networks originated in the 1980s and have been widely used in the field of safety engineering after many years of development. For example, traffic safety [13,14], construction safety [15,16], power security [17,18], and multimodal transport network security [19] have been studied. Meanwhile, a large number of studies have shown that complex networks can more clearly explain the correlation between things from a system perspective [20]. Therefore, complex networks become an effective method to analyze the complex relationships between things. In particular, the use of complex networks has good application prospects in the field of traffic safety [21].

In this paper, we propose a network model to identify the key causal factors of ship collisions based on complex networks. We analyze the model from the perspective of protection and quantify each causal factor. The quantified results allow us to identify the key factors of ship collisions and provide theoretical guidance for ship safety management.

3. Cascading Failure Theory

A cascading failure is a phenomenon of anomalous information propagation in a network. In some real networks, a failure in one or a few nodes, followed by a failure in other nodes through the coupling between nodes, can cause successive failures in other nodes, which can create a chain reaction that eventually leads to the collapse of a significant portion or even the entire network [22]. There are many models for analyzing cascade failures in complex networks in which each node is assigned an initial load and load capacity, and when the load of a node exceeds its capacity, that node will fail, and its load will be distributed to its neighboring nodes. At this point, the initial load of the neighboring node will also change as the neighboring node accepts the additional load. Consequently, if the load of the neighbor node exceeds its capacity, a new failed node will appear, which in turn will lead to a new round of load distribution.

The load capacity model is one of the common models of cascading failures [23]. Generally, the load capacity model assigns the initial load and capacity to each node in the network and determines whether the fault spreads by comparing the load value and capacity value of the node. According to the basic concept of cascading failures [24], the initial node load, node capacity, and some other parameters are defined as follows:

(1)   Initial load

In 2003, Motter et al. [25] proposed a load capacity model in which the information and energy of nodes in a network are propagated according to the shortest paths between pairs of nodes. Therefore, the initial load size of a node can be expressed in terms of the number of shortest paths through the node. In a complex network, the number of shortest paths through a node can be expressed in terms of the betweenness. Therefore, in this paper, the initial load of a node is expressed by its betweenness, and the formula for calculating the betweenness of a node is shown in Equation (1).

$$D_i(0) = \sum_{i,j \in V (j \neq k)} \frac{n_{jk}(i)}{n_{jk}} \tag{1}$$

where n_{jk} represents the number of shortest paths connecting points j and k; $n_{jk}(i)$ represents the number of shortest paths connecting points j and k and passing through point i.

(2)   Node capacity

In the load capacity model, node capacity refers to the maximum load value that a node can carry. Node capacity is the threshold value that determines fault propagation, and when the node load is greater than its capacity, the node will fail and propagate the load to adjacent nodes. Node capacity is generally proportional to its initial load, and its calculation formula is shown in Equation (2).

$$C_i = (1 + \lambda)D_i(0) \tag{2}$$

where λ is the tolerance factor.

4. Method

4.1. Ship Collision Causation Network Structure

The key step in constructing the causal network model of ship collision is to determine the nodes in the network and the rules of the connected edges between the nodes. In a real network, the objects of study are generally abstracted as nodes, and the interrelationships between the objects of study are used as edge rules. In the same way, we abstracted the causal factors of ship collision as network nodes and the interrelationship between the causal factors as the connected edges between nodes. The specific network construction steps were as follows:

Step 1: Collect reports of ship collisions;

Step 2: Analyze the ship collision report and extract the causal factors from it;

Step 3: Count the causal factors contributing to the same accident;

Step 4: Construct a causal network of ship collision accidents. The causal factors were used as network nodes, and edges were defined as the interre-lationship between the causal factors that appeared in the same acci-dent.

4.2. Successive Security Evolutionary Processes

Based on the constructed network model and according to the idea of cascading failures, we propose a successive safety evolution process for ship collisions. In this study, the load value of each node in the network model is defined as the safety protection strength of the causative factor, the initial load of the node is considered as the initial protection value, and the capacity of the node is defined as the safety protection threshold. If the safety protection strength of a node is greater than the safety threshold of that node, we consider the node as a safe node, i.e., the node will not cause an accident, and thus it is removed from the network. However, due to the complexity and variability of the ship's navigation environment and the limitation of human and material resources in ship safety management, sufficient protection strength cannot be given to each causative factor. As a result, the initial protection value of each node is often lower than its safety threshold.

In this paper, the concept of successive safety is proposed based on the theory of cascading failures. Therefore, the initial load of the node is the initial protection value of the node, and the calculation formula is shown in Equation (3).

$$P_i(t) = \sum_{i,j \in V(j \neq k)} \frac{n_{jk}(i)}{n_{jk}} \tag{3}$$

The safety protection threshold for the node is $C_i = (1 + \lambda)P_i(0)$.

When the protection value of node i exceeds the corresponding safety threshold, the node will be in a secure state, at which time any additional protection beyond the safety threshold of this node will be equally divided among its neighbors j, calculated as follows:

$$\Delta P_{ij} = \frac{P_i(t) - C_i}{d_i(t)} (j \in \Gamma_i) \tag{4}$$

where ΔP_{ij} denotes the protection value passed from node i in a secure state to neighbor node j; $P_i(t)$ denotes the protection value of node i at moment t; $d_i(t)$ denotes the number of neighbor nodes of node i at moment t; C_i denotes the node safety threshold; Γ_i denotes the set of neighbor nodes of node i.

After calculation, the neighboring nodes of node i obtain additional protection values, and therefore their own protection values will change as follows:

$$\begin{cases} P_j(t) = P_j(t-1) + \Delta P_{ij} = P_j(t-1) + \frac{P_i(t-1) - C_i}{d_i} \\ P_j(t+1) = \begin{cases} P_j(t) & (P_j(t) \leq C_i) \\ C_i & (P_j(t) > C_i) \end{cases} \end{cases} \tag{5}$$

The evolution process of "successive safety" of the causal network model of ship collision proposed in this paper is shown in Figure 1. The number in the circle in Figure 1 indicates the protection rate of the node, i.e., protection rate = protection value of the node/safety threshold of the node × 100%. When the protection rate reaches 100%, the node is in a safe state, and the node will not cause an accident. When the protection rate exceeds 100%, the node maintains a safe state while assigning the additional protection value above the safety threshold to its neighboring nodes. For example, when $t = 0$, the protection rate of each node in the network does not reach 100%; at this time, each node is in a dangerous state, and these nodes may cause an accident at any time. When $t = 1$, additional protection measures are applied to node 4, so that the protection value of node 4 exceeds its safety threshold. At this point, the protection rate of node 4 exceeds 100%, so node 4 has additional protection values to assign to its neighboring nodes (such as node 5 and node 6). At $t = 2$, the protection values of node 5 and node 6 are updated as node 5 and node 6 receive additional protection values, and the protection rates of both node 5 and node 6 exceed 100% after the update. At this time, node 5 and node 6 have additional protection values to assign to their neighboring nodes (e.g., node 1 and node 7). At t = 3, after node 1 and node 7 receive additional protection values, their respective protection values are updated, and the protection rate of node 7 exceeds 100% after the update. At this point, node 7 has additional protection values to distribute to its neighboring nodes (e.g., node 3). At $t = 4$, node 3 receives the additional load, and its protection rate exceeds 100%. At this point, node 3 allocates the additional protection value to node 2. At t = 5, the "successive safety process" in the network ends because no new node has a protection rate exceeding 100%.

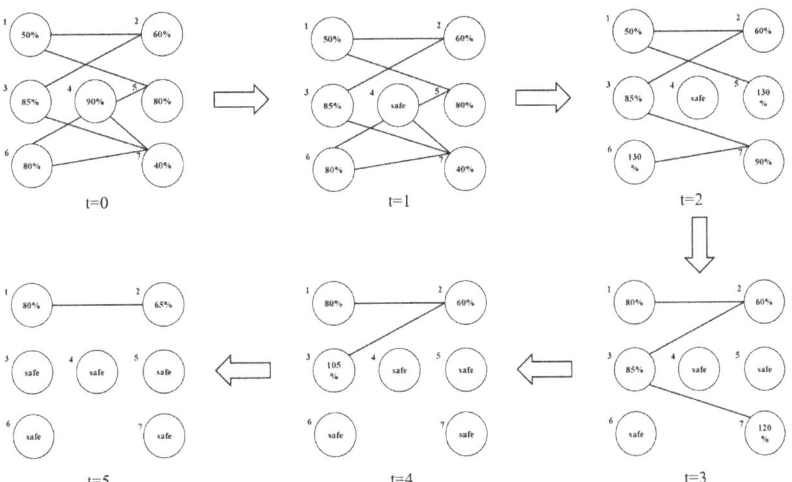

Figure 1. Successive safety evolution process.

4.3. Network Efficiency

According to the successive safety process, the initial node entering the secure state triggers the successive safety evolution process of other nodes in the network. Therefore, in order to evaluate the degree of impact on the network as a whole after the initial node triggers the successive safety process, we introduce the evaluation index of network efficiency. The network efficiency is calculated as follows:

$$E(i) = \frac{N'(i)}{N} \tag{6}$$

where $N'(i)$ is the number of nodes remaining in the network after successive failures of node i, and N is the initial number of nodes in the network.

5. Numerical Case Study

5.1. Constructing a Causal Network for Ship Collisions

In this study, we collected 300 reports of ship collisions that occurred in Chinese waters during a 20-year period from 1999 to 2018. From the reports, we extracted 98 causal factors (see Table A1). According to the network model construction method in Section 4.1, we successfully constructed a causal network of ship collisions, as shown in Figure 2.

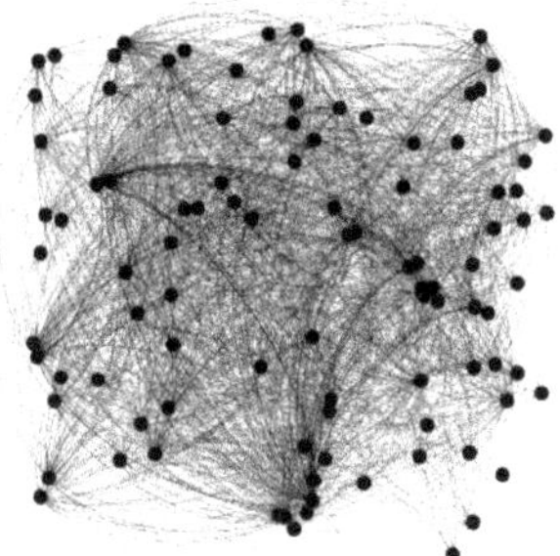

Figure 2. Ship collision causation network model.

5.2. Successive Safety-Triggering Processes

In complex networks, the node with a large degree value is called a hub node, which is coupled with more nodes in the network. In a ship collision causation network, the hub node represents the factor that causes more accidents, and more human and material resources need to be invested in the protection of this type of node to prevent its failure. Therefore, the hub node has a higher initial protection value and safety threshold than other nodes. By analyzing the successive safety processes triggered by hub nodes, it is possible to recognize the important role played by hub nodes in the network, which is of great significance for accident prevention.

The degree values and initial protection values of each node in the ship collision causation network model were calculated, and the results are shown in Figures 3 and 4.

From Figures 3 and 4, it can be seen that the degree value and initial protection value of node 1 (improper lookout) are the largest, so node 1 is used as the hub node in the network. Next, we take this node as an example to analyze the successive safety process of this node in the network.

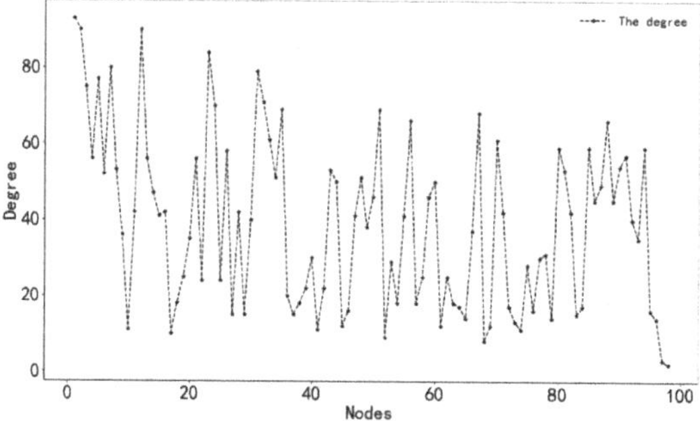

Figure 3. The degree value of each node.

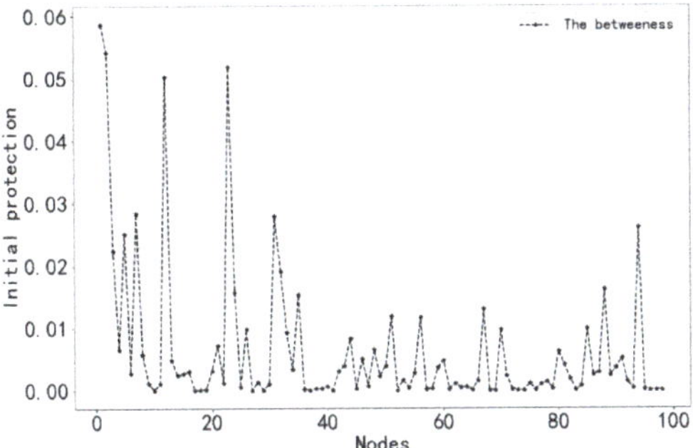

Figure 4. The initial protection value of each node.

Before analyzing the successive safety process of the hub node, we briefly highlight the following assumptions:

Assumption 1: The tolerance factor is set to 0.5, which means that the protection of the causal factor reaches 1.5 times its safety threshold to ensure that the causal factor is in a safe state, i.e., the node will not cause a ship collision to occur.

Assumption 2: In order to trigger the successive safety propagation process, additional protection values need to be assigned to the hub node. In this paper, the hub node is assigned an additional protection value equal to three times its own protection value, so that it is in a fully safe state. At this point, the protection value of the hub node is greater than its own safety threshold, so it will promote the successive safety process.

Assumption 3: When a node's protection value exceeds its safety threshold, the part of the node's protection value that exceeds the safety threshold is assigned to a neighboring node, and the node is removed from the network. When the protection values of all nodes in the network are not affected by other nodes, the process of successive safety propagation ends. At this point, the protection values of all nodes in the network are less than their corresponding safety thresholds.

Based on the above assumptions, we take the hub node as an example to analyze the successive safety evolution process of the causative network of ship collisions, and the specific evolution process is explained in what follows.

$t = 0$: As the initial protection value of node 1 does not reach the safety threshold, node 1 is assigned an additional protection value equal to three times its own protection value to make it in a fully secure state. At this point, the protection value of node 1 is greater than its own safety threshold, and the successive safety evolution starts.

$t = 1$: The protection values of neighboring nodes are updated through the successive safety evolution of the protection value of node 1. The updated protection values of neighboring nodes are compared with their own safety thresholds, as shown in Figure 5.

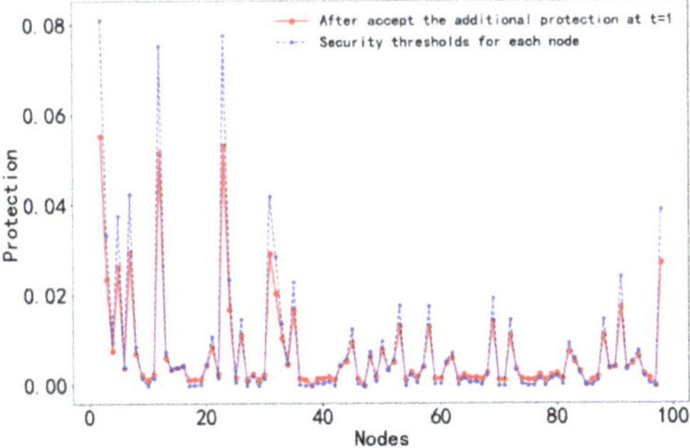

Figure 5. The protection value and safety threshold of each node at $t = 1$.

From Figure 5, the nodes that enter a fully secure state in the network after passing the first successive safety evolution were determined, and the results are shown in Table 1.

Table 1. Node numbers in a fully secured state.

Node Numbers in Fully Secured State									
8	9	10	13	16	17	18	21	24	26
48	50	53	54	55	58	59	62	63	64
65	66	67	69	70	72	73	74	75	76
77	78	79	80	83	85	86	88	91	94
95									

t = 2: Repeating the above successive safety evolution process, the protection value of neighboring nodes at the moment t = 2 is compared with its own safety threshold, as shown in Figure 6.

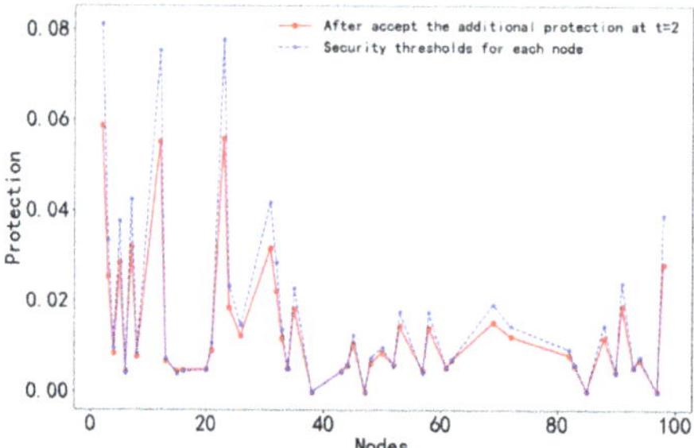

Figure 6. The protection value and safety threshold of each node at t = 2.

The nodes that enter a fully secure state in the network after passing the second successive safety evolution were derived from Figure 6, and they are shown in Table 2.

Table 2. Node numbers in a fully secured state.

Node Numbers in Fully Secured State				
5	14	15	19	33
51	56	89	92	96

t = 3: Similarly, the above successive safety evolution process is repeated to obtain a comparison of the safety threshold of the neighboring node protection with its own at the moment t = 3, as shown in Figure 7.

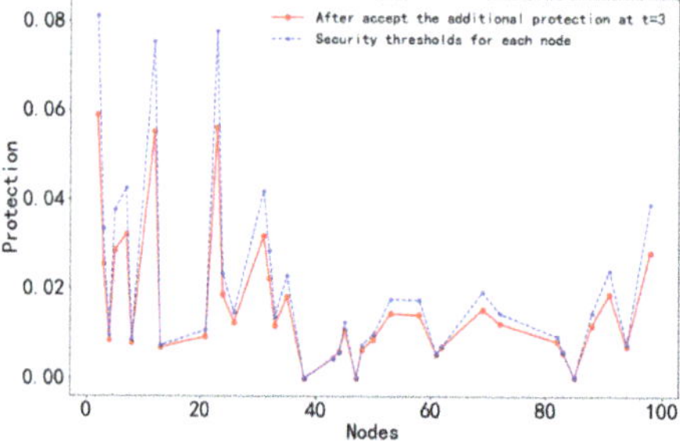

Figure 7. The protection value and safety threshold of each node at t = 3.

As can be seen from Figure 7, only node 84 is in a fully secure state after the third successive safety evolution.

t = 4: Continuing the above process of successive safety evolution, a comparison of the t = 4 neighbor node protection with its own safety threshold is obtained, as shown in Figure 8.

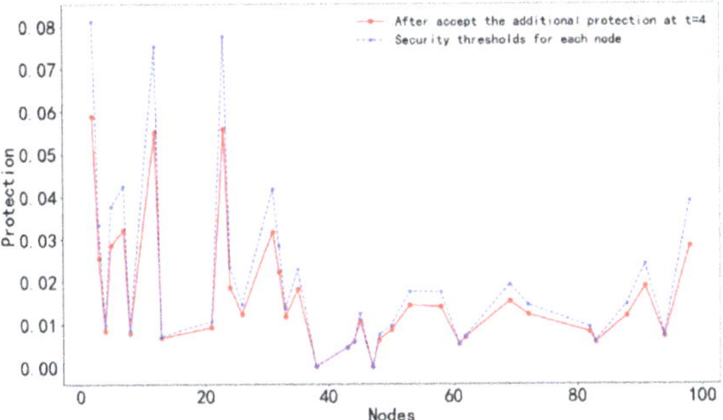

Figure 8. The protection value and safety threshold of each node at t = 4.

As can be seen from Figure 8, at this time, the protection values of the remaining nodes in the network are less than their own safety threshold, and the successive safety evolution process ends.

Throughout this process, the successive evolution toward a secure state triggered by node 1 leads to some nodes entering the secure state at each step in the network. In order to more intuitively reflect the changes in the nodes of the network, we plotted the changes in the number of nodes at each step of the successive safety evolution process, and the node changes are shown in Figure 9.

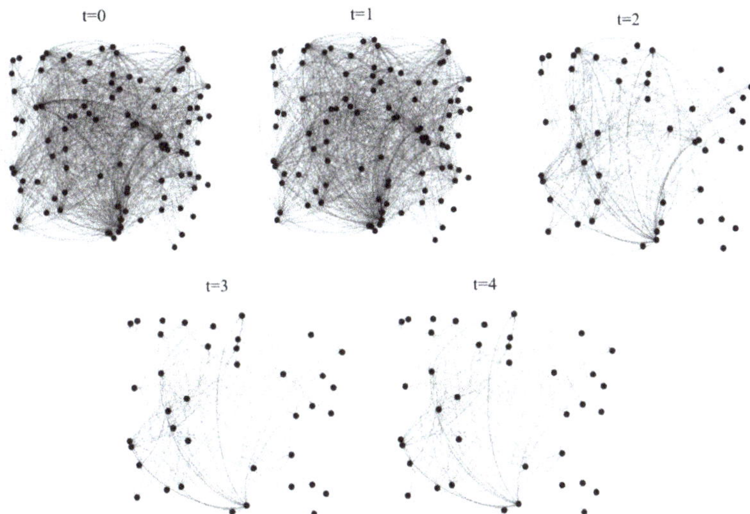

Figure 9. The evolution of the causal network of ship collisions.

As can be seen from Figure 9, the network gradually becomes sparse after each evolutionary step. This means that the successive safety processes triggered by node 1 have more causative nodes entering a secure state. The reduction in the causative nodes in the network can directly indicate that the protection of node 1 can reduce the probability of accidents.

According to Section 4.3, we calculated the network efficiency values after each successive safety evolution, and the calculation results are shown in Table 3. It can be seen that as the successive safety evolution process continues, the network efficiency decreases. This indicates that there are fewer and fewer factors in the causal network that can induce ship collisions, and the probability of ship collisions is increasingly minimized.

Table 3. Network efficiency changes.

Time	t = 0	t = 1	t = 2	t = 3	t = 4
N′	98	97	46	36	35
E(1)	1.0000	0.9899	0.4694	0.3673	0.3571

In summary, the analysis of the successive safety evolution process triggered by node 1 shows that the hub node ends after four successive safety evolution processes. The causative nodes in the network are reduced from 98 to 35, and the network efficiency is reduced from 1.0000 to 0.3571. This indicates that the protection of node 1 can greatly reduce the incidence of accidents.

5.3. The Successive Safety Evolution Process of Each Node

In order to evaluate the importance of each node in the network, we performed the above steps for each node so that each node triggered the successive safety process. Finally, the network efficiency of each node at the end of the successive safety process was obtained (see Table 4).

Table 4. Network efficiency value of each node.

ID	Network Efficiency	ID	Network Efficiency	ID	Network Efficiency	ID	Network Efficiency
1	0.3571	26	0.8775	51	0.9591	76	1.0000
2	0.3775	27	1.0000	52	0.9489	77	0.9591
3	0.6734	28	0.9795	53	0.8367	78	0.9897
4	0.9081	29	1.0000	54	1.0000	79	0.9897
5	0.5918	30	0.9897	55	0.9591	80	0.9693
6	0.9693	31	0.6122	56	0.9693	81	0.9795
7	0.6020	32	0.7448	57	0.9489	82	0.9489
8	0.9183	33	0.8775	58	0.8265	83	0.9693
9	0.9897	34	0.9387	59	0.9897	84	0.9795
10	1.0000	35	0.7857	60	0.9897	85	0.9795
11	0.9897	36	0.9897	61	0.9489	86	0.9897
12	0.4183	37	1.0000	62	0.9285	87	0.9591
13	0.9183	38	1.0000	63	0.9897	88	0.9081
14	0.9897	39	0.9795	64	0.9693	89	0.9897
15	0.9489	40	0.9795	65	0.9897	90	0.9693
16	0.9489	41	0.9897	66	0.9795	91	0.8061
17	1.0000	42	1.0000	67	0.9897	92	0.9693
18	0.9897	43	0.9693	68	0.9897	93	0.9693
19	0.9897	44	0.9285	69	0.8571	94	0.9285
20	0.9693	45	0.9081	70	1.0000	95	0.9795
21	0.9387	46	0.9897	71	1.0000	96	0.9897
22	0.9489	47	1.0000	72	0.8265	97	1.0000
23	0.3979	48	0.9285	73	0.9693	98	0.8061
24	0.8265	49	0.9897	74	0.9795		
25	0.9795	50	0.8979	75	1.0000		

In order to analyze the above results more clearly, we drew a network efficiency diagram at the end of the successive safety processes of each node, as shown in Figure 10. At the same time, the probability of each causative factor triggering the occurrence of 300 accidents was counted, and the probability statistics are shown in Figure 11.

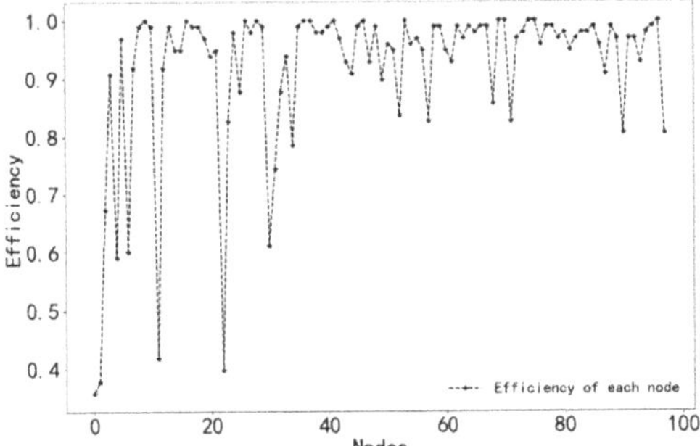

Figure 10. The network efficiency of each node.

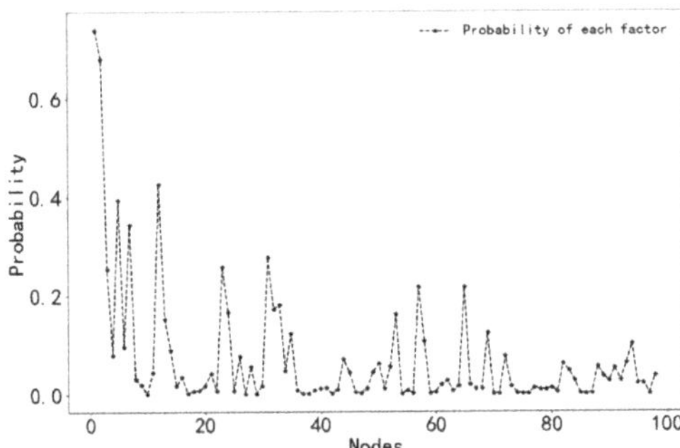

Figure 11. The probability of induced accidents at each node.

In the existing studies, the number of accidents caused is mostly used directly as an important indicator to evaluate the degree of influence of causal factors on ship collisions. Here, we used network efficiency to measure the importance of accident causation. The comparison shows that the evaluation results of these two indicators on the importance of accident causation are basically the same, but there are also some differences. For example, the probability of accidents caused by node 22 is smaller than that of nodes 21 and 23, but the reduction in network efficiency caused by the control of node 22 is much higher than that of nodes 21 and 23, which means that although the number of ship collision accidents caused by the causal factor represented by node 22 is smaller, the incidence of ship collision accidents can be effectively reduced by investing human and material resources in its prevention and control, so the degree of influence of this factor on the incidence of ship collision accidents is much higher than that of nodes 21 and 23. Therefore, the influence of this factor on the occurrence rate of ship collisions is also greater. Secondly, the objects of these two evaluation indexes are different. The calculation of accident probability is influenced by the number of collected accidents, which is more contingent and not general. On the other hand, the network efficiency is determined from the perspective of the protection of causal factors and enables the analysis of the influence of causal factors on the occurrence rate of ship collision accidents under the same level of protection, which is objective and not influenced by the number of collected accidents. With the improvement in ship intelligence, the human and material resources invested in ship safety management are more valuable, and the reasonable allocation of human and material resources in ship safety management has an important influence on the reduction in ship collision rates. Thus, the method proposed in this paper can provide a valuable reference for the reasonable allocation of human and material resources.

In addition, as new routes are opened (e.g., Arctic routes), new causal factors of ship accidents will emerge. However, the ship collision causation network designed in this paper is an open network, which is updateable. For new causes of accidents, we can update the network at any time according to the construction steps of the network described in Section 4.1. By analyzing the new network, we can assess the influence of new causal factors on accident occurrence. This is more in line with the realistic needs resulting from variability in the ship navigation environment.

6. Conclusions

In this paper, a network model of ship collision accidents was established based on the complex network theory. In order to evaluate the importance of causative nodes in the network model, we proposed a successive safety analysis method. The concepts of the initial protection value, the safety threshold, and the protection rate of each node were introduced to help us control the spread of ship collisions by triggering a successive safety evolution process, and each causative factor was quantified according to network efficiency. Lastly, the key causative factors of ship collisions were identified based on the quantified results.

Numerical case studies show that improper lookout is the key cause of accidents, and the probability of accidents will be reduced to less than 40% by taking protective measures against this cause of accidents. Therefore, in ship navigation, the watchkeeping officershould consciously abide by the terms of lookout procedures, use a combination of visual and radar observations for lookout, strictly follow policies while on duty, and refrain from any behavior that affects the driver's formal lookout, such as drunk driving and fatigue driving.

Finally, compared with most of the other existing methods, our proposed method concerns accident protection and enables the quantification of individual causal factors of accidents, and the quantified results are more generalized regardless of the number of collected accidents. In addition, the network model we constructed is an open model, and we will apply this model to special navigation environments (e.g., Arctic routes) in future research to enrich the database of the model and further expand its application scenarios.

Author Contributions: Conceptualization, J.L.; methodology, J.L.; software, J.L.; validation, J.L.; formal analysis, J.L.; investigation, H.Z.; resources, H.Z.; data curation, T.C.; writing—original draft preparation, J.L.; writing—review and editing, J.L. and T.C.; visualization, C.Y.; supervision, T.C.; project administration, T.C. All authors have read and agreed to the published version of the manuscript.

Funding: This research is funded by the Natural Science Foundation of Fujian Province (Grant No. 2019J01326).

Institutional Review Board Statement: Not applicable.

Informed Consent Statement: Not applicable.

Data Availability Statement: Not applicable.

Conflicts of Interest: The authors declare no conflict of interest.

Appendix A

Table A1. The causal factors of ship collisions.

Cause ID	Cause Name
1	Improper lookout
2	Inappropriate assessment of the situation of the risk of collision
3	Did not make a sound signal as per guidelines
4	Did not exhibit lights and shapes as per guidelines
5	Did not navigate at a safe speed

Table A1. *Cont.*

Cause ID	Cause Name
6	Did not take actions such as slacking a vessel's speed, stopping or reversing her propulsion in ample time to avoid close quarters
7	Did not take actions such as slacking a vessel's speed, stopping or reversing her propulsion in ample time to avoid close quarters
8	The person on duty was not on the bridge
9	VHF was not on duty
10	Misunderstood the information from the VHF
11	Took action blindly
12	Did not take effective action in good time to avoid collision (miss the best time to take effective action)
13	The give-way vessel did not carry out the duty to keep out of the way
14	The stand-on vessel took action in error
15	The stand-on vessel took action in error
16	Deviation from specified course
17	Violation of the regulation of no-drinking when on duty
18	Dozed off on duty
19	The person on duty was engaged in something irrelevant to navigation
20	Did not keep a safe distance from the anchoring ship
21	Did not obtain sufficient information about the surrounding navigation environment
22	Did not monitor own ship's position sufficiently
23	Did not meet the requirements of good seamanship and seafarers' usual practice
24	Inadequate use of radar/ARPA
25	Identified the information of radar target in error
26	Inadequate use of AIS (including not installing or turning on AIS)
27	The captain did not give any orders for night navigation
28	The captain failed to command regarding the situation of the bridge as required (e.g., foggy weather, narrow water channel, and traffic-dense area)
29	The navigation alarm on the bridge was not on
30	The steering device was not used at the proper time
31	Did not see other ships as early as possible
32	Failure to track or misjudge the dynamics of other ships
33	Did not take coordinated turning actions in an urgent situation
34	Took improper emergency measures to avoid collision
35	Did not take the most helpful actions to avoid collision in an emergency risk
36	Overtook blindly without other ships' approval
37	Communication failure between the officer on duty and the sailor on duty
38	Communication error between the bridge control and the engine room
39	The officer is not familiar with the rules of the COLREGs
40	The officers' shift error
41	The inexperience of the person on duty
42	The person is sitting when he is on duty
43	The officer is not familiar with the maneuverability of the ship
44	Small alterations in course to avoid collision
45	Underestimated the impact of wind, wave, and current
46	No tug assistance was applied when berthing or unberthing (no application of tug assisting when berthing and unberthing)
47	Operated the tug improperly when berthing and unberthing
48	The remaining speed was too fast when berthing and unberthing
49	Did not check the effectiveness of the action to avoid collision
50	Violation of navigation regulations of the water area (including regulations on ship routing system)
51	Not following VTS advice or traffic control
52	Failure to comply with narrow channel navigation rules
53	The incompetence of the crew member
54	The officer's overfatigue
55	Pilot operation error
56	The officer left the ship too early
57	Failure to obey report obligation
58	Insufficient crew number

Table A1. *Cont.*

Cause ID	Cause Name
59	Machine failure without repair guarantee
60	Ships sailing beyond the approved navigation area
61	Anchoring in waterway, customary route, or dense-traffic area
62	Violation of mooring duty requirements
63	Improper anchoring method
64	No effective monitoring of anchorage position during anchoring
65	No effective measures were taken after anchor dragging
66	Significantly affected by the wind
67	Significantly affected by the wave
68	Significantly affected by the current
69	Poor visibility
70	The effect of navigational obstructions
71	The effect of the bend of the channel
72	Navigation impact in traffic-dense areas (complicated navigation environment)
73	The impact of collision avoidance by third-party vessels
74	The impact of narrow waterways
75	The impact of shallow water
76	The VHF communication channel was too noisy
77	The main engine broke down
78	Failure of the steering gear
79	AIS fault
80	Other facilities' failure
81	The influence of the blind area of the bow
82	The ship was not in a seaworthy condition
83	The Vessel certificate expired or undocumented
84	No ship inspection was conducted as required
85	The ship failed to correct safety defects or faults before sailing
86	No VTS was established
87	VTS supervision error
88	The shipping company failed to fulfill the main responsibility of safety production
89	The shipping company commanded the ship to operate on sea illegally
90	The shipping company did not provide enough qualified crew for the ship
91	The shipping company did not establish SMS or the requirements of SMS were not implemented
92	The training and assessment of the crew members by the shipping company were insufficient
93	(Improper arrangement of persons on duty) insufficient staff on duty
94	No additional lookout staff
95	The shipowner did not sufficiently know about the information and competency of the crewmembers
96	The shipping company did not monitor the ship dynamically
97	The shipping company did not fully grasp the navigation management regulations important to ship safety
98	Defects in bridge resource management

References

1. Chen, J.; Bian, W.; Wan, Z.; Yang, Z.; Zheng, H.; Wang, P. Identifying factors influencing total-loss marine accidents in the world: Analysis and evaluation based on ship types and sea regions. *Ocean Eng.* **2019**, *191*, 106495. [CrossRef]
2. Wang, Y.-F.; Wang, L.-T.; Jiang, J.-C.; Wang, J.; Yang, Z.-L. Modelling ship collision risk based on the statistical analysis of historical data: A case study in Hong Kong waters. *Ocean Eng.* **2020**, *197*, 106869. [CrossRef]
3. Roca, E.; Julià-Verdaguer, A.; Villares, M.; Rosas-Casals, M. Applying network analysis to assess coastal risk planning. *Ocean Coast. Manag.* **2018**, *162*, 127–136. [CrossRef]
4. Provan, D.J.; Woods, D.D.; Dekker, S.W.A.; Rae, A.J. Safety II professionals: How resilience engineering can transform safety practice. *Reliab. Eng. Syst. Saf.* **2020**, *195*, 106740. [CrossRef]
5. Zhang, D.; Yang, X.; Fan, X. Research on Risk Assessment of Vessel-Bridge Collision Accident In Inland Waterway. *Appl. Mech. Mater.* **2013**, *256*, 2790–2793. [CrossRef]
6. Ugurlu, H.; Cicek, I. Analysis and assessment of ship collision accidents using Fault Tree and Multiple Correspondence Analysis. *Ocean Eng.* **2022**, *245*, 110514. [CrossRef]

7. Afenyo, M.; Khan, F.; Veitch, B.; Yang, M. Arctic shipping accident scenario analysis using Bayesian Network approach. *Ocean Eng.* **2017**, *133*, 224–230. [CrossRef]
8. Chai, T.; Zhu, H.; Peng, L.; Wang, J.; Fan, Z.; Xiao, S.; Xie, J.; Hu, Y. Constructing and analyzing the causation chain network for ship collision accidents. *Int. J. Mod. Phys. C* **2022**, *33*, 2250118. [CrossRef]
9. Zhen, R.; Shi, Z.; Liu, J.; Shao, Z. A novel arena-based regional collision risk assessment method of multi-ship encounter situation in complex waters. *Ocean Eng.* **2022**, *246*, 110531. [CrossRef]
10. Wu, B.; Yip, T.L.; Yan, X.; Guedes Soares, C. Fuzzy logic based approach for ship-bridge collision alert system. *Ocean Eng.* **2019**, *187*, 106152. [CrossRef]
11. Huang, Y.; van Gelder, P. Time-Varying Risk Measurement for Ship Collision Prevention. *Risk Anal.* **2020**, *40*, 24–42. [CrossRef]
12. Chen, P.; Li, M.; Mou, J. A Velocity Obstacle-Based Real-Time Regional Ship Collision Risk Analysis Method. *J. Mar. Sci. Eng.* **2021**, *9*, 428. [CrossRef]
13. Li, K.; Pan, Y. An effective method for identifying the key factors of railway accidents based on the network model. *Int. J. Mod. Phys. B* **2020**, *34*, 2050192. [CrossRef]
14. Li, M.; Wang, H.; Wang, H. Resilience Assessment and Optimization for Urban Rail Transit Networks: A Case Study of Beijing Subway Network. *IEEE Access* **2019**, *7*, 71221–71234. [CrossRef]
15. Guo, S.; Zhou, X.; Tang, B.; Gong, P. Exploring the behavioral risk chains of accidents using complex network theory in the construction industry. *Phys. A Stat. Mech. Its Appl.* **2020**, *560*, 125012. [CrossRef]
16. Zhou, C.; Ding, L.; Skibniewski, M.J.; Luo, H.; Jiang, S. Characterizing time series of near-miss accidents in metro construction via complex network theory. *Saf. Sci.* **2017**, *98*, 145–158. [CrossRef]
17. Deng, X.; Wang, S.; Wang, W.; Yu, P.; Xiong, X. Optimal defense strategy for AC/DC hybrid power grid cascading failures based on game theory and deep reinforcement learning. *Front. Energy Res.* **2023**, *11*, 247. [CrossRef]
18. Guo, H.; Yu, S.S.; Iu, H.H.C.; Fernando, T.; Zheng, C. A complex network theory analytical approach to power system cascading failure-From a cyber-physical perspective. *Chaos* **2019**, *29*, 053111. [CrossRef]
19. He, Z.; Guo, J.-N.; Xu, J.-X. Cascade Failure Model in Multimodal Transport Network Risk Propagation. *Math. Probl. Eng.* **2019**, *2019*, 3615903. [CrossRef]
20. Valente, T.W. Network interventions. *Science* **2012**, *337*, 49–53. [CrossRef]
21. Ruths, J.; Ruths, D. Control profiles of complex networks. *Science* **2014**, *343*, 1373–1376. [CrossRef] [PubMed]
22. Buldyrev, S.V.; Parshani, R.; Paul, G.; Stanley, H.E.; Havlin, S. Catastrophic cascade of failures in interdependent networks. *Nature* **2010**, *464*, 1025–1028. [CrossRef] [PubMed]
23. Valdez, L.D.; Shekhtman, L.; La Rocca, C.E.; Zhang, X.; Buldyrev, S.V.; Trunfio, P.A.; Braunstein, L.A.; Havlin, S.; Estrada, E. Cascading failures in complex networks. *J. Complex Netw.* **2020**, *8*, cnaa013. [CrossRef]
24. Kinney, R.; Crucitti, P.; Albert, R.; Latora, V. Modeling cascading failures in the North American power grid. *Eur. Phys. J. B* **2005**, *46*, 101–107. [CrossRef]
25. Motter, A.E.; Lai, Y.C. Cascade-based attacks on complex networks. *Phys. Rev. E Stat. Nonlin. Soft Matter Phys.* **2002**, *66*, 065102. [CrossRef]

Journal of
Marine Science and Engineering

Review

Maritime Anomaly Detection for Vessel Traffic Services: A Survey

Thomas Stach *, Yann Kinkel , Manfred Constapel and Hans-Christoph Burmeister

Sea Traffic and Nautical Solutions, Fraunhofer Center for Maritime Logistics and Services CML, Blohmstraße 32, 21079 Hamburg, Germany; yann.kinkel@cml.fraunhofer.de (Y.K.); manfred.constapel@cml.fraunhofer.de (M.C.); hans-christoph.burmeister@cml.fraunhofer.de (H.-C.B.)
* Correspondence: thomas.stach@cml.fraunhofer.de; Tel.: +49-40-2716461-1513

Abstract: A Vessel Traffic Service (VTS) plays a central role in maritime traffic safety. Regulations are given by the International Maritime Organization (IMO) and Guidelines by the International Association of Marine Aids to Navigation and Lighthouse Authorities (IALA). Accordingly, VTS facilities utilize communication and sensor technologies such as an Automatic Identification System (AIS), radar, radio communication and others. Furthermore, VTS operators are motivated to apply Decision Support Tools (DST), since these can reduce workloads and increase safety. A promising type of DST is anomaly detection. This survey presents an overview of state-of-the-art approaches of anomaly detection for the surveillance of maritime traffic. The approaches are characterized in the context of VTS and, thus, most notably, sorted according to utilized communication and sensor technologies, addressed anomaly types and underlying detection techniques. On this basis, current trends as well as open research questions are deduced.

Keywords: maritime surveillance; vessel traffic service; VTS; monitoring; anomaly detection; decision support tool; DST

Citation: Stach, T.; Kinkel, Y.; Constapel, M.; Burmeister, H.-C. Maritime Anomaly Detection for Vessel Traffic Services: A Survey. *J. Mar. Sci. Eng.* **2023**, *11*, 1174. https://doi.org/10.3390/jmse11061174

Academic Editor: Claudio Ferrari and Mihalis Golias

Received: 12 April 2023
Revised: 29 May 2023
Accepted: 30 May 2023
Published: 3 June 2023

1. Introduction

Maritime transportation is the backbone of global trade, in value as well as in volume [1]. While navigable waters on the oceans are wide, the bottlenecks of maritime transportation are normally narrow straits, channels and port approaches themselves. Most ports do only provide one approach from port to sea, making the approach itself a critical infrastructure without redundancy [2]. In channels and straits, vessels must follow the water way in a structured manner so as not to expose themselves into risk of hazards, e.g., collisions. This is aggravated by the various meteorological and hydrological conditions under which maritime traffic operates. Thus, accurate traffic information and traffic coordination is needed in those areas, to ensure safe and smooth traffic flows. This task is, nowadays, supported by Vessel Traffic Services (VTS) [3,4]. VTS make use of various communication and sensor technologies to establish an extensive situation awareness. Familiar examples are Automatic Identification System (AIS), radar or radio communication in the very high frequency (VHF) range. Decision Support Tools (DST) help the VTS operators to outsource significant workloads so that tasks such as the anchor watch can be performed by machines instead of humans. Assuming that normal traffic flow is safe and smooth, an anomaly-detecting DST could indicate traffic events that impair traffic flow, in this regard [5].

Anomaly detection is a well-known research field in the scientific community [6] and also in the maritime context specifically [5,7]. Specifically, from the perspective of VTS, this research field seems promising due to the manifold of technologies applicable as data sources. However, as VTS operates within a safety-critical environment, the recommendations and solutions proposed by any DST must be explainable and reliable [5,8]. Further, the addressed anomaly types must suit the VTS traffic scenarios, too. Lastly, requirements related to VTS operation, such as real-time capability, must be taken into account as well.

In this survey, we review the recent literature on maritime anomaly-detection approaches from the point of view of VTS operations. We do this by collecting literature, selecting relevant publications and classifying them corresponding to data source, detection technique and addressed-anomaly types. These and other classifying dimensions are chosen based on the capabilities and requirements of VTS. To the best of our knowledge, this is the first survey on maritime anomaly detection in which the literature is reviewed according to VTS-relevant properties. This survey may be of relevance for any researcher in the field of surveillance or monitoring of maritime traffic or closely related topics. Closely related topics may be any intermediate step of maritime anomaly detection such as traffic extraction or representation.

This survey is organized as follows: In Section 2, this survey is compared with related work. It is shown how comparable surveys can be distinguished and on what other authors focus. Subsequently, in Section 3, the reader is introduced to the context and scope of this survey. As a result, the tasks and services, as well as the technological capabilities and application of DST, are explained. The Section closes with an outline of anomaly types and detection techniques. The literature-review process comprising the collection, selection and classification of the literature in this survey is explained in Section 4. Section 5 presents the results of the literature review. Finally, the survey is concluded in Section 6.

2. Related Work

Within recent years, some surveys reviewing anomaly-detection articles in the maritime context have been published. Here, we focus only on recent surveys starting from 2017 (cf. Section 4).

Sidibé and Shu [9] provide a summary of approaches for anomaly detection for the period 2011–2016. This summary comprises approaches which utilize AIS data only. Further, the authors characterize the approaches by their techniques and applied AIS data attributes.

In [10], an overview of AIS-based anomaly-detection techniques is presented as a part of a broader survey. This survey also deals with related topics such as AIS data providers and methods on route estimation or collision risk assessment. The overview on the detection techniques strongly focuses on technical backgrounds.

Riveiro et al. [7] give a holistic overview of anomaly-detection techniques. The authors cover over two decades of literature, i.e., 1996 to 2017. The reviewed approaches are characterized with various properties such as the utilized data, normalcy extraction and representation, the detection technique itself and anomaly types. Particularly worth mentioning is that this survey does not focus on a specific data source such as AIS.

In a study by Yan and Wang, an overview of AIS-based data-driven detection techniques is given [11]. Further the authors distinguish among purely statistical or Machine-Learning-based as well as hybrid techniques.

In [12], the authors introduce a distinction between the detection of events and anomalies. For both, the reviewed approaches are characterized according to their underlying detection techniques. All reviewed approaches rely either on AIS data solely or in combination with other data, e.g., environmental data.

Similarly to the majority of the previously mentioned related work, Dogancay et al. provide an overview of the techniques of anomaly-detection approaches [13]. However, this survey focuses on AIS-based approaches only.

The most recent survey was published by Wolsing et al. in 2022 [14]. Even though this survey focuses on AIS data only, the noteworthy key result of this work is a tabular overview which sorts anomaly-detection approaches from the years 2007 to 2021 by detection techniques, anomaly types, utilized AIS data attributes and more.

Therefore, to the best of our knowledge, this work represents the first survey which reviews maritime anomaly-detection approaches w.r.t. VTS context—more specifically, the Information Service of VTS—and sorts the reviewed literature, accordingly, by relevant properties.

3. Situation Awareness in Vessel Traffic Services

In this section, the reader is introduced to VTS and how its operators build up and maintain their situation awareness about maritime traffic flow. It starts with a brief summary of the historic origin and development of VTS and moves to the service goals and levels. Then, the technological capabilities are briefly introduced and how DST build on them. Finally, the topic of maritime anomaly detection within the scope of this survey is introduced.

3.1. Tasks and Services

After the radio had been employed for a some time already, it was complemented by a novel technology when, in 1948, one of the first harbour surveillance radars was introduced in Liverpool [3,15]. Both technologies, radio and radar, not only offered extended communication and surveillance capabilities but also ensured these during adverse weather conditions. Although initially adapted for efficiency reasons, e.g., to reduce congestion, it was acknowledged that the utilization of these technologies also increased traffic safety. Since the advent of these informal VTS, not only has ship traffic been increasing but also the technological capabilities of VTS. In 2021, the latest guidelines for VTS were outlined in resolution A.1158(32) [4]. According to this resolution, vessel traffic service shall serve three purposes, which are:

1. "providing timely and relevant information on factors that may influence ship movements and assist onboard decision-making";
2. "monitoring and managing ship traffic to ensure the safety and efficiency of ship movements";
3. "responding to developing unsafe situations".

Thereby, VTS operation itself is no longer a pure observer and information provider over a traffic coordinator; rather, it is now a direct shore-based supporter as indicated by their service levels [3,16]:

1. Information Service (INS);
2. Traffic Organisation Service;
3. Navigation Assistance Service.

This study specifically focuses on the INS of VTS.

3.2. Technologies and Their Operations

In order to fulfil the above-listed services and purposes, VTS can make use of various complementary communication and sensing technologies. They are specified by the International Association of Marine Aids to Navigation and Lighthouse Authorities (IALA) in Guideline G1111 [17]. Table 1 gives an overview of these technologies.

The data and information of the utilized communication and sensing technologies are fused together with the aim of providing holistic situation awareness. Software is used which visualizes the situation awareness in an electronic navigational chart.

3.3. Decision Support

In order to facilitate and enhance the situation awareness further, VTS operators are encouraged to utilize decision support tools, which deliver more elaborate functionalities. Examples of applications of DST are listed in IALA Guideline G1110 [5]. The complexity of DST ranges from simple functionalities such as geographically confined vessel-speed alerts to anomalous-behaviour alerts. Accordingly, the complexity of the technical implementation varies from threshold-based alerts to traffic models build on historical data.

To our knowledge, VTS operators most commonly use rather simple DST such as vessel-speed or -number alerts, anchor watch or geofence-based monitoring of undesired entering or leaving of specified zones. However, these alert systems are highly customizable, so that, e.g., thresholds and geographically confined areas can be defined freely based on situation and experience. More complex DST that are, for example, based on statistical models or Machine-Learning-based systems have not found their way into VTS systems yet.

Table 1. Overview of communication and sensor technologies of VTS as defined by IALA in Guideline G1111 [17].

Technology	Description
Radar	A radar system emits electro-magnetive waves and detects the echo signal of reflected waves by targets such as vessels [18]. Direction and distance to the target as well as its motion direction and velocity can be deduced.
AIS	AIS is a standardized, automatic communication system which is used over transceivers [17,19]. It is used by vessels, VTS and for (virtual) aids to navigation. Depending on the message type, the message contains static, dynamic and voyage-related information about the sender.
Environmental monitoring	Various relevant environmental conditions can be monitored [17]. Common is the measurement and monitoring of hydrological (e.g., height of tide, current speed or ice coverage) and meteorological (e.g., wind speed, wind direction or visibility) conditions.
Electro-optical systems	Electro-optical systems refer to imaging devices that can be, for example, daylight or night-vision camera surveillance [17]. Usually, the field of view of the utilized cameras is adjustable.
Radio communications	Spoken communication takes place over VHF radio communication systems [17]. VHF is used to enable real-time situation assessment.
Radio direction finders	A radio direction-finder device is able to deduce the bearing to a VHF emitting station. The bearing can be associated with an AIS target in the vicinity.
Long-range sensors	For situation awareness beyond the operation range of short-range sensors (e.g., radar or AIS), long-range sensors can be applied [17]. Common examples are the so-called long-range identification and tracking system or satellite-based AIS.

3.4. Anomaly Types and Detection Techniques

An anomalous pattern, i.e., an anomaly, inherently requires an understanding of the normalcy from which it deviates. In the context of maritime traffic, VTS operators expect certain traffic patterns which they perceive as normal. Therefore, deviating traffic patterns may appear anomalous. Humans perform this reasoning through different approaches such as using formalized rules or simply by experience. Similarly, anomalies can be detected by a machine. The mechanism performing this is called anomaly detection. In the following, we will first describe a variety of anomaly types and subsequently outline briefly anomaly-detection techniques.

In this review, we focus on five generic anomaly types, which are listed in Table 2. This selection is based on our talks with experts in the field of VTS development and operation, other studies [12,20,21] and a preliminary, exploratory investigation of the reviewed literature. These anomaly types, on the one hand, are sufficiently generalized to form the basis for more specific or complex anomaly scenarios and, on the other hand, cover frequent anomaly scenarios which can be addressed by DST [17].

Note that an anomaly detection must not be restricted to the detection of anomalies in the present situation but may also predict upcoming anomalies and their probabilities of occurrence.

The complexity of the an anomaly-detection technique can range from a rule-based system to a system based on a neural network. Generally, multiple techniques can serve an anomaly type. The choice of the technique can be driven by various factors such as the available type and amount of data or computational power. In safety-critical environments such as VTS, in particular, reliability and explainability of the technique play important roles.

Table 2. Considered generic anomaly types in the review classification scheme.

Anomaly Type	Description
Kinematic deviation	Deviation in a single kinematic parameter, e.g., speed over ground or course over ground.
Route deviation	Deviation in a route due to deviation in the sequence of positions.
Collision risk	Close approach between vessels or vessels and (abstract) objects. Objects can be visible on water (e.g., bouys) as well as regulatoric (e.g., traffic separation schemes as abstract objects) or physical (e.g., shallow water or coastlines) confinements of the waterway.
Zone entry	Penetration of regulatorily or physically defined zones.
Inconsistency	Information inconsistency in the situation awareness either due to sensors providing contrasting information or one sensor providing false data.

In our review, we identified five groups of techniques which are applied for the problem of anomaly detection (cf. Table 3). We define a techniques as groups of methods. It is important to note that a detection approach does not stick to one specific method but can use multiple methods from one or several of the following techniques.

Table 3. Considered anomaly-detection techniques in the review classification scheme.

Detection Technique	Description
Descriptive Statistics	Detection techniques based on descriptive statistics are data-driven. They rely on data sets which are used to derive a statistical distribution to model behaviour patterns [6,22]. Here, normalcy is defined as any pattern that is close to the mean behaviour pattern. An anomalous behaviour pattern would be anything that deviates too far from the mean, thus the term outlier. The degree or threshold of an anomaly is given by the distance from the mean.
Stochastic Processes	Behaviour patterns can be described as stochastic processes. In this case, a model is created which is able to describe the change in a pattern (or a state) over time randomly or due to influential conditions. Generally known approaches that fall into this technique are Markov models or models based on Bayesian statistics [22].
Clustering	The technique of clustering comprises approaches for the creation of clusters and which are based on Machine Learning (ML). There exists a variety of clustering algorithms which all serve specific purposes [23].
Classification	Classification can be performed through various approaches. In this review, only ML-based classification approaches are counted as classification approaches. Similarly to clustering, there is a broad variety of classification algorithms [23].
Neural network-based	Neural networks (NN) are universal function approximators and, thus, theoretically can address every problem type. Any NN-based detection approach is considered under this technique [24].
Rule-based	Rule-based systems build upon human-made rules. The other aforementioned techniques rely on sets and interactions of rules, too; however, this technique crucially differs in its simpler technical structure and the intuitive explainability of decision processes [12].

In case an anomaly-detection approach does not fall under the introduced techniques, it is classified as other

4. Literature Review

In this section, the literature-review process is briefly explained. This includes the collection, selection and classification of the literature. The whole process is depicted in Figure 1.

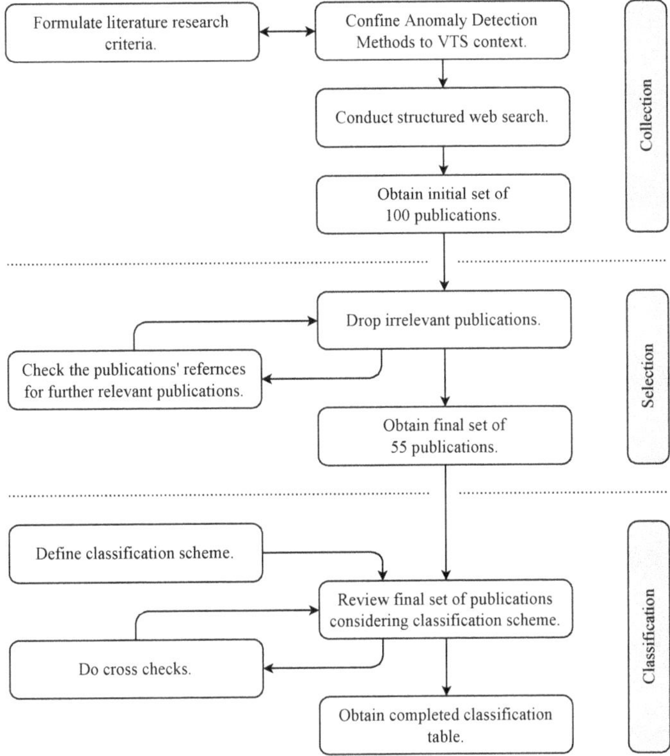

Figure 1. Literature-review process comprising collection, selection and classification steps.

4.1. Collection

The focus of this literature review lies in the approaches of anomaly detection in the context of maritime traffic surveillance conducted by VTS, specifically, INS. Beyond that, the review is not constrained to specific anomalies, a data source or other dimensions as is done in some related works (cf. Section 2). Accordingly, the scope is set by the definition of the search terms:

- maritime;
- (surveillance OR VTS OR monitoring);
- anomaly detection.

Here, each term is connected by a Boolean AND operator and the words of the VTS-specific term are connected by Boolean OR operators. By doing this, we aim to retrieve a broad variety of literature within our scope. To focus on more recent approaches, the search is confined to literature from the years 2017 to 2022. Initially, the first 100 search results on Google Scholar were collected.

4.2. Selection

Irrelevant publications, i.e., topic out of scope, were dropped. Kept literature was checked for potentially relevant references. These steps were repeated iteratively, as can be seen in Figure 1. This way, we made sure to collect publications which had not been listed in the initial set of 100 search results but had been cited by other publications.

Thesis works were not kept; however, the references on which they were based were kept. Other surveys were not kept either; however, their references were checked for relevance, too. This selection process of the literature was cross-checked internally.

4.3. Classification

The finally obtained publications were compared. To compare the introduced context of the INS of VTS, corresponding classification dimensions were defined. Initially, the following set of classification dimensions was formulated:

- Data source (i.e., communication and sensor technologies, cf. Table 1);
- Anomaly types (cf. Table 2);
- Area types (i.e., physically, regulatory or not constrained)
- Ship types;
- Detection techniques (cf. Table 3).

The literature review was then performed using these classification dimensions. During the review process and at its end, the filled classification table was cross-checked internally to make sure that the classification was being performed under common understanding and to check for any flaws.

5. Results

The number of publications which were retrievable with the defined search phrase (cf. Section 4) has increased significantly within recent years, as can be seen in Figure 2. To put this into context: more publications are retrievable from the year 2022 than from the years 2010–2015 combined.

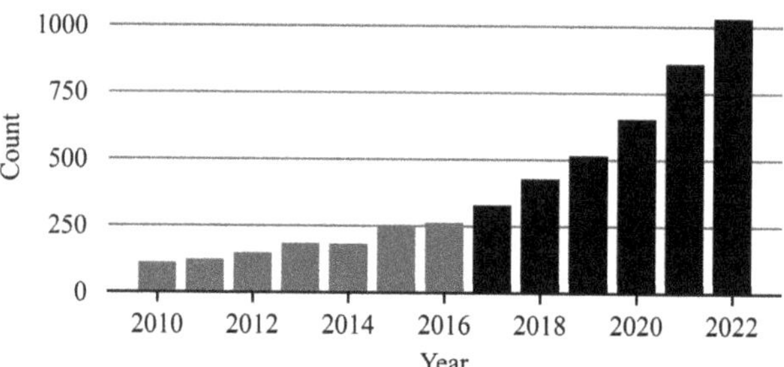

Figure 2. Counts of publications retrievable with the defined search phrase via Google Scholar. The deep black bars are the years that are within the scope of this survey.

Following the review methodology depicted in Figure 1, in total, 136 publications were collected and screened. From those, 55 passed the selection criteria and cross-checks. These publications were sorted according to the classification explained in Section 4. By doing this, this survey addresses topics in or close to the field of maritime traffic surveillance, monitoring or VTS operations, specifically.

The resulting classification of the publication can be examined in Table 4 for the AIS data source alone and Table 5 for all other data sources and their combinations (with AIS). Multiple publications of the same author and approach are merged into one entry and classified according to the latest publication. As can be seen in the tables, and as stated in Section 4, the classification dimensions area types and ship types were dropped. This is due to the fact that only a small minority of the publications stated that their approaches were clearly addressing specific (or all) area or ship types. The scalability of presented approaches was rarely indicated either. Given that geographical features and maritime traffic can strongly differ among VTS regions, the choice of suitable anomaly-detection approaches is hampered.

Table 4. Screened publications grouped by data sources and denoted according to detection techniques and anomaly types. Data sources are abbreviated as follows: AIS (A), camera (C), radar (R) and other (O). Abbreviations are concatenated when data sources are combined. Filled circle (●) when a technique or feature is used in publication, otherwise empty circle (○).

Data Source	Author	Year	Descriptive Statistics	Stochastic Processes	Clustering	Classification	NN-Based	Rule-Based	Other	Route Deviation	Kinematic Deviation	Collision Risk	Zone Entry	Inconsistency
A	Abreu et al.	2021 [25,26]	●	○	○	○	○	○	○	○	●	○	○	○
	Cai et al.	2018 [27]	●	○	○	○	○	○	○	●	○	○	○	○
	Chatzikokolakis et al.	2019 [28]	○	○	○	○	○	●	●	●	○	●	○	●
	Chen et al.	2022 [29]	○	○	○	○	○	●	○	○	○	○	●	○
	Daranda and Dzemyda	2020 [30]	○	○	●	○	○	○	○	●	○	○	○	○
	Eljabu et al.	2021 [31]	○	○	○	○	●	○	○	●	○	○	○	○
	Filipiak et al.	2018 [32] 2019 [33]	●	○	○	○	○	●	○	○	●	○	○	●
	Ford et al.	2018 [34]	●	○	○	○	○	○	○	○	○	○	○	●
	Forti et al.	2018–2022 [35–39]	○	●	○	○	○	○	○	●	○	○	○	○
	Fu et al.	2017 [40]	○	○	●	○	○	○	○	●	●	○	○	○
	Goodarzi and Shaabani	2019 [41]	○	○	●	○	○	○	○	●	○	○	○	○
	Guo et al.	2021 [42]	○	○	●	○	○	○	●	○	●	○	○	●
	Han et al.	2020 [43]	○	○	●	○	○	○	○	●	●	○	○	○
	Hu et al.	2022 [44]	○	○	○	○	●	○	○	○	●	○	○	○
	Karatas et al.	2021 [45]	○	○	●	○	●	○	○	●	○	○	○	○
	Keane	2017 [46]	●	○	○	○	○	○	○	●	●	○	○	○
	Kontopoulos et al.	2020 [47]	○	○	○	○	○	●	○	○	○	○	○	●
	Kontopoulos et al.	2020 [48]	●	○	●	○	○	○	○	●	●	○	○	○
	Krüger	2019 [49]	○	○	○	●	○	●	○	○	○	○	○	●
	Nguyen et al.	2018–2021 [50–53]	●	○	○	○	●	○	○	●	●	○	○	○
	Patroumpas et al.	2017 [54]	●	○	○	○	○	●	○	○	●	●	●	○
	Roberts	2019 [55]	○	○	●	○	○	○	○	●	○	○	○	○
	Rong et al.	2020 [56]	●	○	●	○	○	○	○	●	○	○	○	○
	Singh and Heymann	2020 [57]	○	●	○	●	●	○	○	○	○	○	○	●
	Singh and Heymann	2020 [58]	○	○	○	○	●	○	○	○	○	○	○	●
	Tyasayumranani et al.	2022 [59]	●	○	○	○	○	●	○	●	●	●	○	○
	Wang et al.	2020 [60]	○	○	○	●	○	○	○	●	●	○	○	○
	Wang	2020 [61]	○	○	○	○	●	○	○	○	●	○	○	●
	Xia and Gao	2020 [62]	○	○	○	○	●	○	○	●	●	○	○	○
	Yan et al.	2022 [63]	●	○	○	●	○	○	○	○	○	○	○	●
	Zhao and Shi	2019 [64]	○	○	●	○	●	○	○	●	●	○	○	○
	Zhen et al.	2017 [65]	○	●	●	○	○	○	○	●	○	○	○	○
	Zhou et al.	2019 [66]	●	○	○	○	○	○	○	●	○	○	○	●
	Zissis et al.	2020 [67]	○	○	●	○	○	●	●	●	●	○	○	○
	Zor and Kittler	2017 [68]	○	●	○	○	○	○	○	●	●	○	○	○

Table 5. Screened publications grouped by data sources and denoted according to detection techniques and anomaly types. Data sources are abbreviated as follows: AIS (A), camera (C), radar (R) and other (O). Abbreviations are concatenated when data sources are combined. Filled circle (●) when a technique or feature is used in publication, otherwise empty circle (○).

Data Source		Publication		Technique							Anomaly				
		Author	Year	Descriptive Statistics	Stochastic Processes	Clustering	Classification	NN-Based	Rule-Based	Other	Route Deviation	Kinematic Deviation	Collision Risk	Zone Entry	Inconsistency
AO		Coleman et al.	2020 [69]	○	○	●	●	○	○	○	○	○	○	○	●
		Mazzarella et al.	2017 [70]	●	○	○	●	○	○	○	○	○	○	○	●
		Ray	2018 [71]	○	○	○	○	○	●	○	○	○	○	○	●
ARO		Thomopoulos et al.	2019 [72]	●	○	○	○	○	●	○	○	●	●	○	●
AR		d'Afflisio et al.	2018 [73,74] 2021 [75,76]	○	●	○	○	○	○	○	○	○	○	○	●
R		Bauw et al.	2020 [77]	○	○	○	●	●	○	○	○	○	○	●	○
		Van Loi et al.	2020 [78]	●	○	●	○	○	○	○	○	●	○	○	○
C		Fahn et al.	2019 [79]	○	○	○	●	○	●	○	○	○	●	○	○

In the following subsections, the approaches are presented sequentially from the perspective of the utilized data source, the underlying detection technique and the applicable anomaly type.

5.1. Data Sources

The frequency distribution of the utilized communication and sensor technologies as data sources is depicted in Figure 3. As can be seen, the majority of the proposed approaches utilize AIS solely as data source; however, some make use of AIS and other complementary data sources (cf. Tables 4 and 5). This may be due to the fact that AIS data is easily accessible, available in large quantities and consistently structured, which makes the research and development of novel anomaly detection techniques feasible [12,14]. Another reason may be that, with AIS, data anomalies can be detected to a large extent.

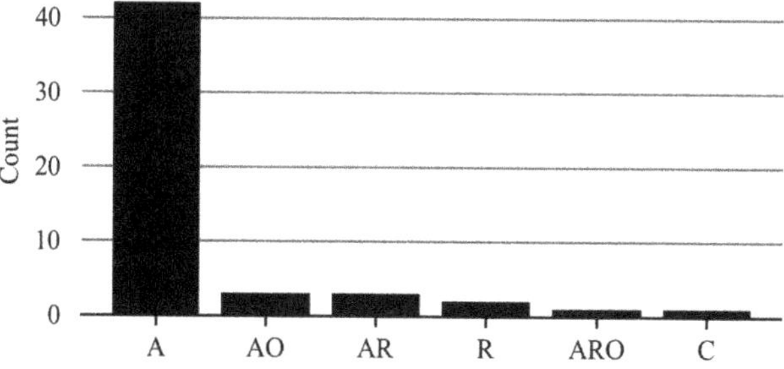

Figure 3. Counts of utilized data sources, i.e., communication and sensor types, as data sources in reviewed approaches. Data sources are abbreviated as follows: AIS (A), camera (C), radar (R) and other (O). Abbreviations are concatenated when data sources are combined.

Out of the 55 screened publications, 42 presented approaches relying on AIS solely. This is hardly surprising, as explained initially. A recent literature review specifically on the application of AIS-based maritime anomaly detection is presented by Wolsing et al. [14].

Some approaches combine AIS and other sources. D'Afflisio et al. propose an approach which relies on positional and kinematic data, which can come from AIS solely or complemented by radar [73–76]. In [72], Thomopoulos et al. present an anomaly-detection toolbox which fusions vessel data based on AIS, radar and Vessel Monitoring System [72]. In the concept by [69], it is proposed to enhance a vessel's own situation awareness by complementing ordinary AIS data with additional target vessels' data, such as from temperature sensors. Another data source is exploited by Mazzarella et al. [70] and Ray [71], who utilize the received signal strength indicator (RSSI) of terrestrial AIS base stations. As VTS maintain their own AIS base stations, the RSSI (or a similar dimension) is a potential data source for anomaly detection. The proposed data sources by [69–71] are listed under other in Tables 4 and 5.

Out of the screened publications, only three utilized approaches are described that rely on data sources other than AIS. This is very striking, due to the availability of other communication and sensor technologies. The approaches from Bauw et al. [77] and Van Loi et al. [78] are both tested on real coastal-surveillance radar datasets. Bauw et al. further specifies that one-dimensional high-resolution range profiles are used in their study. High-resolution satellite-based image data, which includes seashores, rivers and islands, is tested in [79]. Remarkably, no shore-based image data has been applied to the screened approaches. Similarly to the situation with radar-based detection techniques, this is striking, as camera systems are widespread at VTS sites, too.

Czapelewski et al. [80] apply purely synthetic and simplified image data from an aerial view. They intend, however, to extend their approach and experiments on (synthetic) radar data. Due to the very conceptual character of the approach and its current application of synthetic data only, this publication is not listed in the completed classification schemes.

The review from the perspective of data sources indicates, from the variety of communication and sensor technology, that VTS includes (cf. Table 1) a remarkable minority of methods, i.e., only AIS and radar, is considered in maritime anomaly-detection approaches. Notably, no approach was screened within this review which was based on VHF, despite the fact that this communication technology is used widely and provides context information which cannot be obtained by other communication or sensor technology. In addition, that is despite current research and development progress on natural language processing, on the one hand, and the established Standard Marine Communication Phrases (SMCP), on the other hand, which is followed by Gözalan et al. in [81].

5.2. Anomaly Types

The screened publications were classified according to the anomaly types (cf. Table 2) which they aim to detect. The majority of the publications, viz. 33, covers one specific anomaly type. The remaining 22 publications cover two or three anomaly types (cf. Tables 4 and 5). The most common combinations of addressed anomaly types cover, at least, either route deviation, kinematic deviation or both. The frequency distribution of the addressed anomaly types is depicted in Figure 4.

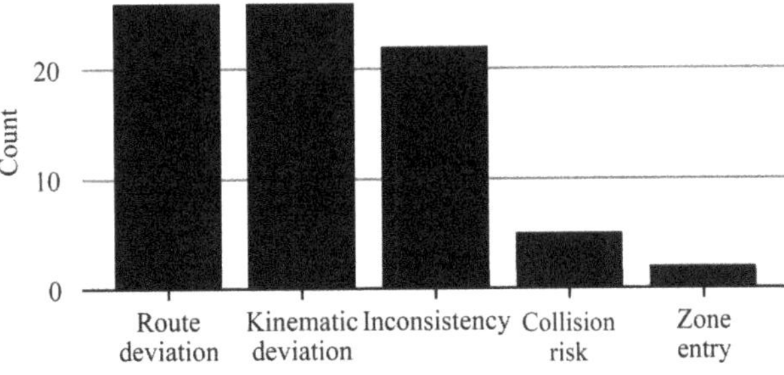

Figure 4. Counts of addressed anomaly types in reviewed approaches.

Tyasayumranani describes, in [59], an approach to detect whether a ship is being steered under the influence of alcohol. This complex detection is based on the detection of kinematic deviations, route deviations and collision risk. In [28,34], the authors present an approach to detect information inconsistency in the form of AIS reporting gaps due to potentially intentional AIS switch-off. Singh and Heymann describe approaches which shall be able to classify whether AIS reporting gaps are intentional or due to power outages [57,58]. Utilizing the RSSI through an AIS base station (cf. Section 5.1), Mazzarella et al. [70] and Ray [71] demonstrate the detection of AIS switch-off. Furthermore, Mazzarella et al. [70] and Thomopoulos et al. [72] outline a logic-based formalism which is able to detect falsified AIS messages. AIS spoofing is another scenario of information inconsistency and is covered extensively by the work of d'Afflisio et al. [73–76]. In [49], the author compares the outcome of different techniques (cf. Section 5.3) to detect spoofing committed by fishing vessels. In some approaches, inconsistency refers to invalid positional or kinematic data based on simple projections [32,33,42], taking a vessel's manoeuvrability into account [29] or detecting implausible value changes [61]. Another form of inconsistency detection is described by Yan et al. [63] and Zhou et al. [66]. Based on different approaches, both compare the detected ship type with the actually reported one and, so, potentially detect information inconsistency. Close-approaches scenarios are defined by [28] as imminent collisions between vessels or the grounding of a vessel. Patroumpas et al. formalize scenarios such as fast or close approaches to detect suspicious interactions such as package picking [54]. Based on radar-range profiles, Bauw et al. demonstrate, in [77], the detection of the visit of unusual ship types, e.g., fishing vessels, in areas where, normally, tankers and container and cargo vessels operate. This is the only publication which covers the anomaly type of zone entry in our review.

It seems that there is already a variety of anomaly scenarios covered by the reviewed literature. The complexity level, ranging from the detection of simple kinematic deviation to that of hiding activities via switched-off AIS, varies, too. Taking into account the differences in the areas and traffic patterns that VTS oversee, most probably some addressed anomaly scenarios may be used universally, where others may need more customized detections, e.g., specific environmental or geographical conditions.

5.3. Anomaly-Detection Techniques

In Section 3.4, the techniques are explained according to the screened publications that have been sorted into each technique. Figure 5 illustrates the frequency distribution of the techniques applied to solve the different problem types of anomaly detection. It shall be noted, however, that the counts of the usages of specific technique types may be biased since some technique types may cover more potential approaches than others.

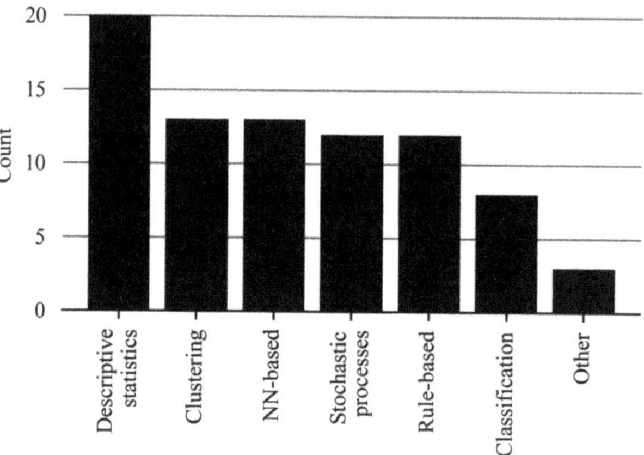

Figure 5. Counts of underlying techniques of reviewed approaches.

Remarkably, descriptive statistics is used most often followed by approaches which are based on clustering or neural networks. In [45,49,57,70,77], the authors compare the outcome of various techniques within specific anomaly-detection scenarios. Bauw et al. applies unsupervised approaches, notably, Support Vector Machine (SVM), Isolation forest, Local Outlier Factor, Convolutional Autoencoder and a semi-supervised NN-based classification approach proposed by Ruff et al. [77,82]. In the context of route deviation, Karataş compares the precision of numerous approaches, e.g., based on decision tree, random forest, and density-based spatial clustering of applications with noise (DBSCAN) in combination with classification and long short-term memory (LSTM), and states that decision-tree-based approaches perform the best in his context [45]. Krüger investigates the precision of AIS-spoofing detection with classification methods such as random forest, decision tree or k-nearest neighbours (kNN) algorithms and a fuzzy rules-based method [49]. In the context of detection of intentional AIS on–off switching, Mazzarella et al. compare the detection precision using SVM and a historically based spatial distribution to model the AIS-base-station RSSI [70]. In a similar context, Singh and Heymann compare the effectiveness at detecting whether, more specifically, AIS on–off switching occurs intentionally or due to a power outage by applying approaches such as Naive Bayes, SVM, kNN, decision tree, random forest and linear regression and NN [57].

As can be seen in Figure 6, the applied technique type is associated with the addressed anomaly type. The anomaly-types route deviation and kinematic deviation are covered by the commonly applied techniques descriptive statistics, clustering or NN-based classification.

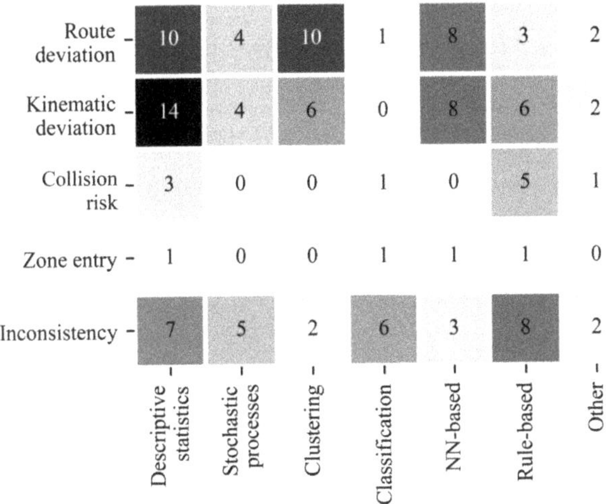

Figure 6. Counts of detection techniques combined with anomaly types.

Most detection techniques which are based on descriptive statistics consider, somehow, outliers in the (spatio-temporal) distribution of the vessels' attributes [27,32,33,46,59,72]. Outliers can be defined with, e.g., distance to the mean or standard deviations. Commonly used attributes are, particularly, the vessels' position, speed and course. Abreu et al. propose an approach based on visual analytics [25,26]: For visually selected spatial subtrajectories, a score indicating the degree of anomaly is calculated. Analogously to the approaches mentioned before, the calculation is based on the statistical deviation from the mean of given attributes. Implementation of anomaly-detection techniques based on descriptive statistics are very obvious in the age of big data. VTS are able to record data which is produced by their communication and sensor technologies. This data can serve as a solid basis for the aforementioned and other data-driven techniques.

Numerous approaches apply clustering, another data-driven technique, before proceeding with other techniques such as descriptive statistics or NN-based classification. This way, most authors try to reduce the amount of data while still keeping information about relevant traffic patterns, e.g., stopping or turning points. In [30,40–43,48,56,64], the authors apply DBSCAN as a clustering algorithm to vessels' positional or kinematic AIS data. Zhao et al. use these clusters as training data for a Recurrent Neural Network (RNN), i.e., an LSTM [64]. The authors highlight that this method requires high data quality, which is a challenging task in a real-time environment such as VTS. It shall be mentioned that adjustment of the parameters of the DBSCAN clustering algorithm is known to be arbitrary [53]. Against the background of scalability and explainability of anomaly-detection techniques in safety-critical environments, the application of the DBSCAN clustering algorithm, however, is disputable.

Since NN are universal function approximators, they can process various data types and address a broad range of problem types in general. Due to the advent of big data, especially the availability of AIS data, NN can be trained intensely. A commonly addressed anomaly type is route deviation; however, the precise NN classes applied vary. Eljabu et al. trained a Graph Convolution Network which models the spatio-temporal vessel traffic network and demonstrates that it is able to identify anomalous route behavior [31]. In order to detect kinematic anomalies, Hu et al. apply a Variational Autoencoder and show that it outperforms other approaches, e.g., decision tree, SVM and LSTM [44]. In the work of Nguyen et al., an RNN is trained to detect kinematic and route deviations. It does so by learning the stochastic representation of given routes and detecting improbable events which can be anomalies [50–53]. A similar approach applying a Bayesian RNN is followed by Xia et al. [62]. Both working groups state that RNN are able to learn vessel behaviour patterns based on AIS data without preliminary traffic extraction and claim promising results. The literature shows that the results of NN-based anomaly detection are promising. Be that as it may, the decision finding process of NN-based systems intrinsically lack explainability [12,83,84]. Due to this, the application of NN in safety-critical environments such as VTS is problematic.

Forti et al. and d'Afflisio et al. base their work on the Ornstein–Uhlenbeck mean-reverting stochastic process [22]. However, both working groups have extended their approaches differently. Forti et al. cover route-deviation detection and demonstrate the capabilities of their approach [35,36,39] by applying it on the real-world scenario of the blocking of the Suez canal. They state that their approach would have been capable of predicting this anomaly early [38]. D'Afflisio, on the other hand, integrate a hypothesis testing framework and they aim at detecting AIS spoofing and stealth deviations [73–76]. Beyond that, inconsistency anomalies can be detected by other techniques. Some examples are rule-based approaches with predetermined thresholds [28,29,33,47,71,72], statistical approaches with data-driven thresholds [33,70,72] or those based on cross-checking data values using pre-trained classification models, e.g., SVM or random forest [63,70], or even NN, as performed by Wang [61]. The latter author further proposes a rare behaviour factor which can be used by (VTS) operators to tune the sensitivity of the anomaly detection [60]. The rule-based or statistically driven techniques offer a comprehensible decision process of anomaly detection. However, these techniques can only uncover anomalies which are predefined by experts and covered by statistical data. Furthermore, it shall be mentioned that anomaly-detection techniques based on statistical data assume that statistically frequent values (or events) correspond to normalcy. However, statistically rare values (or events) can be normal, too, and vice versa.

Two anomaly-detection techniques are sorted under other in the tabular overview (cf. Tables 4 and 5). A unique approach within the scope of this review is followed by Guo et al. who deduce a vessel's manoeuvrability from its trajectory through kinematic interpolation of the AIS track [42]. An AIS data point that does not follow the ship's manoeuvrability is a potential anomaly. Another unconventional approach is the utilization of convex hulls, which is carried out by [28,67].

6. Conclusions

To ensure safe and smooth maritime traffic, its surveillance is performed by VTS for specific regions such as port or channel areas. VTS operators make use of various DST in order to reduce their workload and increase maritime-traffic safety. The DST themselves can utilize a variety of communication and sensor technologies and resulting data at VTS sites. One application of DST is the detection of anomalies in maritime traffic.

In this survey, the state-of-the-art of anomaly-detection approaches are presented. For this purpose, 136 publications from the years 2017 to 2022 were collected and screened for relevance. Therefrom, 55 publications were classified as relevant (cf. Section 4) and, subsequently, further investigated. Noteworthily, the number of relevant publications increases significantly for later years, as can be seen in Figure 2.

The proposed approaches are presented from the perspectives of the data sources (cf. Section 5.1), detectable anomaly types (cf. Section 5.2) and anomaly-detection techniques (cf. Section 5.3). The results can be summarized as follows:

- The primary data source used as a base for anomaly detection is AIS. The widespread application of this data source can be explained by its ubiquity, standardized structure and sufficient coverage of relevant maritime traffic information. Only few approaches utilize other data sources. However, despite the contextual information, VHF has not been used in any approach.
- The served anomaly scenarios can be grouped into five generic anomaly types. Most approaches aim at detecting either route or kinematic deviations, or information consistency. However, the precisely served anomaly scenario can be very specific and, for example, cover detection of intentional AIS switch-off, unusual ship visits or a ship being steered under the influence of alcohol.
- The underlying detection techniques are manifold and can range from rather simple and transparent approaches such as measures from descriptive statistics or rule-based systems to more complex and intransparent approaches, e.g., based on NN. The applied detection techniques are connected to the aimed anomaly detection type. For example, approaches for the detection of implausible kinematic or route deviation are often based on measures from descriptive statistics.

Finally, it is worth mentioning that almost all publications focus on specific anomaly scenarios. The approaches presented by Nguyen et al. [53] and Thomopoulos et al. [72] are part of holistic systems. However, none of the investigated approaches are applied operatively at VTS sites, yet.

7. Discussion

Despite all the approaches outlined in science, monitoring a VTS area is still a manual task for the VTS operators and, subsequently, connected with challenges to human observation capabilities and the ever-increasing volume of communication and sensor data. DST are under development, as outlined before; however, they are not used widely spread in current VTS operations. The authors assume that this is due to a variety of reasons, namely:

1. Transparency of and trust in DST decision-making;
2. Real-time capabilities;
3. Scattered national solutions;
4. Geographical scalability of approaches and training data sets.

As a first step, the potential lack of trust of the VTS operators in DST decision-making capabilities must be addressed. Specifically, with approaches not based on descriptive statistics or rule-based systems, the decision-finding is not clear to the operator due to a lack of explainability. Humans, then, tend not to trust the respective DST. Most of the solutions discussed above are concepts, prototypes or case studies which have never been integrated or tested within the operators' daily working systems. This advanced prototyping and testing, which could help to close that trust gap, is also challenging, as not all VTS are computationally equipped to allow for higher levels of anomaly detection in real-time, given the existing computational power installed. Further, some interfaces to third-party computational systems cannot be realized due to security considerations, since VTS are considered critical infrastructure. As VTS systems are often national solutions for one coastal state resulting nationally in a bilateral monopoly structure as well as different interpretations of normality based on the geographical region, the scalability of research and development efforts is not a given. This makes development more challenging and rather expensive. Furthermore, the step from descriptive or rule-based DST towards other techniques outlined in this survey would require data-quality measures such as properly annotated training data sets, which go beyond pure AIS traffic tracking. According to the best knowledge, those set are not publicly available and significant efforts are needed to generate such sets, even though initial approaches to automatically annotate AIS data sets exists, such as, e.g., by Constapel et al. in [85] for COLREGs information.

Despite those implementation hurdles, DST capabilities for VTS operations are continuously progressing and have reached a state that allows a constantly more in-depth monitoring of maritime traffic than can be realistically achieved by human observation only. Safe but realistic test environments are needed to ensure early human-oriented testing by VTS operators, such as, e.g., achieved by the European Maritime Simulator Network within the STM Validation project [86]. Such environments can help to safely test DST to achieve trust and acceptance by the VTS operators in advance, to overcome one of the implementation hurdles for innovative DST in VTS. Additionally, integration of VHF information into DST must be further investigated, as there is still relevant contextual information missing for the DST given the currently considered data sources, which is AIS mainly.

In summary, the following future work is required, according to the authors' opinion, to facilitate DST in daily VTS operations:

1. Integration of VHF data into DST to bridge the missing data gap;
2. Generation and availability of annotated and approved training data to bridge the training and testing gap;
3. System integration of realistic test beds for operators under operational conditions to bridge the human–machine trust gap.

This would help DST to reach their full potential in assisting the VTS operators in their tasks by relieving them from routine, but still complex, monitoring tasks. They could then focus on ensuring safe and smooth maritime traffic in critical situations.

Author Contributions: Conceptualization, T.S., M.C. and H.-C.B.; methodology, T.S., Y.K. and M.C.; literature review, T.S., Y.K. and H.-C.B.; supervision, M.C.; validation, T.S.; visualization, T.S.; writing—original draft preparation, T.S.; writing—review and editing, M.C. and H.-C.B.; discussion H.-C.B. All authors have read and agreed to the published version of the manuscript.

Funding: This research is part of the project LEAS (https://www.cml.fraunhofer.de/en/research-projects/LEAS.html, accessed on 1 June 2023). The project is funded by the German Federal Ministry of Education and Research (BMBF) within the programm "Research for Civil Security 2018-2023" under grant number 13N16246 managed by VDI Technologiezentrum.

Institutional Review Board Statement: Not applicable.

Informed Consent Statement: Not applicable.

Data Availability Statement: No new data were created or analyzed in this study. Data sharing is not applicable to this article.

Conflicts of Interest: The authors declare no conflict of interest.

Abbreviations

The following abbreviations are used in this manuscript:

AIS	Automatic identification system
COLREGs	Convention on the International Regulations for Preventing Collisions at Sea
DBSCAN	Density-based spatial clustering of applications with noise
DST	Decision support tool
IALA	International Association of Marine Aids to Navigation and Lighthouse Authorities
IMO	International Maritime Organization
INS	Information Service
kNN	k-nearest neighbours
LSTM	Long short-term memory
ML	Machine learning
NN	Neural network
RNN	Recurrent neural network
RSSI	Received signal strength indicator
SMCP	Standard marine communication phrases
SVM	Support vector machine
VHF	Very high frequency
VTS	Vessel traffic service

References

1. Wan, Z.; Chen, J.; Makhloufi, A.E.; Sperling, D.; Chen, Y. Four Routes to Better Maritime Governance. *Nature* **2016**, *540*, 27–29. [CrossRef] [PubMed]
2. Burmeister, H.C. Ensuring Navigational Safety and Mitigate Maritime Traffic Risks While Designing Port Approaches and Ship Maneuvering Areas. In *Handbook of Terminal Planning*; Böse, J.W., Ed.; Springer International Publishing: Cham, Switzerland, 2020; pp. 269–286. [CrossRef]
3. *VTS Manual*; Technical report; International Association of Marine Aids to Navigation and Lighthouse Authorities: Saint-Germain-en-Laye, France, 2021.
4. *Resolution A.1158(32), Guidelines for Vessel Traffic Services*; Technical report; International Maritime Organization: London, UK, 2021.
5. *G1110 Use of Decision Support Tools for VTS Personnel*; Technical report; International Association of Marine Aids to Navigation and Lighthouse Authorities: Saint-Germain-en-Laye, France 2022.
6. Chandola, V.; Banerjee, A.; Kumar, V. Anomaly Detection: A Survey. *ACM Comput. Surv.* **2009**, *41*, 15:1–15:58. [CrossRef]
7. Riveiro, M.; Pallotta, G.; Vespe, M. Maritime Anomaly Detection: A Review. *Wiley Interdiscip. Rev. Data Min. Knowl. Discov.* **2018**, *8*, e1266. [CrossRef]
8. *G1141 Operational Procedure for Delivering VTS*; Technical report; International Association of Marine Aids to Navigation and Lighthouse Authorities: Saint-Germain-en-Laye, France, 2022.
9. Sidibé, A.; Shu, G. Study of Automatic Anomalous Behaviour Detection Techniques for Maritime Vessels. *J. Navig.* **2017**, *70*, 847–858. [CrossRef]
10. Tu, E.; Zhang, G.; Rachmawati, L.; Rajabally, E.; Huang, G.B. Exploiting AIS Data for Intelligent Maritime Navigation: A Comprehensive Survey from Data to Methodology. *IEEE Trans. Intell. Transp. Syst.* **2017**, *19*, 1559–1582. [CrossRef]
11. Yan, R.; Wang, S. Study of Data-Driven Methods for Vessel Anomaly Detection Based on AIS Data. In *Smart Transportation Systems 2019*; Springer: Singapore, 2019; pp. 29–37. [CrossRef]
12. May Petry, L.; Soares, A.; Bogorny, V.; Brandoli, B.; Matwin, S. Challenges in Vessel Behavior and Anomaly Detection: From Classical Machine Learning to Deep Learning. In *Advances in Artificial Intelligence, Proceedings of the 33rd Canadian Conference on Artificial Intelligence, Canadian AI 2020, Ottawa, ON, Canada, 13–15 May 2020*; Lecture Notes in Computer Science (including subseries Lecture Notes in Artificial Intelligence and Lecture Notes in Bioinformatics); Springer International Publishing: Cham, Switzerland, pp. 401–407.
13. Dogancay, K.; Tu, Z.; Ibal, G. Research into Vessel Behaviour Pattern Recognition in the Maritime Domain: Past, Present and Future. *Digit. Signal Process.* **2021**, *119*, 103191. [CrossRef]
14. Wolsing, K.; Roepert, L.; Bauer, J.; Wehrle, K. Anomaly Detection in Maritime AIS Tracks: A Review of Recent Approaches. *J. Mar. Sci. Eng.* **2022**, *10*, 112. [CrossRef]
15. Vessel Traffic Services. Available online: https://www.imo.org/en/OurWork/Safety/Pages/VesselTrafficServices.aspx (accessed on 1 June 2023).
16. *G1089 Provision of a VTS*; Technical report; International Association of Marine Aids to Navigation and Lighthouse Authorities: Saint-Germain-en-Laye, France, 2022.
17. *G1111 Preparation of Operational and Technical Performance Requirements for VTS Systems*; Technical report; International Association of Marine Aids to Navigation and Lighthouse Authorities: Saint-Germain-en-Laye, France, 2022.

18. Skolnik, M.I. *Introduction to Radar Systems*, 3rd ed.; McGraw-Hill Electrical Engineering Series; McGraw Hill: Boston, MA, USA; Burr Ridge, IL, USA; Dubuque, IA, USA, 2001.
19. *Resolution MSC.74(69), Adoption of New and Amended Performance Standards*; Technical report; International Maritime Organization:London, UK, 1998.
20. Lane, R.O.; Nevell, David, A.; Hayward, S.D.; Beaney, T.W. Maritime Anomaly Detection and Threat Assessment. In Proceedings of the 2010 13th International Conference on Information Fusion, Edinburgh, UK, 26–29 July 2010; pp. 1–8. [CrossRef]
21. van Laere, J.; Nilsson, M. Evaluation of a Workshop to Capture Knowledge from Subject Matter Experts in Maritime Surveillance. In Proceedings of the 2009 12th International Conference on Information Fusion, Seattle, WA, USA, 6–9 July 2009; pp. 171–178.
22. d'Afflisio, E. Maritime Anomaly Detection Based on Statistical Methodologies: Theory and Applications. Ph.D. Thesis, Università degli Studi di Firenze, Florence, Italy, 2022.
23. Russell, S.; Norvig P. *Artificial Intelligence, A Modern Approach*, 4th ed.; Pearson series in artificial intelligence; Pearson: Hoboken, NJ, USA, 2021.
24. Alzubaidi, L.; Zhang, J.; Humaidi, A.J.; Al-Dujaili, A.; Duan, Y.; Al-Shamma, O.; Santamaría, J.; Fadhel, M.A.; Al-Amidie, M.; Farhan, L. Review of Deep Learning: Concepts, CNN Architectures, Challenges, Applications, Future Directions. *J. Big Data* **2021**, *8*, 53. [CrossRef]
25. Abreu, F.H.O.; Soares, A.; Paulovich, F.V.; Matwin, S. A Trajectory Scoring Tool for Local Anomaly Detection in Maritime Traffic Using Visual Analytics. *ISPRS Int. J. Geoinf.* **2021**, *10*, 412. [CrossRef]
26. Abreu, F.H.; Soares, A.; Paulovich, F.V.; Matwin, S. Local Anomaly Detection in Maritime Traffic Using Visual Analytics. In Proceedings of the CEUR Workshop Proceedings, Nicosia, Cyprus, 23–26 March 2021; Volume 2841.
27. Cai, C.; Chen, R.; Liu, A.D.; Roberts, F.S.; Xie, M. iGroup Learning and iDetect for Dynamic Anomaly Detection with Applications in Maritime Threat Detection. In Proceedings of the 2018 IEEE International Symposium on Technologies for Homeland Security (HST), Woburn, MA, USA, 23–24 October 2018; IEEE: Piscataway, NJ, USA, 2018; pp. 1–6.
28. Chatzikokolakis, K.; Zissis, D.; Vodas, M.; Tsapelas, G.; Mouzakitis, S.; Kokkinakos, P.; Askounis, D. BigDataOcean Project: Early Anomaly Detection from Big Maritime Vessel Traffic Data. In Proceedings of the 18th International Conference on Computer and IT Applications in the Maritime Industries, Tullamore, Ireland, 25–27 March 2019.
29. Chen, S.; Huang, Y.; Lu, W. Anomaly Detection and Restoration for AIS Raw Data. *Wirel. Commun. Mob. Comput.* **2022**, *2022*, 5954483. [CrossRef]
30. Daranda, A.; Dzemyda, G. Navigation Decision Support: Discover of Vessel Traffic Anomaly According to the Historic Marine Data. *Int. J. Comput. Commun. Control* **2020**, *15*, 1–9. [CrossRef]
31. Eljabu, L.; Etemad, M.; Matwin, S. Anomaly Detection in Maritime Domain Based on Spatio-Temporal Analysis of AIS Data Using Graph Neural Networks. In Proceedings of the 2021 5th International Conference on Vision, Image and Signal Processing (ICVISP), Kuala Lumpur, Malaysia, 18–20 December 2021; IEEE: Piscataway, NJ, USA, 2021; pp. 142–147. [CrossRef]
32. Filipiak, D.; Stróżyna, M.; Krzysztof, W. Anomaly Detection in the Maritime Domain: Comparison of Traditional and Big Data Approach. In Proceedings of the NATO IST-160-RSM Specialists' Meeting on Big Data & Artificial Intelligence for Military Decision Making, Bordeaux, France, 30 May–1 June 2018; pp. 1–13. [CrossRef]
33. Filipiak, D.; Stróżyna, M.; Węcel, K.; Abramowicz, W. Big Data for Anomaly Detection in Maritime Surveillance: Spatial AIS Data Analysis for Tankers. *Marit. Tech. J.* **2019**, *215*, 5–28. [CrossRef]
34. Ford, J.H.; Peel, D.; Kroodsma, D.; Hardesty, B.D.; Rosebrock, U.; Wilcox, C. Detecting Suspicious Activities at Sea Based on Anomalies in Automatic Identification Systems Transmissions. *PLoS ONE* **2018**, *13*, e0201640. [CrossRef] [PubMed]
35. Forti, N.; Millefiori, L.M.; Braca, P. Hybrid Bernoulli Filtering for Detection and Tracking of Anomalous Path Deviations. In Proceedings of the 2018 21st International Conference on Information Fusion (FUSION), Cambridge, UK, 10–13 July 2018; pp. 1178–1184. [CrossRef]
36. Forti, N.; Millefiori, L.M.; Braca, P. Unsupervised Extraction of Maritime Patterns of Life from Automatic Identification System Data. In Proceedings of the OCEANS 2019-Marseille, Marseille, France, 17–20 June 2019; pp. 1–5. [CrossRef]
37. Forti, N.; Millefiori, L.M.; Braca, P.; Willett, P. Random Finite Set Tracking for Anomaly Detection in the Presence of Clutter. In Proceedings of the IEEE National Radar Conference - Proceedings, Florence, Italy, 21–25 September 2020; Volume 2020. [CrossRef]
38. Forti, N.; d'Afflisio, E.; Braca, P.; Millefiori, L.M.; Willett, P.; Carniel, S. Maritime Anomaly Detection in a Real-World Scenario: Ever Given Grounding in the Suez Canal. *IEEE Trans. Intell. Transp. Syst.* **2021**, *23*, 13904–13910. [CrossRef]
39. Forti, N.; Millefiori, L.M.; Braca, P.; Willett, P. Bayesian Filtering for Dynamic Anomaly Detection and Tracking. *IEEE Trans. Aerosp. Electron. Syst.* **2022**, *58*, 1528–1544. [CrossRef]
40. Fu, P.; Wang, H.; Liu, K.; Hu, X.; Zhang, H. Finding Abnormal Vessel Trajectories Using Feature Learning. *IEEE Access* **2017**, *5*, 7898–7909. [CrossRef]
41. Goodarzi, M.; Shaabani, M. Maritime Traffic Anomaly Detection from Spatio-temporal AIS Data. In Proceedings of the Second International Management Conference and Fuzzy Systems, 2019; pp. 1–8.
42. Guo, S.; Mou, J.; Chen, L.; Chen, P. An Anomaly Detection Method for AIS Trajectory Based on Kinematic Interpolation. *J. Mar. Sci. Eng.* **2021**, *9*, 609. [CrossRef]
43. Han, X.; Armenakis, C.; Jadidi, M. DBscan Optimization for Improving Marine Trajectory Clustering and Anomaly Detection. *Int. Arch. Photogramm. Remote Sens. Spat. Inf. Sci.* **2020**, *43*, 455–461. [CrossRef]

44. Hu, J.; Kaur, K.; Lin, H.; Wang, X.; Hassan, M.M.; Razzak, I.; Hammoudeh, M. Intelligent Anomaly Detection of Trajectories for IoT Empowered Maritime Transportation Systems. *IEEE Trans. Intell. Transp. Syst.* **2022**, *24*, 2382–2391. [CrossRef]

45. Karataş, G.B.; Karagoz, P.; Ayran, O. Trajectory Pattern Extraction and Anomaly Detection for Maritime Vessels. *IEEE Internet Things J.* **2021**, *16*, 100436. [CrossRef]

46. Keane, K.R. Detecting Motion Anomalies. In Proceedings of the 8th ACM SIGSPATIAL Workshop on GeoStreaming, IWGS'17, Redondo Beach, CA, USA, 7–10 November 2017; Association for Computing Machinery: New York, NY, USA, 2017; pp. 21–28. [CrossRef]

47. Kontopoulos, I.; Chatzikokolakis, K.; Zissis, D.; Tserpes, K.; Spiliopoulos, G. Real-Time Maritime Anomaly Detection: Detecting Intentional AIS Switch-Off. *IJBDI* **2020**, *7*, 85. [CrossRef]

48. Kontopoulos, I.; Varlamis, I.; Tserpes, K. Uncovering Hidden Concepts from AIS Data: A Network Abstraction of Maritime Traffic for Anomaly Detection. In *Multiple-Aspect Analysis of Semantic Trajectories: First International Workshop, MASTER 2019, Held in Conjunction with ECML-PKDD 2019, Würzburg, Germany, 16 September 2019*; Lecture Notes in Computer Science (Including Subseries Lecture Notes in Artificial Intelligence and Lecture Notes in Bioinformatics); Springer International Publishing: Cham, Switzerland, 2020; Volume 11889, LNAI, pp. 6–20. [CrossRef]

49. Krüger, M. Detection of AIS Spoofing in Fishery Scenarios. In Proceedings of the 2019 22th International Conference on Information Fusion (FUSION), Ottawa, ON, Canada, 2–5 July 2019; pp. 1–7. [CrossRef]

50. Nguyen, D.; Vadaine, R.; Hajduch, G.; Garello, R.; Fablet, R. Multi-Task Learning for Maritime Traffic Surveillance from AIS Data Streams. In Proceedings of the IEEE 5th International Conference on Data Science and Advanced Analytics (DSAA), Turin, Italy, 1–3 October 2018; pp. 1–3.

51. Nguyen, D.; Vadaine, R.; Hajduch, G.; Garello, R.; Fablet, R. A Multi-Task Deep Learning Architecture for Maritime Surveillance Using AIS Data Streams. In Proceedings of the 2018 IEEE 5th International Conference on Data Science and Advanced Analytics, DSAA 2018, Turin, Italy, 1–4 October 2018; pp. 331–340.

52. Nguyen, D.; Simonin, M.; Hajduch, G.; Vadaine, R.; Tedeschi, C.; Fablet, R. Detection of Abnormal Vessel Behaviours from AIS Data Using GeoTrackNet: From the Laboratory to the Ocean. *arXiv* **2020**.

53. Nguyen, D.; Vadaine, R.; Hajduch, G.; Garello, R.; Fablet, R. GeoTrackNet–A Maritime Anomaly Detector Using Probabilistic Neural Network Representation of AIS Tracks and A Contrario Detection. *IEEE Trans. Intell. Transp. Syst.* **2021**, *23*, 5655–5667.

54. Patroumpas, K.; Alevizos, E.; Artikis, A.; Vodas, M.; Pelekis, N.; Theodoridis, Y. Online Event Recognition from Moving Vessel Trajectories. *GeoInformatica* **2017**, *21*, 389–427. [CrossRef]

55. Roberts, S.A. A Shape-Based Local Spatial Association Measure (LISShA): A Case Study in Maritime Anomaly Detection. *Geogr. Anal.* **2019**, *51*, 403–425. [CrossRef]

56. Rong, H.; Teixeira, A.; Guedes Soares, C. Data Mining Approach to Shipping Route Characterization and Anomaly Detection Based on AIS Data. *Ocean Eng.* **2020**, *198*, 106936. [CrossRef]

57. Singh, S.K.; Heymann, F. On the Effectiveness of AI-Assisted Anomaly Detection Methods in Maritime Navigation. In Proceedings of the 2020 IEEE 23rd International Conference on Information Fusion (FUSION), Rustenburg, South Africa, 6–9 July 2020; IEEE: Piscataway, NJ, USA, 2020; pp. 1–7. [CrossRef]

58. Singh, S.K.; Heymann, F. Machine Learning-Assisted Anomaly Detection in Maritime Navigation Using AIS Data. In Proceedings of the 2020 IEEE/ION Position, Location and Navigation Symposium (PLANS), Portland, OR, USA, 20–23 April 2020; IEEE: Piscataway, NJ, USA, 2020; pp. 832–838. [CrossRef]

59. Tyasayumranani, W.; Hwang, T.; Hwang, T.; Youn, I.H. Anomaly Detection Model of Small-Scaled Ship for Maritime Autonomous Surface Ships' Operation. *J. Int. Marit. Saf. Environ. Aff. Shipp.* **2022**, *6*, 224–235. [CrossRef]

60. Wang, F.; Lei, Y.; Liu, Z.; Wang, X.; Ji, S.; Tung, A.K. Fast and Parameter-Light Rare Behavior Detection in Maritime Trajectories. *Inf. Process. Manag.* **2020**, *57*, 102268. [CrossRef]

61. Wang, Y. Application of Neural Network in Abnormal AIS Data Identification. In Proceedings of the 2020 IEEE International Conference on Artificial Intelligence and Computer Applications (ICAICA), Dalian, China, 27–29 June 2020; IEEE: Piscataway, NJ, USA, 2020; pp. 173–179. [CrossRef]

62. Xia, Z.; Gao, S. Analysis of Vessel Anomalous Behavior Based on Bayesian Recurrent Neural Network. In Proceedings of the 2020 IEEE 5th International Conference on Cloud Computing and Big Data Analytics (ICCCBDA), Chengdu, China, 10–13 April 2020; pp. 393–397. [CrossRef]

63. Yan, Z.; Song, X.; Zhong, H.; Yang, L.; Wang, Y. Ship Classification and Anomaly Detection Based on Spaceborne AIS Data Considering Behavior Characteristics. *Sensors* **2022**, *22*, 7713. [CrossRef]

64. Zhao, L.; Shi, G. Maritime Anomaly Detection Using Density-based Clustering and Recurrent Neural Network. *J. Navig.* **2019**, *72*, 894–916. [CrossRef]

65. Zhen, R.; Jin, Y.; Hu, Q.; Shao, Z.; Nikitakos, N. Maritime Anomaly Detection within Coastal Waters Based on Vessel Trajectory Clustering and Naïve Bayes Classifier. *J. Navig.* **2017**, *70*, 648–670. [CrossRef]

66. Zhou, Y.; Wright, J.; Maskell, S. A Generic Anomaly Detection Approach Applied to Mixture-of-unigrams and Maritime Surveillance Data. In Proceedings of the 2019 Symposium on Sensor Data Fusion: Trends, Solutions, Applications, SDF 2019, Bonn, Germany, 15–17 October 2019. [CrossRef]

67. Zissis, D.; Chatzikokolakis, K.; Spiliopoulos, G.; Vodas, M. A Distributed Spatial Method for Modeling Maritime Routes. *IEEE Access* **2020**, *8*, 47556–47568. [CrossRef]

68. Zor, C.; Kittler, J. Maritime Anomaly Detection in Ferry Tracks. In Proceedings of the 2017 IEEE International Conference on Acoustics, Speech and Signal Processing (ICASSP), New Orleans, LA, USA, 5–9 March 2017; pp. 2647–2651. [CrossRef]
69. Coleman, J.; Kandah, F.; Huber, B. Behavioral Model Anomaly Detection in Automatic Identification Systems (AIS). In Proceedings of the 2020 10th Annual Computing and Communication Workshop and Conference (CCWC), Las Vegas, NV, USA, 6–8 January 2020; pp. 0481–0487. [CrossRef]
70. Mazzarella, F.; Vespe, M.; Alessandrini, A.; Tarchi, D.; Aulicino, G.; Vollero, A. A Novel Anomaly Detection Approach to Identify Intentional AIS On-off Switching. *Expert Syst. Appl.* **2017**, *78*, 110–123. [CrossRef]
71. Ray, C. Data Variety and Integrity Assessment for Maritime Anomaly Detection. In Proceedings of the International Conference on Big Data and Cyber-Security Intelligence, Las Vegas, NV, USA, 12–14 December 2018.
72. Thomopoulos, S.C.A.; Rizogannis, C.; Thanos, K.G.; Dimitros, K.; Panou, K.; Zacharakis, D. OCULUS Sea™ Forensics: An Anomaly Detection Toolbox for Maritime Surveillance. In *Lecture Notes in Business Information Processing*; Springer: Cham, Switzerland, 2019; Volume 373, pp. 485–495. [CrossRef]
73. d'Afflisio, E.; Braca, P.; Millefiori, L.M.; Willett, P. Detecting Anomalous Deviations from Standard Maritime Routes Using the Ornstein–Uhlenbeck Process. *IEEE Trans. Signal Process.* **2018**, *66*, 6474–6487. [CrossRef]
74. d'Afflisio, E.; Braca, P.; Millefiori, L.M.; Willett, P. Maritime Anomaly Detection Based on Mean-Reverting Stochastic Processes Applied to a Real-World Scenario. In Proceedings of the 2018 21st International Conference on Information Fusion (FUSION), Cambridge, UK, 10–13 July 2018; IEEE: Piscataway, NJ, USA, 2018; pp. 1171–1177. [CrossRef]
75. d'Afflisio, E.; Braca, P.; Willett, P. Malicious AIS Spoofing and Abnormal Stealth Deviations: A Comprehensive Statistical Framework for Maritime Anomaly Detection. *IEEE Trans. Aerosp. Electron. Syst.* **2021**, *57*, 2093–2108. [CrossRef]
76. d'Afflisio, E.; Braca, P.; Chisci, L.; Battistelli, G.; Willett, P. Maritime Anomaly Detection of Malicious Data Spoofing and Stealth Deviations from Nominal Route Exploiting Heterogeneous Sources of Information. In Proceedings of the 2021 IEEE 24th International Conference on Information Fusion (FUSION), Sun City, South Africa, 1–4 November 2021; IEEE: Piscataway, NJ, USA, 2021; pp. 1–7. [CrossRef]
77. Bauw, M.; Velasco-Forero, S.; Angulo, J.; Adnet, C.; Airiau, O. From Unsupervised to Semi-Supervised Anomaly Detection Methods for HRRP Targets. In Proceedings of the 2020 IEEE Radar Conference (RadarConf20), Florence, Italy, 21–25 September 2020; IEEE: Piscataway, NJ, USA, 2020; pp. 1–6.
78. Van Loi, N.; Kien, T.T.; Hop, T.V.; Van Khuong, N. Abnormal Moving Speed Detection Using Combination of Kernel Density Estimator and DBSCAN for Coastal Surveillance Radars. In Proceedings of the 2020 7th International Conference on Signal Processing and Integrated Networks (SPIN), Noida, India, 27–28 February 2020; IEEE: Piscataway, NJ, USA, 2020; pp. 143–147. [CrossRef]
79. Fahn, C.; Ling, J.; Yeh, M.; Huang, P.; Wu, M. Abnormal Maritime Activity Detection in Satellite Image Sequences Using Trajectory Features. *IJFCC* **2019**, *8*, 29–33. [CrossRef]
80. Czaplewski, B.; Dzwonkowski, M. A Novel Approach Exploiting Properties of Convolutional Neural Networks for Vessel Movement Anomaly Detection and Classification. *ISA Trans.* **2022**, *119*, 1–16. [CrossRef]
81. Gözalan, A.; John, O.; Lübcke, T.; Maier, A.; Reimann, M.; Richter, J.G.; Zverev, I. Assisting Maritime Search and Rescue (SAR) Personnel with AI-Based Speech Recognition and Smart Direction Finding. *J. Mar. Sci. Eng.* **2020**, *8*, 818. [CrossRef]
82. Ruff, L.; Vandermeulen, R.A.; Görnitz, N.; Binder, A.; Müller, E.; Müller, K.R.; Kloft, M. Deep Semi-Supervised Anomaly Detection. In Proceedings of the International Conference on Learning Representations, Addis Ababa, Ethiopia, 26–30 April 2020. [CrossRef]
83. Samek, W.; Müller, K.R. Towards Explainable Artificial Intelligence. In *Explainable AI: Interpreting, Explaining and Visualizing Deep Learning*; Springer: Cham, Switzerland, 2019; Volume 11700, pp. 5–22.
84. Wang, Y.; Chung, S.H. Artificial Intelligence in Safety-Critical Systems: A Systematic Review. *Ind. Manag. Data Syst.* **2021**, *122*, 442–470. [CrossRef]
85. Constapel, M.; Koch, P.; Burmeister, H.C. On the Implementation of a Rule-Based System to Perform Assessment of COLREGs Onboard Maritime Autonomous Surface Ships. *J. Phys. Conf. Ser.* **2022**, *2311*, 012033. [CrossRef]
86. Burmeister, H.C.; Scheidweiler, T.; Reimann, M.; Jahn, C. Assessing Safety Effects of Digitization with the European Maritime Simulator Network EMSN: The Sea Traffic Management Case. *TransNav. Int. J. Mar. Navig. Saf. Sea Transp.* **2020**, *14*, 91–96. [CrossRef]

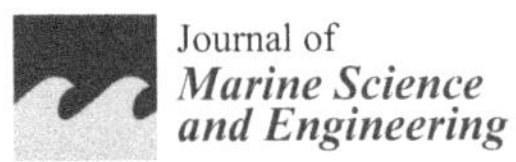

Journal of
*Marine Science
and Engineering*

MDPI

Brief Report

Improving Maritime Domain Awareness in Brazil through Computer Vision Technology

Matheus Emerick de Magalhães [1,2,*], Carlos Eduardo Barbosa [1,2], Kelli de Faria Cordeiro [2], Daysianne Kessy Mendes Isidorio [1] and Jano Moreira de Souza [1]

[1] Program Systems Engineering and Computer Science,
Alberto Luiz Coimbra Institute for Graduate Studies and Research in Engineering (Coppe),
Universidade Federal do Rio de Janeiro (UFRJ), Rio de Janeiro 21941-972, Brazil;
eduardo@cos.ufrj.br (C.E.B.); daysiannekessy@metalmat.ufrj.br (D.K.M.I.); jano@cos.ufrj.br (J.M.d.S.)
[2] Centro de Análises de Sistemas Navais (CASNAV), Brazilian Navy, Rio de Janeiro 20091-000, Brazil;
kelli@marinha.mil.br
* Correspondence: emerick@cos.ufrj.br

Abstract: This article discusses the Brazilian maritime authority's efforts to monitor and control vessels in specific maritime areas using data from the naval traffic control system. Anomalies in vessel locations can signal security threats or illegal activities, such as drug trafficking and illegal fishing. A reliable Maritime Domain Awareness (MDA) is necessary to reduce such occurrences. This study proposes a data-driven framework, CV-MDA, which uses computer vision to enhance MDA. The approach integrates vessel records and camera images to create an annotated dataset for a Convolutional Neural Network (CNN) model. This solution supports detecting, classifying, and identifying small vessels without trackers or that have deliberately shut down their tracking systems in order to engage in illegal activities. Improving MDA could enhance maritime security, including identifying warships invading territorial waters and preventing illegal activities.

Keywords: maritime domain awareness; data integration; computer vision

Citation: Emerick de Magalhães, M.;
Barbosa, C.E.; Cordeiro, K.d.F.;
Isidorio, D.K.M.; Souza, J.M.d.
Improving Maritime Domain
Awareness in Brazil through
Computer Vision Technology. *J. Mar.
Sci. Eng.* **2023**, *11*, 1272. https://
doi.org/10.3390/jmse11071272

Academic Editors: Marko Perkovic,
Lucjan Gucma, Sebastian Feuerstack
and Mihalis Golias

Received: 6 May 2023
Revised: 26 May 2023
Accepted: 9 June 2023
Published: 23 June 2023

1. Introduction

The Maritime Domain Awareness (MDA) concept and solutions were reformulated after the 9/11 terrorist attacks [1]. The MDA concept can be described as understanding the elements, mainly vessels and other assets, in a given region.

The expansion of the MDA is necessary for maritime authorities to monitor and control the strategic planning of maritime traffic and to support naval actions. These actions are mainly related to safety, security, and economic activities such as oil extraction, fishing, and freight transport.

Several solutions can be related to MDA activities. In summary, the main activities of trajectory analysis used to increase the MDA are vessel density [2], waterway, maritime waypoint [3], maritime route extraction [4], anomalous vessel detection [5], vessel collision avoidance [6], and others. In almost all solutions, vessel data is the central resource. Furthermore, the challenges found in extracting data knowledge lead to the use of multiple data sources to reach the maritime domain. Handling the challenges encountered in acquiring data knowledge and using multiple data sources is crucial for obtaining better insights [7,8].

In recent years, the maritime authority has faced several MDA challenges related to the knowledge and intelligence needed for maritime security [9]. Increasing automatic, self-reported, and computational vision data leads to a data-driven solution that increases maritime knowledge and provides intelligent solutions. Furthermore, many countries have proposed specific MDA frameworks to improve the MDA, such as Greece [10], Canada [8], and the Philippines [11]. On the other hand, the literature presents exhaustive and unreliable MDA solutions using generic methods, such as movement extraction, data aggregation, data analysis, computer vision, and deep learning, through automatic or synthetic data [12].

In 2021, the United States and Brazil signed a cooperation agreement to best bid for addressing regional and global threats. The maritime domain threats can be hitched to various scenarios, such as drug trafficking, illegal fishing, piracy, oil spills, and illegal immigration [13–15]. Bad actors can also falsify AIS messages to cover illegal activities [16].

This research identifies several challenges related to the absence, veracity, and completeness of the navigation records in monitoring and controlling the activities of vessels in territorial waters. Detection and classification of ships in MDA are crucial, but it requires data integration to feed the Convolutional Neural Network (CNN). The absence of veracity and data completeness in monitoring and controlling the activities of vessels in territorial waters, and an interest in regional cooperation and interoperability presented in this new scenario, motivated this work [17,18]. Our research contributes to the advancement of the field and helps push the boundaries of knowledge and foster innovation.

The objectives of this work are related to the creation of datasets of vessel images using navigation records from various data sources. The aim is to increase data labeling integrity and accuracy while supporting the development of neural network models for vessel detection solutions. Additionally, we cover different approaches, methods, technologies, and definitions presented in the literature related to this work. Therefore, this work aims at creating a navigation dataset and vessel records through the integration of heterogeneous data sources, the formation of an automated structure for the annotation of identified objects through the integration with navigation and vessel records, and the creation of a data-driven model for the classification of vessels that are present in Brazilian inland waters using a solution based on computer vision.

2. Vessel Detection Issues

The vessel-detection problem refers to identifying and locating maritime vessels in large bodies of water using various remote sensing techniques. This problem has gained increasing attention in recent years due to the growing importance of maritime transportation and the need for effective surveillance and monitoring of sea traffic. Various methods have been developed to address this problem, including satellite imagery, radar systems, and acoustic sensors. However, vessel detection remains a complex task due to factors such as adverse weather conditions, vessel size and speed, clutter, and other sources of interference. Advances in computer vision are helping to improve the accuracy and efficiency of vessel detection systems, but this remains an ongoing area of research and development.

Figures 1 and 2 present images captured from the Tactical Image Console with Augmented Reality (Console de Imagens Táticas com Realidade Aumentada—CITRA). CITRA [19] is a system developed by the Brazilian Navy that allows an operator to visualize real images captured by a video surveillance camera combined with synthetic elements from sensors in the maritime Command and Control systems, using Augmented Reality techniques. In this visualization, CITRA combines the real images from the connected cameras with metadata obtained from a Command and Control system, functioning as a video situational awareness module. However, CITRA has limitations related to inconsistent data between the live feed from the camera and the vessel's last known position in the Command and Control system.

Figure 1. No Sensor Data to Vessel Identification.

Figure 2. Low Sensor Data Veracity.

In Figure 1, we highlight the limitation of a vessel that appears in the camera and is missing in the command and control system. In contrast, in Figure 2, we highlight a record in the command and control system representing an absent vessel—probably due to the latency of the positioning update. Therefore, this work proposes key steps for creating a data-driven framework to support increasing MDA detection and classification of vessel types using a computer vision-based solution, detecting and reducing these inconsistencies.

3. Maritime Domain Awareness (MDA)

3.1. Data Integration in MDA

Information on the behavior of vessels is the main situational element maritime authorities use to increase domain awareness. Several data sources provide the vessels' movement records, including records from Automatic Identification Systems and other sources, such as Long-Range Identification and Tracking (LRIT) [20]. In general, movement records are self-reported by vessels or automatically identified using satellites. Maritime data integration aims to consolidate vessels' movement data and information [21]. The importance of integrating data sources to increase the MDA is represented in the works of Bannister and Neyland [7] and Battistello et al. [8].

Consolidating movement and vessel data involves multiple data-processing activities [22]. The main activities are data collection and movement records associated with the various vessel identifiers, organization, temporal alignment, and data smoothing [23].

The heterogeneous data from multiple sources can provide a more complete picture of vessel movements and a better understanding of their behavior. A more complete understanding of vessel trajectory can overcome errors in vessel movement records, such as data loss, self-information error, malfunctioning radar systems, and problems that can occur while processing the data from these maritime traffic systems [24].

Developing heterogeneous data is geared towards developing domain, situation, and impact analysis strategies. Integrating and combining different maritime data sources create an efficient system of responding to the various data sources [11]. Table 1 presents the data sources in the Brazilian Maritime Area.

Table 1. Data sources in the Brazilian Maritime Area.

Data Source	Description	Sensor Type	Privacy
AIS	Automatic Identification System used for short-range coastal monitoring on ships	Terrestrial	Public
SISTRAM	Information Systems on Maritime Traffic in Brazil	Information System	Private
SIMMAP	Maritime Monitoring System to support oil-related activities	Terrestrial	Private
LRIT	Long-Range Global Monitoring System	Satellite	Public
PREPS	Program for Tracking Fishing Vessels via Satellite	Satellite	Private
VRMTC (TRMN)	Virtual Regional Maritime Traffic	Terrestrial	Private
MSSIS	The Maritime Safety and Security Information System	Terrestrial	Private
PRENAV	Program for Tracking Vessels operating in navigations regulated by the National Waterway Transport Agency (Agência Nacional de Transportes Aquaviários—ANTAQ)	Satellite	Private
CRTAMAS	Regional Traffic Center for the South Atlantic area (Brazil, Argentina, Uruguay, and Paraguay)	Terrestrial	Public
MILITARES	Brazilian Navy Ships	Radio	Private
ORBCOMM	Orbcomm's second and third generation of satellite constellation, operating in Low Earth Orbit	Satellite	Private
Tietê-Paraná	AIS Tietê-Paraná for navegable rivers.	Terrestrial	Public

We aim to use computer vision on the CITRA camera view, which extracts potential vessels in the field of view. Then, the detected vessels are combined with the suitable data source (listed in Table 1) to increase the MDA.

3.2. Computer Vision in MDA

Computer vision appears as a solution to support human visual tasks to interpret and understand the maritime domain using Deep Learning Neural Network (DNN) models. The DNN models accurately identify and classify objects [12,25]. Yet, the DNN solutions have been used in several MDA challenges, including detecting and classifying vessel types using synthetic or real-world images.

A specific type of DNN, the Convolutional Neural Network, has revolutionized the art of object detection and recognition, achieving faster and more accurate results [26].

CNN uses far fewer weights in our deep network, allowing for significantly faster training. One of the most accurate, fast, and precise CNN available is the state-of-the-art framework You Only Look Once (YOLO) [27]. The YOLO meets the requirements in some proposal works (Redmon et al. [28] and Redmon and Farhadi [29]) as real-time processing and is robust to changes in lighting in images and, in non-synthetic images, keeping a simple neural network, using a single GPU with a smaller mini-batch size to train a model.

The literature presents other object detection algorithm ancestors and alternatives of the YOLO framework, such as Histogram of Oriented Gradients (HOG) [30], Fast R-CNN [31], Faster R-CNN [32], Single Shot Detector (SSD) [33], and RetinaNet [34].

CNN can adopt an approach based on transfer learning [35]. It starts with a pre-trained model for generic feature detection, usually using a generic dataset such as the Microsoft Common Objects in Context (MS COCO) [36]. The model is specialized with a fine-tuning strategy using a data-driven perspective.

4. The MDA Problem in Brazil

In Brazil, the need to monitor vessels and assets in territorial water such as the Blue Amazon (Amazônia Azul), shown in Figure 3, has also motivated the development of novel and specific solutions intended to increase the MDA [37]. In Figure 3, the lighter blue region represents Brazil's Exclusive Economic Zone of 200 nautical miles. In contrast, the darker blue represents Brazil's proposal to the UN, which extends the zone to the Continental Shelf, up to 350 nautical miles [38].

Figure 3. Blue Amazon—Adapted from Thompson and Muggah [38].

Monitoring territorial waters is necessary for efficient maritime policymaking. Maritime policymaking encompasses military and economic actions related to national security [37]. The maritime authority expresses great concern regarding incidents related to foreign vessel intrusions, drug trafficking, protected biological areas, oil spills, search and rescue operations, and illegal fishing. These situations highlight the importance of implementing effective monitoring protocols within the maritime domain. Tracking vessel activities across various maritime zones is crucial to protect marine environments.

Bunholi et al. [39] explained Brazil's biological importance in summarizing its data collection system for all of its fisheries. In Brazil, more information is needed for the expansion of policies and actions in order to inspect illegal fishing activities, where the current lack of data makes it difficult to control the maritime domain better [40].

From a Brazilian maritime perspective, the security environment in the South Atlantic focuses on threats such as piracy, drug trafficking, and other forms of transnational crime [41,42]. In the military domain, the disclosure of information regarding maritime security is subject to restrictions and limitations due to the classified nature of military information. However, the intrusion of foreign entities into territorial waters remains a constant concern for maritime authorities [43].

5. CV-MDA: A Computer Vision-Based Framework to Improve Maritime Domain Awareness

This section proposes three layers in an architectural framework for broad MDA, called CV-MDA, presented in detail in Figure 4. The Data Acquisition Layer conducts a unique vessel identification parameter and the vessel's motion image data acquisition. The Integration Layer establishes the vessel movement dataset, and the Detection and Classification Layer proposes a fine-tuned transferred learning approach to a detection and classification solution. The following subsections describe the tools chosen for developing the framework.

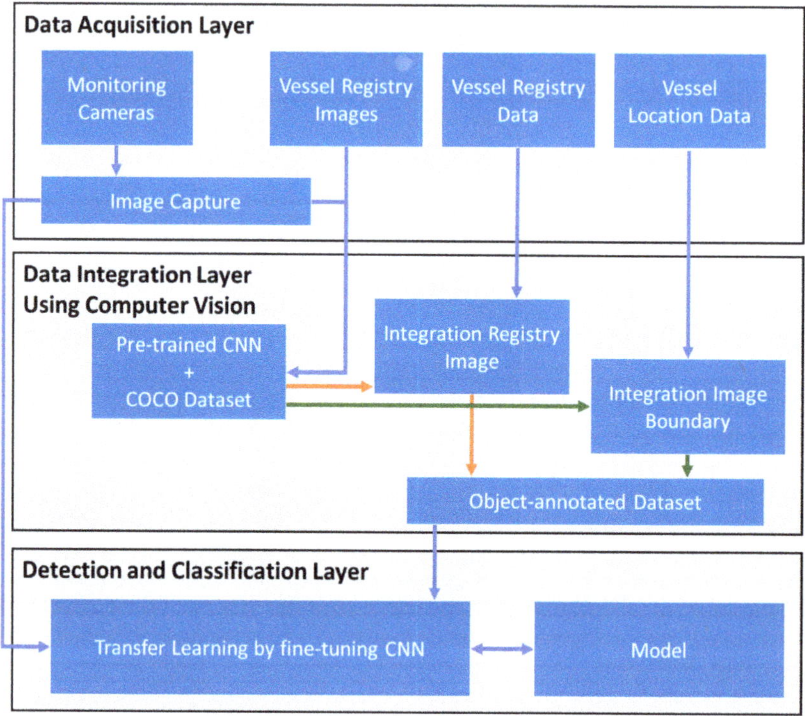

Figure 4. The CV-MDA Framework.

5.1. Data Acquisition Layer

This work used traffic open data sources from Brazilian Maritime Traffic and tracked image data from Console Tactical Images. A heterogeneous data source can adopt specific vessel identification parameters in such sensor systems. We conducted a study with 25,618,290 navigation records from eight data sources in a 60-day horizon in order to understand and establish a unique vessel identification parameter.

The three main identifiers used for vessel identification are International Maritime Organization Number (IMO) [40], International Radio Call Sign (IRIN) [44], and Maritime Mobile Service Identity (MMSI) [45]. In our data, several vessels had more than one identifier, as shown in Figure 5. However, 82.6% had all three identifiers. Furthermore, we associated them in order to obtain a unique vessel identification.

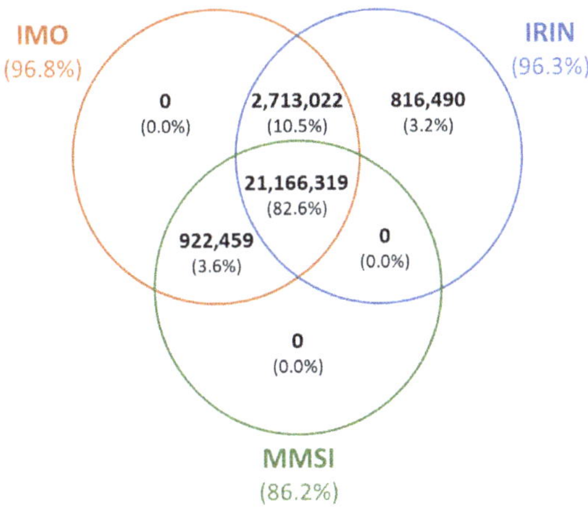

Figure 5. Vessel Identifications in Movement Data.

We collected images of the Brazil vessel registry and integrated them using the new vessel identification to provide some ground truth input to our object-annotated dataset. We also developed a specific tool to capture the vessel images from the maritime area, presented in Figure 6. The main objective was to capture and store real-world vessel images in a specific Brazilian area to build a dataset and train our CNN model.

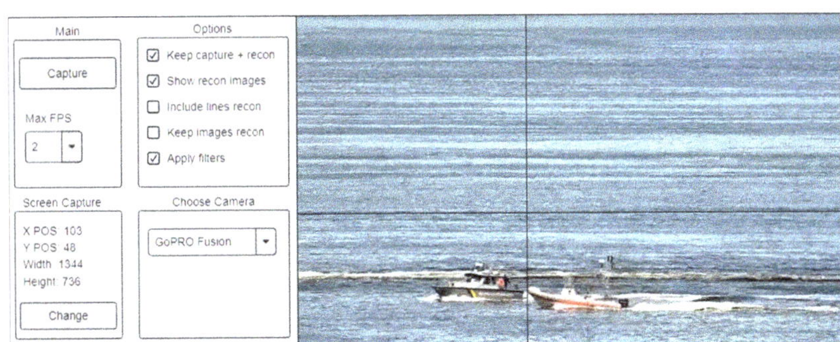

Figure 6. Tool for capturing vessel images.

We were able to use a data augmentation solution to improve our object-annotated dataset. Data augmentation is a strategy that enables slight variations to improve the diversity and volume of the image dataset [46].

5.2. Data Integration Layer Based on Computer-Vision

In this section, we propose correlating vessel registry, movement data, and real maritime images to label a training dataset to build a fine-tuned CNN approach model to detect and classify vessels and support a wide MDA. The proposed model—shown in Figure 7—provides a maritime data integration solution to build an automatically annotated Brazilian vessel dataset. Once the images were captured, we applied different computer vision approaches for enhancing the detection and pre-classification of vessels on the sea. We proposed using YOLO in the current implementation with an open-source CNN framework architecture called Darknet.

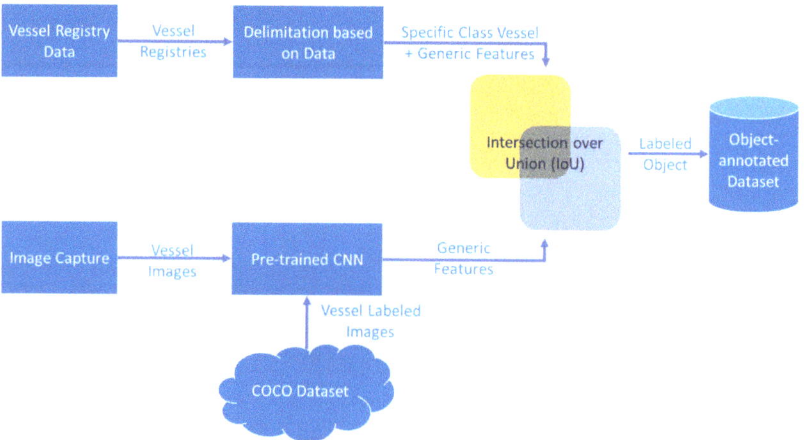

Figure 7. Using Computer-Vision on Maritime Data Integration Steps.

After submitting the images to the model, the pre-trained CNN returned an object's detection boundary box properties (location) and a class (boat) confidence result in a found object, as presented in Figure 8.

Figure 8. Detection Boundary and Pre-Classification.

On the other hand, we correlated the movement data sources (self-automatic and automatic) with vessel registries. We created stay points corresponding with the movement spatial coordinates and built a boundary box around them. Therefore, vessel type information was associated with delimited boxes.

Both boundary boxes, from image and movement data, were used to create an annotated dataset. We compared the bounding boxes in a specific time window using an Intersection over Union (IoU) approach to label the images with vessel type from a movement data source. The IoU model is shown in Figure 9.

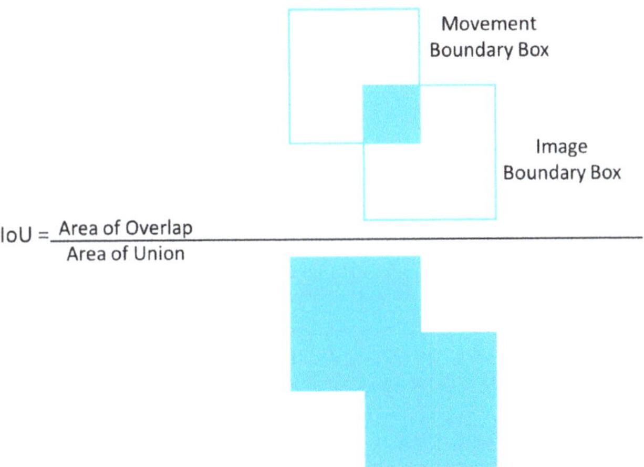

Figure 9. IoU Boundary Boxes.

An IOU threshold determined a correspondence between the boundary boxes to create an annotated image with vessel type. The flow of successful correspondent boundary boxes of a specific vessel was used to build a sequenced ground truth vessel labeled dataset in the Brazilian area. As a result, the accuracy of the combination could provide veracity and completeness to vessel label images. This data was used as an input source for a fine-tuned CNN approach to discovering and classifying vessel data in the next step.

This section may be divided by subheadings. It should provide a concise and precise description of the experimental results, their interpretation, and the experimental conclusions that we can draw.

5.3. Detection and Classification Layer

In this subsection, we propose transfer learning by using a fine-tuning strategy, shown in Figure 10. We chose this strategy to improve CNN generalization and specialization to build a fully connected layer to classify the object (vessel).

A generalization in transfer learning implies that we are taking the generic feature learned from a pre-trained network and applying it to another similar network. In the initial phase, we addressed this approach choice using image data consisting of patterns, common inputs, and generic feature detections. On the other hand, specialization referred to a fine-tuning phase to specialize the detection and obtain specific features.

In transfer learning, we used a pre-trained CNN to acquire low-level features. We could also use the CNN YOLOv5 with a trained vessel MS COCO dataset to obtain the basics proprieties (such as color, patterns, shapes, and edges), results of detection, and classification tasks. Next, in fine-tuning, we froze these low-level trains on the MS COCO and only retrained the high-level features with our labeled dataset. Finally, we replaced the classification layer with our vessel type classes.

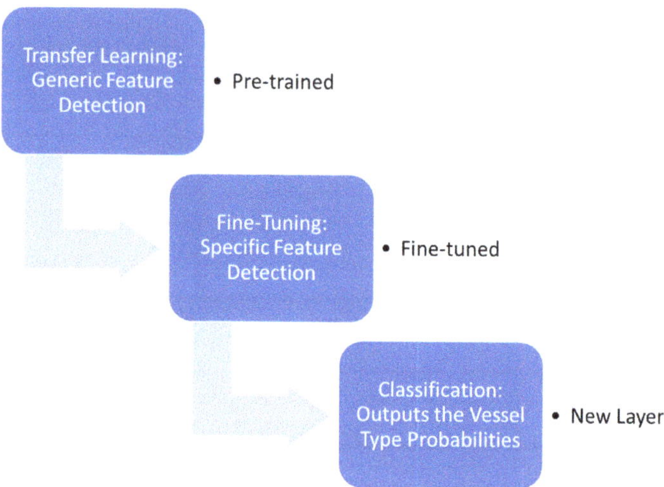

Figure 10. CNN Transfer Learning by Fine-tuning Approach.

We were able to measure the model's performance using a mean Average Precision (mAP) metric. A threshold determined if the object detection and classification task was valid or not and was used to evaluate the performance of an object detection model. After all stages, we started the final classification step for enhancing the real-time captured images into a CNN.

6. Experiment

To showcase the outcomes of our framework, we conducted a straightforward experiment. Firstly, we employed vessel registry images and a pre-trained model to detect and outline the boat class within these images, as shown in Figure 8. Subsequently, we integrated the outlined images with the corresponding navigation records, incorporating the spatial coordinates and timestamps as parameters. This process is depicted in Figure 11, showcasing the integration.

Performing this integration, as demonstrated in Figure 12, we successfully assigned specific classes to the objects. Utilizing the model derived from annotated images using navigation records, we obtained images from a camera and subjected them to the vessel detection model. The labels "Ship1, ... , ShipN" represent the identified vessel types. Upon examining the image, it becomes evident that the proposed model cannot detect certain vessels, while others may be misclassified.

Figure 13 presents an instance illustrating the process of automatic annotation used for labeling vessel classes. Some vessels could not be visually detected, resulting in their exclusion from the labeling process due to insufficient data sources. The colors assigned to the classes represent the various data sources employed to label each vessel. A challenge arises when attempting to detect vessels lacking sensor data, hindering the automatic annotation process.

We can address these limitations by incorporating a crowdsourcing layer into the CV-MDA framework. This additional layer aims to evaluate the detection results and assist in annotating the classes of objects that lack vessel records and, therefore, cannot be annotated based on navigation records, such as small vessels. By introducing crowdsourcing, it becomes possible to gather collective insights and annotations, improving the accuracy and completeness of the vessel classification process.

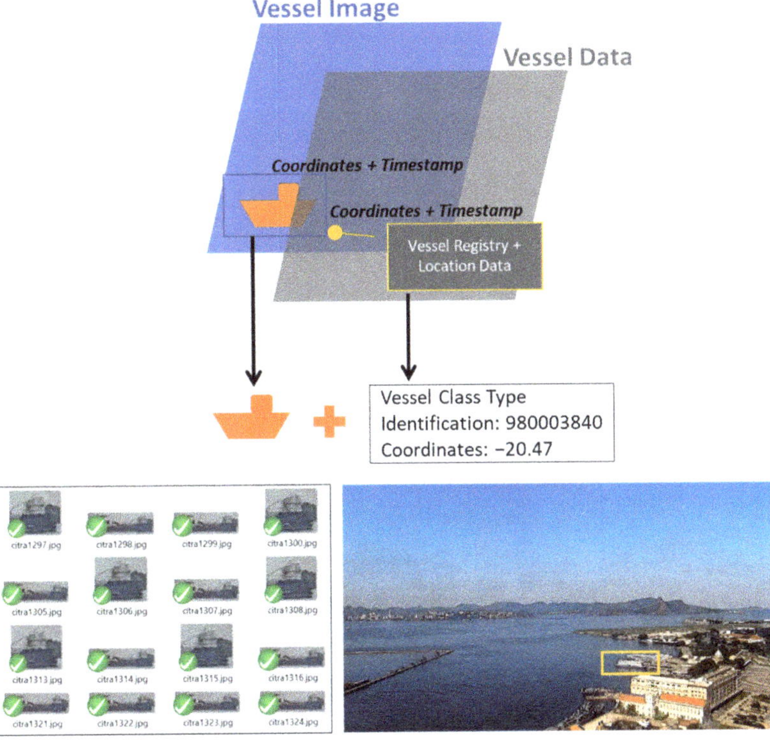

Figure 11. Vessel's image outline process.

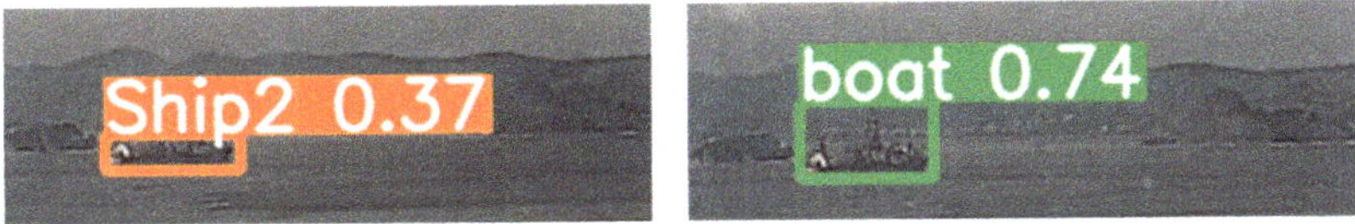

Figure 12. Object Annotated with Vessel Registry (**left**) and Object Pre-Trained Annotated (**right**).

Using multiple integrated navigation data sources for automatically annotating vessel images in the Brazilian maritime region is a novelty and a significant contribution to the field. Combining navigation data from different sensors and navigation identifiers allows for more precise and efficient automatic annotation of vessel images. This automatic annotation surpasses the manual or semi-automatic approach commonly found in the literature, where human experts annotate. Automating the annotation process reduces the dependency on manual annotations, which can be time-consuming and prone to errors. Additionally, integrating multiple sources of navigation data provides a more comprehensive view of maritime activities in the region, enabling a more accurate understanding of the context and navigation patterns.

Our approach also considers the reality in the Brazilian maritime context, which is a relevant contribution. Each maritime region has its specific characteristics and challenges, and adapting the solution to the local needs and particularities is crucial for obtaining more effective results. Integrating sensor data, such as AIS, radar data, and other relevant sources, allows for obtaining more comprehensive and reliable information about vessels in the Brazilian maritime region. This data integration has many practical applications, such as maritime traffic monitoring, border control, maritime security, and strategic planning. The automatic annotation approach for vessel images, combining navigation data from multiple sources, represents a significant advancement compared to previous works focusing on manual or semi-automatic annotation. Our approach can provide higher efficiency, scalability, and accuracy in the detection and annotation of vessels.

Figure 13. Object detection by the fine-tuned CNN.

7. Final Remarks

In conclusion, the Brazilian maritime authority relies on vessel monitoring and control activities to detect anomalies in the maritime domain, which could signify potential security risks or illegal activities such as drug trafficking and illegal fishing. This issue is particularly pressing in the Blue Amazon. In response, we have proposed a novel solution for Maritime Domain Awareness (MDA) based on computer vision. The maritime authority has deemed our technique efficient and valuable for detecting and classifying vessel types, thus supporting a comprehensive MDA.

This work has presented a solution to enhance Maritime Domain Awareness (MDA) by developing a framework for creating datasets of vessel images using navigation records from various data sources. Creating datasets is an important step in increasing the completeness and accuracy of data labeling. The proposed approach strengthens the capability to develop models applied in neural network solutions for vessel detection. Navigation records as the foundation for labeling make obtaining precise and comprehensive information about vessel types and movements possible. This precise information improves the efficiency of detection models, provides a solid foundation for decision-making, and enhances maritime domain awareness. By integrating diverse data sources, a more comprehensive and reliable understanding of the maritime scenario can be achieved, aiding in detecting anomalies, preventing illegal activities, and ensuring the security and surveillance of territorial waters. We have focused on computer vision applied to near-shore camera data. However, we can extend the framework with a few adaptations to real-time satellite images to achieve similar results in a larger area.

Moreover, we can adapt the proposed approach for monitoring other objects, supporting autonomous driving, detecting wildlife, and facilitating intelligent drone surveillance. However, creating annotated datasets can be challenging for small vessels such as fishing and sports vessels with limited movement records. In these cases, manual or semi-automatic labeling is still necessary to create annotated datasets.

In summary, incorporating multiple data sources alongside AIS offers several advantages over relying solely on AIS for ship-related information. These advantages include enhanced redundancy, as incorporating diverse data sources mitigates limitations and improves ship information's reliability and availability. Additional data points from various sources provide supplementary information beyond AIS, enabling a broader range of ship-related data for more detailed analysis and decision-making. Moreover, different data sources capture information from unique sensors and technologies, offering diverse perspectives on maritime activities and facilitating better situational awareness. Specialized data streams cater to specific needs, such as environmental monitoring or security assessments, while cross-verifying and validating data from multiple sources improves the accuracy and reliability of data analysis. By leveraging these advantages, organizations can comprehensively and accurately understand maritime operations, improving safety, security, and efficiency.

Author Contributions: Conceptualization, M.E.d.M. and C.E.B.; methodology, M.E.d.M.; software, M.E.d.M.; formal analysis, M.E.d.M.; investigation, M.E.d.M.; resources, J.M.d.S.; data curation, D.K.M.I.; writing—original draft preparation, M.E.d.M.; writing—review and editing, M.E.d.M., C.E.B., K.d.F.C. and D.K.M.I.; visualization, M.E.d.M.; supervision, C.E.B., K.d.F.C. and J.M.d.S.; project administration, M.E.d.M. and J.M.d.S.; funding acquisition, M.E.d.M. and J.M.d.S. All authors have read and agreed to the published version of the manuscript.

Funding: This study was financed in part by the Coordenação de Aperfeiçoamento de Pessoal de Nível Superior—Brasil (CAPES)—Finance Code 001.

Institutional Review Board Statement: Not applicable.

Informed Consent Statement: Not applicable.

Data Availability Statement: Data are strictly used within the project but could be available upon request.

Conflicts of Interest: The authors declare no conflict of interest.

References

1. National Commission on Terrorist Attacks. *The 9/11 Commission Report: Final Report of the National Commission on Terrorist Attacks Upon the United States (Authorized Audio Edition, Abridged)*; Norton, W.W., Ed.; National Commission on Terrorist Attacks Upon the United States: New York, NY, USA, 2011; ISBN 978-0-393-34215-4.
2. Millefiori, L.M.; Zissis, D.; Cazzanti, L.; Arcieri, G. Scalable and Distributed Sea Port Operational Areas Estimation from AIS Data. In Proceedings of the 2016 IEEE 16th International Conference on Data Mining Workshops (ICDMW), Barcelona, Spain, 12–15 December 2016; pp. 374–381.
3. Liu, L.; Shibasaki, R.; Zhang, Y.; Kosuge, N.; Zhang, M.; Hu, Y. Data-Driven Framework for Extracting Global Maritime Shipping Networks by Machine Learning. *Ocean Eng.* **2023**, *269*, 113494. [CrossRef]
4. Jousselme, A.-L.; Iphar, C.; Pallotta, G. Uncertainty Handling for Maritime Route Deviation. In *Guide to Maritime Informatics*; Artikis, A., Zissis, D., Eds.; Springer International Publishing: Cham, Switzerland, 2021; pp. 263–297, ISBN 978-3-030-61852-0.
5. Iphar, C.; Ray, C.; Napoli, A. Data Integrity Assessment for Maritime Anomaly Detection. *Expert Syst. Appl.* **2020**, *147*, 113219. [CrossRef]
6. Rong, H.; Teixeira, A.P.; Guedes Soares, C. Ship Collision Avoidance Behavior Recognition and Analysis Based on AIS Data. *Ocean Eng.* **2022**, *245*, 110479. [CrossRef]
7. Bannister, N.P.; Neyland, D.L. Maritime Domain Awareness with Commercially Accessible Electro-Optical Sensors in Space. *Int. J. Remote Sens.* **2015**, *36*, 211–243. [CrossRef]
8. Battistello, G.; Gonzalez, J.; Ulmke, M.; Koch, W.; Mohrdieck, C. Multi-Sensor Maritime Monitoring for the Canadian Arctic: Case Studies. In Proceedings of the 2016 19th International Conference on Information Fusion (FUSION), Heidelberg, Germany, 5–8 July 2016; pp. 1881–1888.
9. Zheng, D.X.; Sun, X.D. A Knowledge Acquisition Model in Maritime Domain Based on Ontology. In Proceedings of the Proceedings 2014 IEEE International Conference on Security, Pattern Analysis, and Cybernetics (SPAC), Wuhan, China, 18–19 October 2014; pp. 372–375.
10. Giannakopoulos, T.; Vetsikas, I.A.; Koromila, I.; Karkaletsis, V.; Perantonis, S. AMINESS: A Platform for Environmentally Safe Shipping. In Proceedings of the 7th International Conference on PErvasive Technologies Related to Assistive Environments, Island of Rhodes, Greece, 27–30 May 2014; ACM: New York, NY, USA, 2014; pp. 45:1–45:8.
11. Vicente, R.; Tabanggay, L.; Rayo, J.F.; Mina, K.; Retamar, A. Earth Observations for Goal 14: Improving Maritime Domain Awareness Using Synthetic Aperture Radar Imaging with Automatic Identification System in the Philippines. *ISPRS—Int. Arch. Photogramm. Remote Sens. Spat. Inf. Sci.* **2020**, *XLIII-B3-2020*, 215–219. [CrossRef]
12. Breitinger, A.; Clua, E.; Fernandes, L.A. An Augmented Reality Periscope for Submarines with Extended Visual Classification. *Sensors* **2021**, *21*, 7624. [CrossRef]

13. Bosilca, R.-L. The Use of Satellite Technologies for Maritime Surveillance: An Overview of EU Initiatives. *Incas Bull.* **2016**, *8*, 153.

14. Lindley, J.; Percy, S.; Techera, E. Illegal Fishing and Australian Security. *Aust. J. Int. Aff.* **2019**, *73*, 82–99. [CrossRef]

15. Sakhuja, V. Indian Ocean and the Safety of Sea Lines of Communication. *Strateg. Anal.* **2001**, *25*, 689–702.

16. Iphar, C.; Napoli, A.; Ray, C. An Expert-Based Method for the Risk Assessment of Anomalous Maritime Transportation Data. *Appl. Ocean Res.* **2020**, *104*, 102337. [CrossRef]

17. Soares, A.; Dividino, R.; Abreu, F.; Brousseau, M.; Isenor, A.W.; Webb, S.; Matwin, S. CRISIS: Integrating AIS and Ocean Data Streams Using Semantic Web Standards for Event Detection. In Proceedings of the 2019 International Conference on Military Communications and Information Systems (ICMCIS), Budva, Montenegro, 14–15 May 2019; pp. 1–7.

18. Emmens, T.; Amrit, C.; Abdi, A.; Ghosh, M. The Promises and Perils of Automatic Identification System Data. *Expert Syst. Appl.* **2021**, *178*, 114975. [CrossRef]

19. Centro de Análises de Sistemas Navais Console de Imagens Táticas com Realidade Aumentada (CITRA). Available online: https://web.archive.org/web/20230520030321/https://www.marinha.mil.br/casnav/?q=node/181 (accessed on 20 May 2023).

20. Verma, B. Long Range Identification and Tracking (LRIT) Apropos Global Maritime Security. *Marit. Aff. J. Natl. Marit. Found. India* **2009**, *5*, 39–56. [CrossRef]

21. El Faouzi, N.-E.; Klein, L.A. Data Fusion for ITS: Techniques and Research Needs. *Transp. Res. Procedia* **2016**, *15*, 495–512. [CrossRef]

22. Steinberg, A.N.; Bowman, C.L. Revisions to the JDL Data Fusion Model. In *Handbook of Multisensor Data Fusion*; CRC Press: Boca Raton, FL, USA, 2017; pp. 65–88.

23. García, J.; Guerrero, J.L.; Luis, A.; Molina, J.M. Robust Sensor Fusion in Real Maritime Surveillance Scenarios. In Proceedings of the 2010 13th International Conference on Information Fusion, Edinburgh, UK, 26–29 July 2010; pp. 1–8.

24. Claramunt, C.; Ray, C.; Salmon, L.; Camossi, E.; Hadzagic, M.; Jousselme, A.-L.; Andrienko, G.; Andrienko, N.; Theodoridis, Y.; Vouros, G. Maritime Data Integration and Analysis: Recent Progress and Research Challenges. *Adv. Database Technol.-EDBT* **2017**, *2017*, 192–197.

25. Kim, J.-H.; Kim, N.; Park, Y.W.; Won, C.S. Object Detection and Classification Based on YOLO-V5 with Improved Maritime Dataset. *J. Mar. Sci. Eng.* **2022**, *10*, 377. [CrossRef]

26. Simonyan, K.; Zisserman, A. Very Deep Convolutional Networks for Large-Scale Image Recognition. *arXiv* **2014**, arXiv:1409.1556.

27. Bochkovskiy, A.; Wang, C.-Y.; Liao, H.-Y.M. Yolov4: Optimal Speed and Accuracy of Object Detection. *arXiv* **2020**, arXiv:2004.10934.

28. Ribeiro, R.; Cruz, G.; Matos, J.; Bernardino, A. A Data Set for Airborne Maritime Surveillance Environments. *IEEE Trans. Circuits Syst. Video Technol.* **2019**, *29*, 2720–2732. [CrossRef]

29. Redmon, J.; Farhadi, A. Yolov3: An Incremental Improvement. *arXiv* **2018**, arXiv:1804.02767.

30. Surasak, T.; Takahiro, I.; Cheng, C.; Wang, C.; Sheng, P. Histogram of Oriented Gradients for Human Detection in Video. In Proceedings of the 2018 5th International Conference on Business and Industrial Research (ICBIR), Bangkok, Thailan, 17–18 May 2018; pp. 172–176.

31. Girshick, R. Fast R-CNN. In Proceedings of the 2015 IEEE International Conference on Computer Vision (ICCV), Santiago, Chile, 7–13 December 2015; pp. 1440–1448.

32. Jiang, H.; Learned-Miller, E. Face Detection with the Faster R-CNN. In Proceedings of the 2017 12th IEEE International Conference on Automatic Face Gesture Recognition (FG 2017), Washington, DC, USA, 30 May–3 June 2017; pp. 650–657.

33. Liu, W.; Anguelov, D.; Erhan, D.; Szegedy, C.; Reed, S.; Fu, C.-Y.; Berg, A.C. SSD: Single Shot MultiBox Detector. In Proceedings of the Computer Vision—ECCV 2016, Amsterdam, The Netherlands, 11–14 October 2016; Leibe, B., Matas, J., Sebe, N., Welling, M., Eds.; Springer International Publishing: Cham, Switzerland, 2016; pp. 21–37.

34. Wang, Y.; Wang, C.; Zhang, H.; Dong, Y.; Wei, S. Automatic Ship Detection Based on RetinaNet Using Multi-Resolution Gaofen-3 Imagery. *Remote Sens.* **2019**, *11*, 531. [CrossRef]

35. Ahmed, I.; Ahmad, M.; Ahmad, A.; Jeon, G. Top View Multiple People Tracking by Detection Using Deep SORT and YOLOv3 with Transfer Learning: Within 5G Infrastructure. *Int. J. Mach. Learn. Cybern.* **2021**, *12*, 3053–3067. [CrossRef]

36. Lin, T.-Y.; Maire, M.; Belongie, S.; Hays, J.; Perona, P.; Ramanan, D.; Dollár, P.; Zitnick, C.L. Microsoft COCO: Common Objects in Context. In Proceedings of the Computer Vision—ECCV 2014, Zurich, Switzerland, 6–12 September 2014; Fleet, D., Pajdla, T., Schiele, B., Tuytelaars, T., Eds.; Springer International Publishing: Cham, Switzerland, 2014; pp. 740–755.

37. Boraz, S.C. Maritime domain awareness: Myths and realities. *Nav. War Coll. Rev.* **2009**, *62*, 137–146.

38. Thompson, N.; Muggah, R. The Blue Amazon: Brazil Asserts Its Influence across the Atlantic 2015. Available online: https://igarape.org.br/the-blue-amazon-brazil-asserts-its-influence-across-the-atlantic/ (accessed on 21 May 2023).

39. Bunholi, I.V.; da Silva Ferrette, B.L.; De Biasi, J.B.; de Oliveira Magalhães, C.; Rotundo, M.M.; Oliveira, C.; Foresti, F.; Mendonça, F.F. The Fishing and Illegal Trade of the Angelshark: DNA Barcoding against Misleading Identifications. *Fish. Res.* **2018**, *206*, 193–197. [CrossRef]

40. International Maritime Organization IMO Ship Identification Number Scheme. Available online: https://www.imo.org/en/OurWork/IIIS/Pages/IMO-Identification-Number-Schemes.aspx (accessed on 27 April 2022).

41. Duarte, É.E.; Kenkel, K.M. Contesting Perspectives on South Atlantic Maritime Security Governance: Brazil and South Africa. *South Afr. J. Int. Aff.* **2019**, *26*, 395–412. [CrossRef]

42. Medeiros, S.E.; Moreira, W.D.S. Maritime Co-Operation among South Atlantic Countries and Repercussions for the Regional Community of Security Practice. *Contexto Int.* **2017**, *39*, 281–304. [CrossRef]
43. Phillips, C.E.; Ting, T.C.; Demurjian, S.A. Information Sharing and Security in Dynamic Coalitions. In Proceedings of the Seventh ACM Symposium on Access Control Models and Technologies, Monterey, CA, USA, 3–4 June 2002; Association for Computing Machinery: New York, NY, USA, 2002; pp. 87–96.
44. International Telecommunication Union International Call Sign Series. Available online: https://www.itu.int:443/en/ITU-R/terrestrial/fmd/Pages/call_sign_series.aspx (accessed on 28 April 2022).
45. U.S. Coast Guard Navigation Center Maritime Mobile Service Identity. Available online: https://www.navcen.uscg.gov/?pageName=mtmmsi# (accessed on 28 April 2022).
46. Mikołajczyk, A.; Grochowski, M. Data Augmentation for Improving Deep Learning in Image Classification Problem. In Proceedings of the 2018 International Interdisciplinary PhD Workshop (IIPhDW), Swinoujscie, Poland, 9–12 May 2018; pp. 117–122.

Journal of
Marine Science and Engineering

Article

Exploring the Pirate Attack Process Risk along the Maritime Silk Road via Dynamic Bayesian Network Analysis

Xiaoyue Hu, Haibo Xia, Shaoyong Xuan and Shenping Hu *

Merchant Marine College, Shanghai Maritime University, Shanghai 201316, China;
huxiaoyue958@gmail.com (X.H.); hbxia@shmtu.edu.cn (H.X.); syxuan@shmtu.edu.cn (S.X.)
* Correspondence: sphu@shmtu.edu.cn

Abstract: The Maritime Silk Road (MSR) is an important channel for maritime trade between China and other countries in the world. Maritime piracy has brought huge security risks to ships' navigation and has seriously threatened the lives and property of crew members. To reduce the likelihood of attacks from pirates, it is necessary to study the risk to a ship exposed to attacks from pirates on the MSR. Firstly, risk factors were established from three risk component categories (hazard, mitigation capacity, and vulnerability and exposure) and the risk index system of piracy and armed robbery events was founded. Secondly, the dynamic Bayesian network (DBN) method was introduced to establish a pirate attack risk assessment model ad to conduct a quantitative analysis of the process risk of a ship being attacked by pirates. Finally, combined with the scene data of the MSR, the process risk of a ship being attacked by pirates was modeled and applied as an example. The results showed that the overall risk of a ship being attacked by pirates is the lowest in July and the highest in March. In the whole route, when the ship was in the Gulf of Guinea, the Gulf of Aden–Arabian Sea, and the Strait of Malacca, the risk of pirate attack was the highest. This dynamic network model can effectively analyze the level of risk of pirate attacks on ships, providing a reference for the safety decision-making of ships on ocean routes.

Keywords: Maritime Silk Road; pirate attacks; process risk; dynamic Bayesian network; risk assessment

Citation: Hu, X.; Xia, H.; Xuan, S.; Hu, S. Exploring the Pirate Attack Process Risk along the Maritime Silk Road via Dynamic Bayesian Network Analysis. *J. Mar. Sci. Eng.* **2023**, *11*, 1430. https://doi.org/10.3390/jmse11071430

Academic Editors: Sebastian Feuerstack, Marko Perkovic and Lucjan Gucma

Received: 16 May 2023
Revised: 21 June 2023
Accepted: 26 June 2023
Published: 17 July 2023

1. Introduction

In the context of economic globalization, over 90% of trade goods are transported by sea [1]. The 21st-century MSR proposed by China in 2013 is a new type of trade corridor, which follows the new development of global politics and trade mode, connecting China and the world [2]. However, frequent pirate attacks and hijackings have become a serious threat to the MSR trade [3]. Surveys have shown that frequent incidents of piracy can lead to a significant reduction in traded goods and an increase in trade costs. Piracy causes approximately USD 25 billion in losses to the world economy every year, as reported by the International Maritime Organization (IMO). This underscores the urgent need to strengthen ship security measures and combat maritime terrorism, as the maritime community acknowledges the profound impact of terrorism on maritime transportation [4]. Therefore, we analyzed the risk of a ship being attacked by pirates in the MSR transportation process, and relevant strategies were put forward according to the results, which is of great significance to the establishment and improvement of ship security systems.

In the field of research on ship security risks, as early as 2003, Wang and Gong [5] explored ship security risks in terms of ship alarms. Subsequently, Lin [6] proposed the security assessment risk decision-making method. Due to the introduction of risk assessment in the International Ship and Port Facility Security Code (ISPS), Juan [7] proposed a defend-attack-defend model starting in 2011 to conduct an adversarial analysis of the attack risk of Somali pirates. Afterward, Struwe [8] conducted feasibility discussions about the measures taken by private security companies in response to frequent piracy incidents at sea. In 2014, Vanek et al. [9] improved the number and time of ship groupings for the Gulf of Aden collective transit plan. Subsequently, researchers increasingly conducted in-depth quantitative research on ship security risks under different algorithms and models. Bouejla et al. [10] used Bayesian networks for parameter management of risk factors for pirate attacks on oilfield facilities. Lewis [11] used a multinomial logistic model to study how crews' actions and naval actions affect the probability of pirate attacks or hijackings. However, both of them lacked descriptions of the entire risk indicator systems for pirate attacks. Therefore, Pristrom et al. [4] began to build a Bayesian network model to assess the risk of pirate hijacking and verified the feasibility of the model. However, they ignored the risk control after the risk assessment was completed and did not consider the consequences of the risk. In 2019, Jin et al. [12] used a binary logistic regression model to estimate the probability of a ship being attacked by pirates and the success of the attack, but they did not consider the characteristics of spatial changes. Given the previous lack of research on the risks of pirate attacks in ship security risk, Li [13] proposed using the Bayesian network model and the risk matrix method to assess the probability and consequences of the piracy hijacking risk in the sea area. However, they did not consider the limitations of data and the insufficient identification of risk factors. In addition, the literature above mainly studied the risk to a ship exposed to attacks from pirates under static factors without considering that the risk to a ship exposed to attacks from pirates during sea transportation is constantly changing with time and space; these factors are called the dynamic characteristics of risk.

At the same time, domestic and foreign scholars have conducted analysis and provided applications from different perspectives with respect to model and algorithm problems in risk analysis to make up for the shortcomings of specific methods such as evidential reasoning (ER), fuzzy logic, and other methods. Although these methods have been widely used to address the issue of incomplete data on maritime safety, they have not fully analyzed the causal relationships between various influencing factors. With the accumulation of big data and the improvement in mathematical algorithms, database networks have been used for risk assessment prediction and diagnostic analysis. Jiang et al. [2], Wang et al. [14], and Kabir and Papadopoulos [15] found the Bayesian network to be superior in data application. Based on identifying the influencing factors, they innovatively proposed the Bayesian confidence network model in risk analysis. Wang and Yang [16] proposed that the Bayesian network be used to analyze the causality of various influencing factors of maritime accident risk. In addition, Deng et al. [17] proposed an N-K model to reveal the risk coupling characteristics of maritime accidents, while Hsu et al. [18] proposed a continuous risk matrix model to determine the risk level during navigation. However, in reality, risks are constantly changing. To study real-time risks more accurately, Bi et al. [19] used dynamic irregular grids to analyze and evaluate navigation safety. Li et al. [20,21] and Guo et al. [22] proposed using DBN to study risk evolution. Therefore, the introduction of the DBN method not only solved the uncertainty measurement problem based on risk information but also facilitated the analysis of risk characteristics in the spatial and temporal dimensions.

Therefore, this study aims to enhance the risk system of pirate attacks on ships by examining the process risk of pirate attacks in MSR transportation. This research not only contributes to the academic field but also provides practical insights for decision-makers, ship operators, and security agencies in developing effective measures to mitigate the risk of pirate attacks and enhance the security of maritime trade along the MSR. Based on the analysis of the historical database of pirate attack risk, an index system of pirate and armed robbery events based on the interaction between factors was established. Considering the dynamic characteristics of risk factors, a DBN risk analysis model of network topology jurisdiction was established. In combination with the MSR scenario conditions, the risk level of a ship being attacked by pirates under different environmental conditions was studied.

2. Problem Description

2.1. Pirate Attack Process Risk along the Maritime Silk Road

The MSR is a maritime trade route that starts from Guangzhou, Quanzhou, and other cities on the southeast coast of China and ends at several major ports on the east coast of Southeast Asia, South Asia, the Middle East, and Africa, including Indonesia, India, Sri Lanka, countries along the Persian Gulf, Egypt and other places [23]. As shown in Figure 1, the 21st-century MSR has main sea routes divided into various branches. This paper focused on the western route from Guangzhou to the Mediterranean Sea, which includes key nodes: Malacca Strait, the Gulf of Aden, and the Gulf of Guinea. Due to the important trade position of the MSR, pirate accidents frequently occur. Therefore, it is of great significance to explore the pirate attack risk along the MSR.

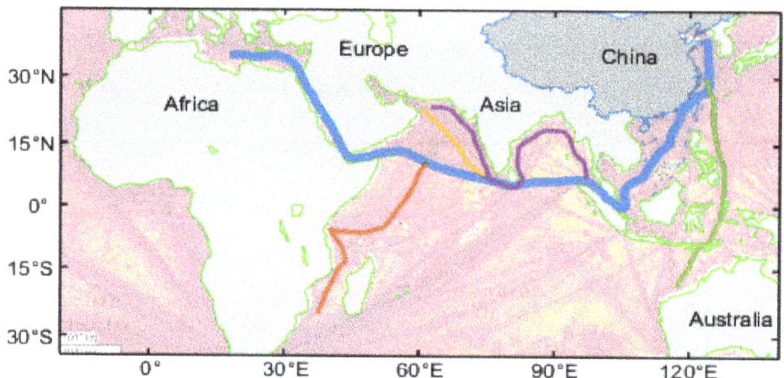

Figure 1. The route along the "21st-century" Maritime Silk Road.

The risk of pirate attack falls under the category of ship security risks. Ship security risk refers to the potential threats faced by a specific ship, including its personnel, cargo, equipment, and operations, stemming from illegal acts and terrorism, and the severity of their consequences [24]. Pirate attack risk is the combination of the possibility of a ship and its personnel, cargo, equipment, and operations being threatened by terrorism during transportation and the severity of its consequences. It is the expression of the interactive evolution of dynamic factors and static factors. The risk of a ship being attacked by pirates in the process of transportation is continuous in time and space, and shows the dynamic evolution process of the risk. Therefore, the mathematical model expressing the risk of pirate attacks can be represented by Equations (1) and (2) [20,25].

$$P(X_n) = \prod_{X_n \in X} P(X_n | Pa(X_n)), \tag{1}$$

$$P(X, Y, Z) = P(Y_t | Y_{t-1}) \prod_{t=0}^{T-1} P(Z_t \backslash Y_t) P(X, Y), \tag{2}$$

where: X is the static factor of the pirate attack risk; Y is the dynamic factor of the pirate attack risk; Z is the risk of the ship being attacked by pirates; $P(X_n)$ is the conditional probability of the factor X_n; and $P(Y_t | Y_{t-1})$ is the transfer probability matrix of the dynamic factor Y of the ship at a time t.

2.2. Risk Mechanism of a Ship Being Attacked by Pirates

In this study, we explored the risks of piracy and armed robbery against ships navigating along established routes. We also examined the critical factors that influence these risks. According to the basis that was discussed by previous scholars, risk is defined as the probability and consequences of unforeseen events occurring during a ship's voyage [26]. A comprehensive risk equation was developed to enhance risk evaluation, incorporating influencing factors from three risk component categories (hazard, mitigation capacity, vulnerability, and exposure) [27–29], and the equation is calculated as follows:

$$\text{Risk} = \text{Hazard} \times \text{Vulnerability} \times \text{Exposure} / \text{Mitigation capacity}, \tag{3}$$

Additionally, by utilizing the existing database, we established a risk index system for piracy and armed robbery incidents during a ship's navigation. The details of this system are shown in Table 1. Within this system, harmfulness mainly refers to the degree to which a system may suffer damage when confronted with specific events or behaviors. Vulnerability and exposure mainly refer to the degree to which a system is susceptible to damage or paralysis when facing external pressure or interference. mitigation capacity describes the ability and effectiveness of disaster prevention and reduction measures, including emergency response, disaster warning, building seismic fortification, and other capabilities. A system with good mitigation capacity can mitigate the impact of risks on the system when facing risks.

Table 1. Risk indicator system for piracy and armed robbery incidents.

	Indicator Name	Explanations	Indicator State	References
Level 1 indicator	Hazard	The higher the system hazard, the greater the risk of pirate attacks.	high; medium; low;	[27]
	Vulnerability and exposure	The higher the system vulnerability and exposure, the greater the risk of pirate attacks	high; medium; low;	[12,27]
	Mitigation capacity	The higher the level of mitigation capacity, the lower the risk of pirate attacks	good; medium; poor;	[27]
Level 2 indicator	Natural conditions	Under adverse natural conditions, the initiative of pirate attacks tends to decrease.	favourable; normal; bad;	[1,4,13,30,31]
	Human-induced hazards	The situation of local pirates affects the level of risk.	high; medium; low;	[12,30]
	Ship condition	When a ship malfunctions, it is more likely to become a target of pirate attacks.	good; moderate; poor;	[2,32]
	Ship's own risk	When the ship itself has sufficient attractiveness to pirates, the risk of being attacked by pirates increases compared to other ships.	high; medium; low;	[1,2,4,12,33]
	The anti-piracy capability of the ship	When a ship has strong anti-piracy capabilities, the risk of being attacked by pirates will be reduced.	good; moderate; poor;	[12,33]
	Naval support	This variable stands for the military response time (i.e., the time necessary to render assistance to the ship under threat).	t15; t30; morethant30;	[11,12,27,34]
Level 3 indicator	Wave	When the waves are larger, they will restrict the operation of boats used by pirates. The risk of a ship being attacked by pirates will be significantly reduced at this time.	normal; moderate; rough;	[2,26,30,31,33]
	Visibility	When the visibility is poor, ship lookouts may not detect pirate boats promptly, which makes it easier for pirates to approach and attack the ship.	good; moderate; poor;	[2,4,30]
	Pirate Capability	The stronger the pirate's capabilities, the greater the risk of being attacked and hijacked.	Strong; General; weak;	[12]
	The situation of surrounding countries	Generally speaking, in a turbulent zone, many criminal factors will breed, and the occurrence likelihood of attacks from pirates will also increase greatly.	high; medium; low;	[10–12,35]

Table 1. *Cont.*

	Indicator Name	Explanations	Indicator State	References
Level 3 indicator	Ship maintenance degree	Ship maintenance is closely related to the safety and stability of vessels during their navigation on water. If effective management and maintenance are not carried out, it can lead to significant consequences. Therefore, ship maintenance is closely tied to the condition of the vessel. Once a vessel experiences malfunctions due to maintenance issues within pirate-infested areas, the risk of pirate attacks significantly increases compared to other vessels [36]	good; moderate; poor;	Experts
	Ship Age	The older the ship is, the more passive and risky it is when facing pirate attacks.	less than 6; between 6 and 15; fifteen years and over;	[2,32]
	Ship Type	Pirates prefer to attack high-value ships because they can bring higher profits, such as bulk carriers, oil tankers, and so on.	high; medium; low;	[2,10–12,32]
	Freeboard	The lower the freeboard is, the lower the safety is, and the easier it is for pirates to board the ship, which increases the risk.	high; medium; low;	[1,4,12]
	Speed	The probability of successful pirate attacks decreases significantly and the risk is lower when the ship can sail at a speed of 15 knots or higher.	fifteen knots and over; less than fifteen; at anchor;	[2,12,33]
	Emergency management	The higher the emergency management capability of a ship is, the better its ability to handle pirate attacks in an orderly manner. Sometimes, the timely summoning of the crew or sounding alarms can reduce the risk of pirate attacks.	good; moderate; poor;	Experts
	Anti-piracy measures	The timely implementation of anti-piracy measures, when a pirate attack occurs, plays a significant role in preventing pirate intrusion.	good; moderate; poor;	[1,4,33]
	Armed security	When armed guards are present on board, pirates are more likely to abort their attacks, resulting in a lower risk of pirate attacks.	armed; unarmed; noGuards;	[8,11,33]
Level 4 indicator	Number of pirates	The greater the number of individuals involved in a pirate attack, the higher the risk posed to the targeted ship.	less than 5; between 5 and 10; 10 persons and over;	[10,12,33]
	Pirates' weapons	The degree of advancement of pirate weapons determines their capabilities. Generally, the more advanced the weapons, the stronger the pirate's abilities, and the greater the threat to ships.	guns and rocket-propelled grenades; knives; other;	[10,12,33]
	Annual average times of pirate attacks in surrounding areas	The annual average times of pirate attacks in the surrounding area reflects the degree of piracy prevalence. The more frequent pirate attacks, the higher the risk of pirate attacks.	less than 5; between 5 and 30; 30 times and over;	[30,35]
	Political situation in neighboring countries	In politically unstable countries, law enforcement may have loopholes, leading to an increase in criminal activities, which may increase the probability of pirate incidents to some extent.	extremely unstable; unstable; stable;	[10–12,30,35]
	Economic situation of surrounding countries	Coastal countries are prone to developing forms of robbery similar to piracy in economically underdeveloped situations.	GDP less than 2500; GDP between 2500 and 6000; GDP 6000 and over;	[10–12,35]

Table 1. *Cont.*

	Indicator Name	Explanations	Indicator State	References
Level 4 indicator	Anti-piracy drill	Regularly organizing anti-piracy exercises for the crew can enhance their ability to respond to pirate incidents, thereby enabling them to promptly take effective measures to resist piracy.	good; moderate; poor;	[12,33]
	Crew's awareness of anti-piracy	Having a strong awareness of anti-piracy measures can effectively reduce the risk of pirate attacks.	good; moderate; poor;	Experts
	Self-defense equipment and communication facilities	Ships generally equip themselves with certain self-defense devices such as water cannons, foam guns, alarm systems, etc., which can to some extent slow down or prevent pirate attacks and boarding.	good; moderate; poor;	[12,33]
	Monitoring intensity	Frequent observation can facilitate the early detection of potential pirate threats, enabling ships to take preemptive measures.	frequent; moderate; infrequent;	[11]

3. Model and Method

3.1. The Network Structure of Risk Analysis

In the risk indicator system for piracy and armed robbery, the risk of a ship being attacked is determined by the interaction of three risk component categories: hazard, mitigation capacity, and vulnerability and exposure. Moreover, it is also formed by the evolution of dynamic and static indicators.

For example, when the times of pirate attacks in the surrounding area is frequent and the surrounding countries are in a stage of political chaos and economic decline all year round, there is a higher possibility of the "situation of surrounding countries" indicator being "high risk", which will promote a higher possibility of the "human-induced hazards" indicator being "high risk". It will increase the possibility of the " hazard" indicator being "high" and ultimately lead to an increase in the risk of pirate attacks. However, if there are high wave levels during this time, it affect the "natural conditions" indicator as "bad", and ultimately work together with "human-induced hazards" to "hazard", reducing the risk of pirate attacks to a certain extent based on the original increase. The same goes for other indicators. From this perspective, the relationship between indicators is network oriented, thus forming a complete pirate attack risk assessment network.

Further analysis of the factors influencing the risk of pirate attacks on ships reveals that these factors exhibit characteristics of directed acyclic relationships. Bayesian networks utilize Directed Acyclic Graphs (DAGs) to represent dependencies between variables, while other risk analysis methods may employ different model representations such as decision trees or regression models. The graph structure of Bayesian networks captures causal relationships and probabilistic dependencies, providing a more intuitive and clear model representation. Moreover, Bayesian networks are probabilistic graphical models that incorporate probability distributions to describe relationships and uncertainties. In contrast, other risk analysis methods may employ deterministic models, disregarding the impact of uncertainty. The probabilistic modeling in Bayesian networks allows for comprehensive consideration of uncertainty and provides accurate probability inference [37]. Therefore, a Bayesian network can be introduced to conduct a quantitative analysis of the risk of pirate attacks.

3.2. Dynamic Bayesian Network Model

3.2.1. Bayesian Network

Bayesian networks, also known as belief networks, are DAG models [13], which comprise nodes representing variables and directed edges connecting these nodes. Nodes represent random variables and directed edges between nodes represent the relationships between them (from parent nodes to child nodes). The relationship between parent and child nodes is expressed through conditional probability tables, while prior probabilities are used for root nodes to convey information. The variables represented by the nodes can be abstractions of any problem and the entire network comprises multiple nodes and directed edges, showing the causal relationships between variables and the conditional independence relationships between nodes. Generally, the Bayesian network structure can be determined in the following ways: (1) by obtaining the Bayesian network structure based on the database; (2) the network structure being established and adjusted according to previous literature and expert suggestions, resulting in the formation of the Bayesian network structure.

The process risk of a ship being attacked by pirates is a continuous process under a time series. However, the traditional Bayesian network ignores the correlation and interaction of factors at different times, which easily leads to misjudgment of the final result. It is difficult to meet the prediction and assessment requirements after the dynamic evolution of factors in a complex dynamic environment. DBN is an extension of the Bayesian network in time series. It extends the Bayesian network by introducing relevant temporal dependencies to model the dynamic behavior of random variables. A DBN consists of a series of time slices and time links, where each slice represents a static Bayesian network that describes the variables at the corresponding time step. The links between variables across different time slices represent temporal probability dependencies [22]. The conditional probability of each variable in a DBN can be calculated independently, which facilitates the interpretation of DBNs.

By analyzing the risk of a ship being attacked by pirates and combining the description of DBN in the previous section, it is found that DBN can describe the process risk of a ship being attacked by pirates more scientifically, intuitively, and accurately. Therefore, by considering influence factors such as waves, visibility, the number of pirates, pirates' weapons, annual average times of pirate attacks in surrounding areas, the political situation in neighboring countries, the economic situation of surrounding countries, and naval support as dynamic variables in the model, the dynamic property of the system's behavior can be reflected. Based on the study of the risk mechanism of pirate attacks, a process risk analysis model for pirate attacks based on DBN was established, as shown in Figure 2.

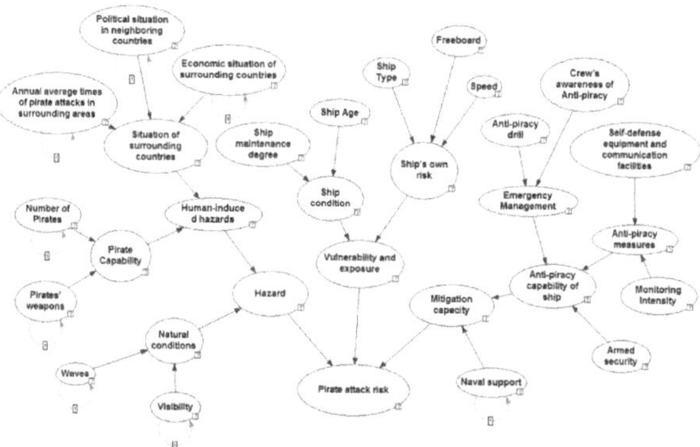

Figure 2. Diagram for assessing the risk of pirate attacks.

3.2.2. Model Parameter

In the DBN model, there are two types of nodes: one type has one or more parent nodes, and its state is influenced by the combination of its parent nodes, requiring the input of conditional probabilities, which can be obtained through training sample data or expert questionnaires. The other type of node is a root node, requiring the input of prior probabilities, which can be obtained through event case statistics or expert questionnaires.

The transition matrix is usually obtained through expert questionnaires or by combining the Markov model with probability distribution assumptions. DBN is represented as an even pair $\langle B_0, B_\rightarrow \rangle$, where B_0 is the initial BN that defines the probability distribution of $P(Y)$ at the initial time and is a 2-TBN containing two-time slices that define the conditional probability distribution between the variables of two adjacent time slices; the state transition probability is expressed as Equation (4).

$$P(Y_t|Y_{t-1}) = \prod_{i=1}^{n_t} P(Y_t^i|Pa(Y_t^i)),\tag{4}$$

where: $P(Y)$ is a set of variables, Y_t^i is the i-th node in the time slice t; $Pa(Y_t^i)$ is the parent node of Y_t^i, n_t is the number of nodes in the t-th time slot.

3.3. Risk Value

Because the process risk is dynamic, the pirate attack risk will also change in different environments in the same place. To identify the specific size of the risk, it is necessary to calculate the average risk on the route in the process of risk analysis, which is expressed in Equations (5) and (6) [20].

$$\overline{R} = \frac{1}{N}\sum_{t=1}^{N} R(t),\tag{5}$$

$$R(t) = \sum_{i=1}^{n} R(t_i)\gamma_i,\tag{6}$$

where: $\overline{R}$ is the average risk of the route, $R(t)$ is the comprehensive risk at time t, $R(t_i)$ is the likelihood of state I at time t, and is the weight of state i. This article divides the risk status of the final pirate attacks into high risk, medium risk, and low risk, with weights of 6, 3, and 1, respectively.

4. Simulation and Results

4.1. Scenario Descriptions

The ship's route was to depart from Lagos Port in Nigeria in July 2021 to Guangzhou Port; the route was chosen to shorten the voyage and reduce sailing time while at the same time considering commercial, economic, and navigational safety requirements, and included transit through the Suez Canal, the Gulf of Aden, Sri Lanka, the Strait of Malacca, and the South China Sea. The entire route was approximately 11,657 nautical miles, with a sailing time of about 45 days. Due to the presence of pirate attack risks at multiple locations along the entire route, this study integrated a previously proposed process risk model to determine the average risk for the entire voyage.

Due to the relative length of the entire route, to improve the accuracy of risk description, this route is divided into three segments for process risk calculation considering that the dynamic Bayesian transfer matrix is homogeneous. One data sampling point is a time slice.

Seven-time slices were set in the Lagos-Suez Canal route, ten time slices were set in the Suez Canal–Sri Lanka route, and ten time slices were set in the Sri Lanka–Guangzhou Port route. The last time slice of each segment serves as the first time slice of the next segment. The entire route comprises 25 (0~24) time slices.

4.2. Information Acquisition and Parameter Determination

(1) This article incorporates environmental observation data to obtain model parameters of environmental impact factors. The visibility data along the route were gathered from literature sources [4] and transformed to determine the parameter status values. This article introduced environmental observation data to obtain model parameters of environmental impact factors. The visibility data on the route was collected through literature [4] and converted to determine the parameter status values. The sea wave data on the route were

obtained from a shared meteorological information website. The data were extracted from the daily variation of sea wave meteorological information. We selected "Significant height of combined wind waves and swell" from the website as the data for the waves in the model [38], for example, as is shown in Figure 3. The daily average wave data from March 1 to March 6 were extracted. Based on expert experience, the wind and wave range were set [0, 3] as normal, [4, 6] as moderate, and level 7 and above as rough. Using interval division to determine the state description data of environmental data, for example, at t = 0, we took these six small pictures as an example in which the daily average wave level of 6 days is in the range of [0, 3]. Therefore, the probability that the wave variable is "normal" at t = 0 is 1, and other data can be obtained in the same way.

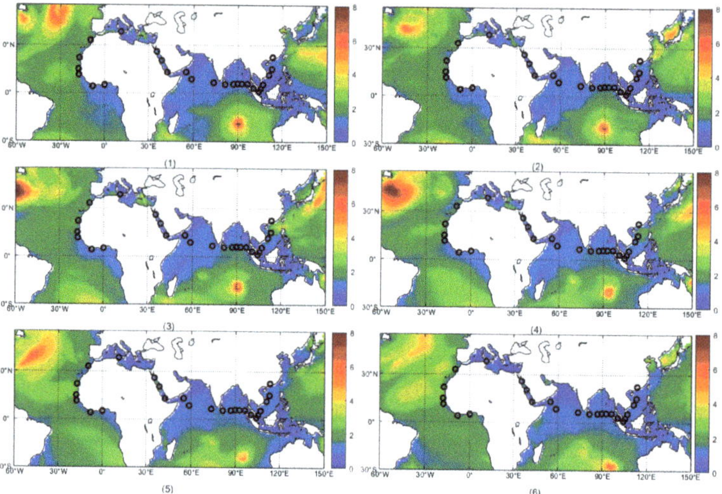

Figure 3. Daily average wave data from March 1 to March 6 and 25 data collection points (Data Source: https://cds.climate.copernicus.eu/cdsapp#!/search?type=dataset. URL (accessed on 20 May 2023)).

(2) Acquisition of prior probability and conditional probability. Historical database information was introduced to determine the conditional probability, prior probability, and other parameters of the network model. Based on the database of pirate and armed robbery incidents from 1994 to 2022 obtained from the IMO website, taking the ship type as an example and combining the statistical results of the database, Figure 4 was obtained. Finally, the prior probability of the ship type was determined as follows: high risk: 0.698; medium risk: 0.221; and low risk: 0.081. The prior probabilities of other non-environmental root nodes were obtained by similar statistical methods.

(3) Acquisition of transition probability. The acquisition of transition probabilities for dynamic nodes is a challenging task in the establishment of the entire model. Through the process risk analysis of pirate attacks, ten variables, including waves, visibility, the numbers of pirates, pirates' weapons, annual average times of pirate attacks in surrounding areas, the political situation in neighboring countries, the economic situation of surrounding countries, and naval support, were defined as dynamic variables. The transition matrices of different variables were different. Basic transition probabilities of dynamic nodes were established by analyzing the 20-year historical data in the sample database. Taking pirate personnel as an example, the transition matrix is

$$P_0 = \begin{bmatrix} 0.989 & 0.951 & 0.872 \\ 0.0092 & 0.032 & 0.085 \\ 0.0018 & 0.017 & 0.043 \end{bmatrix}$$

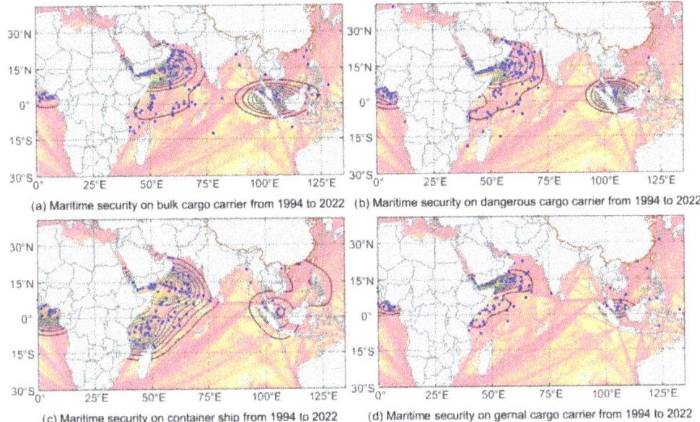

Figure 4. The security incidents distribution of different types of ships.

Due to the unavailability of certain parameters in the database, such as the political situation in neighboring countries, the economic situation of surrounding countries, naval support, and so on, parameter values were obtained through questionnaires and expert consultations. The experts involved in this study include professors who have been engaged in this field for many years, as well as captains with extensive navigation experience. For instance, at t = 0, there were three possible states for the political situation in neighboring countries: extremely unstable, unstable, and stable. The probabilities of these three states were determined through questionnaires and expert consultations, with the probabilities of extremely unstable, unstable, and stable being 0.786, 0.201, and 0.013, respectively.

4.3. Process Risk Analysis of Pirate Attacks

(1) Based on the database of pirate attacks and armed robberies obtained from the IMO website, it is evident that there are significant differences in the times of pirate attacks that occur each month out of the 12 months in a year. Therefore, in the following analysis, this paper took the month of July 2021 as an example to assess the risk of pirate attacks, as illustrated in Figure 5a.

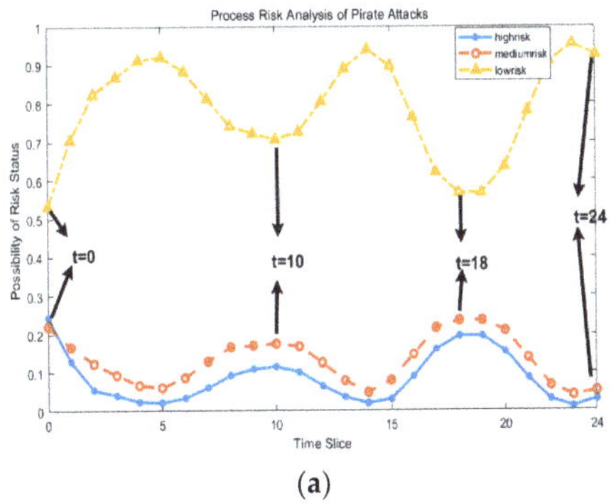

(a)

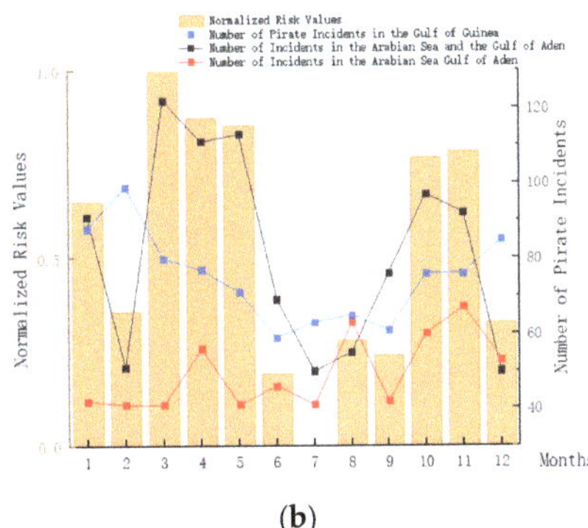

(b)

Figure 5. Results of dynamic Bayesian model simulation. (**a**) Results of the dynamic Bayesian model. (**b**) Normalized risk values vs. pirate attack incidents chart.

(2) Through the analysis of wave information, significant changes in wave conditions could be observed. Therefore, in the analysis of the process risk of pirate attacks, it was necessary to consider the monthly effects of wave factors on pirate attack risks. To this end, we calculated the average risk value for the 12 months using Equations (2) and (3). For a

better analysis of risk characteristics, we normalized the values, as is shown in Figure 5b. The results indicated that the trend of pirate accidents is consistent with the trend of wave effects on pirate risk. From March to May, the times of pirate events were relatively high compared to other months, and the risk values were relatively high. From June to September, the times of pirate events and risk values were relatively lower compared to other months. Therefore, we analyzed the months of March with the highest risk values and July with the lowest risk values.

4.4. Sensitivity Analysis

In order to verify the validity of the model, 50 pieces of data in March 2021 were randomly extracted from the database of piracy incidents, and some subjective data were graded by experts with rich navigation experience. By inputting 50 groups of data into the established dynamic Bayesian network, the results of 24 time slices can be obtained. The results are shown in Figure 6, in which the dashed line represents the change of process risk and the solid line represents the results of 50 randomly selected samples input into the model.

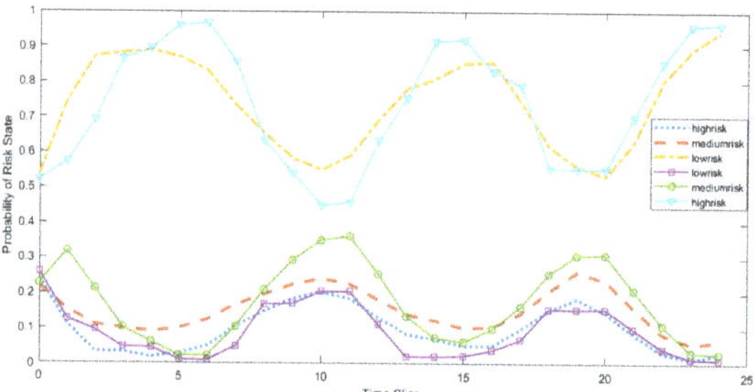

Figure 6. Process risk verification chart for pirate attacks.

The results indicated that the change curve of process risk is roughly consistent with the change curve trend of risk obtained by random sampling, which shows that the model is scientific and effective. In the dynamic Bayesian model, the node "pirate attack risk" was set as the target, and Bayesian network inference could be used to analyze the main sensitive nodes leading to pirate incidents.

As the model is a dynamic network structure, the first time slice was selected for model sensitivity testing to identify the most critically influential factors affecting the risk of pirate attacks. The sensitivity test results for the state "high risk" are shown in Figure 6, where "A" represents the node "pirate attack risk", "B1" represents "hazard", "B2" represents "vulnerability and exposure", "B3" represents "mitigation capacity", "C1" represents "natural conditions", "C2" represents "human-induced hazards", "C6" represents "naval support", "D3" represents "pirate capability", "D4" represents "situation of surrounding countries", "F4" represents "waves", "F5" represents "pirates' weapons", and "F7" represents "annual average times of pirate attacks in surrounding areas". The green color indicates a positive effect on "high risk", while the red color indicates a negative impact.

From Figure 7, we can obtain the following conclusions. Under the comprehensive influence of hazard, vulnerability and exposure, and mitigation capacity, "B1 = high | C1 = favourable, C2 = highrisk", "A = highrisk | B2 = high, B3 = good, B1 = high", and so on, have positive impacts on the high risk of pirate attacks, while "B1 = low | C1 = favourable, C2 = low risk", "F5 = Other", and so on, have negative impacts on the high risk of pirate attacks. Therefore, it can be concluded that "natural conditions" and "human-induced hazards" are the nodes that have the greatest impact on high-risk pirate attacks. "Hazard", "vulnerability and exposure", "mitigation capacity", "navy support", "pirate capability", "situation of surrounding countries", "wave", "pirates' weapons", and "annual average times of pirate attacks in surrounding areas" have a significant impact on high-risk pirate attacks.

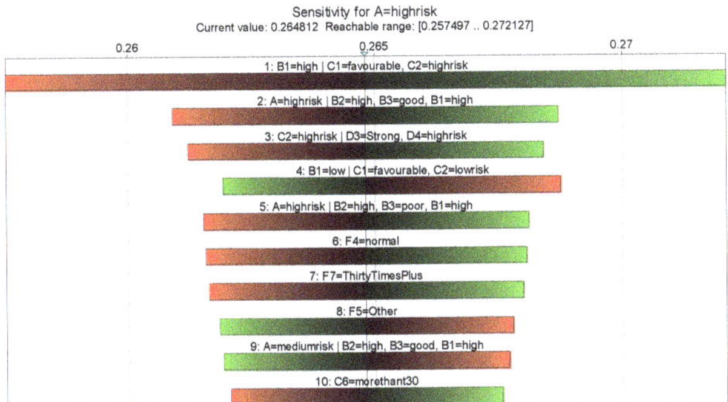

Figure 7. Sensitivity analysis chart.

5. Discussion

5.1. Process Risk of a Ship Being Attacked by Pirates along the Maritime Silk Road

To investigate the impact of the environment on the risk of pirate attacks further, the months of March (a) and July (b) were taken as examples, as shown in Figure 8.

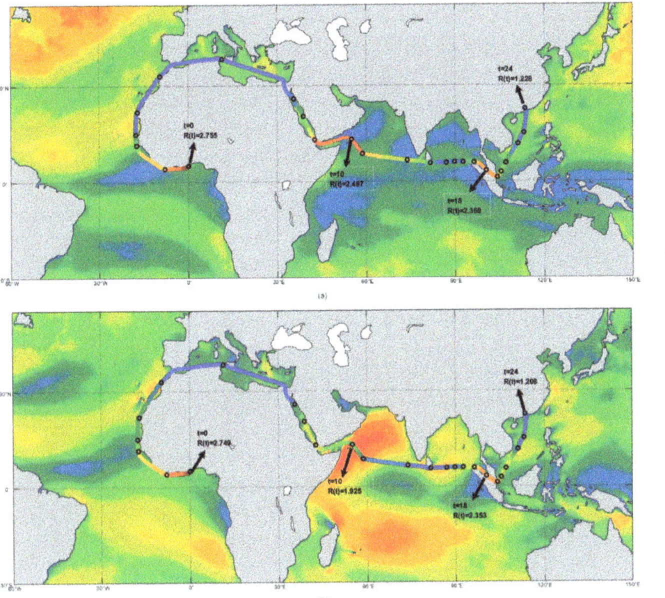

Figure 8. Diagram of Pirate Attack Risk for March (**a**) and July (**b**).

The Gulf of Guinea (t = 0~1) is located near the equator and experiences relatively stable wave conditions, so there was not much difference in pirate activities between March and July. showing a high intensity. In the region of the Gulf of Aden–Arabian Sea (t = 9~11), March is a non-monsoon period with favorable environmental conditions, resulting in high pirate attack risks. In contrast, during the monsoon period in July, the environment becomes harsh and pirate activities become difficult, leading to a significant decrease in attack risks compared to March. In the Malacca Strait (t = 17~19), the environmental conditions remain stable throughout the year due to its geographical features, resulting in stable pirate attack risks. It could be concluded that pirate attack risks were high when the environment is favorable and decreased with changes in environmental conditions. Based on the analysis, it could be concluded that the average risk of pirate attacks was the lowest in July and the highest in March along the route. Throughout the voyage, the risk of pirate attacks was the highest in the Gulf of Guinea, followed by the Gulf of Aden–Arabian Sea region and the Malacca Strait.

In addition to environmental factors, other dynamic factors also affected the process risk of a ship being attacked by pirates. For example, as shown in Figure 8, the March (a) chart indicates that when the ship is at t = 0, the risk of pirate attacks is high. At this time, the probabilities of the nodes "annual average times of pirate attacks in surrounding areas", "number of pirates", and "pirates' weapons" being in a state of "thirty times and over", "ten persons and over", and "guns and rocket-propelled grenades" are 0.528, 0.111, and 0.317, respectively. When the ship reaches t = 7, the risk of pirate attacks is low, with probabilities for the aforementioned nodes of 0.0085, 0.00202, and 0.0091, respectively. At t = 13, the risk of pirate attacks decreases, with probabilities for the nodes of 0.095, 0.007011, and 0.00456. At t = 18, the risk increases, with probabilities for the nodes of 0.44, 0.1712, 0.312. At t = 24, the risk reaches its lowest point, with probabilities for the nodes of 0.0023, 0.00082, and 0.00108. From this analysis, it could be concluded that the risk of pirate attacks increased when there were high times of pirate incidents in the past, and when there were many pirates with advanced weapons, coupled with poor political and economic conditions in the surrounding areas. When such factors were absent or certain measures were taken, the risk of pirate attacks decreases. Moreover, when naval support was prompt or the times of pirates were low, the risk of pirate attacks also decreased.

5.2. The Realistic Situation of Pirate Attacks Risk along the Maritime Silk Road

When a ship is navigating through pirate-prone areas, the likelihood of a ship being attacked by pirates was high. However, encountering adverse weather conditions could significantly reduce the likelihood of such attacks, regardless of the area. In regions where pirate incidents were frequent, the level of piracy capabilities, such as the number of pirates and the weapons used, could also impact the risk to a ship exposed to attacks from pirates. When the number of pirates was small, such as one or two, and when they had no weapons or only used simple tools, the risk of pirate attacks was greatly reduced due to the system's ability to resist risks. However, if the number of pirates exceeded ten and they used advanced weapons and equipment, the risk increased. Additionally, ships carrying high-value goods such as petroleum were more likely to be targeted by pirates.

In reality, the Gulf of Guinea is frequented by pirates due to a lack of naval support, political instability in surrounding countries, and economic recession, among many other factors that trigger pirate attacks. As a vital channel connecting the Red Sea and the Indian Ocean, the Gulf of Aden witnesses a large number of high-value ships, such as oil tankers and cargo ships, pass through annually. The geographical location of the Gulf of Aden also holds significant military strategic importance, contributing to the surrounding countries' political instability and poor economic conditions, leading to numerous severe pirate incidents. The Strait of Malacca, as a critical trading route in Asia, has an enormous amount of maritime transportation. Due to the vast wealth involved in trade and the significant wealth gap in nearby regions, some areas experience political instability and economic stagnation. Consequently, some individuals engage in illegal activities with support from coastal villages. Governments making efforts to combat piracy also face certain challenges. These combined factors contribute to a higher risk of piracy in the waters of the Strait of Malacca.

5.3. Space-Time Characteristics of Pirate Attacks along the MSR

Moreover, the presence of naval and coastal guard personnel has a tremendous impact on deterring piracy. The analysis of data on piracy and armed robbery incidents from 2003 to 2023 provided by the IMO website illustrated the spatiotemporal characteristics of pirate attacks on MSR coastal ships, as shown in Figure 9.

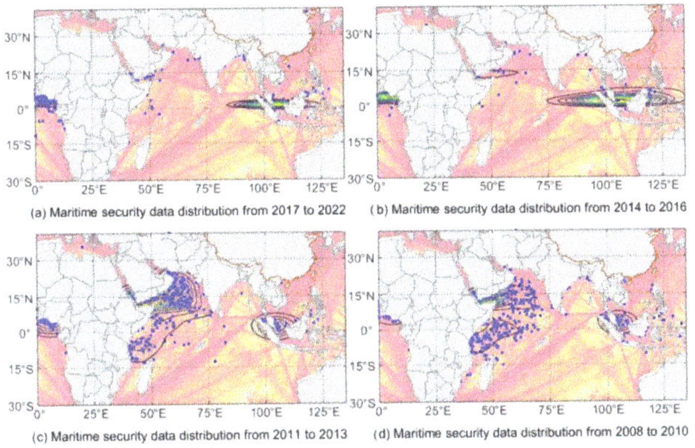

Figure 9. The security incidents distribution from 2008 to 2022.

Through analysis of event data, it could be found that from 2008 to 2013, piracy incidents in the Malacca Strait and the Gulf of Guinea began to gradually emerge in the public eye. Of the 2464 piracy incidents that occurred globally during this period, 1205 incidents occurred in the waters of the Gulf of Aden–Arabian Sea, accounting for 48.89% of the global total. The notorious Somali pirates also originated from this area, and much of the reason for their existence can be attributed to the economic situation and political chaos of the surrounding countries [34]. Due to the severe global impact of Somali pirate incidents, the United Nations called on countries to send naval escorts. Starting in 2014, piracy incidents in the Gulf of Aden–Arabian Sea began gradually decreasing each year, with no piracy incidents occurring from January to June 2015. Thus, the presence of pirate groups in the Gulf of Aden–Arabian Sea, represented by Somali pirates, has rapidly declined under the joint global crackdown. However, the Malacca Strait, as a vital maritime transportation route, has a large wealth gap among its people and a certain pirate culture and tradition due to historical reasons [39]. It became a high-risk area for piracy incidents from 2014 to 2016 and quickly gained widespread global attention. However, due to its complex geographical location, it is difficult to combat piracy and, thus, there were still piracy incidents in the region from 2017 to 2022. The Gulf of Guinea, due to political and economic problems and weak naval forces, has gradually become another high-risk area for piracy incidents in succession to the Malacca Strait region [40].

By incorporating a transition matrix, the changes in risk over time can be represented, allowing for risk assessment. Ship navigation is a dynamic process, and the factors contributing to the risk of pirate attacks are correlated within the changing time series. By using Bayesian inference, which combines subjective and objective data, the probability of target events occurring can be calculated through network inference. Therefore, in this study, a dynamic transition matrix and nodes in a Bayesian network evaluation model were utilized to obtain risk values at different time slices, enabling the assessment of process risk. The findings indicated that the risk of pirate attacks was relatively high in the Gulf of Guinea, the Gulf of Aden, and the Malacca Strait along the route from Lagos to Guangzhou Port. The lowest risk was in July, while the highest risk was observed in March. "Natural conditions" and "human-induced hazards" were the most influential factors contributing to high-risk pirate attacks.

6. Conclusions

This paper focused on researching the risk of pirate attacks along shipping routes and developed a process risk analysis model for pirate attacks based on DBN, which incorporates the standards of the three risk component categories. Thirty-one factors, including hazard, mitigation capacity, and vulnerability and exposure, among others, were considered to enhance the risk assessment system for pirate events, enabling the assessment of the process risk of a ship being attacked by pirates along the route. The results indicated the effectiveness of process risk analysis for pirate attacks based on dynamic Bayesian networks.

Based on the environmental characteristics, physical features of pirate incidents, and logical characteristics of pirate incidents on shipping routes, this article analyzed the process risk of a ship being attacked by pirates and identified the risk characteristics of pirate attacks in different spatial and temporal contexts. The results indicated that the overall average risk of a ship being attacked by pirates on the entire route was the lowest in July and highest in March. Among the entire route, the risk values in the waters of the Gulf of Guinea and the Malacca Strait were the highest, while the risk values in the waters of the Gulf of Aden–Arabian Sea were unstable, with the lowest risk in July and the highest risk in March, demonstrating an overall high risk. Furthermore, the study highlighted the significant impact of natural conditions, human-induced hazards, the situation of surrounding countries, and navy support on the risk of a ship being attacked by pirates.

The risk of a ship being attacked by pirates in a sea route represents a complex and multi-factorial system. Pirate attacks can have varying impacts, including economic, personnel, and environmental consequences. This study specifically focused on assessing pirate attack risks along the western route of the Silk Road, and as such, it may have limitations when evaluating risks in other scenarios. Additionally, during the construction of the model, certain environmental factors, such as wind speed and precipitation [41,42], were not thoroughly explored. Therefore, accurately identifying and fully exploring the environmental information, analyzing and controlling risks, and quantifying consequences scientifically are important areas for further study in this field.

Author Contributions: X.H., methodology, software, validation, formal analysis, investigation, resources, data curation, visualization, writing—original draft preparation, writing—review and editing; H.X., conceptualization, formal analysis, writing—review and editing, supervision, project administration; S.X., methodology, validation, formal analysis, investigation, resources; S.H., conceptualization, methodology, validation, formal analysis, visualization, supervision, writing—original draft, writing—review and editing, funding acquisition. All authors have read and agreed to the published version of the manuscript.

Funding: The project is supported by the National Natural Science Foundation of China (Grant No. 52272353). This work was also supported by funding from the National Key Research and Development Program of China (Grant No. 2021YFC2801005).

Institutional Review Board Statement: Not applicable.

Informed Consent Statement: Not applicable.

Data Availability Statement: Data sharing not applicable. No new data were created or analyzed in this study. Data sharing is not applicable to this article.

Acknowledgments: We thank all those who helped us in the writing review and editing of this paper.

References

1. Jiang, M.; Lu, J. The analysis of maritime piracy occurred in Southeast Asia by using the Bayesian network. *Transp. Res. Part E Logist. Transp. Rev.* **2020**, *139*, 101965. [CrossRef]
2. Jiang, M.Z.; Lu, J.; Yang, Z.L.; Li, J. Risk analysis of maritime accidents along the main route of the Maritime Silk Road: A Bayesian network approach. *Marit. Policy Manag.* **2020**, *47*, 815–832. [CrossRef]
3. Yang, Y.; Liu, W. Resilience Analysis of Maritime Silk Road Shipping Network Structure under Disruption Simulation. *J. Mar. Sci. Eng.* **2022**, *10*, 617. [CrossRef]
4. Pristrom, S.; Yang, Z.L.; Wang, J.; Yan, X.P. A novel flexible model for piracy and robbery assessment of merchant ship operations. *Reliab. Eng. Syst. Saf.* **2016**, *155*, 196–211. [CrossRef]

5.   Wang, M.; Gong, S. Functions and Risks of Ship Security Alarm Systems. *China Ship Surv.* **2003**, *11*, 84–86.
6.   Ling, Z. Implementation of Ship Security Assessment. *Transp. Sci. Technol.* **2006**, *1*, 96–98.
7.   Sevillano, J.C.; Rios Insua, D.; Rios, J. Adversarial Risk Analysis: The Somali Pirates Case. *Decis. Anal.* **2012**, *9*, 86–95. [CrossRef]
8.   Struwe, L.B. Private Security Companies (PSCs) as a Piracy Countermeasure. *Stud. Confl. Terror.* **2012**, *35*, 588–596. [CrossRef]
9.   Vanek, O.; Hrstka, O.; Pechoucek, M. Improving Group Transit Schemes to Minimize Negative Effects of Maritime Piracy. *IEEE Trans. Intell. Transp. Syst.* **2014**, *15*, 1101–1112. [CrossRef]
10. Bouejla, A.; Chaze, X.; Guarnieri, F.; Napoli, A. A Bayesian network to manage risks of maritime piracy against offshore oil fields. *Saf. Sci.* **2014**, *68*, 222–230. [CrossRef]
11. Lewis, J.S. Maritime piracy confrontations across the globe: Can crew action shape the outcomes? *Mar. Policy* **2016**, *64*, 116–122. [CrossRef]
12. Jin, M.J.; Shi, W.M.; Lin, K.C.; Li, K.X. Marine piracy prediction and prevention: Policy implications. *Mar. Policy* **2019**, *108*, 10. [CrossRef]
13. Li, J.; Liu, Y.; Zhu, L. Risk Assessment of Pirate Hijacking Based on Bayesian Network. *J. Saf. Environ.* **2021**, *21*, 7.
14. Wan, C.; Yan, X.; Zhang, D.; Qu, Z.; Yang, Z. An advanced fuzzy Bayesian-based FMEA approach for assessing maritime supply chain risks. *Transp. Res. Part E Logist. Transp. Rev.* **2019**, *125*, 222–240. [CrossRef]
15. Kabir, S.; Papadopoulos, Y. Applications of Bayesian networks and Petri nets in safety, reliability, and risk assessments: A review. *Saf. Sci.* **2019**, *115*, 154–175. [CrossRef]
16. Wang, L.; Yang, Z. Bayesian network modelling and analysis of accident severity in waterborne transportation: A case study in China. *Reliab. Eng. Syst. Saf.* **2018**, *180*, 277–289. [CrossRef]
17. Deng, J.; Liu, S.; Xie, C.; Liu, K. Risk Coupling Characteristics of Maritime Accidents in Chinese Inland and Coastal Waters Based on N-K Model. *J. Mar. Sci. Eng.* **2022**, *10*, 4. [CrossRef]
18. Hsu, W.-K.K.; Chen, J.-W.; Huynh, N.T.; Lin, Y.-Y. Risk Assessment of Navigation Safety for Ferries. *J. Mar. Sci. Eng.* **2022**, *10*, 700. [CrossRef]
19. Bi, J.; Gao, M.; Zhang, W.; Zhang, X.; Bao, K.; Xin, Q. Research on Navigation Safety Evaluation of Coastal Waters Based on Dynamic Irregular Grid. *J. Mar. Sci. Eng.* **2022**, *10*, 733. [CrossRef]
20. Li, Z.; Hu, S.; Gao, G.; Yao, C.; Fu, S.; Xi, Y. Decision-making on process risk of Arctic route for LNG carrier via dynamic Bayesian network modeling. *J. Loss Prev. Process Ind.* **2021**, *71*, 104473. [CrossRef]
21. Li, Z.; Hu, S.; Zhu, X.; Gao, G.; Yao, C.; Han, B. Using DBN and evidence-based reasoning to develop a risk performance model to interfere ship navigation process safety in Arctic waters. *Process Saf. Environ. Prot.* **2022**, *162*, 357–372. [CrossRef]
22. Guo, Y.L.; Jin, Y.X.; Hu, S.P.; Yang, Z.L.; Xi, Y.T.; Han, B. Risk evolution analysis of ship pilotage operation by an integrated model of FRAM and DBN. *Reliab. Eng. Syst. Saf.* **2023**, *229*, 18. [CrossRef]
23. Yang, Y.B.; Liu, W.; Xu, X. Identifying Important Ports in Maritime Silk Road Shipping Network from Local and Global Perspective. *Transp. Res. Record* **2022**, *2676*, 798–810. [CrossRef]
24. Zhu, Q. Research on the Application of Risk Management in Ship Security. *J. Saf. Sci. Technol.* **2012**, *8*, 5.
25. Lu, X.; Xu, C.; Hou, B.; Du, L.; Li, L. Risk Assessment of subway tunnel Construction Based on Dynamic Bayesian network. *Ocean Eng.* **2022**, *208*, 12.
26. Hu, S. Monte Carlo simulation of maritime traffic Systemic risk. *J. Shanghai Marit. Univ.* **2011**, *32*, 7–11.
27. Zhou, X.; Cheng, L.; Li, M.C. Assessing and mapping maritime transportation risk based on spatial fuzzy multi-criteria decision making: A case study in the South China sea. *Ocean. Eng.* **2020**, *208*, 12. [CrossRef]
28. Bruno, M.F.; Motta Zanin, G.; Barbanente, A.; Damiani, L. Understanding the Cognitive Components of Coastal Risk Assessment. *J. Mar. Sci. Eng.* **2021**, *9*, 780. [CrossRef]
29. Bruno, M.F.; Saponieri, A.; Molfetta, M.G.; Damiani, L. The DPSIR Approach for Coastal Risk Assessment under Climate Change at Regional Scale: The Case of Apulian Coast (Italy). *J. Mar. Sci. Eng.* **2020**, *8*, 531. [CrossRef]
30. Jiang, M.; Lu, J. Maritime accident risk estimation for sea lanes based on a dynamic Bayesian network. *Marit. Policy Manag.* **2020**, *47*, 649–664. [CrossRef]
31. Dabrowski, J.J.; de Villiers, J.P. Maritime piracy situation modelling with dynamic Bayesian networks. *Inf. Fusion* **2015**, *23*, 116–130. [CrossRef]
32. Ung, S.T. Navigation Risk estimation using a modified Bayesian Network modeling-a case study in Taiwan. *Reliab. Eng. Syst. Saf.* **2021**, *213*, 107777. [CrossRef]
33. Shane, J.M.; Magnuson, S. Successful and Unsuccessful Pirate Attacks Worldwide: A Situational Analysis. *Justice Q.* **2014**, *33*, 682–707. [CrossRef]
34. Coito, J.C. Pirates vs. Private Security: Commercial Shipping, the Montreux Document, and the Battle for the Gulf of Aden. *Calif. Law Rev.* **2013**, *101*, 173–226.
35. Gong, X.X.; Lu, J. Strait/canal security assessment of the Maritime Silk Road. *Int. J. Shipp. Transp. Logist.* **2018**, *10*, 281–298. [CrossRef]
36. Chen, C. Exploring how to do a good job in the maintenance and upkeep of ship machinery and equipment. *Mod. Manuf. Technol. Equip.* **2020**, *4*, 204–208. [CrossRef]
37. Miao, Y.F.; Wang, L.; Zhang, G.A.; Su, Q.H.; Li, X.L. Weather threat assessment based on dynamic bayesian network. *Appl. Ecol. Environ. Res.* **2019**, *17*, 9391–9400. [CrossRef]

38. Wang, Z.; Zhang, R.; Ge, S.; Ju, Y.; Cao, Z. Natural Environmental Risk Zoning of the Arctic Northeast Passage: Taking the Northern Sea Area of Russia as an Example. *Ocean. Eng.* **2017**, *35*, 61–70. [CrossRef]
39. Rediker, M. Pirates, Privateers, and Rebel Raiders of the Carolina Coast (Book). *J. South. Hist.* **2002**, *68*, 152. [CrossRef]
40. Onuoha, F. Piracy and Maritime Security in the Gulf of Guinea: Trends, Concerns, and Propositions. *J. Middle East Afr.* **2013**, *4*, 267–293. [CrossRef]
41. Dzvonkovskaya, A.; Nikolic, D.; Orlic, V.; Peric, M.V.; Tosic, N. Remote Observation of a Small Meteotsunami in the Bight of Benin Using HF Radar Operating in Lower HF Band. *IEEE Access.* **2019**, *7*, 88601–88608. [CrossRef]
42. Peel, M.C.; Finlayson, B.L.; McMahon, T.A. Updated world map of the Koppen-Geiger climate classification. *Hydrol. Earth Syst. Sci.* **2007**, *11*, 1633–1644. [CrossRef]

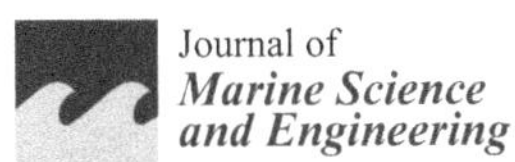

Journal of
Marine Science and Engineering

MDPI

Article

An Analytic Model for Identifying Real-Time Anchorage Collision Risk Based on AIS Data

Zihao Liu [1,*], Dan Zhou [2], Zhongyi Zheng [1], Zhaolin Wu [1] and Longhui Gang [1]

[1] Navigation College, Dalian Maritime University, Dalian 116026, China; dlzzyi@sina.com (Z.Z.); wuzl2010@126.com (Z.W.); ganglh@dlmu.edu.cn (L.G.)
[2] Intelligent Shipping Center, China Waterborne Transport Research Institute, Beijing 100088, China; zhoudan@wti.ac.cn
* Correspondence: zihaoliu0407@gmail.com

Abstract: With the increasing volume of ship traffic, maritime traffic safety is facing a great challenge because the traffic in port becomes more and more crowded and complicated, which will make ship collisions more likely to happen. As a special water area of the port, the anchorage is also threatened by collision risk all the time. For accurately assessing the collision risk in anchorage and its adjacent waters in real time, this paper proposed an analytic model based on Automatic Identification System (AIS) data. The proposed anchorage collision risk model was established in microscopic, macroscopic, and complexity aspects, which considered ship relative motion, anchorage characteristics, and ship traffic complexity, respectively. For validation, the AIS data of the anchorages near the Shandong Peninsular were used to carry out a series of experiments. The results show that the proposed model can identify the anchorage collision risk effectively and has an advantage in dealing with complicated scenarios. The proposed anchorage collision risk model can help maritime surveillance better monitor and organize the ship traffic near the port and provide mariners with a reference about the collision risk situation of the anchorage on their route, which are important to improving maritime traffic safety.

Keywords: anchorage; collision risk; relative motion; safe navigable waters; traffic complexity

Citation: Liu, Z.; Zhou, D.; Zheng, Z.; Wu, Z.; Gang, L. An Analytic Model for Identifying Real-Time Anchorage Collision Risk Based on AIS Data. *J. Mar. Sci. Eng.* **2023**, *11*, 1553. https://doi.org/10.3390/jmse11081553

Academic Editors: Marko Perkovic, Lucjan Gucma and Sebastian Feuerstack

Received: 3 July 2023
Revised: 1 August 2023
Accepted: 3 August 2023
Published: 5 August 2023

1. Introduction

After entering the 21st century, with the continuous development of the global economy and shipping industry, ship traffic around the world has increased rapidly, especially in areas with frequent ship activities, such as port areas [1,2]. The increase in ship traffic will complicate ship traffic, making the already busy port waters more crowded and increasing the possibility of ship collisions [3]. In port ship traffic, according to the arrangement of the port authority or the plan of the ship itself, some ships choose to anchor in the port anchorage so as to prepare for their next activities [4]. For the port waters with heavy and complicated ship traffic, the collision risk exists not only in the channel of the ship but also in the anchorage. Anchorage is a special navigable functional water area, which was paid less attention to in the studies of ship collision risk in the past compared with common waters. For some anchorage collision risk studies, there are still some limitations. The characteristics of anchorage and the ships within were not well considered. More importantly, the impact of traffic complexity on collision risk in anchorage was not incorporated into the model. This not only made the collision risk in the anchorage difficult to be accurately identified, endangering the navigation and anchoring safety of ships in the anchorage and its vicinity but also hindered the effective supervision of traffic in busy and complex port waters by relevant personnel, affecting their monitoring and analysis of ship traffic risks in the anchorage and even port waters. Therefore, it is crucial to propose a collision risk identification model which can assess the real-time danger in anchorage accurately, especially under complicated traffic situations. This paper proposed an analytic model for identifying real-time anchorage collision risk based on AIS data, which not only considered the microscopic relative motion between ships and calculated the microscopic collision risk, but also considered the macroscopic characteristics of anchorage through a special ship domain arena. Moreover, a higher safety criterion was adopted through the introduction of an arena and the setting of a dynamic variable range of anchorage boundaries. More importantly, this paper considered the impact of traffic complexity on anchorage collision risk from the perspective of ship position in establishing the model, which was more helpful in accurately identifying the collision risk in anchorage under complex traffic scenarios. The reminders of this paper were arranged as follows. In Section 2, the literature on collision risk identification was reviewed. Section 3 described the modeling of anchorage collision risk, which considered the microscopic, macroscopic, and complexity aspects. In Section 4, some experimental case studies were carried out to validate the proposed model. Some discussions were made to explain the advantages and limitations of the proposed model in Section 5. At last, the conclusion was drawn in Section 6, and some future works on the anchorage collision risk model were outlooked.

2. The Literature Review

Collision risk identification is one of the hot issues in the maritime research field. To enhance navigational safety, many scholars have been researching how to accurately quantify the collision risk between ships or in a specific water area. In terms of the research scope, collision risk identification research can be divided into two categories, which are microscopic collision risk, which is to quantify the collision risk between ships, and macroscopic collision risk, which is to quantify the collision risk of a specific water area. The two kinds of collision risk have different uses in the maritime field. Both collision risks are researched extensively by maritime scholars.

The microscopic collision risk refers to the collision risk between two ships or multi ships. It is usually used as a safety criterion in collision avoidance or relevant decision-making. Generally, the methods of modeling microscopic collision risk can be divided into three categories, which are the analytic method, fuzzy method, and machine learning method. The analytic method is to model the microscopic collision risk by analytic expression. At an early stage, Kearon [5] proposed an analytic expression to calculate the collision risk index, namely, the microscopic collision risk. In this analytic expression, two crucial parameters in collision avoidance, Distance to the closest point of approach (DCPA) and Time to the closest point of approach (TCPA), were used. Since then, some scholars have improved or proposed analytical expressions of microscopic collision risk based on other parameters [6–10]. The advantage of the analytical method is that it can simply and clearly quantify the relationship between the collision risk and the input variables, which is convenient to use and has strong objectivity. However, the input variables considered in this method are limited, and a lot of models only involve DCPA and TCPA variables, which is slightly insufficient in representing collision risk. To overcome the limitation of the analytic method, some scholars began to model the microscopic collision risk by the theory of fuzzy mathematics, including fuzzy inference and fuzzy comprehensive assessment. Most of the early studies on microcosmic collision risk calculation by fuzzy inference only included DCPA and TCPA as input variables [11]. Gradually, scholars began to involve more influencing factors [12] or combined this method with other approaches, such as ship domain [13] and neural network [14]. Although the fuzzy method can overcome the disadvantage of the analytic method in considering factors to some extent, it also has drawbacks. It is limited in strong subjectivity because the fuzzy method should rely on the knowledge of an expert. There were also some scholars modeling microscopic collision risk with a machine learning approach, mainly by the neural network. As a black box, the neural network cannot model microscopic collision risk directly but needs to rely on the sample training data derived from other microscopic collision risk models. Earlier studies mostly used DCPA and TCPA as network inputs [15]. Considering the computing efficiency, some scholars began to use the raw data of the ship's movement as input variables [16].

The advantage of the neural network method is that it can establish the nonlinear mapping relationship between the input variable and the output variable and obtain the result of the collision risk quickly. However, as a black box to simulate the human brain, the mechanism has not been clarified, so it is difficult to clearly explain this nonlinear mapping relationship.

The macroscopic collision risk refers to the collision risk in a specific water area, and it is normally used in maritime surveillance. As the water area near the port is the busiest and the most crowded water area, many scholars pay attention to the study of collision risk in port waters. At an early stage, Fujii and Shiobara [17] and Macduff [18] proposed a framework to identify the collision risk in the water area, which multiplied the geometric collision numbers or probability with causation collision probability. The framework was widely used in regional collision risk identification by probability means [19–21]. Under the framework, a famous model proposed by the International Association of Lighthouse Authorities (IALA) is the International Association of Lighthouse Authorities Waterways Risk Assessment Program (IWRAP) model [22]. Silveiria et al. [23] developed an algorithm to assess the relative importance and risk profile of routes associated with ports and proposed a method to identify the collision risk by estimating future distances between ships off the coast of Portugal. Zhen et al. [24] proposed a novel framework to identify and analyze the collision risk of the west coastal waters of Sweden. The collision risk near the port area can be obtained in an analytic way using DCPA and TCPA. To reduce the collision probability in Istanbul Strait, which is busy with the crossing traffic from port to port and passing traffic, Korçak and Balas [25] defined the hot spots for encounters and calculated the collision probability between ships. Breithaupt et al. [26] plotted the ship routes between ports along the Atlantic coast of the United States, which can indicate the distribution of collision risk in the waters to some extent. Li et al. [27] proposed an integrated method for regional collision risk analysis. The random forest was used to integrate the accident risk model and non-accident critical events risk model. Lan et al. [28] proposed a data-driven method integrating association rule mining, complex network, and random forest to explore the correlation among collision risk factors. Via this method, the critical factor can be found and used to assess the severity of collision accidents.

Although the studies on macroscopic collision risk were extensively carried out, there was little research on collision risk in anchorage. Burmeister et al. [29] utilized the famous IWRAP model and improved it to MKII to assess the collision risk between the underway ship and the anchored ship in the anchorage. However, the influencing factors of collision risk were not considered sufficiently, and the model was limited in real-time collision risk identification. Weng and Xue [30] evaluated the ship collision frequency in port fairways. Through some case studies in Singapore Strait, they found the hot spots of different dangerous encountering types. Debnath and Chin [31] utilized a binomial logistic model to determine the relationship between the anchorage collision risk and various characteristics by the Navigation Traffic Conflict Technique proposed by them [32]. This method can be used to identify the anchorage collision risk, but the model relatively relied on expert judgment. Liu et al. [33] proposed an anchorage collision risk model. This model assessed the collision risk in anchorage by calculating the collision risk between any two ships within anchorage and the safe room in anchorage for navigation. However, this model was limited in expressing the impact of the complication of maritime traffic.

Compared with the study of macroscopic collision risk, the study of microscopic collision risk was more extensive. These microscopic collision risk studies mainly focused on the microscopic factors that affect the collision risk between ships, such as the relative motion parameters of ships. Therefore, they had a wide range of applications in the actual collision avoidance of ships and were helpful as a safety criterion to assist the decision-making of collision avoidance. However, because it was limited to the relative motion between ships and took little consideration of the overall influencing factors of collision risk, it was difficult to be directly applied to evaluate the macroscopic collision risk of the water area, and it was difficult to assess the overall collision risk of the special water area, such as the collision risk of anchorage. At present, the studies of macroscopic collision risk mainly focus on some open waters or straits and pay less attention to special navigational waters, such as anchorage. Although some scholars have researched the collision risk of anchorage, there were still some limitations in these works, such as a lack of consideration of the characteristics of anchorage and difficulty in obtaining instantaneous anchorage collision risk. More importantly, under the increasingly complicated ship traffic, no matter the macroscopic collision risk study or the anchorage collision risk study, complexity was not considered as a factor affecting the possibility of ship collision, and the consideration of safety criterion was insufficient, so it was difficult to identify the potential collision risks in anchorage accurately.

3. The Anchorage Collision Risk Model

The anchorage collision risks modeled in this paper refer to the global risk levels of ship collision within the scope of the anchorage and its adjacent waters. For identifying the anchorage collision risks more efficiently under the complicated traffic situations nowadays, the anchorage collision risk model was established in three aspects in this paper, which are collision risk in microscopic, macroscopic, and spatial complexity, respectively.

The microscopic collision risk refers to the collision risk objectively existing between two ships in the anchorage waters due to their relative motion. As a ship can exist in anchorage either in an underway state or anchored state, there are three different situations for a pair of two ships in the anchorage, which are two underway ships, one underway ship, one anchored ship, and two anchored ships. Since anchored ships have no sailing speed, they were considered fixed objects in this paper. For the three situations mentioned above, the microscopic collision risk between two ships can be calculated as follows.

For two underway ships, the method to calculate the microscopic collision risk between them is the same as the calculation of the collision risk between two ships in open water, which is obtained in an analytic way. For them, one ship is considered as its own ship, and the other ship is considered as a target ship. Then, the relative motion relationship between these two ships can be established in a Cartesian coordinate system, as shown in Figure 1.

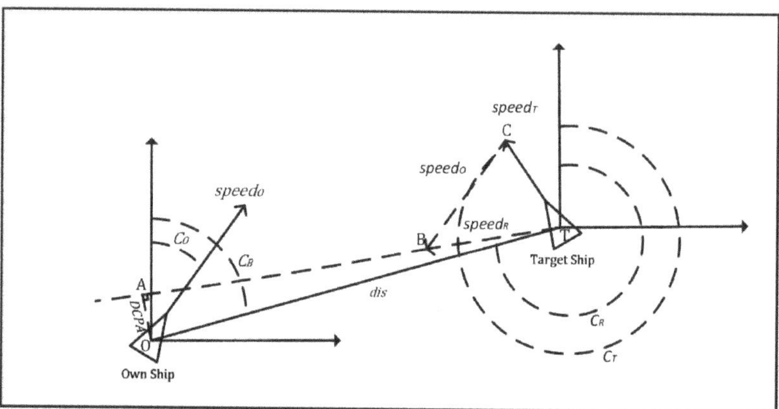

Figure 1. The relative motion between two ships in a Cartesian coordinate system.

Using the theory of analytic geometry, the two crucial collision risk parameters, DCPA and TCPA, can be obtained in the coordinate system as follows:

$$DCPA = dis \times |sin(C_R - C_B - \pi)| \tag{1}$$

$$TCPA = \frac{dis}{speed_R} \times cos(C_R - C_B - \pi) \tag{2}$$

where *dis* refers to the distance in space between two ships. C_B refers to the bearing of the target ship relative to its own ship, which can be obtained by the arctangent trigonometric function according to the coordinates of the two ships.

$$C_B = \begin{cases} arctan\frac{long_T - long_O}{lat_T - lat_O} & lat_T > lat_O \\ arctan\frac{long_T - long_O}{lat_T - lat_O} + \pi & lat_T \leq lat_O \end{cases} \tag{3}$$

$speed_R$ refers to the speed of the target ship relative to the own ship, which can be calculated by the speed and course information of the two ships.

$$speed_R = \sqrt{speed_O^2 + speed_T^2 - 2speed_O speed_T cos(course_T - course_O)} \tag{4}$$

C_R refers to the course of the target ships relative to the own ship, which can be calculated based on $speed_R$.

$$C_R = course_O + \pi \pm arccos\frac{speed_O^2 + speed_R^2 - speed_T^2}{2speed_O speed_R} \tag{5}$$

After obtaining the *DCPA* and *TCPA* between two ships, referred to the analytic expression of collision risk index proposed by Kearon [5] which considered *DCPA* and *TCPA*, and the negative exponential function used in Zhen [24] when established the relationship between collision risk and *DCPA*/*TCPA*, the microscopic collision risk between two ships in anchorage can be expressed as follows:

$$CR_{micro} = \sqrt{\alpha_{DCPA}e^{-\beta_{DCPA} \cdot DCPA^2} + \alpha_{TCPA}e^{-\beta_{TCPA} \cdot TCPA^2}} \tag{6}$$

where α and β are the parameters in a negative exponential function, which can be obtained by setting two extreme scenarios between *DCPA* and collision risk and between *TCPA* and collision risk.

For an underway ship and an anchored ship, the calculation of the microscopic collision risk is similar to that of two underway ships. The only difference is that one ship was considered a fixed object with no speed. In other words, the relative motion between the two ships can be seemed as the relative motion of the underway ship relative to a fixed object, as shown in Figure 2.

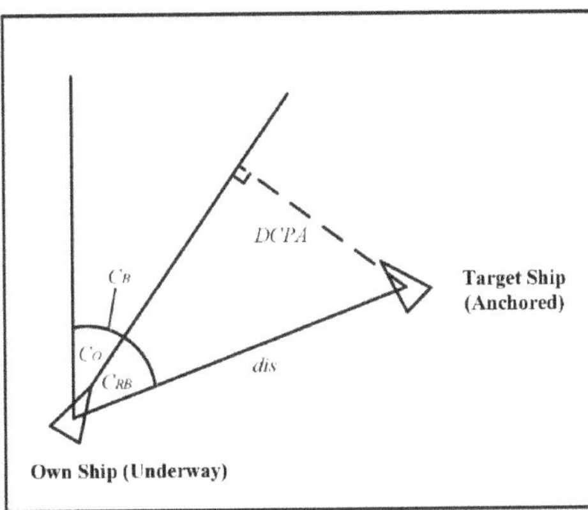

Figure 2. The relative motion between an underway ship and an anchored ship.

Therefore, the two crucial collision risk parameters, *DCPA* and *TCPA*, can be calculated as follows:

$$DCPA = dis \times sinC_{RB} \tag{7}$$

$$TCPA = \frac{dis}{speed_O} \times cossinC_{RB} \tag{8}$$

where C_{RB} refers to the bearing of the target ship relative to the own ship, which can be obtained via Equation (9).

$$C_{RB} = C_B - C_O \tag{9}$$

Similarly, after obtaining *DCPA* and *TCPA*, the microscopic collision risk between them can be obtained via Equations (7) and (8). For two anchored ships, as an anchored ship was considered a fixed object, the relative motion between them does not exist. Therefore, the microscopic collision risk between them does not exist. After identifying the microscopic collision risk between two ships in the three situations mentioned above, for the entire anchorage, the global microscopic collision risk can be obtained by average processing.

The macroscopic collision risk of anchorage is calculated by considering the characteristics of the anchorage itself and from the perspective of the safe navigable waters in the anchorage and its adjacent waters so as to evaluate the overall collision risk of the anchorage that ships sailing to or through the waters. For a ship sailing to or through the waters, the smaller the safe navigable waters, the more difficult it is to conduct collision avoidance maneuvers, and, thus, the higher the possibility of a collision accident. Therefore, the key to calculating macroscopic collision risk is to identify the size of safe navigable waters.

To identify the size of safe navigable waters, the first step is to determine the scope of the studied waters. Considering that the macroscopic collision risk of anchorage is for ships sailing to or through the waters, this study's area includes not only the anchorage area but also the area around the anchorage area. This paper extended the anchorage outward according to the average ship domain scale of the ships in the anchorage, as shown in Figure 3. The extended range is a variable dynamic range and will vary depending on the size and number of ships within the anchorage.

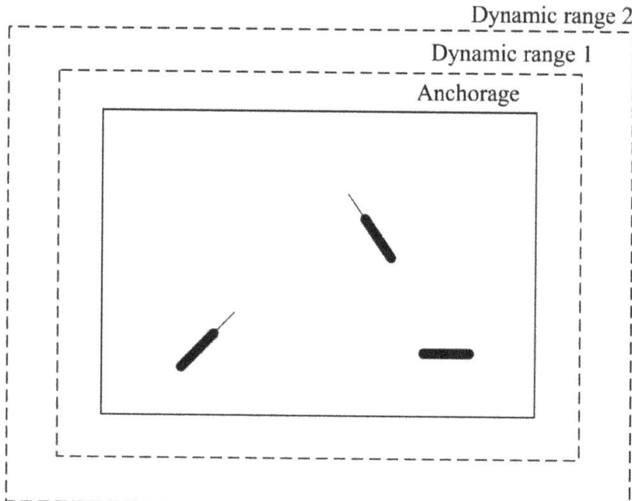

Figure 3. The dynamic extended range of an anchorage.

In addition, it is necessary to calculate the overall area of the anchorage and its surrounding waters $Area_{anc}$, according to the coordinates of the anchorage edges, and the size of safe navigable waters $Area_{safe}$, based on the area of all arenas within anchorage. The size of the arena of the ship was determined according to the relationship between the arena and ship domain summarized by Davis et al. [34]. The safe navigable waters in this paper are shown in Figure 4.

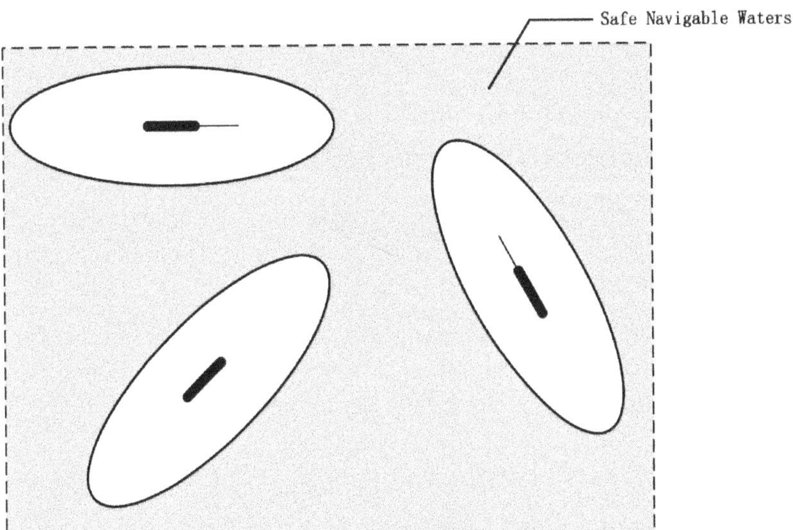

Figure 4. The safe navigable waters in an anchorage.

Then, the macroscopic collision risk can be obtained by calculating the ratio of the safe navigable waters $Area_{safe}$ to the total area of the study waters $Area_{anc}$ and combining with the negative exponential function in Equation (6), which can be expressed as Equation (10).

$$CR_{macro} = \alpha_{macro}e^{-\beta_{macro}\cdot\frac{Area_{safe}}{Area_{anc}}} \tag{10}$$

In addition to the microscopic and macroscopic collision risk, this paper also considered the complexity of ship traffic in the anchorage and its adjacent waters in the identification of the anchorage collision risk, so that the collision risk of the anchorage can be identified more accurately and effectively under the increasingly complex ship traffic situation. The consideration of the complexity of ship traffic in the anchorage in this paper is based on the compactness of the spatial distribution of ships in the anchorage and its adjacent waters, which can be obtained by applying the radial distribution function in statistical mechanics [35,36].

In the radial distribution model, the ships in anchorage were considered particles, and all ships within the boundary of anchorage were considered a particle system. In other words, to calculate the spatial complexity of the ships within anchorage, the first step is to calculate all distances between any two ships, as shown in Figure 5.

Then, the radial distribution function was brought in by considering the anchorage and its adjacent waters as the distribution space. It should be noted that due to the characteristic of ship traffic, which sails on a two-dimensional plane, the distribution space is also a two-dimensional space, so the corresponding radial distribution function needs to be transformed from three-dimensional to two-dimensional. As the ship is regarded as a particle, its radial distribution in the two-dimensional space of the studied water is described, and the radial distribution is integrated within a certain threshold to obtain the spatial complexity according to the characteristic of the radial distribution function, as expressed in Equations (11) and (12).

$$comp = \int_0^{\frac{R_{anc}}{2}} \sum_i^N \frac{N_i(r,\Delta r)}{\lambda N \rho S(r,\Delta r)} dr \tag{11}$$

$$CR_{comp} = \alpha_{comp}e^{-\beta_{comp}\cdot comp} \tag{12}$$

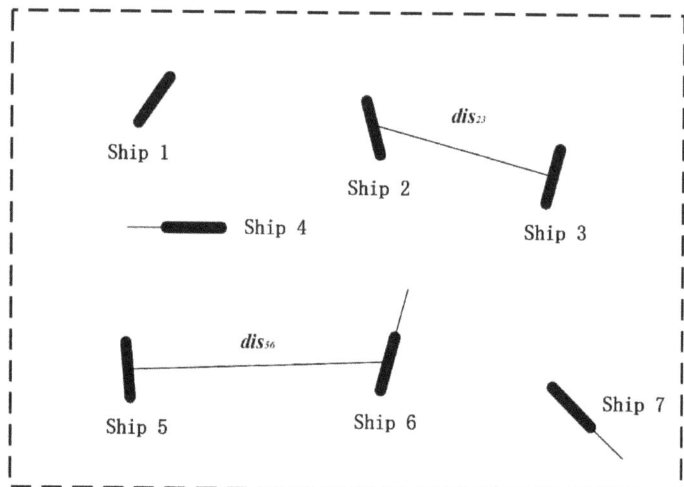

Figure 5. The ship distribution in an anchorage.

After obtaining the collision risk in microscopic aspect CR_{micro}, macroscopic aspect CR_{macro}, and spatial compactness CR_{comp}, referring to the analytic expression proposed by Kearon [5], the anchorage collision risk can be identified by the analytical expression in Equation (13).

$$CR_{anchorage} = \sqrt{a \times CR_{micro}^2 + b \times CR_{macro}^2 + c \times CR_{comp}^2} \tag{13}$$

where a, b, and c are the weight coefficients of the collision risk in each aspect. The three coefficients are set equal for normal situations and can be decided by maritime experts according to the characteristics of the studied water area, traffic situation, and identification purpose. For example, in the water area with extremely complicated ship traffic, the coefficient c, which represents the importance of the collision risk in the complexity aspect, needs to be increased appropriately because the traffic complexity has a significant impact on collision risk in such water area. The anchorage collision risk $CR_{anchorage}$ can reflect the risk level of collision within the anchorage and its adjacent waters considering the relative motion, anchorage characteristics, and spatial complexity, which can provide mariners and maritime surveillance operators with reference to the collision risk level within the anchorage waters so as to improve the maritime traffic safety.

4. Case Study

To validate that the proposed model can identify the collision risk of anchorage effectively, some experimental case studies were carried out using the real AIS data in some of the anchorages of the Northern Yellow Sea. The studied anchorages are on the north side of the Shandong Peninsula. There are some busy ports nearby, such as Weihai Port, Yantai Port, and Penglai Port. The volume of ship traffic here is relatively big, so some of the ships have to wait in the anchorage according to the schedule. Although anchoring in the anchorage can relieve the traffic pressure to a certain extent, when the number of ships in the anchorage increases, the possibility of collision accidents will increase both inside and outside the anchorage. The studied anchorages are illustrated in Figure 6.

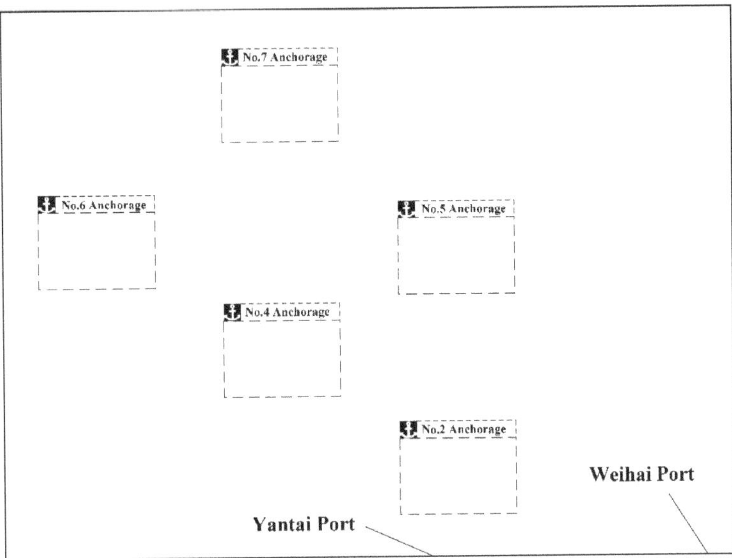

Figure 6. The anchorages studied in this article.

In this article, the AIS data used in case studies is in June 2022. Before experiments, the source data should be encoded first. The encoded AIS data were stored in a database. Then, the data should be filtered to exclude the invalid one. Because the receiving time of AIS data of ships is different, the time labels of the ship navigation information obtained from the AIS data of different ships will be inconsistent. Therefore, it is necessary to interpolate the data according to the time labels before experiments to make the data have the same time labels.

The first case study is to validate the proposed model in a spatial aspect. Anchorage No. 2, Anchorage No. 4, Anchorage No. 5, Anchorage No. 6, and Anchorage No. 7 were selected in this case study. After inputting the required information into the proposed model, the collision risk of these anchorages for a designated timing can be obtained. Firstly, the collision risks at 1620 on 18 June 2022 were identified. The results are shown in Table 1. The ship positions in the anchorages are shown as blue dots in Figure 7.

Table 1. The collision risks of anchorage at 1620 on 18 June 2022.

Anchorage	No. 2	No. 4	No. 5	No. 6	No. 7
Collision Risk	0.1116	0.1078	0.1629	0.1320	0.0664

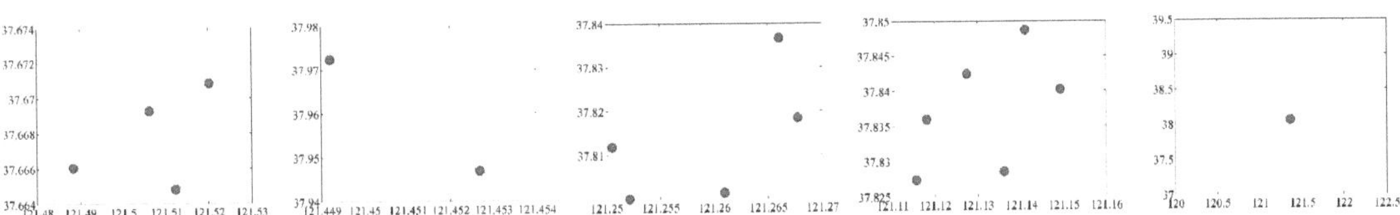

Figure 7. The ship positions in the anchorages at 1620 on 18 June 2022.

For analyzing the results in Table 1, another index, ship traffic density, which can reflect the collision risk to some extent, was adopted. The ship densities for these anchorages at the moment are shown in Table 2.

Table 2. The ship densities of anchorage at 1620 on 18 June 2022.

Anchorage	No. 2	No. 4	No. 5	No. 6	No. 7
Density	4	2	5	6	1

It can be observed that the anchorage with relatively high density also has a larger collision risk value, such as Anchorage No. 5 and No. 6. Anchorage No. 7, which has the least density, also presents the smaller collision risk value. As the ship traffic density can represent the collision risk to some extent, the capability of the proposed model to identify the collision risk of anchorage was validated. Further, another timing, at 2350, on 18 June 2022, was selected to carry out the case study mentioned above again. The collision risks for these anchorages obtained from the proposed model are shown in Table 3, and the ship densities are shown in Table 4. The ship positions in the anchorages are shown in Figure 8.

Table 3. The collision risks of anchorage at 2350 on 18 June 2022.

Anchorage	No. 2	No. 4	No. 5	No. 6	No. 7
Collision Risk	0.1570	0.1015	0.1710	0.3057	0.0700

Table 4. The ship densities of anchorage at 2350 on 18 June 2022.

Anchorage	No. 2	No. 4	No. 5	No. 6	No. 7
Density	6	2	5	7	1

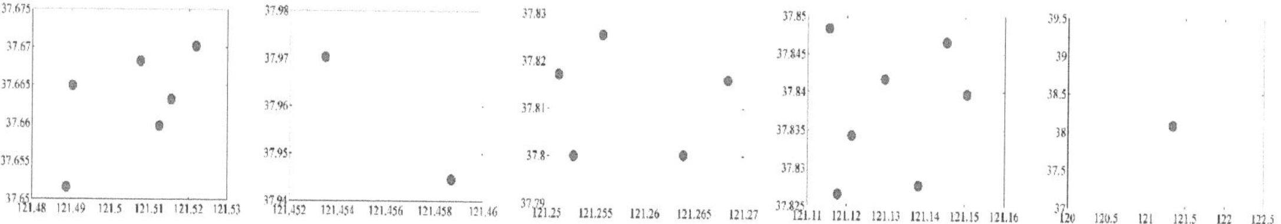

Figure 8. The ship positions in the anchorages at 2350 on 18 June 2022.

Comparing the results in Table 3 to that in Table 4, the positive relationship between the anchorage collision risk and ship traffic density can also be found, where Anchorage No. 6 has the largest collision risk value and ship traffic density and Anchorage No. 7 has the smallest collision risk value and ship traffic density. In addition, comparing the results of 2350 to that of 1620, it can be found that with the increase or decrease in the number of ships in the anchorage, the collision risk in the anchorage will also increase or decrease in most cases. Therefore, the effectiveness of the proposed model can be validated.

Apart from validating the proposed model in the spatial aspect, the temporal experiment was also carried out. Anchorage No. 2 and Anchorage No. 4 were selected in this experiment, where Anchorage No. 2 normally has relatively large numbers of ships. For the two selected anchorages, the collision risks were identified by the proposed model for 24 h by selecting a timing point for each hour. The results are shown in Figure 9.

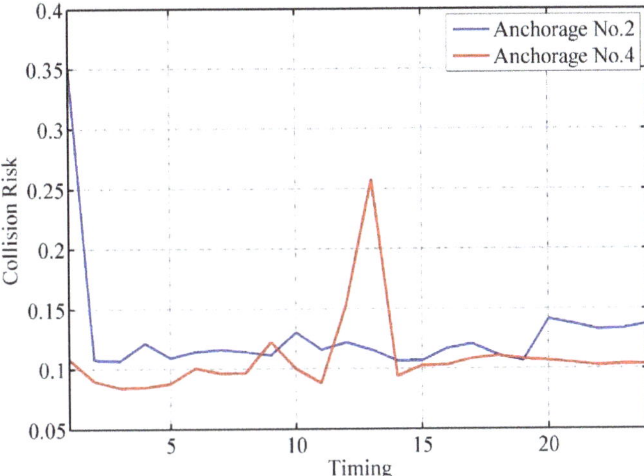

Figure 9. The collision risks of Anchorage No. 2 and No. 4 on 18 June 2022.

It can be observed that the collision risk of Anchorage No. 2 is higher than that of Anchorage No. 4, generally. By calculating, the average value of collision risks in Anchorage No. 2 is 0.1286, and for Anchorage No. 4 is 0.1090. For verifying the results, the ship traffic density was also adopted. The ship densities for the two anchorages in the 24 h are shown in Figure 10.

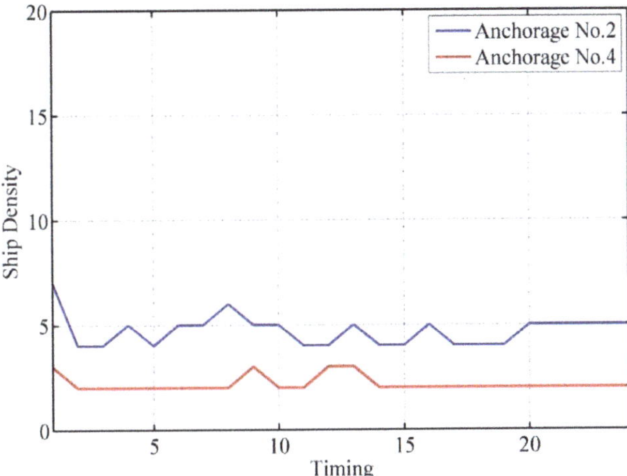

Figure 10. The ship densities of Anchorage No. 2 and No. 4 on 18 June 2022.

It can also be found that the ship traffic density for Anchorage No. 2 is higher than that for No. 4, which reveals that the ships in Anchorage No. 2 are more crowded and have higher possibilities for collision. To check the relationship between the collision risk and ship traffic density for Anchorage No. 2 and No. 4 on 18 June 2022, a Pearson Correlation Analysis (PCA) was conducted, and the correlation result is shown in Tables 5 and 6.

Table 5. The PCA result between the collision risk and ship traffic density for Anchorage No. 2 on 18 June 2022.

Correlation Coefficient	*p*-Value
0.718	<0.001

Table 6. The PCA result between the collision risk and ship traffic density for Anchorage No. 4 on 18 June 2022.

Correlation Coefficient	*p*-Value
0.676	<0.001

The results showed in Tables 5 and 6 both reveal the strong positive correlation between the collision risk and ship traffic density for either Anchorage No. 2 or Anchorage No. 4. As ship traffic density can represent the collision risk to some extent, the proposed model was validated in identifying anchorage collision risk. The numbers of receiving AIS messages within the studied time interval were also identified, and the results are shown in Table 7.

Table 7. The numbers of receiving AIS messages for Anchorage No. 2 and No. 4 on 18 June 2022.

Anchorage	No. 2	No. 4
AIS messages	3029	1579

The number of receiving AIS messages can also represent the busyness of the water area for a time interval and is applied to identify traffic density [37]. It can be found from Table 7 that the number of receiving AIS messages for Anchorage No. 2 is higher than that of Anchorage No. 4, which is in line with the results obtained by the proposed model in revealing collision risk. Therefore, the effectiveness of the proposed model can be validated.

In addition, a dynamic experiment was also carried out to validate the proposed model. Anchorage No. 2 was applied, and the scenario is shown in Figure 11.

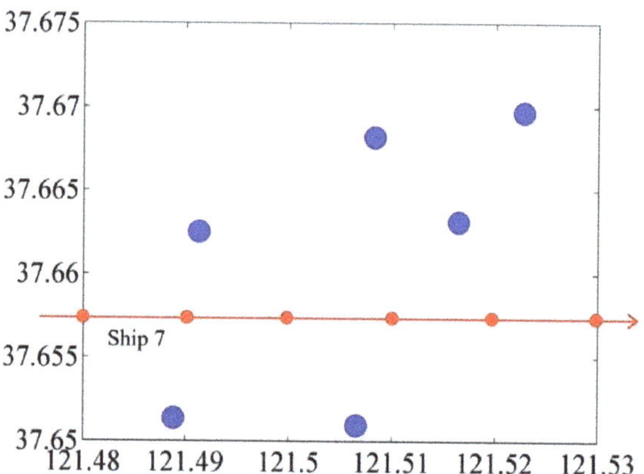

Figure 11. The scenario for the dynamic experiment in Anchorage No. 2.

There are six ships shown as blue dots in Anchorage No. 2, which are all anchored ships. Another ship, which is named Ship 7, was crossing the anchorage from the west edge to the east edge. For this process, six timing moments were selected and marked as red dots in Figure 11. The collision risks for these six timing moments were calculated based on the proposed model, and the results are shown as blue line in Figure 12.

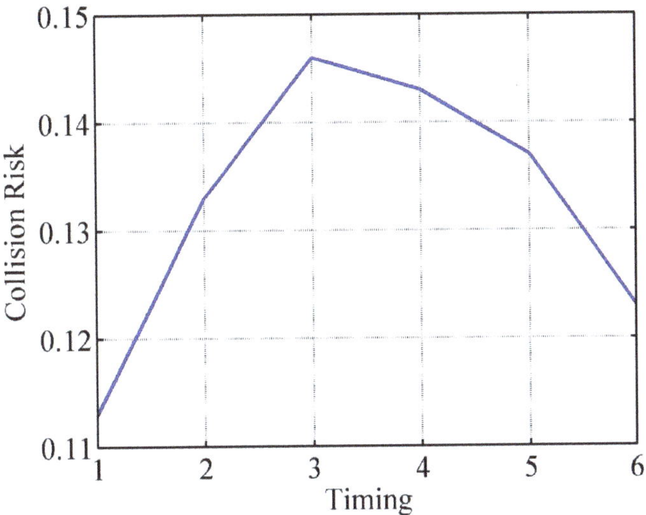

Figure 12. The change in collision risk during the crossing process for Ship 7.

It can be found that the collision risk increased at first and then decreased. This is consistent with the fact that as Ship 7 sailed deeper and deeper into the anchorage during the process of crossing the anchorage, the microscopic collision risk, macroscopic collision risk, and compactness risk between Ship 7 and other ships increased continuously. When this situation reached the threshold, Ship 7 gradually sailed out of the anchorage, and the relevant value began to decrease. Therefore, it can be proved that the proposed model is effective in identifying the collision risk in anchorage.

5. Discussion

In this paper, an anchorage collision risk model was proposed. The proposed model was established in microscopic, macroscopic, and complexity aspects, which considered the relative motion between ships, the characteristic of anchorage, and the spatial complexity of ship distribution. In Section 3, to validate the effectiveness of the proposed model in identifying the collision risk in anchorage, some experimental case studies were carried out. In the spatial and temporal experiments, the ship traffic density was adopted as a validation index as it can evaluate the collision risk through the busyness of anchorage to some extent. However, ship traffic density is still inadequate for assessing the collision risk compared with the proposed model. In other words, the proposed model has the advantage of assessing anchorage collision risk compared with ship traffic density. To explain this, some scenarios with the same ship traffic density were selected. Firstly, for the same anchorage, Anchorage No. 2, we selected five scenarios with a ship traffic density of 5; the scenarios are shown in Figure 13.

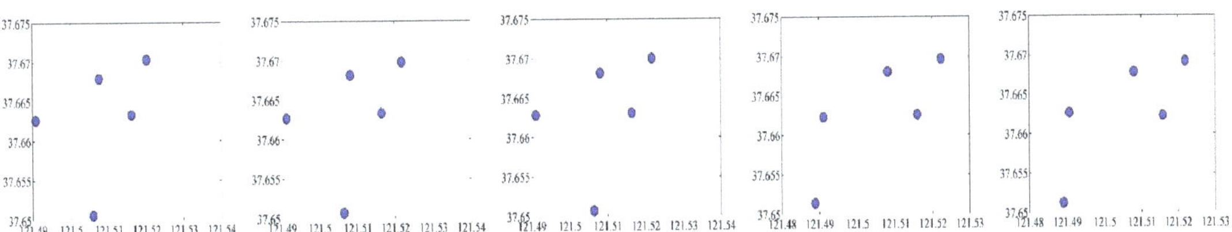

Figure 13. The illustration of the 5 selected scenarios in Anchorage No. 2.

By applying the proposed model, the collision risks can be calculated and are shown in Table 8.

Table 8. The collision risk in Anchorage No. 2 for the 5 selected scenarios.

Scenario	1	2	3	4	5
Collision Risk	0.1214	0.1141	0.1159	0.1113	0.1304

It can be found that although the five scenarios are with the same ship traffic density, the collision risks for them are not equal. This is because the different positions and different motion parameters of the ships in anchorage can lead to differences in micro-collision risk, macro-collision risk, and compactness. The experiment was also conducted in different anchorages. Each scenario in Anchorage No. 2 and No. 5 was selected, as shown in Figure 14, and the ship traffic density is 5 for each scenario.

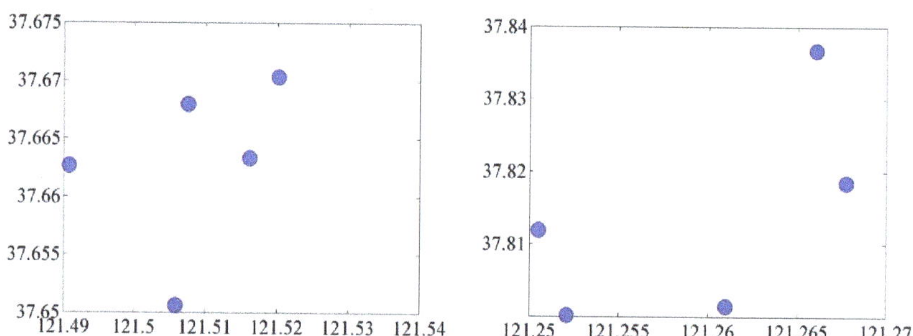

Figure 14. The illustration of the 2 selected scenarios in Anchorage No. 2 and No. 5.

Calculated by the proposed model, the results are shown in Table 9.

Table 9. The collision risk in Anchorage No. 2 and No. 5 for the 2 selected scenarios.

Scenario	1	2
Collision Risk	0.1214	0.1629

Other two scenarios, one is in Anchorage No. 2, and another is in Anchorage No. 6, are shown in Figure 15.

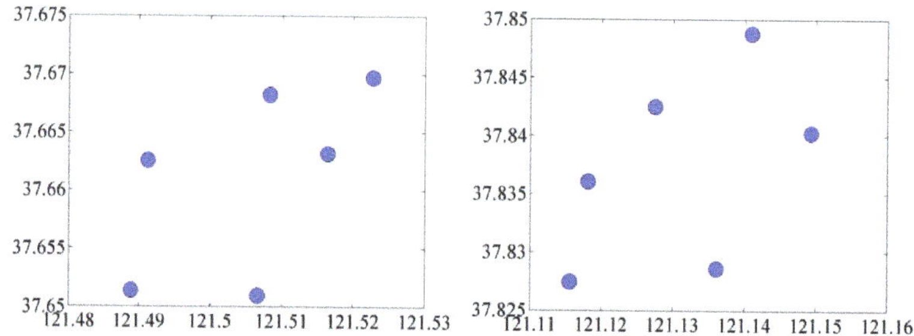

Figure 15. The illustration of the 2 selected scenarios in Anchorage No. 2 and No. 6.

The collision risk results obtained from the proposed model are shown in Table 10.

Table 10. The collision risk in Anchorage No. 2 and No. 6 for the 2 selected scenarios.

Scenario	1	2
Collision Risk	0.1140	0.1320

It can be found in Tables 9 and 10 that even in different anchorages, the collision risks are not equal when the ship densities are the same, which proves the superiority of the proposed model compared with the ship traffic density in assessing anchorage collision risk.

In addition, compared with the anchorage collision risk model [33], the main contribution of the proposed model is the consideration of ship compactness, namely, the spatial complexity. It can make the model more accurate in complex situations. To prove this advantage, a scenario in Anchorage No. 2 was selected, as shown in Figure 16.

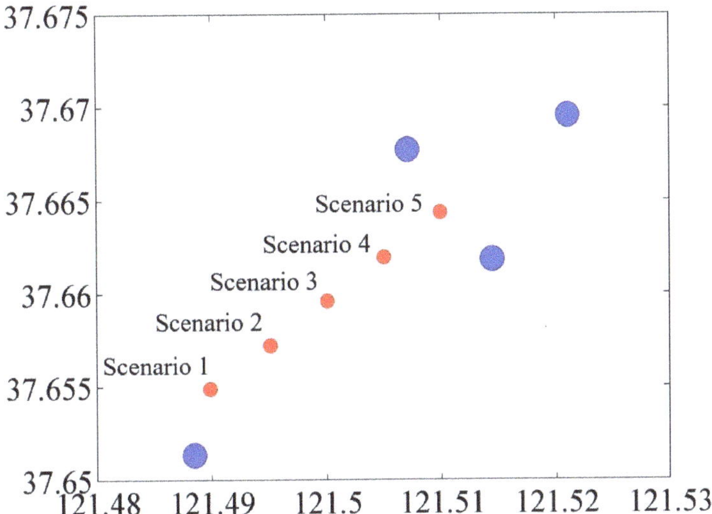

Figure 16. The selected scenario in Anchorage No. 2.

For this scenario, a ship on the southwest edge of the anchorage, which is Ship 1, was simulated to move to different positions in the anchorage, which approach the center of the anchorage gradually.

There is a total of six different positions for Ship 1, and the collision risks of them were calculated using the proposed model. The results are shown in Figure 17.

It can be observed that as the position of Ship 1 gradually approached the center, the collision risk generally increased but fluctuated at some intermediate time points. For analyzing this phenomenon, the collision risk in each aspect was also calculated and shown in Figure 18.

It can be found that the microscopic collision risk was always kept very low and the same because all of the ships in the anchorage in this scenario were anchored ships. However, the collision risks in macroscopic and compactness are different for each moment. For macroscopic collision risk, it increased before Moment 2 and started to decrease after Moment 2. This is because, after Moment 2, the arena of Ship 1 began to overlap with other ships' arenas, which led to the enlargement of safe navigable waters. The collision risk in compactness kept increasing for these moments because Ship 1 was gradually approaching the center of the anchorage. If the compactness was not considered, the collision risks for these moments are shown in Figure 19.

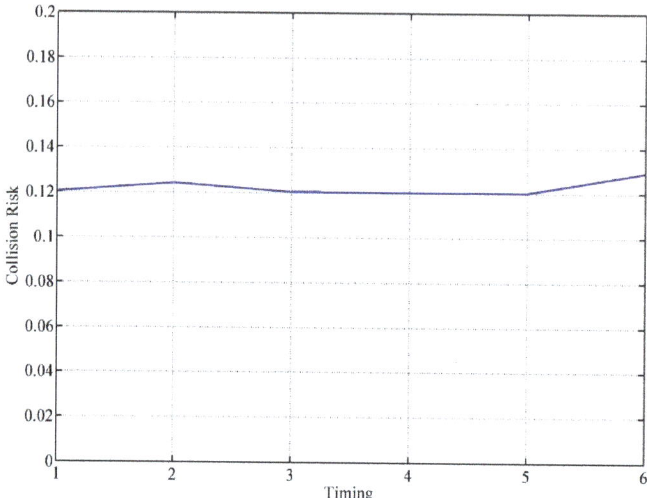

Figure 17. The collision risks in Anchorage No. 2 under the different positions of Ship 1.

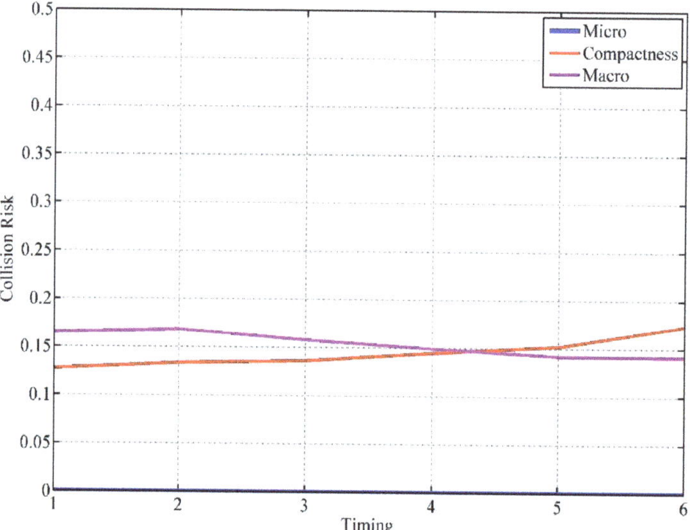

Figure 18. T The collision risk in three sub-aspects in Anchorage No. 2 under the different positions of Ship 1.

It can be found that the trend of the collision risks for these moments is the same as that of macroscopic collision risk (purple line in Figure 18), where the collision risk increased before Moment 2 and started to decrease after Moment 2. However, the results cannot reflect the collision risk in the anchorage sufficiently. Because as Ship 1 continued to approach the center, that is, close to other ships, even though they were anchored ships, the complex ship traffic situation was easier to be formed, which made collision accidents more likely to happen. Therefore, without taking compactness into account, it is difficult to accurately identify the collision risk in such situations. Therefore, since the ship's compactness is considered, the proposed model can identify the anchorage collision risk more accurately in complicated situations.

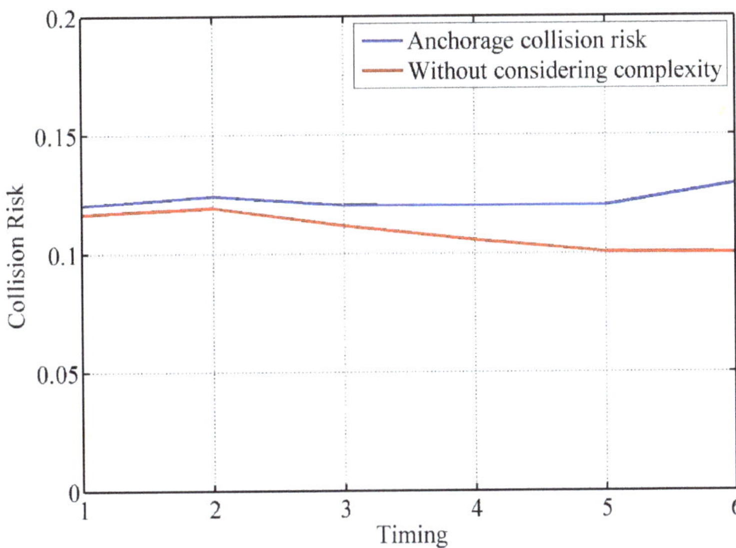

Figure 19. The collision risks for the moments without considering compactness.

In addition, the proposed anchorage collision risk model was compared with other relevant approaches. The compared approaches include Nguyen et al. [8], Debnath and Chin [31], Burmeister et al. [29], Liu et al. [33], Huang et al. [9], Ha et al. [10], which are all the approaches to identifying the collision risks of a water area or between two ships.

Together with the proposed model, these approaches were compared from eight different perspectives, including the ability to assess collision risk regionally, the ability to assess the collision risk in anchorage, whether consider the microscopic factors or not, whether consider the macroscopic factors or not, whether consider the complexity factors or not, the ability to assess collision risk in real time, and the safety criterion. The comparison is shown in Table 11.

Table 11. The comparison for the relevant collision risk identification approaches.

Approaches	Regional	Anchorage	Micro	Macro	Complexity	Real-Time	Safety Criterion
Nguyen et al. (2018) [8],			✓			✓	Relatively High
Debnath and Chin (2016) [31],	✓	✓	✓	✓			Relatively High
Burmeister et al. (2014) [29],	✓	✓		✓			Medium High
Liu et al. (2020) [33],	✓	✓	✓	✓		✓	Relatively High
Huang et al. (2020) [9],			✓			✓	Relatively High
Ha et al. (2021) [10],			✓			✓	Relatively High
The proposed model	✓	✓	✓	✓	✓	✓	High

It can be found that some of the approaches were modeled to assess the collision risk between two ships, such as Nguyen et al. [8], Huang et al. [9], and Ha et al. [10], which pay attention to the microscopic factors of ship collision, such as DCPA, TCPA, etc. These methods can be utilized in automatic collision avoidance or relevant decision-making, but the ability to assess the collision risk regionally was limited. In addition, the characteristics of the anchorage and the ships within anchorage were not well considered, so it is also difficult to assess the collision risk of anchorage accurately by these approaches. For some other approaches which pay attention to the collision risk of anchorage, such as Debnath and Chin [31], Burmeister et al. [29], and Liu et al. [33], there also existed some limitations. Some of the models did not consider the macroscopic factors of anchorage collision risk, such as Burmeister et al. [29], while some of the models were limited in obtaining real-time risk value, such as Debnath and Chin [31]. In addition, all these approaches were limited in considering traffic complexity. The safety criterion for them was not as high as the model proposed in this paper because most of them applied DCPA and TCPA, or the equivalent ship domain, which was treated as the encounter radius in all directions, as the safety criterion. However, the proposed model applied the higher safety criterion, which was the arena. Compared with DCPA or ship domain, the arena is a super domain of the ship and can allow two ships to pass safely at a distance beyond the encounter radius or the DCPA, namely, beyond the scope of the ship domain. Therefore, it has a higher safety criterion and can make more contributions to the safety of navigation.

In addition, some typical approaches were selected to make a numerical comparison with the proposed anchorage collision risk model. The scenario in Figure 16 was used for this comparison experiment. The compared approaches were micro-approach (e.g., Nguyen et al. (2018) [8]), traditional anchorage collision risk model (e.g., Burmeister et al. (2014) [29]), and a recent anchorage collision risk model (Liu et al. (2020) [33]), respectively.

It can be found that the collision risk values obtained via the various approaches for each moment are different. For the micro-approach, since the ships in each moment of the scenario were all anchored ships, there was no relative motion between the ships, and the collision risk calculated based on the relative motion would be almost zero. For the traditional anchorage collision risk models, such as Burmeister et al.'s (2014) [29], the anchorage collision risk was calculated based on the traffic parameters or ship parameters in a period of time, so it was difficult to identify the instantaneous anchorage collision risk value by such approaches; the collision risk values would be N/A, and they are not shown in Table 12. Compared with the anchorage collision risk in [33], it can be found that with the position change of Ship 1, the anchorage collision risk values obtained by [33] are not only smaller than those of the proposed model but also show a downward trend overall. This indicates that compared with the proposed model, the model in [33] did not take into account the impact of traffic complexity on the anchorage collision risk under the situation of relatively complex traffic. At the same time, due to the relatively low safety criterion adopted, the results could also be affected. In other words, the proposed model in this paper can identify the anchorage collision risk more effectively in relatively complex anchorage waters under the premise of considering the traffic complexity and adopting higher safety criteria.

Table 12. The results of the comparison experiment.

Approaches	Moment 1	Moment 2	Moment 3	Moment 4	Moment 5	Moment 6
Micro-approach (e.g., Nguyen et al. (2018) [8])	<0.01	<0.01	<0.01	<0.01	<0.01	<0.01
Liu et al. (2020) [33]	0.1167	0.1192	0.1117	0.1057	0.1006	0.1002
The proposed model	0.1205	0.1243	0.1205	0.1203	0.1204	0.1294

In sum, compared with previous studies on anchorage collision risk, the model has the following advantages. Firstly, the influence of traffic complexity was considered in the modeling of anchorage collision risk, which is important for identifying the collision risk of anchorage under complex traffic conditions accurately. Secondly, this paper adopted a higher safety criterion, which is the arena, instead of the ship domain when calculating the safe navigable waters, which can make potential collision risk in the anchorage easier to spot. In addition, the anchorage boundary in the proposed model was set as a variable dynamic range, which can further identify some potential collision risk at the boundary position in addition to the collision risk in the anchorage.

However, the proposed anchorage collision risk model also has some limitations. Firstly, in modeling the microscopic collision risk, the anchored ship was considered a fixed object with no speed. Actually, some anchored ships have a very low drift speed around the anchor. For the simplification of the model, the effect of this tiny drift speed was not incorporated into the model. For further improving the model, this tiny drift speed is supposed to be considered. Secondly, in modeling the complexity level, this paper only considered the complexity in the spatial aspect, which was obtained by assessing the distribution compactness of ships. Some other factors in the motion aspect should also be taken into account. Thirdly, in the case studies of this paper, the three weight coefficients of the collision risk in microscopic, macroscopic, and complexity aspects were determined equal. To further improve the accuracy of the results, the relationship between the coefficient and the characteristic of anchorage should be investigated deeper.

6. Conclusions

In this paper, an analytic model for identifying the real-time anchorage collision risk was proposed based on AIS data. In modeling the anchorage collision risk, three perspectives were considered, which are microscopic collision risk, macroscopic collision risk, and traffic complexity, respectively. Firstly, the microscopic collision risk was modeled based on the relative motion between ships considering the feature of the anchored ship. Secondly, the macroscopic collision risk was modeled according to the characteristic of anchorage by identifying the ratio of safe navigable waters to total water area within a variable dynamic range of anchorage boundary based on the arena of the ship. Thirdly, the traffic complexity was evaluated from the perspective of the spatial compactness of ships based on the radial distribution function in statistical mechanics. Finally, the three aspects were synthesized to obtain a real-time anchorage collision risk value by a classical analytic expression. Compared with the previous studies on anchorage collision risk, the proposed model additionally considered the influence of traffic complexity, which is important for identifying the anchorage collision risk under complicated traffic situations. In addition, the proposed model adopted a higher safety criterion in identifying collision risk by introducing ship arena and setting a variable dynamic range of anchorage boundary, which is more helpful for identifying the potential collision risk of anchorage. For validating the proposed model, some experimental case studies were carried out using the AIS data in the anchorages off the coast of Shandong Peninsular in China, including spatial experiment, temporal experiment, and dynamic experiment. The experiment results show that the proposed model can effectively identify the collision risk level within anchorage and its adjacent waters and has the advantage of dealing with the scenario with relatively higher complexity. The proposed model is helpful for maritime surveillance operators to monitor the ship traffic in anchorage and can also provide mariners with a cognitive reference about the danger in anchorage on their route, which can both facilitate the enhancement of maritime traffic safety.

Notwithstanding, the proposed anchorage collision risk model still has some limitations which should be overcome in the future. Firstly, in modeling microscopic collision risk, the drift speed of the anchored ship is supposed to be considered in order to fully represent the feature of the anchored ship. Secondly, in modeling complexity, it would be better to incorporate the factors on ship motion to make the consideration of traffic complexity more sufficient. Thirdly, in determining the final collision risk of anchorage, the weights of each aspect should be set according to the characteristic of the water area, which is important to improve the proposed model to a higher level.

Author Contributions: Conceptualization, Z.L.; data curation, Z.L.; formal analysis, Z.L. and D.Z.; funding acquisition, Z.L. and L.G.; investigation, Z.L.; methodology, Z.L.; project administration, Z.L. and L.G.; resources, Z.Z. and Z.W.; software, Z.L.; supervision, Z.Z. and Z.W.; validation, Z.L. and D.Z.; visualization, Z.L.; writing—original draft, Z.L.; writing—review and editing, Z.L., D.Z., Z.Z. and Z.W. All authors have read and agreed to the published version of the manuscript.

Funding: This research was funded by "The Fundamental Research Funds for the Central Universities" (Grant. 3132023145), the National Natural Science Foundation of China (Grant. 52171345) and the Talent Research Start-up Funds of Dalian Maritime University (Grant. 02500128).

Institutional Review Board Statement: Not applicable.

Informed Consent Statement: Not applicable.

Data Availability Statement: Not applicable.

Acknowledgments: We would like to express our gratitude to the editors and reviewers whose valuable comments and suggestions will make an improvement in the quality of this paper.

Conflicts of Interest: The authors declare no conflict of interest.

References

1. Jia, S.; Li, C.L.; Xu, Z. Managing navigation channel traffic and anchorage area utilization of a container port. *Transp. Sci.* **2019**, *53*, 728–745. [CrossRef]
2. Yeo, G.T.; Roe, M.; Soak, S.M. Evaluation of the marine traffic congestion of north harbor in busan port. *J. Waterw. Port Coast. Ocean Eng.* **2007**, *133*, 87–93. [CrossRef]
3. Xin, X.; Liu, K.; Loughney, S.; Wang, J.; Li, H.; Yang, Z. Graph-based ship traffic partitioning for intelligent maritime surveillance in complex port waters. *Exp. Sys. Appl.* **2023**, *231*, 120825. [CrossRef]
4. Kaptan, M. Risk assessment of ship anchorage handling operations using the fuzzy bow-tie method. *Ocean Eng.* **2021**, *236*, 109500. [CrossRef]
5. Kearon, J. Computer program for collision avoidance and track keeping. In Proceedings of the International Conference on Mathematics Aspects of Marine Traffic, London, UK, September 1977.
6. Zhao, J.; Song, S. Measurement of the Mariners' Subjective Collision Risks. *J. Dalian Mar. Coll.* **1990**, *16*, 29–31. (In Chinese)
7. Lisowski, J. Determining the optimal ship trajectory in collision situation. In Proceedings of the IX International Scientific and Technical Conference on Marine Traffic Engineering, Szczecin, Poland, 23 February 2001.
8. Nguyen, M.; Zhang, S.; Wang, X. A Novel Method for Risk Assessment and Simulation of Collision Avoidance for Vessels based on AIS. *Algorithms* **2018**, *11*, 204. [CrossRef]
9. Huang, Y.; van Gelder, P.H.A.J.M. Collision risk measure for triggering evasive actions of maritime autonomous surface ships. *Saf. Sci.* **2020**, *127*, 104708. [CrossRef]
10. Ha, J.; Roh, M.I.; Lee, H.W. Quantitative calculation method of the collision risk for collision avoidance in ship navigation using the CPA and ship domain. *J. Comput. Des. Eng.* **2021**, *8*, 894–909. [CrossRef]
11. Hasegawa, K.; Kouzuki, A.; Muramatsu, T.; Komine, H.; Watabe, Y. Ship auto-navigation fuzzy expert system (SAFES). *J. Soc. Naval Archit.* **1989**, *166*, 445–452. [CrossRef] [PubMed]
12. Bukhari, A.C.; Tusseyeva, I.; Lee, B.; Kim, Y. An intelligent real-time multi-vessel collision risk assessment system from VTS view point based on fuzzy inference system. *Expert. Sys. Appl.* **2013**, *40*, 1220–1230. [CrossRef]
13. Kao, S.; Lee, K.; Chang, K.; Ko, M. A Fuzzy Logic Method for Collision Avoidance in Vessel Traffic Service. *J. Navig.* **2007**, *60*, 17–31. [CrossRef]
14. Ann, J.H.; Rhee, K.P.; You, Y.J. A study on the collision avoidance of a ship using neural networks and fuzzy logic. *Appl. Ocean Res.* **2012**, *37*, 162–173.
15. Yao, J.; Wu, Z.; Fang, X. Adaptive Neural Network of Ship Collision Risk Assessment Method of Fuzzy Inference. *Navig. China.* **1999**, *1*, 14–19. (In Chinese)
16. Yang, G.; Yang, Z. Research of computing the vessel's Collision Risk Index by multiple parameters Based on Neural Network with Genetic Algorithm. In Proceedings of the 2013 Third International Conference On Instrumentation & Measurement, Computer, Communication And Control, Shenyang, China, 21–23 September 2013.
17. Fujii, Y.; Shiobara, R. The Analysis of Traffic Accidents. *J. Navig.* **1971**, *24*, 534–543. [CrossRef]
18. Macduff, T. The probability of vessel collisions. *Ocean Ind.* **1974**, *9*, 144–148.
19. Pedersen, P.T. Collision and grounding mechanics. In Proceedings of the WEMT, Copenhagen, Denmark, 4 November 1995.
20. Kaneko, F. Methods for probabilistic safety assessments of ships. *J. Mar Sci. Technol.* **2002**, *7*, 1–16. [CrossRef]
21. Montewka, J.; Goerlandt, F.; Kujala, P. Determination of collision criteria and causation factors appropriate to a model for estimating the probability of maritime accidents. *Ocean Eng.* **2012**, *40*, 50–61. [CrossRef]
22. International Association of Lighthouse Authorities (IALA). Risk Management Tool for Ports and Restricted Waterways. *IALA Recommendation O-134 Ed 2* **2009**, 18–22.
23. Silveira, P.A.M.; Teixeira, A.P.; Guedes Soares, C. Use of AIS data to characterize marine traffic patterns and ship collision risk off the coast of Portugal. *J. Navig.* **2013**, *66*, 879–898. [CrossRef]
24. Zhen, R.; Riveiro, M.; Jin, Y. A novel analytic framework of real-time multi-vessel collision risk assessment for maritime traffic surveillance. *Ocean Eng.* **2017**, *145*, 492–501. [CrossRef]

25. Korçak, M.; Balas, C.E. Reducing the probability for the collision of ships by changing the passage schedule in Istanbul Strait. *Int. J. Disaster Risk Reduct.* **2020**, *48*, 101593. [CrossRef]
26. Breithaupt, S.A.; Copping, A.; Tagestad, J.; Whiting, J. Maritime route delineation using AIS data from the atlantic coast of the US. *J. Navig.* **2017**, *70*, 379–394. [CrossRef]
27. Li, M.; Mou, J.; Chen, P.; Chen, L.; van Gelder, P.H.A.J.M. Real-time collision risk based safety management for vessel traffic in busy ports and waterways. *Ocean Coast. Manag.* **2021**, *234*, 106471. [CrossRef]
28. Lan, H.; Ma, X.; Qiao, W.; Deng, W. Determining the critical risk factors for predicting the severity of ship collision accidents using a data-driven approach. *Reliab. Eng. Sys. Saf.* **2023**, *230*, 108934. [CrossRef]
29. Burmeister, H.C.; Walther, L.; Jahn, C.; Töter, S.; Froese, J. Assessing the frequency and material consequences of collisions with vessels lying at an anchorage in line with IALA iWrap MkII. *TransNav. Int. J. Mar. Navig. Saf. Sea Transp.* **2014**, *8*, 61–68. [CrossRef]
30. Weng, J.; Xue, S. Ship Collision Frequency Estimation in Port Fairways: A Case Study. *J. Navig.* **2015**, *68*, 602–618. [CrossRef]
31. Debnath, A.K.; Chin, H.C. Navigational traffic conflict technique: A proactive approach to quantitative measurement of collision risks in port waters. *J. Navig.* **2016**, *63*, 137–152. [CrossRef]
32. Debnath, A.K.; Chin, H.C. Modelling collision potentials in port anchorages: Application of the navigational traffic conflict technique (NTCT). *J. Navig.* **2010**, *69*, 183–196. [CrossRef]
33. Liu, Z.; Wu, Z.; Zheng, Z. A novel model for identifying the vessel collision risk of anchorage. *Appl. Ocean Res.* **2020**, *98*, 102130. [CrossRef]
34. Davis, P.V.; Dove, M.J.; Stockel, C.T. A computer simulation of marine traffic using domains and arenas. *J. Navig.* **1980**, *33*, 215–222. [CrossRef]
35. Liu, Z.; Wu, Z.; Zheng, Z.; Wang, X.; Soares, C.G. Modelling dynamic maritime traffic complexity with radial distribution functions. *Ocean Eng.* **2021**, *241*, 109990. [CrossRef]
36. Liu, Z.; Wu, Z.; Zheng, Z.; Yu, X. A Molecular Dynamics Approach to Identify the Marine Traffic Complexity in a Waterway. *J. Mar. Sci. Eng.* **2022**, *10*, 1678. [CrossRef]
37. Wu, L.; Xu, Y.; Wang, Q.; Xu, Z. Mapping Global Shipping Density from AIS Data. *J. Navig.* **2017**, *70*, 67–81. [CrossRef]

Article

A Decision Support System Using Fuzzy Logic for Collision Avoidance in Multi-Vessel Situations at Sea

Tanja Brcko * and Blaž Luin

Faculty of Maritime Studies and Transport, University of Ljubljana, 6320 Portoroz, Slovenia;
blaz.luin@fpp.uni-lj.si
* Correspondence: tanja.brcko@fpp.uni-lj.si

Abstract: The increasing traffic and complexity of navigation at sea require advanced decision support systems to ensure greater safety. In this study, we propose a novel decision support system that employs fuzzy logic to improve situational awareness and to assist navigators in collision avoidance during multi-vessel encounters. The system is based on the integration of the rules of the Convention on International Regulations for Preventing Collisions at Sea (COLREGs) and artificial intelligence techniques. The proposed decision model consists of two main modules to calculate the initial encounter conditions for the target vessels, evaluate the collision risk and navigation situation based on COLREG rules, sort the target vessels, and determine the most dangerous vessel. Fuzzy logic is used to calculate the collision avoidance maneuver for the selected ship, considering the closest point of approach, relative bearing, and the ship's own speed. Simulation tests demonstrate the effectiveness of the fuzzy-based decision model in scenarios with two ships. However, in complex situations with multiple ships, the performance of the model is affected by possible conflicts between evasive maneuvers. This highlights the need for a cooperative collision avoidance algorithm for all vessels in high traffic areas.

Keywords: multi-ship collision avoidance; fuzzy reasoning; decision support model

Citation: Brcko, T.; Luin, B. A Decision Support System Using Fuzzy Logic for Collision Avoidance in Multi-Vessel Situations at Sea. *J. Mar. Sci. Eng.* **2023**, *11*, 1819. https://doi.org/10.3390/jmse11091819

Academic Editor: Chung-yen Kuo

Received: 23 August 2023
Revised: 11 September 2023
Accepted: 17 September 2023
Published: 18 September 2023

1. Introduction

The latest approaches to ensuring greater safety at sea are reflected in the form of decision support systems that harness advanced computer technology to enhance situational awareness and decision-making both onboard the ship and ashore. The main functions of the systems include tools such as prediction of the ship's course, warnings of possible ship collisions, groundings and approach to a guard zone, planning of collision avoidance maneuvers based on COLREG rules (Convention on the International Regulations for Preventing Collisions at Sea, 1972), etc. Such decision support can be found to some extent in ECDIS (Electronic Chart Display and Information System) on a modern ship today.

As the traffic density increases, navigators face a distinct challenge when it comes to avoiding collisions in multi-vessel situations. Existing COLREG rules governing the right-of-way between vessels in close quarter situations do not address such scenarios, leaving the decision-making responsibility in the hands of the navigators. This becomes particularly problematic at sea, where ships, especially in busy areas, lack sufficient time to communicate and coordinate collision avoidance measures. Conversely, the rules regulate collision avoidance for various types of encounters involving two vessels, which can potentially be applied to situations involving multiple vessels. However, to do so effectively, it is necessary to classify and determine the navigational situation. By employing decision-making tools, this classification process becomes simple and efficient, particularly if the tool is integrated into a navigation system utilized daily.

The COLREG rules consider several factors in determining the right of way: the area of navigation where the vessels are located, the relative position of the vessels involved, the type of the navigational situation (head-on, crossing, overtaking), and the navigational status of the vessels (power-driven, fishing boat, sailing boat, restricted in her ability to maneuver, etc.). From a navigator's point of view, the first important step in a situation where several vessels must avoid a collision is to identify the most dangerous vessel among them, followed by the choice of a maneuver that meets the safety requirements of the COLREG regulations.

1.1. Literature Review

Many researchers deal with decision models, some solve them holistically, with heuristic algorithms such as path planning, while others solve individual steps of collision avoidance, i.e., the extent of a speed or course change, which are considered deterministic algorithms [1]. However, not all studies consider COLREG as a part of the decision. The following literature review mainly focuses on three basic components of decision models: the Collision Risk Analysis, the Navigation Situation Classification, and the Collision Avoidance Maneuver.

1.1.1. Collision Risk Calculation Algorithms

The collision risk analysis component includes two functions of the model: namely, detecting the risk of collision with the target vessel and determining the right-of-way between vessels. The quantitative methods for calculating the collision risk could include, first and foremost, the calculation of the DCPA (Distance to the Closest Point of Approach). The Closest Point of Approach (CPA) is the point at which two ships will meet closest to each other. The smaller the distance to this point (DCPA), the greater the risk of collision. This is the first indicator that shows the possibility of collision or crossing the safety domain of one's own vessel.

A vessel's safety domain is an area around a vessel that must remain free of other vessels and fixed installations. In practice, this area takes the form of a circle, the radius of which is subjectively determined or established in advance by the ship's safety management system. The collision avoidance rules basically do not specify how large a vessel's safety domain should be; but, according to Cockcroft [2], it should be limited to two NM (nautical miles) in poor visibility or may be even smaller at low speeds in heavy traffic or when overtaking.

Through various developments in the calculation of the collision risk, the safety domain has also evolved, both in size and shape. Most of the early developments were based on statistical and analytical methods, and the domains were usually oval or elliptical in shape. One of the first was Goodwin [3], who took the COLREG rules as a basis and proposed a division of the navigation area into three sectors corresponding to the angles of the ship's navigation lights. The model was based on a statistical analysis of data from numerous registers and simulations.

Dynamic models of polygonal shapes are found in conjunction with the use of artificial intelligence techniques: Pietrzykowski [4] presented a fuzzy ship domain (combined with a self-learning neural network) as a safety criterion in offshore navigation; a similar method was used by Wang [5], who wrote that fuzzy boundaries in the ship domain are more practical for navigators than well-defined boundaries in assessing navigational safety. Su et al. [6] used variables in their fuzzy system to calculate the size of the safe encounter domain: relative approach speed, size of both ships, and sea state. Later, the development of the domains was dynamically adapted to the different navigation situations: size of vessels, traffic density, relative speeds, type of navigation situation, weather conditions, visibility, and more [7–9]. Du [10], on the other hand, presented an empirically determined ship domain based on a large dataset of ship encounters detected from AIS (Automatic Identification System) data. AIS data today are a very good source for the study of maritime traffic, since it works with dynamic ship data such as position, dimensions, speed, course, etc.

The authors who have dealt holistically with the problem of avoiding collisions at sea have usually used, in the model for the safe area of the ship, a simple radar circle with a radius of up to one NM [11–13]. Li [14] adapted two NM as minimum DCPA and Hu [15] set the value of the safety ship domain 12 to 14 times the ship length. Some dealt with simple ellipses where the size was determined by the length and the width of the ship [16,17]. The quadratic ship domain was used in [18] as the safety area of the ship, with the size determined by four radii (i.e., forward, aft, right, and left). Many authors did not specifically define the size of the domain in the decision models, including [19–21], or they chose a minimum passing distance [22,23]. In the maritime industry ship's domain, research has primarily enabled the application of new domains in existing navigation devices such as radar systems and electronic charts to enable more accurate collision risk calculations.

1.1.2. Algorithms for Determining Collision Risk and Right of Way in Complex Vessel Encounters

Determining the risk of collision and establishing the right of way when two ships encounter, especially in the open sea, is not a significant navigational hurdle. Of greatest importance are the values of the DCPA, the TCPA (Time to Closest Point of Approach), and the relative position of the vessels, as these form the basis for collision avoidance maneuvers.

In areas with dense traffic, there is a higher probability of encountering complex situations involving multiple vessels simultaneously [24]. The more complex the situation, the less situational awareness navigators managing the vessels may have, as their own vessel can be in both a give-way and stand-on position at the same time. This is influenced not only by the vessel's status but also by the navigational area in which the vessels are located (narrow straits, traffic separation scheme, open sea, etc.).

According to COLREG rules 16 and 17, the "Stand-on" vessel must maintain its course and speed, while the "Give-way" vessel must alter its movement to avoid impeding the vessel with the right of way. As mentioned earlier, COLREG rules do not cover complex situations; hence, the need for finding solutions becomes even more significant and demanding. Part of the collision avoidance process in complex scenarios is also the identification of the situation: which vessels are at risk of collision, determining the right of way with each vessel, and to identify the most dangerous vessel among them.

Some authors have dealt with this identification. Zhuo [25] presented an algorithm for calculating the time at which an avoidance maneuver should be initiated and the time frame in which a vessel should take action to identify the most dangerous vessel in a multi-vessel encounter. Elements that influence the time value are DCPA, TCPA, the dynamics of the target vessel, the maneuvering characteristics of the own vessel, the course of the target vessel, and the sea state. To calculate the above values, an adaptive self-learning system was used in combination with neural networks and fuzzy logic techniques. Hasegawa et al. [26], who addressed the problem of a multi-ship collision avoidance, presented the calculation of the collision risk (CR) using fuzzy logic. He used DCPA and TCPA values as input data of the fuzzy inference system and the CR value as output decision. A target ship with the highest CR value was a stand-on ship. Hu [15] followed a similar approach and used, for the input parameters, relative distance, bearing, and speed in addition to DCPA and TCPA. Zhang [27] used the speed ratio between two ships instead of the speed of the ship. Bukhari et al. [28] used DCPA and TCPA values as input data and added the value of VCD (Variation of a Compass Direction), which indicates the change in bearing of a target ship over time. The output decision was the degree of collision risk, and the vessel with the highest value was a stand-on vessel. Wang [29] adopted a basic CR calculation method to construct risk membership functions of DCPA and TCPA, which was also used by Zheng [30]. Ahn et al. [31] focused on situations with limited visibility where he calculated the CR using neural networks. As input variables, he used the speed, the course of the own and the target ship, the distance between the ships, the bearing of the target ship, and the safety range of the ship.

1.1.3. Algorithms for Calculating Collision Avoidance Maneuvers for Multiple Ship Encounters

The next step in the process of collision avoidance in complex scenarios is the planning of collision avoidance maneuvers. The different approaches by researchers to the problem are mainly reflected in the choice of trajectory planning strategies influencing the multi-ship collision avoidance maneuver. Here, we find various algorithms using methods such as deep reinforcement learning, fuzzy logic, artificial potential field, neural networks, swarm intelligence, collision-free trajectory generation, model predictive control, and so on. The main challenge for researchers is to incorporate COLREG rules into decision models, taking into account that they are adapted as a fundamental part of their designs [32]; therefore, studies that do not contain COLREG rules are excluded in the literature review. Liu [33] presented an algorithm for determining the direction of the collision avoidance maneuver, in which a course change amplitude of $10°$ was chosen and the maximum turning course was set at $60°$. Various parameters were considered, such as the distance between the two vessels, true bearings, relative bearings, relative speed, and the heading of the target and the vessel. Lu's [34] work combined the artificial potential field method with a collision avoidance algorithm executed by a particle swarm optimization algorithm, while Miao's [35] work used an improved hybrid A* algorithm that searches for appropriate motion options. Both articles presented a method for calculating the corresponding change in motion of the ships involved in a collision avoidance situation. Zhang [27] introduced an enhanced approach that combines the Velocity Obstacle method, model predictive control, and a ship trajectory prediction model. The objective of this method is to determine a viable space for collision avoidance maneuvers while considering the COLREG, and the constraints imposed by the ship's maneuverability. Authors [20,30], on the other hand, used a proximal policy optimization algorithm for collision avoidance path planning in multi-ship scenarios.

Several authors have also proposed a collision avoidance maneuver based on fuzzy logic. A fuzzy logic algorithm was used as the basis for calculating collision avoidance (course change and speed reduction) using three input parameters: the relative and encounter angle in a ship encounter, and the value of collision risk [36]. Perera [19] gave five input parameters: the region where the target ship is located, the relative course of the target, the degree of encounter risk, the distance to the target ship, and the relative speed of approach. Based on these parameters, the model decided the need to change course or speed based on COLREG rules. The selection of the navigation strategy in the traffic separation scheme, using a decision model based on a fuzzy logic algorithm, was also proposed by Wu [37], who analyzed the dynamic characteristics of the navigation process. Using a similar fuzzy logic approach, the risk of collisions with static and moving objects was calculated by Wu [38] and Hu [15].

In a previous article [39], the authors presented a multi-parameter decision model for collision avoidance at sea using fuzzy logic in the situation of two ships' encounter. In the article, the structure and operation of the fuzzy inference system, decision validation, and examples of model tests were presented. Building upon this prior research, the present article shifts its focus towards the intricacies of ship maneuvering characteristics and collision avoidance strategies in scenarios featuring multiple vessels at sea.

The paper is structured into several sections, each addressing a specific aspect of the research. The first section is the "Introduction," which provides an overview of the latest approaches in ensuring safety at sea using artificial intelligence-based decision support systems. The introduction also addresses existing research on decision models for collision avoidance at sea, algorithms, and methods for calculating collision risks and maneuvers in multiple vessel scenarios. Section 2 is the "Methodology", which outlines the systematic approach used in the paper. Section 3 presents the decision model, which is the core of the research. It explains the simulation of expert decision-making for collision avoidance at sea. The following section provides a practical demonstration of the decision model in a multi-ship scenario. It shows how the model calculates which is the most dangerous ship, determines the appropriate time interval for collision avoidance, and suggests an avoidance maneuver. Section 4 is the "Simulations", where the model's performance is evaluated through simulations in various multi-ship scenarios. The authors utilize a simplified nautical simulator coupled with the fuzzy collision avoidance system to assess the efficiency of the proposed decision model. The fifth section is the "Discussion", where the authors analyze the results of the simulations and discuss the strengths and limitations of the proposed decision model. Conclusions and suggestions for future research are presented in the sixth section.

2. Methodology

The methodology of the paper consists of several approaches:

1. Literature Review: The authors conducted a literature review of scientific papers to gather information on emerging decision models for collision avoidance at sea. Special attention was on articles that addressed multi-vessel situations and incorporated the use of COLREG rules in their models. The review mainly focused on two components of decision models: the collision risk analysis component (CR) and the collision avoidance maneuver component (CA).
2. Algorithm Development: The authors developed algorithms for calculating collision risk and collision avoidance maneuvers in multi-ship encounters. They considered various parameters, such as the DCPA, TCPA, relative bearings, relative speed, ship types, and navigation area, to assess the collision risk and determine the right-of-way between vessels. A classification algorithm was partly presented in the conference paper [40].
3. Fuzzy Logic: Fuzzy logic was utilized as a decision-making tool in the collision avoidance system. The authors implemented fuzzy inference systems with triangular or trapezoidal membership functions to determine the degree to which inputs belonged to different fuzzy sets. The fuzzy logic approach was used to calculate collision avoidance maneuvers in multi-ship encounters.
4. Simulation: A Monte-Carlo class of simulations, involving numerous runs to evaluate the performance of the proposed decision model for collision avoidance in multi-ship scenarios, was conducted. The simulations were carried out using a simplified ship dynamics simulator coupled with the fuzzy collision avoidance system. Different initial positions, speeds, and orientations were considered for each ship to assess collision avoidance performance in various scenarios.

3. Decision Model

The aim of the decision model is to simulate the decision-making of experts in avoiding collisions at sea. The knowledge that the model must contain is summarized in a multi-parameter decision model scheme consisting of two modules (Figure 1):

MODULE 1 "Initial parameter" calculates the initial conditions for the encounter of ships based on the data of ship targets and the ship.

MODULE 2 "Decision Model" is divided into two main components:

- Component 1 "Collision risk assessment and navigation situation analysis",
- Component 2 "Collision course maneuver calculation".

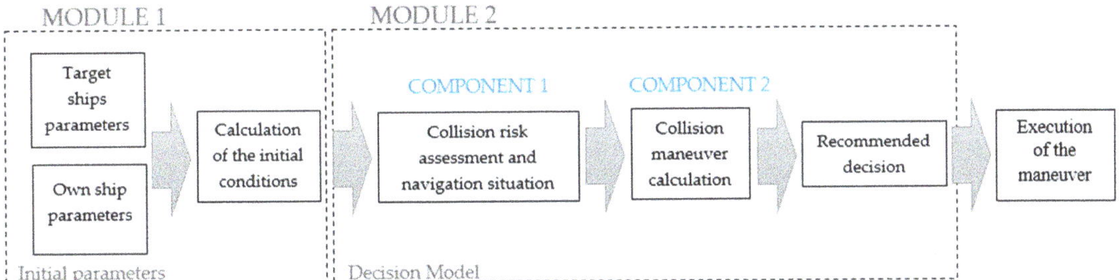

Figure 1. Decision model scheme.

Component 1 is based on the operation of the sorting algorithm and the use of COL-REG rules, and component 2 includes fuzzy reasoning in addition to these rules, which simulates the knowledge of the expert. The operation of both components is explained in more detail below.

3.1. Component 1—Collision Risk Assessment and Navigation Situation

Component 1 contains the algorithm for determining the most dangerous ship (Figure 2). The inputs used are DCPA, TCPA, direction and speed of the target ship, ship type, and COLREG rules eight, nine, 10, 13, 14, 15, 16, 17, 18, and 19. In the "parameter processing" step, the algorithm collects and analyzes the navigation input parameters of the target vessels (initial conditions) within the desired range. The parameters observed are the bearing, range, DCPA, TCPA, and ship's type. In the next step, the algorithm sorts the vessels according to the DCPA value—for further processing, select the vessels whose DCPA value is smaller than the vessel's safety domain (depending on the navigation area and meteorological/oceanographic conditions). If there is only one target in an area, the algorithm determines the right-of-way according to COLREG rules and suggests an evasive maneuver if necessary. If there are multiple vessels, the right-of-way for each of the vessels is determined. The target ships that have right-of-way are sorted by the algorithm according to the minimum TCPA. The ship with the lowest TCPA is selected for avoidance. Two exceptions are used in the algorithm:

- Exception 1—Any target ship within two NM or less, regardless of position or status, has the right of way (as per COLREG rule 17).
- Exception 2—If two vessels in sectors I or IV (see Figure 3) have the right of way, the vessel that is closer should be avoided.

In the algorithm, 90 conditional sentences are used to determine the right-of-way according to the COLREG rules. Their structure is presented in Table 1.

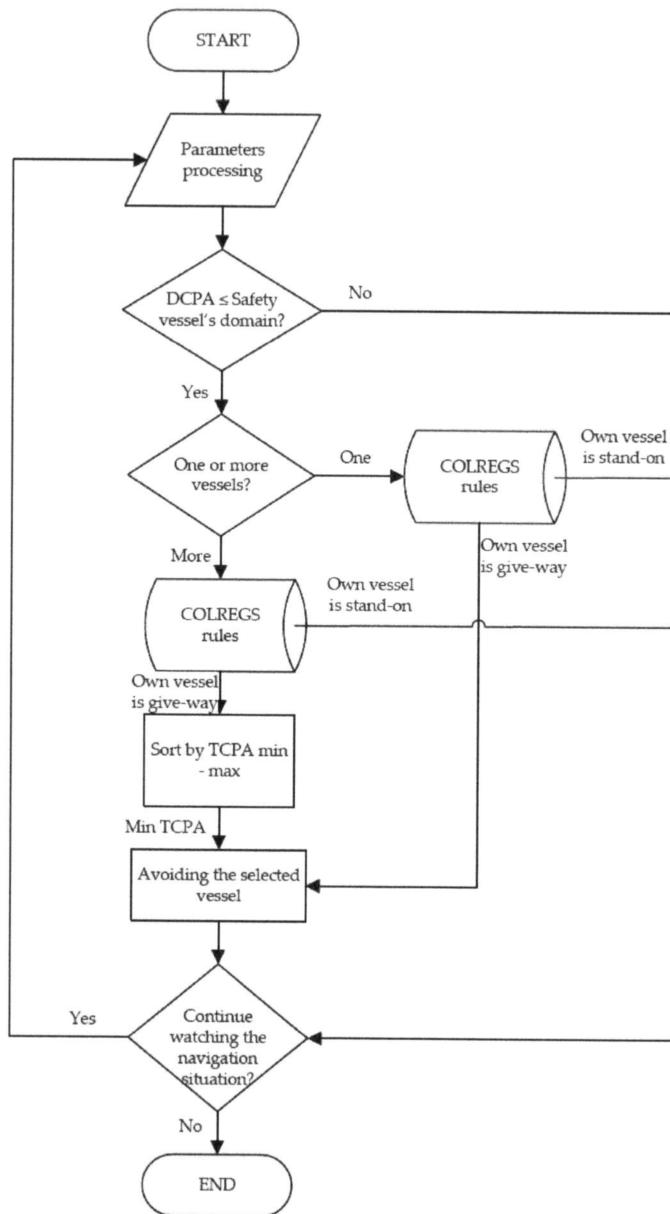

Figure 2. Algorithm for determining the most dangerous ship, adapted from: Brcko [40].

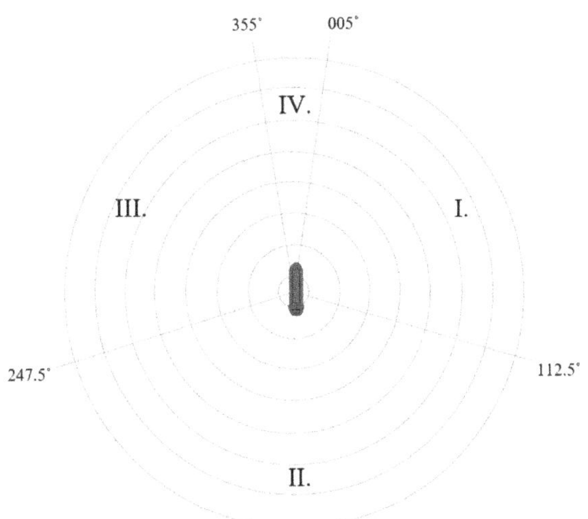

Figure 3. Sectors of the vessel's own domain.

Table 1. Structure of the algorithm.

if	A	Navigation area		Open sea Traffic Separation Scheme Narrow Channel
and	B	Position of the target vessel by sector		Sector I Sector II Sector III Sector IV
and	C	Type of navigation situation		Crossing Head-on Overtaking
and	D	Navigational status of the target vessel		Power-driven vessel Moored Not under command Restricted in her ability to maneuver At anchor Constrained by her draught Aground Engaged in fishing Sailing vessel
then	E	The right of way		The target vessel has the right of way or The own vessel has the right of way

The navigation area is divided into the open sea, the traffic separation scheme, and the narrow navigation channel. From the ship's own perspective, the target ship may be in any of the four sectors (Figure 3), depending on their relative bearings (Equations (1)–(4)):

$$P_t \sec I = 5° \leq RB \leq 112.5° \tag{1}$$

$$P_t \sec II = 112.6° \leq RB \leq 247.5° \tag{2}$$

$$P_t \sec III = 247.6° \leq RB \leq 354.9° \tag{3}$$

$$P_t \sec IV = 355° \le RB \le 004.9° \tag{4}$$

where P_t stands for the target position and RB is the relative bearing of the target. The boundaries of sectors I–III are defined with COLREG rules. The type of navigational situation encountered (rules 13, 14, or 15) is determined by conditional statements. These rules apply only when there is a risk of collision.

According to COLREG rule 18, ships have different priorities over other ships depending on their navigational status. This rule replaces rules 14 and 15, which apply only in situations where there is a risk of collision between power-driven vessels.

The responsibility of one's ship changes with the different navigation conditions. The algorithm provides one of two possible choices: the own ship is a give-way or a stand-by ship. According to the COLREG rules, the give-away ship should maintain its course and speed, while the stand-by ship must perform a collision avoidance maneuver.

3.2. Component 2—Collision Course Manoeuvre Calculation

Collision avoidance is a process that requires planning and observation of a dynamic navigational situation over a reasonable period of time. Path planning is extrapolating the trajectory of ships with a time delay. The paper suggests that the collision avoidance maneuver is calculated based on the data of the most dangerous ship. At the same time, the model checks the risk of collision with all the ships involved and plots the trajectories of the ships. Finally, the model calculates a time frame within which collision avoidance is safe and in accordance with COLREG rules.

Determining the avoidance maneuver is a two-step process. First, the model collects input data of the ship's own and target vessels to assess the collision risk observing minimum acceptable TCPA and DCPA parameters under the assumption speed and the course remains constant. The calculation using relative positions and speeds is defined by the Equations (5) and (6), where Xt and Yt represent the relative position coordinates of the target vessel, while Vrx and Vry denote the components of the relative velocity vector. Vr stands for the relative velocity of the approaching vessels [41].

$$DCPA = \left| \frac{(X_t \cdot V_{ry}) - (Y_t \cdot V_{rx})}{V_r} \right| [M], \tag{5}$$

$$TCPA = - \frac{(Y_t \cdot V_{ry}) + (X_t \cdot V_{rx})}{V_r^2} \cdot 60 \ [min], \tag{6}$$

The collision risk is directly dependent on the DCPA and TCPA and appropriate maneuvers are chosen according to the Fuzzy rules without intermediate quantification of the risk as it was done during the evaluation phase as defined by the Equation (11).

To predict the target vessel position at the time of the collision avoidance maneuver, the time delay calculation is used to obtain the new relative position of a target vessel. At this point four parameters are calculated for further processing as input variables in a fuzzy inference system [39]:

- DCPA—Distance to Closest Point of Approach,
- AP—Action Point distance to the target vessel,
- RB—Relative Bearing of a target vessel,
- Vo—Own vessel Velocity.

The fuzzy inference system (FIS), also known as the rule-based fuzzy system, is the process of formulating the mapping from a given input to an output using fuzzy logic (Figure 4).

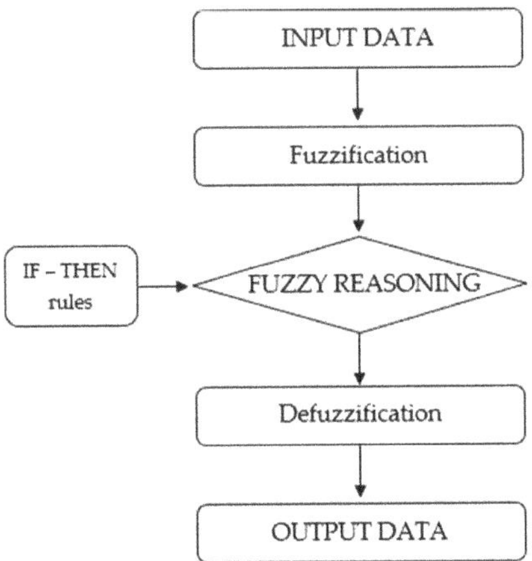

Figure 4. The structure of a fuzzy inference system.

It is a main element of the fuzzy logic system. The FIS formulates rules and based on these rules, the decision is made. The FIS type in this paper is "Mamdani", which is the most used fuzzy method. The first step is to take the inputs and outputs and determine the degree to which they belong to each of the corresponding fuzzy sets using triangular or trapezoidal membership functions (Figures 5–9). A fuzzy system is a set of fuzzy rules that convert fuzzy inputs into fuzzy outputs. It consists of a rule-based system of IF (antecedent)—THEN (consequent). A total of 216 rules forms the IF-THEN statements (Table 2).

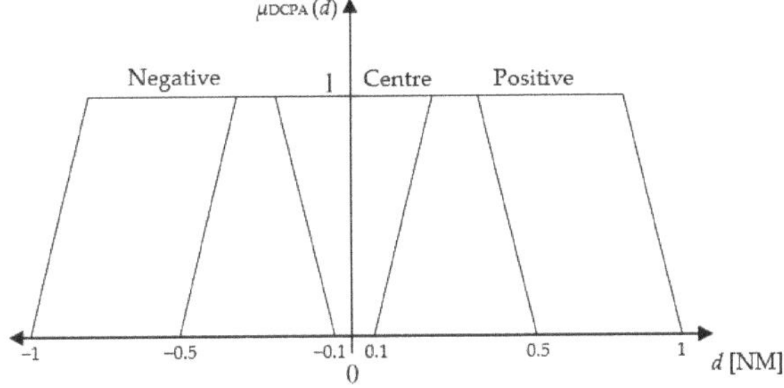

Figure 5. Fuzzy membership functions of DCPA parameter.

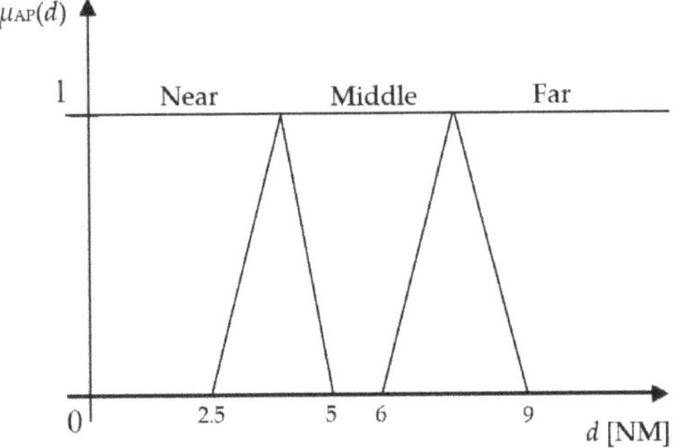

Figure 6. Fuzzy membership functions of AP parameter.

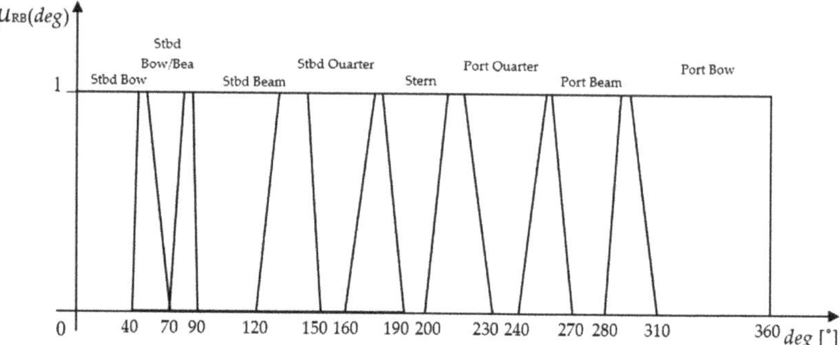

Figure 7. Fuzzy membership functions of RB parameter.

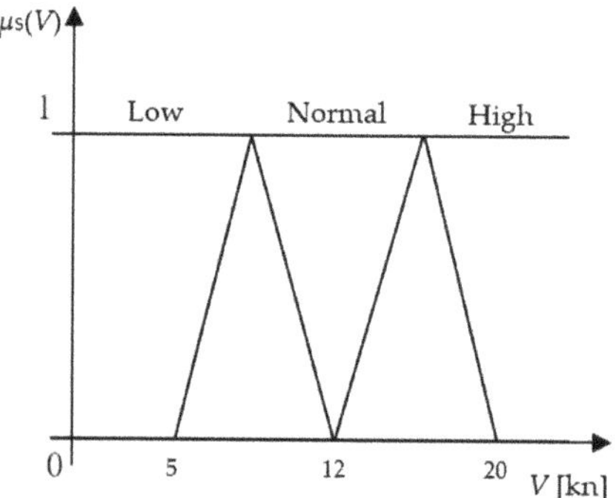

Figure 8. Fuzzy membership functions of the vessel's velocity parameter.

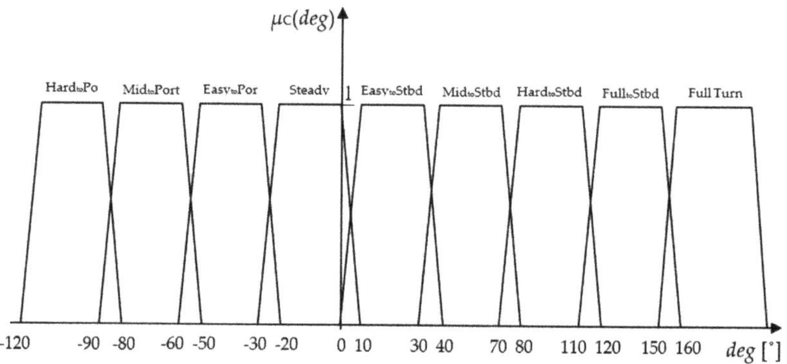

Figure 9. Fuzzy membership functions of the output parameter, the alteration of the vessel's own course.

Table 2. IF–THEN statements.

IF	DCPA	Negative, Positive, Center
AND	AP	Near, Middle, Far
AND	RB	Stbd Bow, Stbd Bow/Beam, Stbd Beam, Stbd Quarter, Stern, Port Quarter, Port Beam, Port Bow
AND	V	Low, Normal, High
THEN	Course alteration	Steady, Easy to port/starboard, Mid to port/starboard, Hard to port/starboard, Full to starboard, Full turn

Source: Adapted from Brcko et al. [39].

The second step involves the intricate computation of the decision or course alteration, employing the advanced principles of fuzzy logic methodology. In this phase, the distinguished techniques of the bisector and centroid are employed to defuzzify the output function derived from the fuzzy system. The bisector method strategically deploys a vertical line, effectively partitioning the region into two distinct sub-regions, each with an equal area. Although not universally so, it often coincides with the centroid line, which holds paramount importance in the realm of the Mamdani's Fuzzy Inference Systems (FIS) technique. The centroid, characterized as the center of gravity, represents the prevailing and most widely adopted approach within Mamdani's FIS method. Its pivotal role in decision-making is derived from its ability to effectively balance the distribution of fuzzy sets, providing a reliable and well-founded basis for subsequent actions.

Upon the completion of this comprehensive decision model, the ultimate outcome materializes as the precise alteration of the vessel's course, precisely expressed in degrees, either towards the port or starboard side.

3.3. Operation of the Model on an Example

In a multi-vessel collision avoidance situation, the model calculates the parameters DCPA, TCPA, position of the CPA point, RB, relative speed, and relative course based on the initial data for each vessel. Then, the right-of-way is determined according to the COLREG rules and the most dangerous vessel (which has the right-of-way) is selected. Based on the parameters of the most dangerous vessel, the collision avoidance course is calculated for each two-minute time delay. During this process, the model observes the DCPA parameters of all ships. Finally, it calculates a time interval in which collision avoidance is recommended, considering all safety parameters and recommending a maneuver. The simulation tests the fuzzy logic response for the encounter situation of three ships in sectors I and III, governed by COLREG rule 15 (crossing). The simulation observes the tuning of the set parameters and rules of the fuzzy inference system that follows COLREG rule 15.

Table 3 shows the initial parameters of the target ships in the radar diagram, where "C" is the ship's course, "V" is the ship's speed, "dt" is the distance to the target ship, and "ωt" is the bearing of the target ship. All ships in the simulation are underway using power.

Table 3. Initial parameters, zero min.

	Own Ship	Target 1	Target 2
C [°]	225	160	338
V [kn]	20	17	13.5
dt [NM]	-	8	7
ωt [°]	-	270	205
Navig. status	Power driven	Power driven	Power driven

In the next step (Table 4), the model calculates the initial navigation conditions, the most important data being DCPA, TCPA, the position of the CPA point with respect to its own ship (Figure 10), and the relative bearing of the target ship. Next is the determination of the right-of-way for each ship (Table 5).

Table 4. Collision risk assessment with targets 1 and 2.

	Target 1	**Target 2**
DCPA [NM]	0.731	0.754
TCPA [min]	23.9	14.8
CPA position	negative	positive
RB [°]	45	340

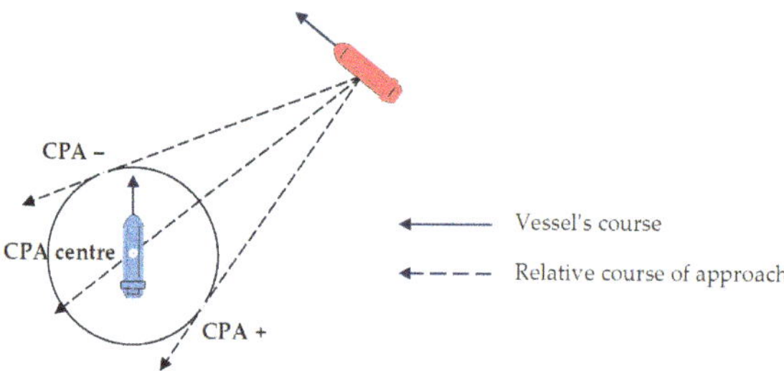

Figure 10. Position of Closest Point of Approach (CPA) for a different relative course of approach, adapted from: Brcko et al. [39].

Table 5. Determination of the right of the way with targets 1 and 2.

	Target 1	**Target 2**
Sector	I	III
Vessel's type	Power driven	Power driven
Nav. Situation	Rule 15	Rule 15
Vessel with the right of the way	Target vessel	Own vessel

For each minute of the time delay, the model calculates the relative position of the target ships. Table 6 shows the position of the ships in the sixth minute of observation. Based on these data, the model calculates the course change shown in Table 7 and reassesses the navigation situation by calculating the DCPA and TCPA for each ship (Table 8).

Table 6. A new relative position of a target vessels for a time delay of six min.

	Target 1	**Target 2**
ωt [°]	268.25	209.13
dt [NM]	6.01	4.21
RB [°]	43.25	344.13

Table 7. Calculated input parameters for the fuzzy inference system (FIS).

DCPA [NM]	−0.73
AP [NM]	6.01
RB [°]	43.25
V [kn]	20.00
Course alteration [°]	47.5

Table 8. Reassessment of the risks of collision.

	Target 1	Target 2
DCPA [NM]	3.4367	1.6389
TCPA [min]	9.6	12.3
Vr [kn]	30.8	18.9
Cr [°]	123.1	52.0

The sample simulation shows a close quarter situation with two ships, where the target ship one has the right of way and the target ship two has to perform a collision avoidance maneuver (Figure 11).

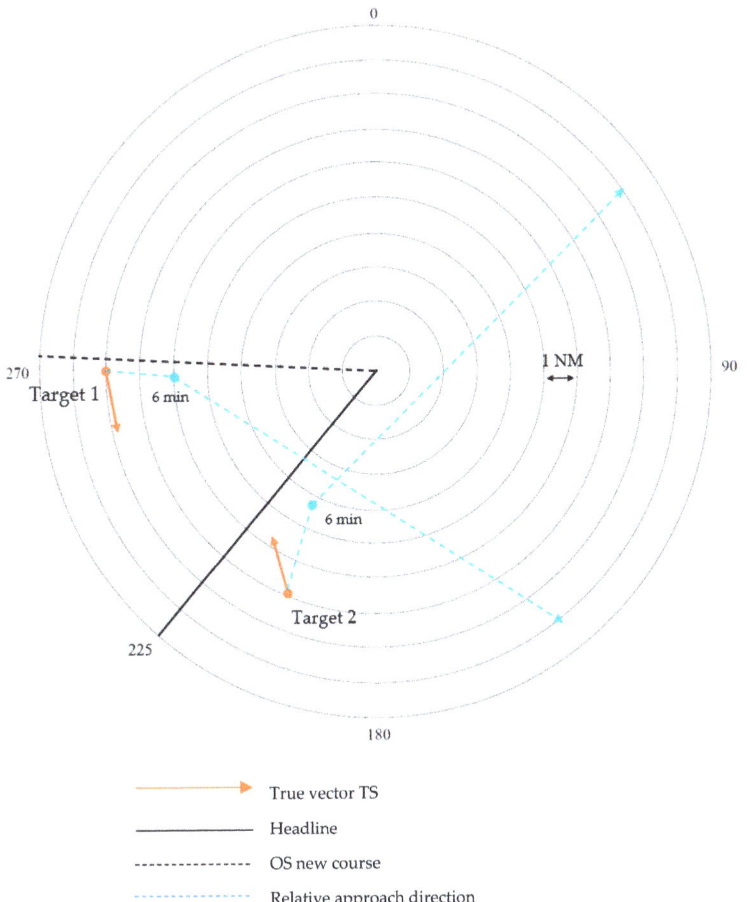

Figure 11. Collision avoidance maneuver in the radar diagram.

The maneuver is calculated based on the parameters of ship one. If ship two violates the COLREG rules, the appropriate time to start the maneuver is between two and nine minutes after the start of the observation. Otherwise, the choice of maneuver is considered appropriate until the 19th minute of the time delay.

4. Simulations

Since the fuzzy system has been well evaluated using two ship scenarios in previous work [39], this article extends its use to cases with multi-ship scenarios. The evaluation of complex navigational situations, when a collision avoidance maneuver may cause a hazardous situation with another vessel in its proximity, can be evaluated by repeating the simulation many times using different initial positions, speeds, and directions. Doing this with conventional nautical simulators would be extremely time-consuming as thousands of simulation runs are required. This approach is known as the Monte-Carlo simulation. To carry it out, a simplified nautical simulator has been developed based on the ship simulator UTSeaSim, Version 1.0, October 2013 [42], modified to include the impact of the rudder direction on longitudinal speed, speed, and direction control, and was tuned for the simulation of larger vessels. It assumes that ship longitudinal motion is described by

$$\ddot{x}m = \sum_i F_i \tag{7}$$

where x is ship position, m is the displacement, and F_i is the forces acting on the ship, such as propulsion and resistance forces. Ship rotational motion is defined by the

$$\alpha I = \sum_i \tau_i \tag{8}$$

where α is angular acceleration, I is the moment of ship inertia, and τ_i are moments acting on the ship due to rudder force and lateral resistance forces. Further details of the model can be found in the publication and code of the model authors [42].

To control the speed and course, PID regulators were used to model the autopilot; therefore, the engine speed command is a function of

$$throttle = c_1\left(v_{target} - v_{ship}\right) + c_2\left(a_{target} - a_{ship}\right) + c_3 v_{target} \tag{9}$$

where *throttle* is the engine speed command based on regulator constants c and differences between the target and actual speed, target acceleration a_{target}, and actual acceleration a_{ship}.

In a similar manner, rudder command is controlled by

$$rudder = d_1\left(\theta_{target} - \theta_{ship}\right) + d_2\left(\dot{\theta}_{target} - \dot{\theta}_{ship}\right) \tag{10}$$

where θ is ship heading and d_i are regulator constants.

The model has been coupled with a fuzzy collision avoidance system that determines avoidance maneuvers for two-ship interactions. When there are more than two ships, the target ship for the avoidance maneuver has been chosen by the lowest TCPA and by observing the priority according to the COLREG rules in the open sea (see Section 3.1). Basically, the evasion of approaching ships coming from the right, overtaking, and head-on avoidance were implemented. In the simulation, all the ships had the same status and none of them had restricted maneuverability.

The approach is illustrated in Figure 12, where the simulator provides the fuzzy collision avoidance system with positions, speeds, and relative bearings of other vessels. Once a need for collision avoidance maneuver arises, the system outputs two commands: heading offset and speed decrease if needed. Afterwards the autopilot system adjusts the heading and speed according to the commands by outputting the desired rudder angle and throttle position which are forwarded to the ship model that is impacted by the wind, sea current, and wave conditions.

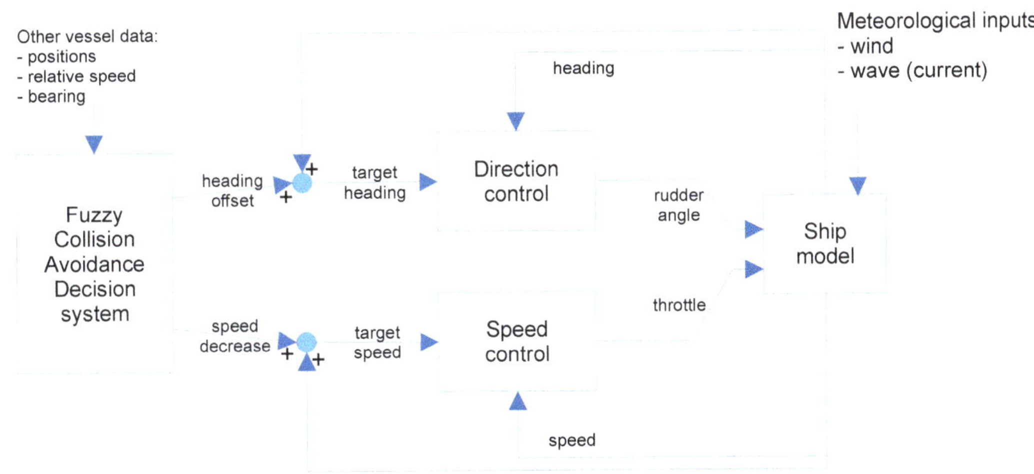

Figure 12. Simulation of the collision avoidance system.

Every ship in the system was controlled by a simulated autopilot coupled to the same fuzzy system as used before. Ship particulars for all ships in all scenarios were:

- Length: 100 m
- Breadth: 18 m
- Displacement: 2500 t
- Power output: 4000 kW

The performance indicators that evaluate the maneuvers are the minimum distance, maximum encounter risk, and maximum average encounter risk of a simulation case. To assess the encounter risk, a modified estimation approach was utilized, building upon the work of [43,44]. This modification ensures that the resulting risk is not influenced by vessel proximity when the distance is already increasing (i.e., when TCPA is negative). Therefore, the encounter risk R can be calculated according to Equation (7):

$$R = \begin{cases} e^{-|DCPA|} \cdot e^{-6TCPA}, & TCPA \geq 0 \\ 0, & TCPA < 0 \end{cases}. \tag{11}$$

In order to focus only on evasion maneuvers, meteorological conditions were set as neutral in all simulation scenarios. The air and water temperature was 20 °C with a zero water current, wave height, and wind velocities. Simulations were carried out by setting the ships' initial positions on a circular ring area with a specified internal and external radius and heading directed towards the center of the ring. An example of three ship scenarios is shown in Figure 13, where the red area marks the possible ship initial positions. They are obtained by generating a random radius between the lower boundary r1 and upper boundary r2 and a random orientation angle α for each ship.

Each of the scenarios was repeated 1000 times to obtain a collision avoidance performance at different initial positions, speeds, and orientations.

The results in Table 9 and Figure 14 indicate that the model performed flawlessly during the reference simulation with only two vessels (case eight), which was carried out to determine the reference risk for the described methodology. During two-ship simulations, no cases of possible collisions were recorded. As the number of ships involved in a simulation increases, more critical conditions arise. One parameter that severely impacted the performance of the maneuvers is the minimum initial distance between two vessels. If there was not such a limitation, initial positions could be set in a way that avoidance could be physically impossible no matter how the system responds.

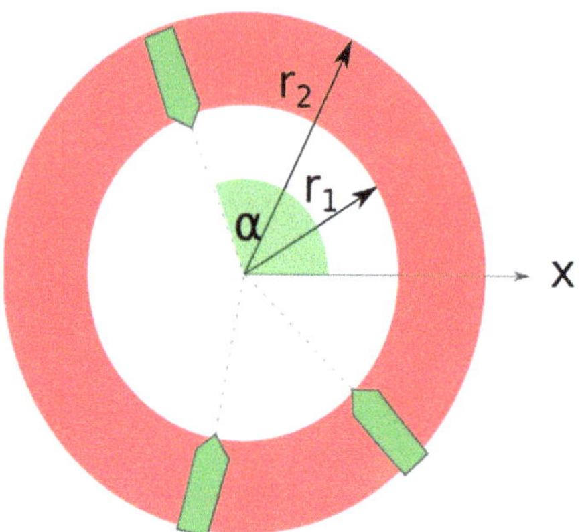

Figure 13. Initial positions at the beginning of a simulation run.

Table 9. Description of the simulation runs.

Case	Simulation Runs	Ships	Initial Radius (NM)		Initial Speed (m/s)		Minimum Initial Distance (NM)
			Min	Max	Min	Max	
1	1000	3	3.2	3.8	5	10	1
2	1000	3	3.2	3.8	5	10	0.8
3	1000	3	3.2	3.8	5	10	0.5
4	1000	3	3.8	3.8	5	10	0.5
5	1000	3	3.8	3.8	6	6	0.5
6	1000	4	3.8	3.8	6	6	0.5
7	1000	5	3.8	3.8	6	6	0.5
8	1000	2	3.2	3.8	6	6	0.5

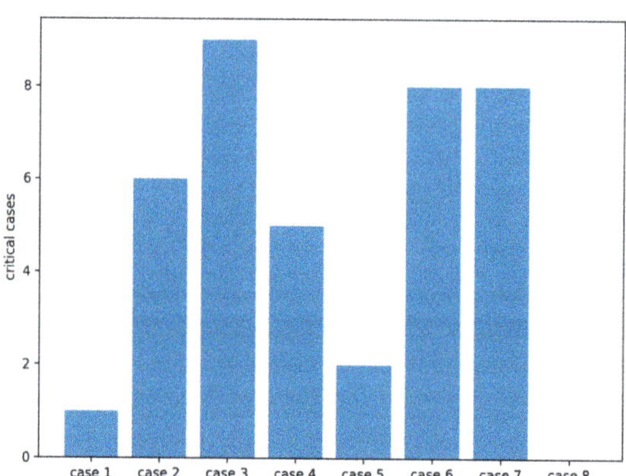

Figure 14. Number of definite collisions.

Cases 1, 2, and 3 show the same scenario using different initial distances, where the number of collisions has increased from one to nine by lowering initial distance from one to 0.5 NM. However, the average minimum distance during the simulation runs does not decrease significantly when lowering to 0.8 NM, but a drop in the average minimum distance is observed at the initial distances of 0.5 NM which is shown in a box blot in Figure 15. Also, the maximum average risk obtained during the simulation runs does not increase significantly, as shown in Figure 16.

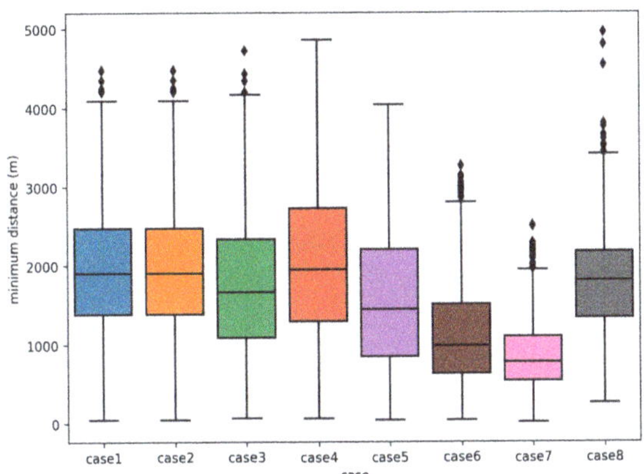

Figure 15. Minimum distances between any pair of ships during the simulation runs.

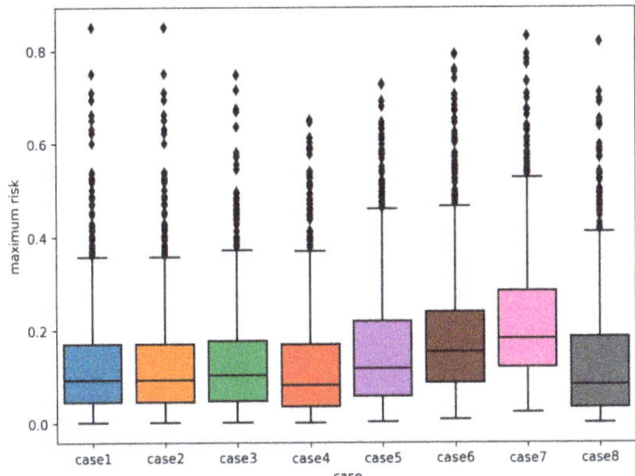

Figure 16. Maximum risk during the simulation runs.

When assessing the simulation results, the number of definite collisions was obtained by checking whether the distance between the vessels dropped below 100 m.

5. Discussion

The overall results of the simulations indicate that the fuzzy algorithm performs ideally with only two ships. It was important to filter out cases when avoidance would have been impossible due to physical restrictions, so the minimum initial distance was always 0.5 NM or more. As the number of ships increases, an in-depth analysis of critical cases has shown that the avoidance of one target caused a critical condition with another. This suggests that relying solely on COLREGS and avoiding a target with the lowest TCPA or highest risk may result in a crash with a third vessel not identified as dangerous before the maneuver. Therefore, in multi-ship scenarios, there is a need for the implementation of a cooperative collision avoidance algorithm that would prescribe mandatory maneuvers for all involved vessels in areas of dense traffic.

The problem could also be solved by determining a time interval within which the ship could safely execute an avoidance maneuver that would comply with COLREG rules. An example of such a time interval is shown in the encounter with five vessels. In the first part, the model calculates the parameters used to analyze the navigation situation and determine the right of way (Tables 10–12). This is followed by the selection of the most dangerous ship. Its parameters determine the time interval of the relevant decisions.

Table 10. Initial parameters.

	Own Vessel	Target 1	Target 2	Target 3	Target 4	Target 5
C [°]	82	288	70	262	181.5	32
V [kn]	12	10	27	15	15	23
dt [NM]		7.9	3.2	9.4	8	7.8
ωt [°]		85	233	79	32	174
Navig. status	Power driven	Power driven	Power driven	Power driven	Power driven	Fishing boat

Table 11. Collision risk assessment with five targets.

	Target 1	Target 2	Target 3	Target 4	Target 5
DCPA [NM]	1.208	0.430	0.492	0.610	0.948
TCPA [min]	21.8	12.3	20.9	23.1	26.0
CPA position	negative	Negative	negative	negative	negative
RB [°]	3	151	357	310	92

Table 12. Determination of the right of the way with five targets.

	Target 1	Target 2	Target 3	Target 4	Target 5
Sector	IV	II	IV	III	I
Vessel's type	Power driven	Power driven	Power driven	Power driven	Fishing boat
Nav. Situation	Rule 14	Rule 13	Rule 14	Rule 15	Rule 15
Vessel with the right of the way	Target ship	Own ship	Target ship	Own ship	Target ship

According to collision risk assessment (Table 11), there is a danger of collision with ships three and five (DCPA < 1 NM), and at the same time, the ships have the right of way. The model for the most dangerous ship chooses target ship five because it is closer (see Section 3.1—exception 2) and calculates a course change based on the parameters of target ship five, for the time interval of two to 10 min. Table 13 shows the DCPA values for each target in the second and fourth minute of the time delay and the calculated course change.

Table 13. DCPA values for each of the target ships in the second and fourth minute of the time delay.

Time Delay [min]	Course Alteration [°]	New Course [°]	DCPA [NM]				
			Target 1	Target 2	Target 3	Target 4	Target 5
2	−40.1	41.9	1.565	1.637	2.116	1.588	3.398
4	−40.1	41.9	1.295	1.392	1.842	1.387	3.189

In the sixth minute of the time delay, the model selects target two as the most dangerous ship based on its distance, since it is 1.68 NM (see Section 3.1—exception 1), and calculates the course change for the sixth, eighth, and 10th minutes (Table 14).

Table 14. DCPA values for each of the target ship in the sixth, eighth and 10th minute of the time delay.

Time Delay [min]	Course Alteration [°]	New Course [°]	DCPA [NM]				
			Target 1	Target 2	Target 3	Target 4	Target 5
6	−48.6	33.4	1.500	1.219	1.985	1.562	3.759
8	−48.6	33.4	1.175	0.948	1.656	1.319	3.466
10	−48.6	33.4	0.850	0.677	1.327	1.076	3.173

Figure 17 shows the change in the DCPA over time. The appropriate time interval for collision avoidance (considering that target 2 violates COLREG rules) is between the second and the seventh minute, when the safety ship domain is one NM.

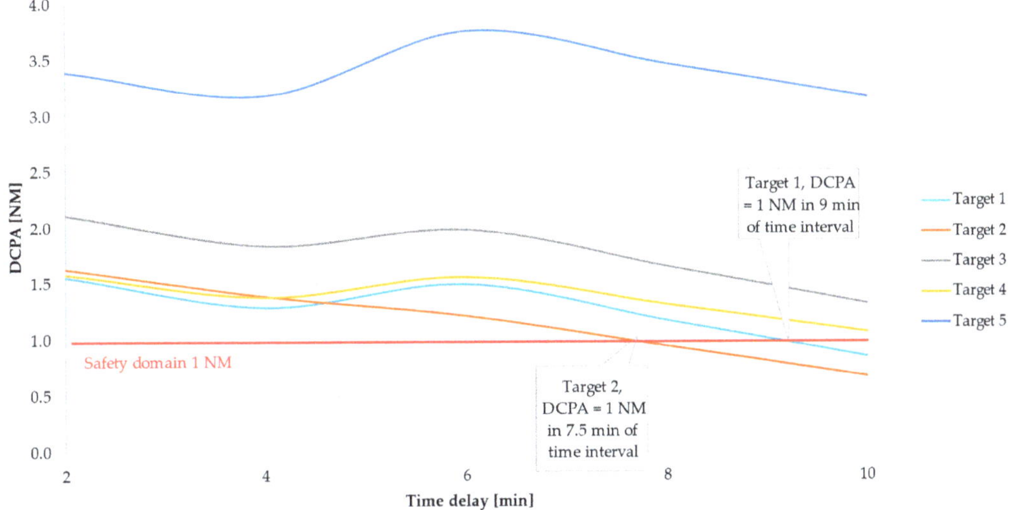

Figure 17. Calculated value of DCPA at different time delays.

6. Conclusions

According to [45], a majority of collisions on open sea happen at night time, mostly due to poor situational awareness, which is primarily the result of a sole lookout, poorer visibility, lack of communicational skills (misspoken, misread, or misheard information), decisions depending mostly on data obtained from navigation devices, etc. To mitigate predominantly human errors, which account for a substantial 78% of maritime accidents, it is imperative to prioritize research endeavors geared towards the development of sophisticated decision systems. These systems hold the promise of assisting seafarers in making optimal judgments precisely when they are most critical. The veracity and effectiveness of such decisions are intricately linked to the robustness and precision of the data upon which navigation devices operate and are disseminated to end-users. Consequently, decision support systems represent a pertinent steppingstone towards the integration of autonomous vessels, given that their transitional management will predominantly rest upon the expertise and competence of seafarers.

This paper highlights pertinent issues that warrant deeper investigation, notably pertaining to the revision of COLREG (Collision Regulations) rules, which govern the avoidance of collisions at sea, particularly in situations involving multiple vessels. An area of utmost significance in this regard is the examination of the feasibility of determining the right of way from a ship's own perspective. Spatial-Based Trajectory Planning emerges as a promising avenue for addressing these challenges and should be subject to thorough exploration and research. By delving into this approach, we may uncover valuable insights into enhancing collision avoidance strategies and maritime safety.

Author Contributions: Conceptualization, T.B. and B.L.; methodology, T.B. and B.L.; software, T.B and B.L.; validation, T.B.; resources, T.B. and B.L.; writing—original draft preparation, T.B.; writing—review and editing, T.B. and B.L. All authors have read and agreed to the published version of the manuscript.

Funding: The authors acknowledge the financial support of the Slovenian Research Agency (research core funding No. P2-0394, Modelling and Simulations in Traffic and Maritime Engineering).

Institutional Review Board Statement: Not applicable.

Informed Consent Statement: Not applicable.

Data Availability Statement: The data presented in this study are available on request from the corresponding author.

Conflicts of Interest: The authors declare no conflict of interest.

References

1. Hu, Y.; Zhang, A.; Tian, W.; Zhang, J.; Hou, Z. Multi-ship collision avoidance decision-making based on collision risk index. *J. Mar. Sci. Eng.* **2020**, *8*, 640. [CrossRef]
2. Cockcroft, A.N.; Lameijer, J.N.F. *Guide to the Collision Avoidance Rules*; Elsevier: Amsterdam, The Netherlands, 2003.
3. Goodwin, E.M. A statistical study of vessel domains. *J. Navig.* **1975**, *28*, 328–344. [CrossRef]
4. Pietrzykowski, Z. Vessel's fuzzy domain—A criterion for navigational safety in narrow fairways. *J. Navig.* **2008**, *61*, 499–514. [CrossRef]
5. Wang, N. An intelligent spatial collision risk based on the quaternion ship domain. *J. Navig.* **2010**, *63*, 733–749. [CrossRef]
6. Su, C.; Chang, K.; Cheng, C. Fuzzy Decision on Optimal Collision Avoidance Measures for Ships in Vessel Traffic Service. *J. Mar. Sci. Technol. Taiwan* **2012**, *20*, 38–48. [CrossRef]
7. Szlapczynski, R.; Szlapczynska, J. Review of ship safety domains: Models and applications. *Ocean Eng.* **2017**, *145*, 277–289. [CrossRef]
8. Szlapczynski, R.; Szlapczynska, J. A ship domain-based model of collision risk for near-miss detection and Collision Alert Systems. *Reliab. Eng. Syst. Saf.* **2021**, *214*, 107766. [CrossRef]
9. Liu, K.; Yuan, Z.; Xin, X.; Zhang, J.; Wang, W. Conflict detection method based on dynamic ship domain model for visualization of collision risk Hot-Spots. *Ocean Eng.* **2021**, *242*, 110143. [CrossRef]
10. Du, L.; Banda, O.A.V.; Huang, Y.; Goerlandt, F.; Kujala, P.; Zhang, W. An empirical ship domain based on evasive maneuver and perceived collision risk. *Reliab. Eng. Syst. Saf.* **2021**, *213*, 107752. [CrossRef]
11. Zhang, J.; Zhang, D.; Yan, X.; Haugen, S.; Soares, C.G. A distributed anti-collision decision support formulation in multi-ship encounter situations under COLREGs. *Ocean Eng.* **2015**, *105*, 336–348. [CrossRef]
12. Szlapczynski, R.; Szlapczynska, J. A Target Information Display for Visualising Collision Avoidance Manoeuvres in Various Visibility Conditions. *J. Navig.* **2015**, *68*, 1041–1055. [CrossRef]
13. Pietrzykowski, Z.; Wołejsza, P.; Borkowski, P. Decision support in collision situations at sea. *J. Navig.* **2017**, *70*, 447–464. [CrossRef]
14. Li, L.; Wu, D.; Huang, Y.; Yuan, Z.M. A path planning strategy unified with a COLREG collision avoidance function based on deep reinforcement learning and artificial potential field. *Appl. Ocean Res.* **2021**, *113*, 102759. [CrossRef]
15. Hu, Y.; Park, G.K. Collision risk assessment based on the vulnerability of marine accidents using fuzzy logic. *Int. J. Nav. Archit. Ocean Eng.* **2020**, *12*, 541–551. [CrossRef]
16. Zhang, J.; Yan, X.; Chen, X.; Sang, L.; Zhang, D. A novel approach for assistance with anti-collision decision making based on the International Regulations for Preventing Collisions at Sea. *Proc. Inst. Mech. Eng. Part M J. Eng. Marit. Environ.* **2012**, *226*, 250–259. [CrossRef]
17. Nguyen, M.; Nguyen, V.; Tamaru, H. Automatic collision avoiding support system for ships in congested waters and at open sea. In Proceedings of the International Conference on Control, Automation and Information Sciences (ICCAIS), Saigon, Vietnam, 26–29 November 2012.
18. Wang, M.; Wang, Y.; Cui, E.; Fu, X. A novel multi-ship collision probability estimation method considering data-driven quantification of trajectory uncertainty. *Ocean Eng.* **2023**, *272*, 113825. [CrossRef]
19. Perera, L.; Carvalho, J.; Soares, C. Fuzzy logic based decision making system for collision avoidance of ocean navigation under critical collision conditions. *J. Mar. Sci. Technol.* **2011**, *16*, 84–99. [CrossRef]
20. Hu, L.; Naeem, W.; Rajabally, E.; Watson, G.; Mills, T.; Bhuiyan, Z.; Salter, I. COLREGs-Compliant Path Planning for Autonomous Surface Vehicles: A Multiobjective Optimization Approach. *IFAC-PapersOnLine* **2017**, *50*, 13662–13667. [CrossRef]

21. Xue, D.; Wu, D.; Yamashita, A.S.; Li, Z. Proximal policy optimization with reciprocal velocity obstacle based collision avoidance path planning for multi-unmanned surface vehicles. *Ocean Eng.* **2023**, *273*, 114005. [CrossRef]
22. Xin, X.; Liu, K.; Yang, Z.; Zhang, J.; Wu, X. A probabilistic risk approach for the collision detection of multi-ships under spatiotemporal movement uncertainty. *Reliab. Eng. Syst. Saf.* **2021**, *215*, 107772. [CrossRef]
23. Chun, D.H.; Roh, M.I.; Lee, H.W.; Ha, J.; Yu, D. Deep reinforcement learning-based collision avoidance for an autonomous ship. *Ocean Eng.* **2021**, *234*, 109216. [CrossRef]
24. Van Westrenen, F.; Baldauf, M. Improving conflicts detection in maritime traffic: Case studies on the effect of traffic complexity on ship collisions. *Proc. Inst. Mech. Eng. Part M J. Eng. Marit. Environ.* **2020**, *234*, 209–222. [CrossRef]
25. Zhuo, Y.; Tang, T. An Intelligent Decision Support System to Ship Anti-Collision in Multi-Ship Encounter. In Proceedings of the 7th World Congress on Intelligent Control and Automation, Chongqing, China, 25–27 June 2008.
26. Hasegawa, K.; Fukuto, J.; Miyake, R.; Yamazaki, M. An Intelligent Ship Handling Simulator with Automatic Collision Avoidance Function of Target Ships. In Proceedings of the International Navigation Simulator Lecturers' Conference 17, Rostock–Warnemünde, Germany, 3–7 September 2012.
27. Zhang, K.; Huang, L.; He, Y.; Wang, B.; Chen, J.; Tian, Y.; Zhao, X. A real-time multi-ship collision avoidance decision-making system for autonomous ships considering ship motion uncertainty. *Ocean Eng.* **2023**, *278*, 114205. [CrossRef]
28. Bukhari, A.C.; Tusseyeva, I.; Kim, Y.G. An intelligent real-time multi-vessel collision risk assessment system from VTS view point based on fuzzy inference system. *Expert Syst. Appl.* **2013**, *40*, 1220–1230. [CrossRef]
29. Wang, S.; Zhang, Y.; Zheng, Y. Multi-ship encounter situation adaptive understanding by individual navigation intention inference. *Ocean Eng.* **2021**, *237*, 109612. [CrossRef]
30. Rongcai, Z.; Hongwei, X.; Kexin, Y. Autonomous collision avoidance system in a multi-ship environment based on proximal policy optimization method. *Ocean Eng.* **2023**, *272*, 113779. [CrossRef]
31. Ahn, J.H.; Rhee, K.P.; You, Y.J. A study on the collision avoidance of a ship using neural networks and fuzzy logic. *Appl. Ocean Res.* **2012**, *37*, 162–173. [CrossRef]
32. Maza, J.A.G.; Argüelles, R.P. COLREGs and their application in collision avoidance algorithms: A critical analysis. *Ocean Eng.* **2022**, *261*, 112029. [CrossRef]
33. Liu, J.; Zhang, J.; Yan, X.; Soares, C.G. Multi-ship collision avoidance decision-making and coordination mechanism in Mixed Navigation Scenarios. *Ocean Eng.* **2022**, *257*, 111666. [CrossRef]
34. Lu, N.; Zhou, W.; Yan, H.; Fei, M.; Wang, Y. A two-stage dynamic collision avoidance algorithm for unmanned surface vehicles based on field theory and COLREGs. *Ocean Eng.* **2022**, *259*, 111836. [CrossRef]
35. Miao, T.; El Amam, E.; Slaets, P.; Pissoort, D. An improved real-time collision-avoidance algorithm based on Hybrid A* in a multi-object-encountering scenario for autonomous surface vessels. *Ocean Eng.* **2022**, *255*, 111406. [CrossRef]
36. Ahmed, Y.A.; Hannan, M.A.; Oraby, M.Y.; Maimun, A. COLREGs compliant fuzzy-based collision avoidance system for multiple ship encounters. *J. Mar. Sci. Eng.* **2021**, *9*, 790. [CrossRef]
37. Wu, B.; Cheng, T.; Yip, T.L.; Wang, Y. Fuzzy logic based dynamic decision-making system for intelligent navigation strategy within inland traffic separation schemes. *Ocean Eng.* **2020**, *197*, 106909. [CrossRef]
38. Wu, B.; Yip, T.L.; Yan, X.; Soares, C.G. Fuzzy logic based approach for ship-bridge collision alert system. *Ocean Eng.* **2019**, *187*, 106152. [CrossRef]
39. Brcko, T.; Androjna, A.; Srše, J.; Boć, R. Vessel multi-parametric collision avoidance decision model: Fuzzy approach. *J. Mar. Sci. Eng.* **2021**, *9*, 49. [CrossRef]
40. Brcko, T. Determining the most immediate danger during a multi-vessel encounter. In Proceedings of the International Conference on Transport Science, Portorož, Slovenia, 14–15 June 2018.
41. Stateczny, A. *Radar Navigation*; GTN: Gdańsk, Poland, 2011.
42. Agmon, N.; Urieli, D.; Stone, P. Multiagent patrol generalized to complex environmental conditions. In Proceedings of the AAAI Conference on Artificial Intelligence, San Francisco, CA, USA, 7–11 August 2011.
43. Koldenhof, Y.; Van der Tak, C.; Glansdorp, C. Risk Awareness—A model to calculate the risk of a ship dynamically. In Proceedings of the XIII International Scientific and Technical Conference on Maritime Traffic Engineering, Malmö, Sweden, 16 October 2009.
44. Jeong, J.S.; Park, G.K.; Kim, K.I. Risk Assessment Model of Maritime Traffic in Time-Variant CPA Environments in Waterway. *J. Adv. Comput. Intell. Intell. Inform.* **2012**, *16*, 866–873. [CrossRef]
45. EMSA. *Safety Analysis of EMCIP Data—Analysis of Navigation Accidents*; European Maritime Safety Agency: Lisbon, Portugal, 2022.

Article

A Contextually Supported Abnormality Detector for Maritime Trajectories

Kristoffer Vinther Olesen [1,*,†], **Ahcène Boubekki** [2], **Michael C. Kampffmeyer** [2], **Robert Jenssen** [2,3,4], **Anders Nymark Christensen** [1], **Sune Hørlück** [5] and **Line H. Clemmensen** [1]

[1] Applied Mathematics and Computer Science, Technical University of Denmark, 2800 Kgs. Lyngby, Denmark; anym@dtu.dk (A.N.C.); lkhc@dtu.dk (L.H.C.)
[2] Machine Learning Group, UiT The Arctic University of Norway, 9019 Tromsø, Norway; ahcene.boubekki@uit.no (A.B.); michael.c.kampffmeyer@uit.no (M.C.K.); robert.jenssen@uit.no (R.J.)
[3] Pioneer Centre for AI, University of Copenhagen, 1350 Copenhagen, Denmark
[4] Norwegian Computing Center, 0373 Oslo, Norway
[5] Terma A/S, 8520 Lystrup, Denmark; snhr@terma.com
* Correspondence: kvol@dtu.dk
† Work done while corresponding author was at UiT.

Abstract: The analysis of maritime traffic patterns for safety and security purposes is increasing in importance and, hence, Vessel Traffic Service operators need efficient and contextualized tools for the detection of abnormal maritime behavior. Current models lack interpretability and contextualization of their predictions and are generally not quantitatively evaluated on a large annotated dataset comprising all expected traffic in a Region of Interest. We propose a model for the detection of abnormal maritime behaviors that provides the closest behaviors as context to the predictions. The normalcy model relies on two-step clustering, which is first computed based on the positions of the vessels and then refined based on their kinematics. We design for each step a similarity measure, which combined are able to distinguish boats cruising shipping lanes in different directions, but also vessels with more freedom, such as pilot boats. Our proposed abnormality detection model achieved, on a large annotated dataset extracted from AIS logs that we publish, an ROC-AUC of 0.79, which is on a par with State-of-the-Art deep neural networks, while being more computationally efficient and more interpretable, thanks to the contextualization offered by our two-step clustering.

Keywords: maritime surveillance; vessel traffic service; AIS; maritime traffic patterns; trajectory clustering; anomaly detection

Citation: Olesen, K.V.; Boubekk, A.; Kampffmeyer, M.; Jenssen, R.; Christensen, A.N.; Hørlück, S.; Clemmensen, L.H. A Contextually Supported Abnormality Detector for Maritime Trajectories. *J. Mar. Sci. Eng.* **2023**, *11*, 2085. https://doi.org/10.3390/jmse11112085

Academic Editors: Sebastian Feuerstack, Marko Perkovic and Lucjan Gucma

Received: 22 September 2023
Revised: 20 October 2023
Accepted: 23 October 2023
Published: 31 October 2023

1. Introduction

According to the International Maritime Organization, international shipping is currently responsible for 80% of global trade and is the most efficient and cost-effective form of long-distance transportation [1]. Despite international efforts to combat maritime piracy, it remains a serious threat to international shipping and is estimated to have a global financial impact of up to 16 billion dollars annually [2]. Accidents such as collisions or groundings can lead to the loss of lives, environmental damage, and disruption of trade routes [3,4]. Furthermore, it is estimated that one fifth of all wild-caught fish is caught illegally or not reported, endangering marine ecosystems and resulting in industry losses upwards of 23.5 billion dollars [5]. Most recently, the sabotage of the Nord Stream gas pipelines in the Baltic Sea [6] has raised the issue of territorial protection and protection of key infrastructure assets. These vulnerabilities in our society highlight the need for maritime security, safety, and threat assessment, to protect the stability of the global supply chain and key infrastructure.

Real-life maritime laws and regulations are complex. While commercial vessels such as cargo and tankers mainly follow well-defined shipping lanes with near-constant speeds, other ship types, such as fishing vessels and sailing ships, have fewer constraints and more complex behaviors. Maritime security requires extensive knowledge of maritime traffic patterns. Research in this field has gained momentum over the past decade, thanks to the advent of the Automatic Identification System (AIS) [7]. The AIS is compulsory for all vessels that exceed a certain tonnage, and it provides hundreds of millions of messages every day on a global scale [8]. The data include static information, such as the unique identifier of the ships (MMSI), size, and dynamic information such as Global Positioning System (GPS) coordinates, speed, course, etc. The AIS forms the basis for modern maritime trajectory data collection, which allows one to model navigational characteristics and rules. The wide variety of possible maritime behaviors means that the most common method of analysis remains clustering, either of individual AIS updates [9,10] or of trajectories compared using specific similarity measures [11,12].

Dense and critical maritime areas are constantly monitored, but the ever-increasing traffic and amount of data call for automatic decision support for Vessel Traffic Service (VTS) operators. In practice, whether or not an event is abnormal is a combination of several factors: the location, the speed, the course, the type of vessel, and the time of day/week/year, etc. For instance, in specific locations, pilot boats steam between harbors and commercial traffic in the shipping lanes, but in other locations or for different ship types, this type of behavior may be highly unexpected. Similarly, it can be expected that diving vessels perform frequent starts or stops to support divers in the water, but if this behavior occurs near major shipping lanes, it can be considered abnormal unless permission has been granted. Automatic detection of abnormal maritime behaviors is thus a difficult, ill-defined problem that requires the disentanglement of multiple possible explanatory factors, of which location, kinematic behavior, and type of ship are the most important.

Previous works have shown that shipping lanes can be identified using only a positional clustering of vessels [9,11,13]. However, to disentangle traffic not constrained to major shipping lanes, the exact route is less important than the local kinematic behavior—that is, changes in speed and course [10,14]. Recently, deep neural network models have been suggested for abnormality detection of multiple ship types [13,15]. These models classify as anomalies the trajectories for which they fail to predict the future position or to make an adequate reconstruction. The main drawback here is the lack of interpretability as to why the networks fail to predict/reconstruct the correct trajectory. According to Riveiro et al. and Stach et al. [16,17], abnormality detection models should offer a large degree of interpretability, to accommodate any skepticism VTS operators may have, and they should promote human–machine interaction [16,17]. At the same time, recent research on dynamical decision making indicates that automated systems for decision making tend to perform poorly, as human operators simply copy the decision of the automated system [18]. This causes a degradation of operator experience and makes human operators less likely to take over manual control when needed. Furthermore, many previous studies simplify their data, to focus on a restricted Region of Interest (ROI) and on a single aspect of maritime traffic—for instance, the behavior of commercial merchant traffic [19] or port entry/exit ways [11,12,20]. However, such restrictions hinder the evaluation of the practical viability of the methods. Stach et al. [17] further highlighted the need for standardized datasets annotated with maritime abnormalities, to bridge the gap between research and practical implementation in VTS operations.

A precise definition of what constitutes a maritime abnormality is difficult to state. To this day, however, many maritime surveillance operations are conducted manually by military or law enforcement. Thus, the reasons for flagging a behavior as abnormal are often classified information, and the specific type of behavior that interests operators is not fully known. For this reason, obtaining a large list of annotated maritime trajectories for training or evaluation is often impossible. Previous works [21,22] have used simulated or self-annotated labels based on extreme values. However, extreme values might not define abnormal trajectories of operational interest for surveillance operators, who are ultimately interested in deterring illegal or dangerous activities, such as the earlier examples. As for unsupervised methods, the most common way to evaluate them is through qualitative examples [15,23,24]. This form of verification illustrates the potential type of abnormal behavior that could be flagged. However, it completely negates the issue of false negatives, which in a military or law enforcement operation may be of greater importance. Ideally, abnormality detection algorithms would be evaluated on datasets with known behavior of operational interest to operators. These datasets should be annotated by subject matter experts, and the annotations should reflect the degree to which the operators find the behavior suspicious or otherwise abnormal.

In this paper, we aim to overcome two gaps in the current research on vessel traffic abnormality detection: the lack of interpretability and contextualization of the predictions and the size and quality of the datasets used for evaluation. First, we introduce an abnormality detection algorithm that provides an explanation of the closest expected/normal behaviors. The normalcy model is learned using a two-step clustering method that disentangles positional and kinematic behavior. The training is performed on historical AIS data and consists of two stages: in the first step, a clustering is learned based on the positional data of the trajectories; in the second step, each positional cluster is refined on the basis of the kinematics of its trajectories—that is, speed and course. The final clustering of the whole data is thus a summary of the typical behavioral patterns in the area. Concurrently, a Local Outlier Factor (LOF) [25] is trained on each positional cluster, also based on the kinematic data, to detect trajectories with abnormal speed and course sequences. In a practical scenario, a new trajectory is first assigned to one of the positional clusters. If flagged as abnormal, using an LOF, it is brought to the attention of an operator who can further assess the situation, using the kinematic clustering of the positional cluster as support.

Next, we present and use for the evaluation two large hand-annotated AIS traffic datasets for abnormal maritime behavior detection that we have created, based on contextual knowledge of the environment and news events. These datasets contain more than 30,000 trajectories from 11 types of vessels and are expected to be of operational interest to operators. Labeled abnormalities cover a full day and include a collision accident, Search and Rescue activity, and deviating commercial traffic. The collision accident is by itself an important test case, but may also serve as a proxy evaluation method for the detection of rendezvous situations that are of interest in finding smuggling events. Similarly, Search and Rescue activity is always of interest, especially as similar behavioral patterns may be seen in more nefarious activities like smuggling and illegal mapping of the seabed or seafloor infrastructure [26]. For the sake of reproducibility and to foster research on methods suitable to a real-world scale, we provide public access to the datasets. Further details are given in Section 4.1.

We summarize our contributions as follows:

- We present a novel method for detecting abnormal maritime trajectories based on two-step clustering, which also provides a contextual decision support tool to help a VTS operator make the final decision.
- We design positional and kinematic similarity measures that focus on different dimensions of maritime trajectories.
- We provide evidence that a multi-step clustering approach can disentangle positional and kinematic information, resulting in a better description of behavioral patterns in a large ROI.
- We provide public access to datasets of preprocessed maritime trajectories in regions of Danish waters, including annotations during a Search and Rescue event.

The paper is organized as follows. In Section 2, we provide an overview of the related work within trajectory clustering. In Section 3, we present our proposed two-step clustering. In Section 4, we give a detailed description of the maritime traffic datasets that we publish and that we also use to show the ability of our proposed method to disentangle maritime traffic patterns, as well as to detect real-life abnormal trajectories from a ship collision. Finally, we present our conclusion in Section 5.

2. Related Work

Clustering of maritime trajectories has been widely employed to extract traffic patterns and find abnormal trajectories. The type of behavior discovered by clustering spatio-temporal trajectories depends heavily on the chosen similarity measure. Laxhammar et al. [27] suggested using the maximum synchronous Euclidean distance between each pair of coinciding points along two trajectories of the same length. The requirements of equal trajectory length and synchronous comparison can be relaxed by using either the Hausdorff distance [12,19,28], Dynamic Time Warping (DTW) [11,29], or the Longest Common Subsequence (LCSS) [24] to measure trajectory similarity. The Hausdorff distance is independent of the time component, which can make trajectories following the same route in opposite directions indistinguishable. In addition, from the clusters reported in [19], we note that the Hausdorff distance may assign a large similarity to significantly different trajectories. Additionally, both of these methods have quadratic time complexity, and several works, therefore, suggest a compression using the Douglas–Peucker (DP) algorithm [30]. Klaas et al. [29] proposed a two-stage DP algorithm: first, reducing the trajectory based on the speed time series and, secondly, based on the position. This two-stage approach was found to better retain periods of acceleration, such as stops.

Several different clustering algorithms have been applied to clustering of trajectories. Methods such as K-means [29] and K-medoids [28] have been utilized in collaboration with different similarity measures. However, density-based clustering techniques have long been the predominant approach to data mining within maritime trajectory analysis. Pallotta et al. [31] proposed the widely used TREAD method, to cluster trajectories into traffic routes, which can then be used for anomaly detection and trajectory prediction. TREAD is a point-based method that extracts the coordinates of new entries, exits, and stops within the ROI. These points are clustered using DBSCAN [32], to form waypoints in which ships enter, exit, or stop within the ROI. A route between waypoints is then formed whenever a certain number of transitions between them have been observed. Several works [11,12,19,24] have combined the idea of a similarity measure and density-based clustering. First, trajectories are simplified using the DP algorithm. The similarities are then computed using the Hausdorff distance, DTW, or LCSS, before being clustered by DBSCAN. Wang et al. [19] considered a hierarchical search over the hyperparameters of DBSCAN, which allowed for groups with different densities, and helped to find clusters in sparsely populated geographical regions.

Recently, trajectory similarities based on deep learning have been suggested. Murray et al. [13] clustered the latent encodings of a Recurrent Variational Autoencoder (RVAE) trained for trajectory reconstruction using hierarchical DBSCAN and found clusters corresponding to the major shipping lanes. The clusters were then used to train neural networks to predict the future position. Luo et al. [33] proposed a graph-based trajectory contrastive learning framework. A Graph Neural Network encoder was trained, using contrastive learning with five different trajectory augmentations. The similarity of two trajectories could then be computed by their distances in the latent space. The method was evaluated by downsampling random trajectories from the training set as test trajectories. The proposed similarity measure was found to perform better than traditional trajectory distance measures.

The abovementioned approaches only considered the positional input, yielding clusters that mostly corresponded to the primary shipping lanes. Zhen et al. [28] introduced the difference of the average course in their similarity measure, and Liu et al. [10] extended the DBSCAN clustering model, to consider not only the geographical distance of the coordinates, but also the difference in speed and course. This allowed them to distinguish between shipping lanes in opposite directions and to find speed differences within the main shipping lanes. However, the work was limited to small geographical areas and a limited number of ship types. Li et al. [23] suggested a similar extension to the DBCSAN algorithm but split the speed and course differences into two different clustering models.

Knowledge about maritime traffic patterns is useful for detecting abnormal activity. Widyantara et al. [24] directly reported outliers from the DBSCAN clustering, but several clustering methods have been extended with a detection step. Often, this step includes knowledge about the kinematic behavior. Pallotta et al. [14] proposed a two-stage anomaly detection scheme, using the routes extracted by TREAD. First, using only the positional inputs, a trajectory was associated with a route. Afterwards, kinematic outliers were found, by comparing the speed and course to the average behavior of the route. Liu et al. [10] proposed to divide the clusters into smaller geographical regions and compute the average kinematic values for each split. These values would then be used to detect abnormalities [20]. Zhao et al. and Li et al. [23,34] used normal trajectories determined from DBSCAN clusters to train deep neural networks for trajectory prediction. Abnormalities were then detected, based on the prediction error.

Recently, an abnormality detection model based purely on deep learning has been suggested. Hu et al. [21] suggested an ensemble of a Variational LSTM AutoEncoder and a Graph Variational AutoEncoder. Each ensemble member was trained to reconstruct the input trajectory, and the reconstruction errors were then combined, to make a final binary prediction of the abnormality. Liu et al. [22] self-annotated training data based on extreme position, speed, or course values and trained a deep neural network to classify abnormalities. Nguyen et al. [15] suggested a Variational Recurrent Neural Network (VRNN) for the detection of abnormalities based on trajectory reconstructions. In this work, Nguyen et al. also suggested an A-Contrario detection methodology, which was supposed to account for regional differences in reconstruction accuracy. Although the results reported using VRNN looked promising, our feedback from VTS operators mentioned the lack of explainability as a key limitation for operational use. The lack of explainability of decision support tools has been identified as a key issue for the automated detection of abnormal maritime behavior in surveys by Riveiro et al. and Stach et al. [16,17].

Table 1 summarizes the normalcy models and the limitations of these normalcy models utilized by the previous research discussed. We present a novel abnormality detection algorithm based on a positional clustering followed by a kinematic clustering of historical maritime trajectories. We rely on an efficient positional similarity measure, which allows us to process a large, complex dataset of maritime trajectories representative of real-life traffic in a reasonable time. The abnormality detection is made with respect to the kinematic of the vessel, for which we design an alternative similarity measure based on DTW. The latter is able to distinguish behaviors within the same positional clustering, giving VTS operators a clear summary of normal behavior when assessing the suggestion of our abnormality detector.

Table 1. Summary of the normalcy models and the limitations of these normalcy models utilized by the previous research.

Normalcy Model	Limitation of Normalcy Model	Works
Clustering of individual updates	Applied on restricted datasets Lack description of kinematic behavior	[20] [9,14]
Clustering of trajectory similarities	Applied on restricted datasets Lack description of kinematic behavior	[12,19,28] [12,19,24,27,29,33]
Deep learning methods	Interpretability	[15,21–23,33,34]

3. Methodology

In this section, we first discuss similarity measures for trajectories and then introduce our abnormality detection algorithm.

3.1. Notations

An AIS trajectory A of length $T_A \in \mathbb{N}$ is a four-dimensional time series $A = \left(a_1, \ldots, a_{T_A}\right)$, where $a_t = (\text{lon}_t, \text{lat}_t, s_t, c_t)$, with each dimension representing, respectively, the longitude, latitude, speed, and course of the vessel as recorded in its AIS message at time t. For legibility, the timestamp t is used indiscernibly as an index of a variable, such that $A(t) = a_t$, or $s(t) = s_t$.

Throughout the section, we consider two AIS trajectories A and B of time duration T_A and T_B, and two timestamps $t \in \{0, \ldots, T_A\}$ and $\tau \in \{0, \ldots, T_B\}$. Also, we assume that the trajectories are regularly sampled without missing data. The function d is a generic distance on $\mathbb{R}$ or $\mathbb{R}^2$, depending on the context.

3.2. Similarity Measures

In the following, we discuss three commonly used trajectory similarity measures: the Hausdorff distance, the average Haversine distance, and Dynamic Time Warping (DTW). We define these similarity measures without specifying which dimensions of the time series are used (positional or kinematics), as this depends on the use case. We also propose a variant of DTW tailored to kinematic data.

3.2.1. Hausdorff

The Hausdorff distance [35] between two trajectories corresponds to the maximum smallest distance realized by any pair of points in each one of the trajectories:

$$\text{Hausdorff}(A, B; d) = \max_{t \in [0, T_A - 1]} \min_{\tau \in [0, T_B - 1]} d\Big(A(t), B(\tau)\Big). \tag{1}$$

The computations require a comparison of all possible pairs of points, resulting in quadratic time complexity. Furthermore, the Hausdorff distance ignores the time component. This means that ships along parallel shipping lanes sailing in opposite directions are not distinguishable. Such a situation is studied in [19].

3.2.2. Average Haversine

The quadratic time complexity and the issues mentioned above make the Hausdorff distance unsuitable for measuring the similarity of many long and complex sequences of geographical coordinates. On the other hand, the Average Haversine distance (AH) proposed in [36] is able to compare the positional evolution of the AIS trajectories in linear time, with respect to the length of the trajectories. It is defined as a continuous distance measure, but it can be approximated using the trapezoidal rule and assuming a regular sampling:

$$\text{AH}(A, B; d_H) = \sum_{t=0}^{T-1} \frac{d_H\Big(A(t), B(t)\Big) + d_H\Big(A(t+1), B(t+1)\Big)}{2T}, \tag{2}$$

where $T = \min(T_A, T_B)$ and d_H is the Haversine distance [36]. This similarity measure computes the geographical distance between the trajectory points one by one in a linear fashion until the length of the shortest trajectory is reached. This means that the measure places an increased weight on the beginning of the trajectories. Thus, we expect the measure to be able to separate trajectories based on their starting location. This is ideal in a real-time operational setting when observing new trajectories, as even short trajectories can very quickly be classified into a subset of historical trajectories with similar behavior.

3.2.3. Dynamic Time Warping

Dynamic Time Warping minimizes the pair-wise distance by re-indexing (alignment of) the data points in the trajectories, according to certain rules. It can be defined as follows:

$$\text{DTW}(A, B; d) = \min_{\pi \in \Pi(T_A, T_B)} \left(\sum_{(i,j) \in \pi} d\Big(A(t_i), B(t_j)\Big) \right), \tag{3}$$

where $\Pi(T_A, T_B)$ is the set of all possible alignments that are sequenced pairs of indices $(i, j) \in [0, T_A - 1] \times [0, T_B - 1]$ satisfying three constraints: (1) the beginning and end of the time series must be matched; (2) the sequence must be monotonically increasing in i and j;

(3) all indices i and j must appear at least once. These ensure that the sequences start and end together and that each point on either sequence is mapped onto at least one point of the other sequence without these mappings crossing in time.

As DTW processes pairs of indices, it also has a quadratic time complexity. The DTW alignment may overestimate the distance of trajectories with similar behavior if this behavior is spread over a large area. For example, consider two trajectories with the same starting point and sailing along the same direction as illustrated in Figure 1. At one point, trajectory A makes a 30 degree turn and continues in this direction, moving away from trajectory B. Later, trajectory B makes a similar 30 degree turn and continues parallel to trajectory A. Both vessels return to their initial course some time later, and the trajectories terminate at the same point. As these two trajectories have the same origin and terminal location and have similar behaviors throughout the journey, we would expect the distance between them to be very small. However, their distance, calculated by DTW on the sequence of geographical coordinates, may be significant. The re-indexing procedure of DTW aligns the course changes between the trajectories. However, due to the spatial nature of the geographical coordinates, DTW calculates the geographical distance between the location where the trajectories changed course. If we instead were to use the time series of the measured angles towards true north, the DTW distance would calculate the difference of the course values. As these values are the same before and after the changes, the DTW distance between the two trajectories would be zero. Using the time series data, we remove the spatial dependence, and DTW can properly calculate the similarity of the course after aligning the changes. Therefore, DTW is a good candidate as the building block of a similarity measure for course and speed time series.

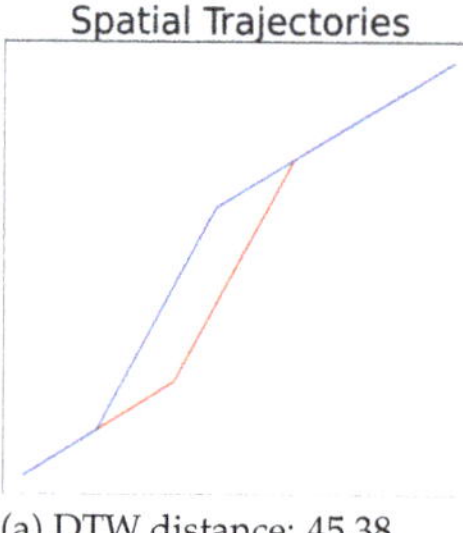

(a) DTW distance: 45.38

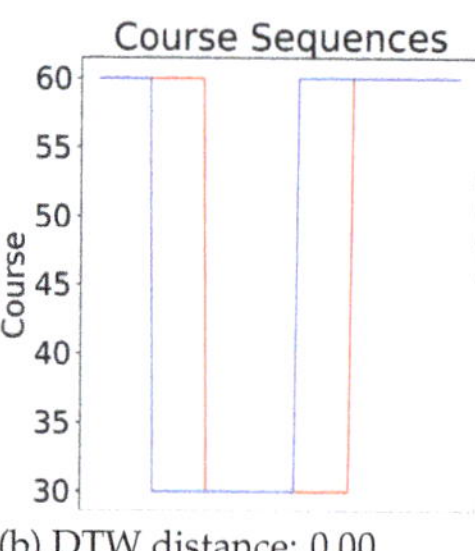

(b) DTW distance: 0.00

Figure 1. (**a**) Two trajectories with the same origin and terminal point and similar behaviors throughout the journey may obtain a large distance calculated by DTW; (**b**) however, the course sequences of the two trajectories may be warped perfectly onto each other and have zero distance between them calculated by DTW.

3.2.4. Kinematic DTW

Following the previous discussion, we propose a variation of DTW for kinematic data, referred to as D_{kin}. The measure is defined as the sum of the DTW of the time series of the speed and of the DTW of the course time series:

$$D_{\mathrm{kin}}(A, B) = \mathrm{DTW}(s_A, s_B; d_{\mathrm{speed}}) + \mathrm{DTW}(c_A, c_B; d_{\mathrm{course}}), \tag{4}$$

where s_A, s_B and c_A, c_B are, respectively, the speed and course sequences of trajectories A and B. The differences in speed and course at each timestamp are measured, respectively, by d_{speed} and d_{course}, which correspond to the standardized absolute difference of the speed and the normalized angular difference in radians, respectively:

$$d_{\mathrm{speed}}(x, y) = \frac{|x - y|}{\Sigma}, \tag{5}$$

$$d_{course}(x,y) = \frac{1}{\pi} \cdot \begin{cases} |x-y| & \text{if } |x-y| \leq \pi, \\ \pi - (|x-y| \bmod \pi) & \text{otherwise,} \end{cases} \tag{6}$$

where Σ is the standard deviation of the speed computed empirically from the speed time series s_A and s_B.

3.3. Two-Step Clustering for Abnormality Detection

Our intention was to design an abnormality detection algorithm to assist VTS operators, which may serve as a contextual decision support tool and let them make the final decision based on the contextual information provided by the algorithm itself. The reason for a trajectory to be flagged as abnormal is that it is either similar to other abnormal trajectories or that it diverges from the most similar non-abnormal trajectories. It is important to state that the notion of the behavior of a vessel is not limited to a sequence of locations, but also includes its speed and course. The similarity measure involved, to compare trajectories, thus needs to take into account both the spatial and the temporal information. Note that the assignment of kinematic clusters gives a context to the prediction of the LOF, as it shows the most similar trajectories. Yet it is not an explanation, as the kinematic clustering is not used by the detector.

We modeled this line of thought as a two-step algorithm:

1. Assign an input trajectory to a cluster, based on its positional dimensions (latitude, longitude, and time).
2. Decide on abnormality, based on the kinematic dimension (speed and course), and provide a context to the decision with the most similar trajectories.

Figure 2 shows the flow of our proposed two-step method.

(1) Positional Clustering: The first step required clustering of a historical database and a classifier, using a fast-to-compute similarity measure, to ensure the reactivity of the system. Hence, the Hausdorff and DTW were excluded. Also, both measures either distort or simply disregard the time component, which is at odds with the rationale exposed above. We chose to rely, both for clustering and classifying, on the average Haversine distance (Equation (2)), which has a linear complexity and compares synchronous positions. The clustering was a hierarchical clustering with average linkage which, once computed, allowed us to easily change the number of clusters and, thus, isolate outliers. During inference, cluster assignment was decided by a K-Nearest Neighbors (KNN) classifier with $k = 3$ trained on the clustering.

(2a) Abnormality Detection: As we did not have access to a large set of labels, we made the assumption that none of the historical trajectories were abnormal. Therefore, none of the positional clusters were considered as abnormal, and abnormality was defined as a divergence from the training set. More precisely, we defined it as a divergence with respect to the trajectories within the assigned positional cluster. As outlier detection, we employed the Local Outlier Factor (LOF) [25] and D_{kin} (Equation (4)) as a similarity measure. Before calculating D_{kin}, we compressed the trajectories, using the two-stage DP compression [29]. As the trajectory had been assigned to a cluster based on its positional information, it could not be an outlier purely based on these data. The divergence needed to be measured on the basis of another aspect of the behavior, namely the variations of the kinematics (speed and course), which could be understood as the derivative of the positional data.

The LOF compares the density of the local neighborhood of a point to that of its KNN. If the density of a point is significantly lower than its neighbors, the point is flagged as an outlier. Following the discussion in [27], we set $k = 5$ nearest neighbors for the LOF algorithm in our experiments. In practice, we did not see large changes in the number of outliers detected when varying that number. However, we recommend a low value to capture information only from the local neighborhood. The LOF also has a hyperparameter, called contamination, related to the expected percentage of outliers. As we expect only a small number of outliers, we recommend again to use small values for the hyperparameter. See Section 4.6.2 for an ablation study.

(2b) Kinematic Clustering: The explanation or context of the prediction of the LOF consists of the most similar historical trajectories. Again, with the aim of speeding up calculations, these trajectories will be extracted from a precomputed hierarchical clustering with average linkage of the trajectories of the assigned positional cluster, using D_{kin} as a similarity measure. The cluster assignment is decided by a KNN classifier with $k = 3$ using D_{kin} and trained on that kinematic clustering.

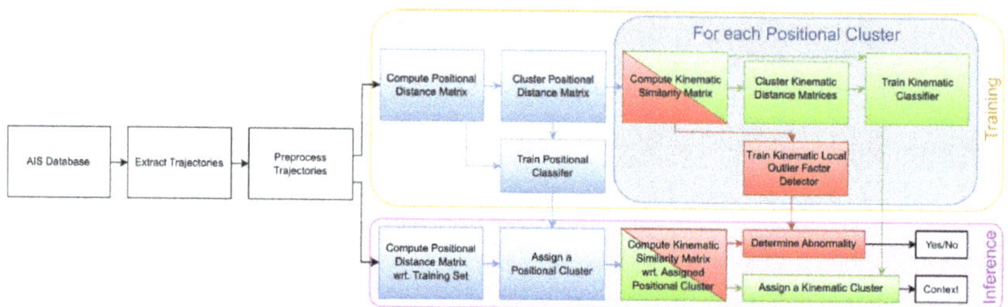

Figure 2. Flowmap of the training and inference of our proposed two-step abnormality detector. In the training phase, the positional distance matrix based on the Haversine distance is computed for the preprocessed training dataset. The matrix is then used to train a hierarchical clustering base and a KNN classifier for that clustering. For each positional cluster, the kinematic similarity matrix based on D_{kin} is computed and then used to train an LOF (red), a hierarchical clustering (green), and a KNN classifier for that clustering. In the inference phase, after preprocessing, the distances between the input trajectory and the training data are used by the positional classifier (blue) to assign a position cluster. In the second step, the kinematic similarities to the trajectories within the same cluster are computed and then used by, on the one hand, the LOF to determine abnormality (red) and, on the other hand, by the kinematic classifier to assign a kinematic cluster (green). The output is thus double: the answer of the LOF and a context, in the form of the trajectories of the kinematic cluster.

4. Experimental Results

In this section, we evaluate the choice of similarity measure and of algorithm for each positional and kinematic clustering. We also compare the final clustering to that of a single-step algorithm. Finally, we discuss the abnormality detection capabilities of our approach and compare it to State-of-the-Art neural-network-based baselines.

4.1. Datasets

For this work, we built two datasets of AIS data from Danish waters that cover large ROIs and contain various types of vessels with different priorities and expected behavior patterns. Both datasets are available for public use, to facilitate reproducibility and to give researchers the ability to evaluate their proposed models on a complex dataset representative of a real-world setting [37] (datasets available at https://data.dtu.dk/collections/ AIS_Trajectories_from_Danish_Waters_for_Abnormal_Behavior_Detection/6287841. Accessed on 20 October 2023). The complete AIS data from all Danish waters are available publicly [38]; however, minor differences between the two sources may occur.

The first dataset covers a rectangular ROI covering the island of Sjælland, bounded by (54.4° N, 10.5° E) to (56.4° N, 13.5° E). The data were collected during November 2021 and contain 18.738 of trajectories from 11 different types of ships, ranging from commercial cargo and tanker ships to private sailing and fishing boats. The second dataset covers a rectangular ROI around Bornholm Island bounded by (54.5° N, 13° E) to (56° N, 16° E). The data were collected during December 2021 and contain 12.591 of trajectories from 8 different types of ships. The speed was limited to 20 m/s, and updates with higher speeds were discarded. For both datasets, if the time interval between two successive AIS messages exceeded 15 min, the trajectory was split into two contiguous trajectories. Trajectories shorter than 10 min were discarded and trajectories exceeding 12 h were divided into smaller trajectories, each between 10 min and 12 h. All trajectories were resampled every

120 s, using linear interpolation. Table 2 shows an example of the information associated with each extracted trajectory.

Table 2. Example information associated with a trajectory from a passenger ship.

MMSI	Timestamp	Latitude	Longitude	Speed	Course
211149000	2021-11-29 22:47:39	54.40	12.16	7.13	25.83
211149000	2021-11-29 22:49:39	54.41	12.17	7.12	27.30
211149000	2021-11-29 22:51:39	54.42	12.18	7.15	27.05
211149000	2021-11-29 22:53:39	54.42	12.18	7.16	27.40
$\vdots$	$\vdots$	$\vdots$	$\vdots$	$\vdots$	$\vdots$
211149000	2021-11-30 09:45:39	54.41	11.91	8.16	122.16
211149000	2021-11-30 09:47:39	54.40	11.90	8.09	233.83

We used the Sjælland data to evaluate the proposed two-step clustering algorithm. The entire dataset was used for training. We evaluated the proposed automated anomaly detection algorithm on the Bornholm dataset. Data from 13 December 2021 was withheld as a test set, the rest serving as a training set. On that day, a collision accident between two ships occurred, causing several abnormal trajectories. Trajectories from this day were manually labeled, resulting in 25 abnormalities out of 521 trajectories. In addition to the colliding vessels, the abnormal trajectories corresponded to commercial traffic, which had to deviate from the planned course, for Search-and-Rescue and law-enforcement vessels responding to the accident, and to any other vessel taking part in the search for the two missing sailors.

4.2. Experimental Setting

Similarity measure baselines include the Hausdorff distance and DTW, both based on Haversine distance, as suggested in [11,12]. In terms of the clustering algorithm, we compared hierarchical clustering and DBSCAN. The linkage distance threshold for hierarchical clustering was decided using the Kneedles algorithm [39], to select the number of clusters. The hyperparameters of the DBSCAN were tuned by creating candidate lists of minimum distances and samples, as suggested in [12]. The optimal value of these candidates was then determined, using the Kneedles algorithm [39]. We have provided quantitative, qualitative, and runtime analyses of the clustering. We quantitatively evaluated the clusterings, using the Silhouette score [40]. The qualitative evaluation was performed by manually gauging the similarity of the behaviors of the extracted clusters while also accounting for the purpose of the two steps. In the positional clustering, the primary purpose was to be a fast clustering of all the trajectories into groups, in which similar behaviors might be discovered. As such, we were looking for trajectories that originated in the same area and shared some common positional evolution. In the kinematic clustering, we were interested in clusters describing uniquely different maritime behavior, i.e., we wished to identify different behavioral patterns across clusters. The baselines for abnormality detection included the State-of-the-Art VRNN [15] and RVAE [13] deep learning architectures. We measured the performance of the detectors, using the area under the receiver operating characteristic (AUC). The code was implemented in Python 3.8, using standard libraries, and it ran on an Intel Xeon Processor 2660v3. Similarity calculations were parallelized across eight cores.

4.3. Positional Clustering

The positional clustering is the basis for the outlier detector and the kinetic clustering. It needs, thus, to separate well-different behaviors, e.g., by distinguishing vessels traveling along shipping lanes in different directions. We considered different combinations of distance measures (Hausdorff, DTW based on Haversine distance, and average Haversine distance computed using Equation (2)) and clustering algorithms (hierarchical, DBSCAN). Recall that our model combines the average Haversine distance with hierarchical clustering.

4.3.1. Quantitative Analysis

In Table 3, we report various statistics on the clustering of each combination of distance measure and algorithm, including hyperparameters selected using the Kneedles algorithm, the number of clusters found, the median number of members in each cluster, and, when DBSCAN was used, the percentage of outliers. The quality of the clusterings was measured in terms of silhouette score (the larger, the better). Our combination of hierarchical clustering with the average Haversine distance achieved the best score.

Table 3. Positional clustering performance, in terms of the silhouette score for various combinations of distance and clustering algorithms, along with the hyperparameters and characteristics of the clusterings. Bold denote the highest recorded silhouette score. Our model corresponds to the last line.

Distance Measure	Clustering Algorithm	Eps-Threshold	MinSamples	# Clusters	Median of # of Members	% Outliers	Silhouette Score
Hausdorff	Hierarchical	9000	-	1515	2	-	0.535
Hausdorff	DBSCAN	12,504	242	15	569	45.7	0.127
Hausdorff	DBSCAN	12,504	25	53	110	10.7	0.376
Hausdorff	DBSCAN	27,000	242	7	1446	12.6	0.265
DTW	Hierarchical	140,000	-	2862	1	-	0.349
DTW	DBSCAN	60,941	91	7	190	68.8	−0.454
Avg. Haversine	DBSCAN	1.07	261	14	485.5	57.8	−0.033
Avg. Haversine	Hierarchical	10	-	52	232	-	**0.651**

The silhouette scores show that DBSCAN generally performed worse than hierarchical clustering. One explanation could be the large number of trajectories flagged as outliers by DBSCAN. For all three similarity measures, DBSCAN considered at least 45% of the data as outliers. This is too many false positives for an automated system to be useful. Despite the better silhouette scores, hierarchical clustering with the Hausdorff distance and DTW suffered from a similar phenomenon. In fact, both combinations produced the largest number of clusters. Most of these clusters contained very few trajectories and, thus, served a similar purpose as the outliers in DBSCAN.

On the other hand, our proposed combination found a reasonable number of 52 clusters with a median number of trajectories per cluster of 232. These were reasonable numbers that allowed for further analysis in the second step. Note that these numbers of trajectories per cluster are comparable to the size of the full datasets used in most other works, such as [11,12].

4.3.2. Qualitative Analysis

Clustering using DTW or Hausdorff distances resulted in clusters corresponding to well-defined shipping lanes, as seen in Figure 3a,b. However, they failed to cluster two types of trajectories: small groups of trajectories on less populated maritime routes and trajectories that shared a segment along a shipping lane but did not follow it; see Figure 3d. These trajectories were marked as outliers by DBSCAN or as single-observation clusters by hierarchical clustering. Fine-tuning the hyperparameters might reduce the number of outliers by slowly admitting trajectories with similar positions into the clusters but with the risk of joining clusters of different shipping lanes, as shown in Figure 3c. This indicates that real, unfiltered trajectory recordings from a diversely populated ROI have too much randomness for these combinations of measures and algorithms to find well-separated clusters using only the latitude, longitude, and the timestamps without flagging the majority of the data as outliers.

In Figure 3e,f, we plotted the four most populated clusters in the ROI around Sjælland, using our positional clustering algorithm. Each cluster shown contained more than 1000 trajectories. We see that in each cluster, the trajectories began in the same geographical area unique to each cluster. This was expected, due to the increased attention by the average Haversine distance to the initial part of the trajectories. The discovered clusters contained trajectories from all different shipping lanes that originated in a given area. However, this was acceptable, as we expected the second-step clustering to separate the shipping lanes based on their common kinematic behaviors and the outlier detector to catch those that did not travel steadily along the lanes.

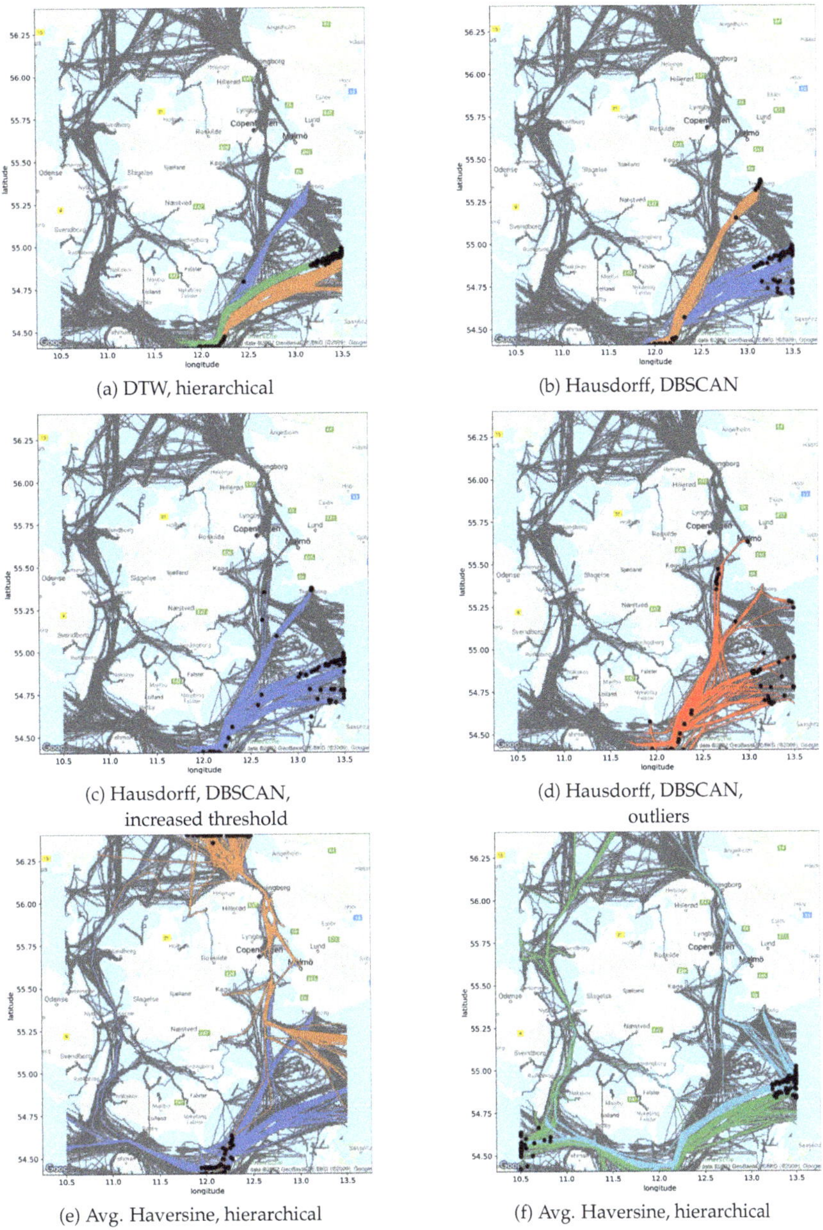

(a) DTW, hierarchical

(b) Hausdorff, DBSCAN

(c) Hausdorff, DBSCAN,
increased threshold

(d) Hausdorff, DBSCAN,
outliers

(e) Avg. Haversine, hierarchical

(f) Avg. Haversine, hierarchical

Figure 3. Examples of clusters and outliers discovered in the first positional step, using different similarity measures and clustering methods. Clustered trajectories are shown in color, historical traffic is shown in gray, and trajectory origins are denoted by a black circle.

Note that the blue cluster of Figure 3e looks similar to that of Figure 3c. However, the difference is major. In Figure 3e, all the trajectories start in the same area, while in Figure 3c there are trajectories starting where others end. This is due to the fact that the Hausdorff distance compares asynchronous pairs of positions and is thus invariant to the direction of travel. This is a potential problem for the real-time classification of incomplete trajectories in the discovered clusters.

4.3.3. Runtime Analysis

In Table 4, we report average runtimes for computing the average Haversine distance, based on Equation (2), DTW, and the Hausdorff distance. The distances were implemented in Python 3.8.11 programming language, using Numpy 1.23.2. The Hausdorff and DTW distances were calculated using the trajectory_distance library (https://github.com/bguillouet/traj-dist accessed on 20 October 2023), implemented in Cython 0.29.24. With its linear time complexity, the average Haversine distance is undoubtedly the fastest to compute: it is 10 times faster than DTW and 100 times faster than the Hausdorff distance.

Table 4. Average time in seconds to compute a pair of trajectory similarities during computation of the distance matrix.

Avg. Haversine, Equation (2)	DTW	Hausdorff	Kinematic, Equation (4)
16.38 µs	107.2 µs	1084 µs	12,788 µs

4.3.4. Discussion of the Positional Clustering

The traditional distance measures DTW and Hausdorff result in many outliers when applied to a complex, unfiltered dataset that resembles trajectories expected in real-life applications. By contrast, the average Haversine distance results in clusters that contain all the different routes that originate in a location that varies between clusters. Additionally, the average Haversine distance is much faster to compute, which allows for real-time assignment of unseen trajectories into precomputed clusters. However, these clusters do contain trajectories from many different maritime routes. Therefore, simply reducing the threshold in the hierarchical clustering does not yield a more detailed clustering describing their global positional or local kinematic behavior. To refine the cluster, a different distance measure must thus be used.

4.4. Kinematic Clustering

The kinematic clustering of the second step serves to provide a context to the outlier detectors prediction. It is expected to refine the gross positional clustering. Therefore, in this section, we study the refinement of the blue positional cluster shown in Figure 3e. The trajectories of this cluster originated at the southeastern edge of the ROI and split into four major shipping lanes—one going west towards the Kieler Channel, allowing passage to the Atlantic, one going north towards the North Sea, one going east towards the Baltic Sea, and one going northeast, terminating in the Swedish port of Trelleborg. In addition to these shipping lanes, the Danish port of Gedser (southern tip of the Lolland island) is a hub for pilot boats, which often have to rendezvous with larger ships passing through the Fehmarn Belt between Denmark and Germany. These pilot boats form a triangle fanning outwards east from the port of Gedser, seen in the bottom of Figure 3e.

We compared the clusters obtained using both kinematic and positional similarity measures. Regarding the positional clustering, based on the results of Section 4.3, we considered only hierarchical clustering combined with the average Haversine distance and the Hausdorff distance. The former produced better groupings, and the latter showed potential to further split clusters. As for the kinematic clustering, we tested our proposed kinematic similarity measure D_{kin}, Equation (4) with hierarchical clustering, and DBSCAN. Finally, to evaluate the benefit of basing D_{kin} on DTW, we also considered the average of the synchronous speed and course distances of Equations (5) and (6):

$$AK(A, B) = \sum_{t=0}^{T-1} \frac{d_{\text{speed}}\left(s_A(t), s_B(t)\right) + d_{\text{course}}\left(c_A(t), c_B(t)\right)}{2T}, \tag{7}$$

where $T = \min(T_A, T_B)$.

We report in Table 5 the hyperparameters and statistics about each clustering. The two positional-based clusterings found fewer clusters and obtained better silhouette scores than our proposed kinematic distance measure, Equation (4). However, if we look at some of the clusters shown in Figure 4a,b, we note that these methods did not produce a more detailed clustering, in terms of the local kinematic behavior of the trajectories. This was expected, as without a refinement in local kinematic behavior a two-step clustering was not relevant, as we would have expected to find the same subclusters if we had accepted more groups in the first step when clustering the whole data.

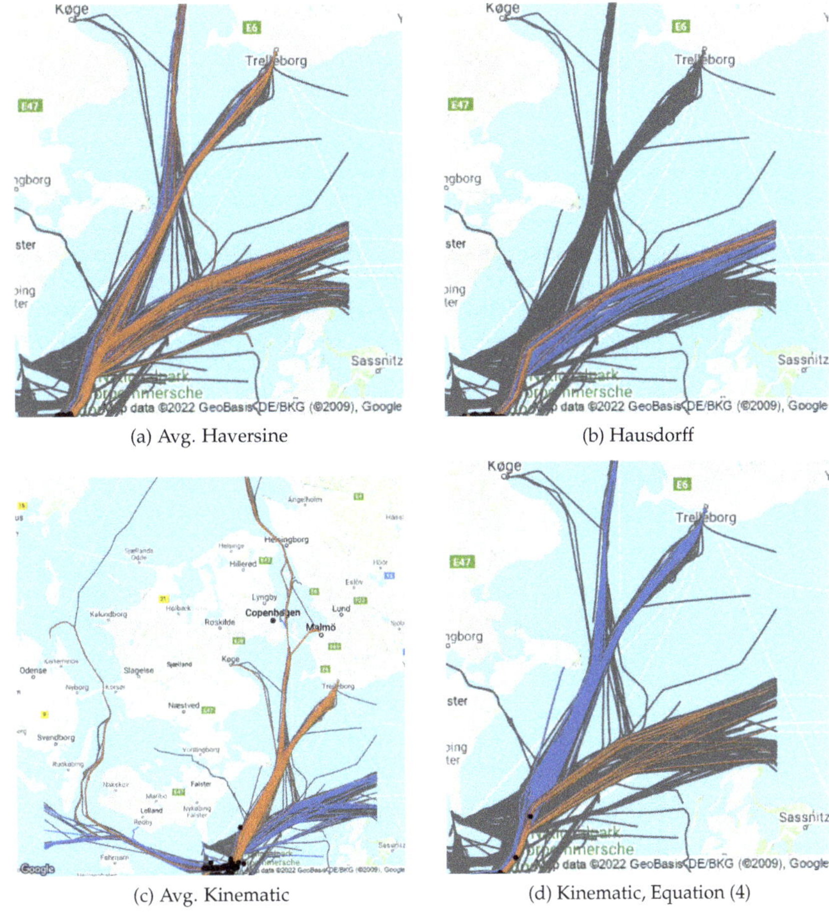

(a) Avg. Haversine

(b) Hausdorff

(c) Avg. Kinematic

(d) Kinematic, Equation (4)

Figure 4. Second-step clusters obtained using hierarchical clustering and different similarity measures. Clustered trajectories are shown in color, trajectories in gray denote trajectories in the same positional cluster, and trajectory origins are denoted by a black circle.

Table 5. Hyperparameter values and clustering results of DBSCAN and hierarchical clustering using distances computed by Hausdorff, the average Haversine distance, Equation (2), or our proposed kinematic distance measure, Equation (4), on trajectories assigned to positional cluster 0 in the first step. Bold denote the highest recorded silhouette score.

Distance Measure	Clustering Method	Eps-Threshold	MinSamples	# Clusters	# Outliers/ Singletons	Silhouette Score
Avg. Haversine, Equation (2)	Hierarchical	0.9	-	49	12	**0.558**
Hausdorff	Hierarchical	6250	-	134	62	0.533
Avg. Kinematic	Hierarchical	1.8	-	34	0	0.126
Kinematic	DBSCAN	12.0	46	2	821	0.036
Kinematic	DBSCAN	12.0	2	21	477	−0.221
Kinematic	Hierarchical	22.5	-	221	147	0.217

4.4.1. Positional Similarity Measures

Clustering based on the average Haversine distance (Figure 4a) was unable to split the shipping lanes. We believe the focus on the initial part was the cause. The Hausdorff distance (Figure 4b) allowed the separation of the maritime routes through the ROI. We also noticed some maritime routes divided into two or more clusters, as seen in Figure 4b. Thus, we gained a more detailed clustering, in terms of describing their global positional behavior. In Figure 5, we show the speed and course of the trajectories assigned to the two clusters of Figure 4b. We see that both clusters are not clearly distinguishable, in terms of speed or course. Note that fast trajectories significantly decreased their speeds while in the shipping lane (blue trajectories with an initial speed of about 12 m/s). We would expect such trajectories to be grouped separately. Looking at the course, we see the two clusters generally had similar course changes, although they happened at different times, due to the time invariance of the Hausdorff distance. Based on the results above, we conclude that using the Hausdorff distance in the second-step clustering resulted in a more detailed clustering regarding the global positional behavior but not the local kinematic behavior.

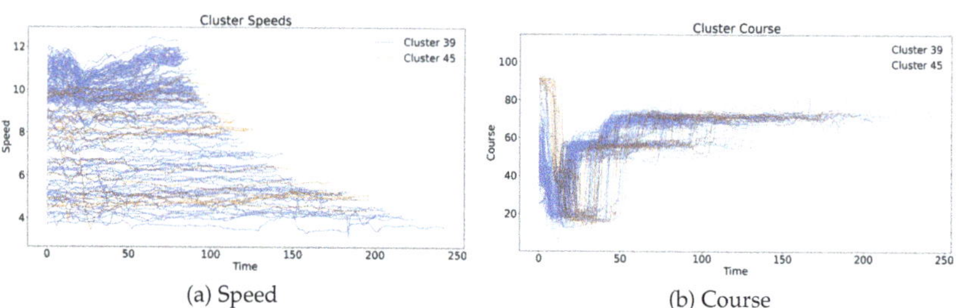

(a) Speed

(b) Course

Figure 5. Kinematic time series of the trajectories assigned to clusters obtained from the Hausdorff clustering shown in Figure 4b.

4.4.2. Kinematic Similarity Measures

We now study the clusterings obtained using kinematic-based similarity measures. Our proposed similarity measure combined with hierarchical clustering obtained a higher silhouette score than the average kinematic distance, yet the latter found fewer clusters.

We found that our choice of hierarchical clustering with average linkage was superior to the DBSCAN variants, as shown in Table 5. DBSCAN returned only two large clusters and most of the trajectories were assigned as outliers. Reducing the minimum number of samples required to define clusters increased the number of clusters to 21, but the vast majority of trajectories were assigned to the same cluster.

We see in Figure 4c that the average kinematic distance groups trajectories followed very different shipping lanes. Using Equation (4) as a similarity measure for hierarchical clustering resulted in 221 different clusters. Most of these clusters were singleton clusters and could themselves be considered outliers. Despite their naturally similar behavior, we found that pilot boats were not clustered together, but belonged to singleton clusters that were closer to one other than to other kinematic clusters. This shows that singleton clusters could occur for normal expected behavior in sparse regions of our feature space. Clusters with more than five trajectories assigned to them are shown in Figure 6. The clusters clearly split the major maritime routes. Looking at the speed trajectories within these clusters (Figure 7a–c), we note that the clusters clearly partitioned the speed behaviors. Although less obvious, this also applied to the trajectories of Figure 6d heading towards the port of Trelleborg. In Figure 7b,d, we can distinguish five unique types of behavior: slow-speed returns south (green), fast-speed return south (red), slow-speed stops in port (orange), fast-speed stops in port (yellow), and fast-speed stops in port with a spike in speed during slowdown (blue). Note that the high-frequency course changes at low speeds in Figure 7d were due to a vessel drifting in port. These random course changes may artificially decrease the similarity between trajectories of the same behavior, but it is expected that two-stage DP compression [29] filters out the majority of these stationary periods at drift. In general, D_{kin} yielded well-separated clusters with consistent kinematic behavior.

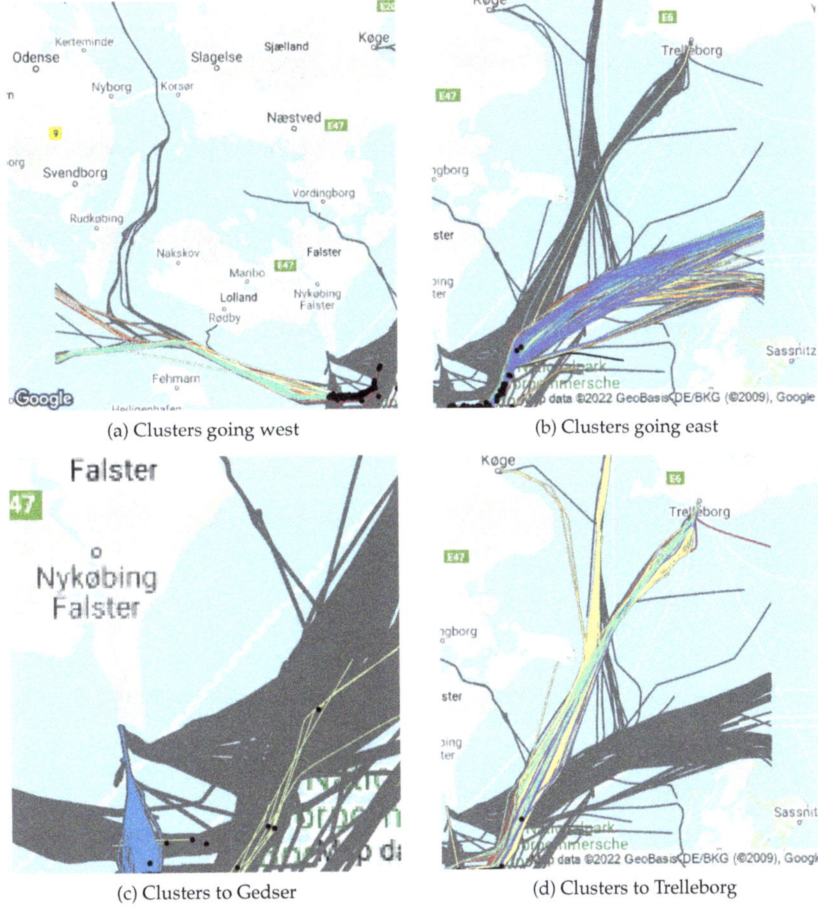

(a) Clusters going west   (b) Clusters going east

(c) Clusters to Gedser   (d) Clusters to Trelleborg

Figure 6. Second-step clusters with more than five assigned trajectories obtained using the kinematic distance matrix. Colors denote different clusters. Trajectories in gray denote trajectories in the same positional cluster.

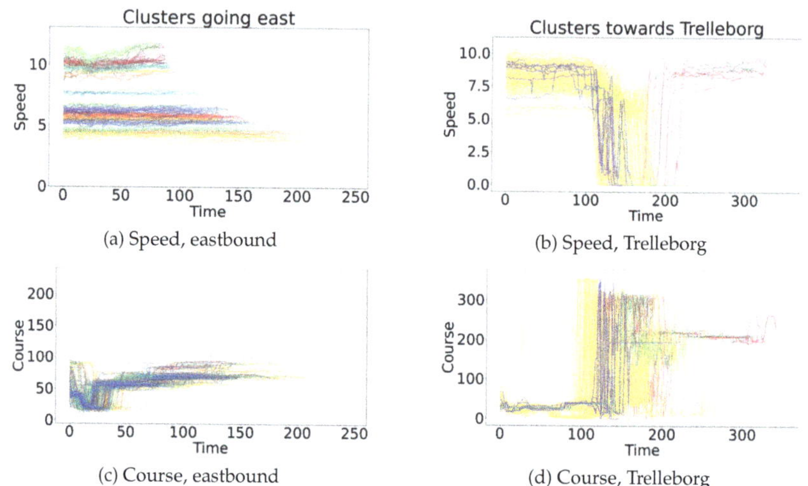

Figure 7. Kinematic time series of all clusters following two different maritime routes; see Figure 6b–d. Different colors represent different clusters.

4.4.3. Discussion on the Kinematic Clustering

Our combination of D_{kin} and hierarchical clustering was able to refine a positional clustering from the first step. This refinement disentangled the positional and kinematic features, which resulted in subgroups with well-defined and unique kinematic behaviors, thus obtaining a more detailed description of maritime behavioral patterns. There existed an inherent trade-off between clustering kinematic behaviors that we knew to be similar but that naturally had a higher distance and clustering different behaviors that naturally were very close to one another. As discussed above, all the pilot boats were clustered into singleton clusters. But had the clustering threshold been increased, all the pilot boats would have formed a single kinematic cluster. However, increasing the threshold would have had the added downside of merging the different speed clusters of Figure 7a.

4.5. Single-Step Clustering

We now compare the clustering obtained from our two-step algorithm with one computed with a single hierarchical clustering using the sum of the average Haversine distance, Equation (2), and of D_{kin} of Equation (4) as the similarity measure. Combining positional and kinematic information into a single similarity measure may hide some variations that may be captured when the dimensions are processed separately, as in our approach. Therefore, the two-step method was expected to return more clusters.

The Kneedles algorithm for the single-step clustering obtained a silhouette score of 0.095 and found 2880 clusters. The two-step clustering approach achieved a silhouette score of 0.162 and a total of 6963 clusters when applied to the entire dataset. In both methods, the majority of the discovered clusters were singleton clusters, which we previously highlighted for the two-step algorithm. The single-step and two-step clustering approaches found 2168 and 6221 singleton clusters, respectively.

Using single-step clustering, trajectories traveling along different shipping lanes could be clustered together (Figure 8a), while our two-step algorithm disentangled them (Figure 8b). A detailed analysis reveals that the single-step approach grouped together (orange cluster of Figure 8a) trajectories with distant initial points and following different shipping lanes because the speed of the trajectories was very similar. The differences in position were compensated by similar speed behaviors. On the other hand, the two-step algorithm split these two routes into multiple clusters, as it processed positional and kinematic information separately.

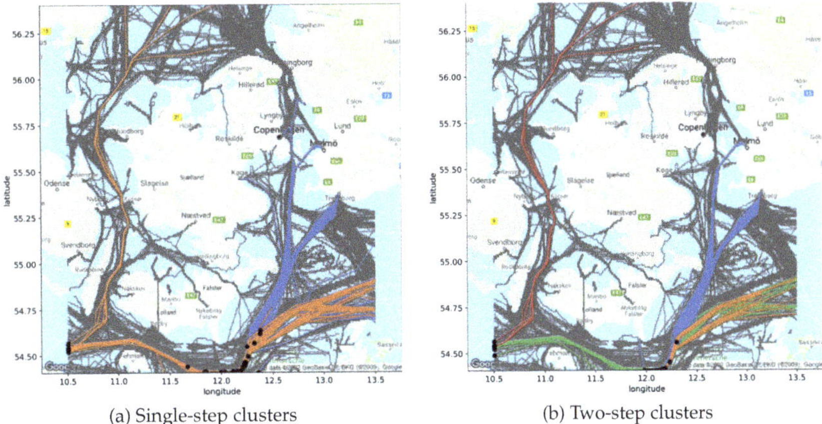

(a) Single-step clusters   (b) Two-step clusters

Figure 8. Trajectory clusters produced by the single-step clustering (**a**) that were split into multiple clusters, (**b**) using the the two-step algorithm. Clustered trajectories are shown in color, trajectories in gray denote trajectories in the same positional cluster, and trajectory origins are denoted by a black circle.

Our proposed two-step approach results in a better disentanglement of position and kinematic behavior. Even though our proposed similarity measure focuses on different aspects of the trajectories, treating them as a sum results in a situation where differences in the positional similarity are canceled by differences in the kinematic similarity. The better disentanglement of the two-step approach results in clusters with more well-defined and unique kinematic behaviors, thus obtaining a more detailed description of maritime behavioral patterns. Better disentanglement also means that the learned normalcy model can distinguish a higher number of possible kinematic behaviors at each location in the ROI. As such, the normalcy model is more useful for supplying contextual information to VTS operators. As discussed previously, the two-step approach has a trade-off between clustering behaviors that we know to be similar but that naturally have a larger distance and clustering different behaviors that naturally are very close to one another. Using a single-step approach seems to push this trade-off towards the latter option, automatically. Additionally, the two-step approach is computationally more efficient than the single-step approach. The computational requirements of the kinematic distance measure shown in Table 4 are large compared to the positional distance measure. Thus, computing the proposed kinematic distance measure on the entire dataset is not feasible for large datasets. Comparatively, the first positional clustering in the two-step approach functions as a filter that reduces the number of trajectory pairs for which to calculate the kinematic distance.

4.6. Outliers and Embedding Analysis

In this section, we use a TSNE representation of the kinematic similarity matrix of the blue positional cluster of Figure 3e, to show how the LOF and the kinematic clustering are related despite being computed independently.

4.6.1. LOF Contamination and Cluster Size

Although the LOF does not use kinematic clustering, it is computed on the same similarity matrix. In this experiment, we use a TSNE representation of the kinematic similarity matrix, to show how the kinematic clustering and the LOF handle outliers.

In Figure 9, we plotted a TSNE embedding of the kinematic distance matrix and the LOF results with different contamination levels of all trajectories of the blue positional cluster of Figure 3e. In the following, we compare four values of the contamination hyperparameter: (a) 0.14, computed as suggested in [25], (b) 0.1, (c) 0.05, and (d) 0.01. Points in clusters with at least five trajectories are colored blue and the others are in yellow. Outliers are denoted by red borders.

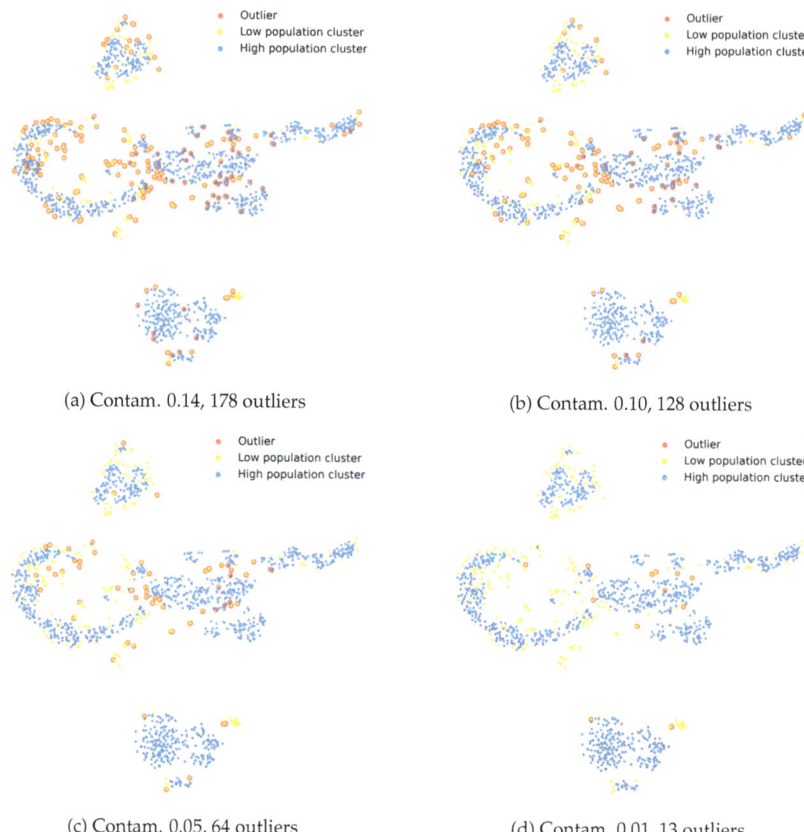

(a) Contam. 0.14, 178 outliers

(b) Contam. 0.10, 128 outliers

(c) Contam. 0.05, 64 outliers

(d) Contam. 0.01, 13 outliers

Figure 9. T-SNE of the kinematic distance matrix of the blue cluster of Figure 3e with varying levels of contamination in the LOF. Crosses denote detected outliers. Colors denote cluster assignments, with gray being clusters with fewer than five members. The contamination level in (a) is determined using [25].

We see that the TSNE projects the trajectories in low populated clusters in the fringe of those in highly populated. Increasing or decreasing the contamination hyperparameter causes more or less of the trajectories along the fringe to be flagged as outliers. With a contamination of 0.10, even trajectories from large clusters are flagged as abnormal. Reducing it to 0.01 leaves almost no outliers. Therefore, we recommend contamination levels of 0.05.

Note that yellow dots that are grouped together in the TSNE projection without necessarily being in the same kinematic cluster better resist the increase of the contamination hyperparameter and remain inliers. An example is the group of yellow dots on the top-right of the large blue cluster at the bottom of the representation.

4.6.2. Outliers and Embedding Analysis

In Figure 10, we plot the same TSNE representation of the kinematic similarity matrix of all trajectories in positional cluster 0, shown in blue in Figure 3e, but the colors denote here cluster assignments of the kinematic clustering. The gray dots are clusters with fewer than five trajectories. The red borders denote outliers flagged by the LOF, using a contamination level of 0.05. The squares denote embeddings of pilot boats, triangle trajectories with U-turns, and crosses trajectories with stops outside a port.

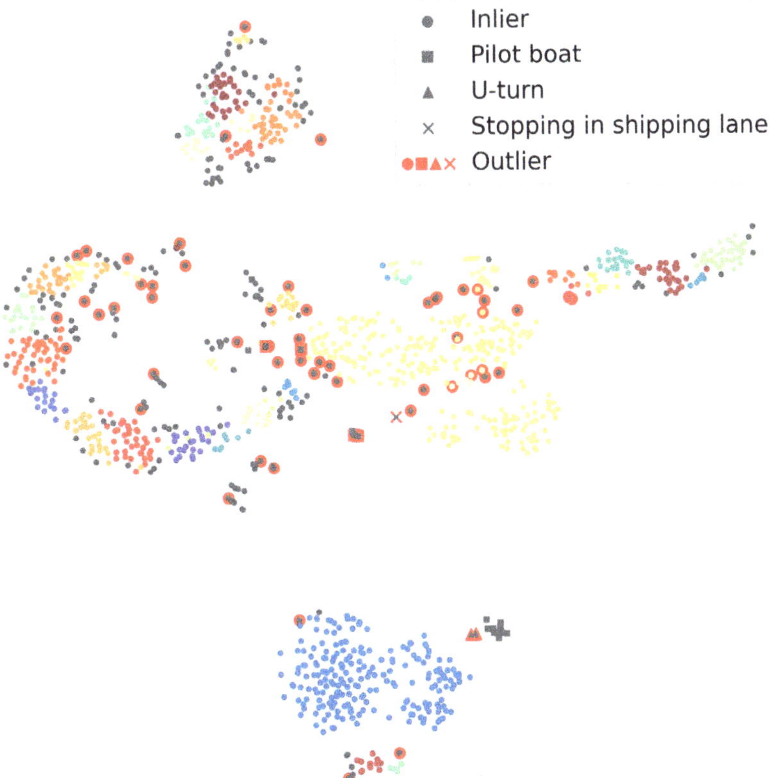

Figure 10. TSNE of the kinematic distance matrix of the blue cluster in Figure 3e. The colors denote the kinematic cluster assignment, with gray being clusters with less than five members. The red borders denote outliers flagged by the LOF. The shapes denote manually identified special trajectories.

As we mentioned previously, the majority of the closest neighbors of the pilot boats were other pilot boats. Most of the pilot boats form a small gray group near a larger cluster (blue) in the bottom of the figure. This blue cluster represents ships coming from the south that docked at the port of Gedser and return back towards the south after docking. In the same area, we find cases of ships (red, green) sailing north and docking at Trelleborg before returning south. In between the pilot boats and the large blue cluster we notice three trajectories discovered to be outliers by the LOF. These three trajectories were boats coming from the south and making a stop and a U-turn in the middle of the shipping lane, very similarly to the pilot boats. However, small differences in the course time series led the three U-turns trajectories to be flagged as abnormal. In Figures 11a,b, we show an example of a U-turn trajectory and that of a pilot boat nearby.

Note the crescent clustered in many different clusters on the left of Figure 10. These clusters correspond to the traffic following the shipping lanes to the east (Figure 6b) at different speeds. Around the right end of this crescent, we find some pilot boats (square) and a highlighted outlier corresponding to a trajectory following the shipping lanes going east with multiple sudden stops in the middle of the shipping lane (cross). This type of behavior is of interest for certain types of ships, such as research vessels and diving vessels. These two trajectories are shown in Figure 11c,d. The large yellow cluster in the center of Figure 10 corresponds to trajectories going north and stopping in the port of Trelleborg. Contrary to the previous smaller red and green clusters in the bottom of Figure 10, these trajectories ended in the port. This means that our proposed kinematic distance measure found the start-stopping behavior to be more akin to the trajectories ending in port or pilot boats; however, not so much as to be part of their cluster, due to differences in the course. This confirms that our proposed distance measure is capable of capturing local kinematic similarities, regardless of the geographical position, and it can serve to flag trajectories with abnormal local kinematic behavior.

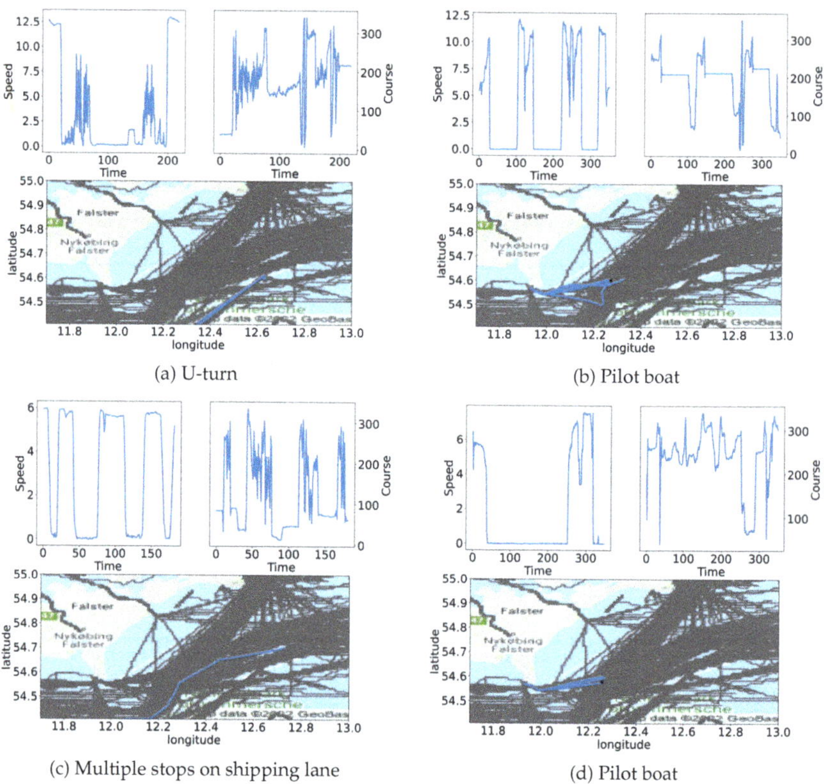

(a) U-turn

(b) Pilot boat

(c) Multiple stops on shipping lane

(d) Pilot boat

Figure 11. Trajectories of abnormal activity, such as (**a**) U-turns and (**c**) stopping in the shipping lane, found to have similarity with the behavior of pilot boats (**b**,**d**). Historical traffic is shown in gray, and trajectory origins are denoted by a black circle.

4.7. Abnormality Detection

We evaluate now the performance of our algorithm for the detection of abnormalities. We first compare it to State-of-the-Art abnormality detectors and then provide an example showing how the context provided by the kinematic clustering helps to understand the prediction of the LOF. For comparison, we also propose an interpretation of the VRNN baseline.

Throughout this section, we use the data from the Bornholm area of December 2021. All the models were trained using all the data, except that from 13 December and tested on that day. As a Search and Rescue operation took place that day, all the trajectories were annotated.

4.7.1. Anomaly Detection

We investigate the precision of our model and discuss which value of contamination parameter to use, based on the receiver operating curve from outlier detection on the 13 December data, as shown in Figure 12. We compare to the A-Contrario outlier detection method [15], using RVAE [13] and VRNN [15].

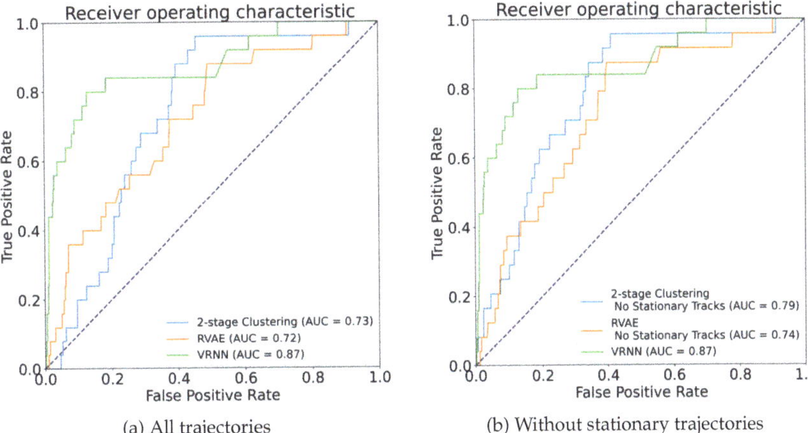

(a) All trajectories　　　　　　　　(b) Without stationary trajectories

Figure 12. ROC of the outlier detection on the Bornholm 13 December data.

We see that our method outperforms detection based on RVAE reconstruction but not VRNN reconstruction. In particular, we see that our method suffers from false positives early in the detection. Looking at these trajectories, we note that the vessels were mostly stationary in port with a highly varying course, resulting in a large directional distance to all other trajectories. By removing trajectories with an average speed of less than 0.3 m/s from the detection, we can reduce the number of early false positives to similar levels as RVAE.

Previously, we counted outliers to infer the contamination parameter. Here, we show how the ROC curve may provide more insight into this hyperparameter. In order to detect 96% of the abnormal trajectories, a false alarm was triggered on 40% of the normal trajectories, which corresponded to a contamination rate of 0.43. A contamination level of 0.05, 0.10, and 0.14 yielded a true positive rate of about 20%, 25%, and 40%, respectively.

The detection performance of our proposed method is below that of the VRNN, but it is much faster to computer. Indeed, in our experiments, the VRNN model and A-Contrario detection had an average evaluation time of 176 seconds per trajectory. By comparison, our two-step algorithm required only 6.8 seconds per trajectory. The large processing time of the VRNN model and A-Contrario detection is a major drawback that questions the viability of both algorithms for a real-life scenario where a quick response is vital, e.g., in the case of a collision or prediction of ongoing piracy. Our algorithm spends most of the 7 s to compute the positional similarity matrix, which compares the input to all the training dataset. This operation could be dramatically sped up, using some heuristics to avoid computations with obviously different trajectories.

4.7.2. Anomaly Detection and Context

We present here an example to highlight how the context provided by the kinematic clustering helps to assess the prediction of the LOF. We consider the case of an abnormal trajectory making a double U-turn in the shipping lane. The trajectory was made by the sister ship of the vessel that caused the collision accident and was returning to the site of the collision, perhaps to transfer crew to the collided vessel.

We fed the algorithm with the trajectory truncated at $t = 205$, which corresponded to when the vessel finished the first U-turn and was traveling in the opposite direction of the shipping lane. The positional trajectory (a), speed (b), and course (c) time series of that trajectory are depicted in red (plain before $t = 205$ and dotted after) in the plots of Figure 13. The five most similar trajectories in the kinetic context are depicted in black. Note that all five trajectories originated in the northeastern part of the ROI, and all five vessels had extended periods of time in which they traveled at reduced speeds.

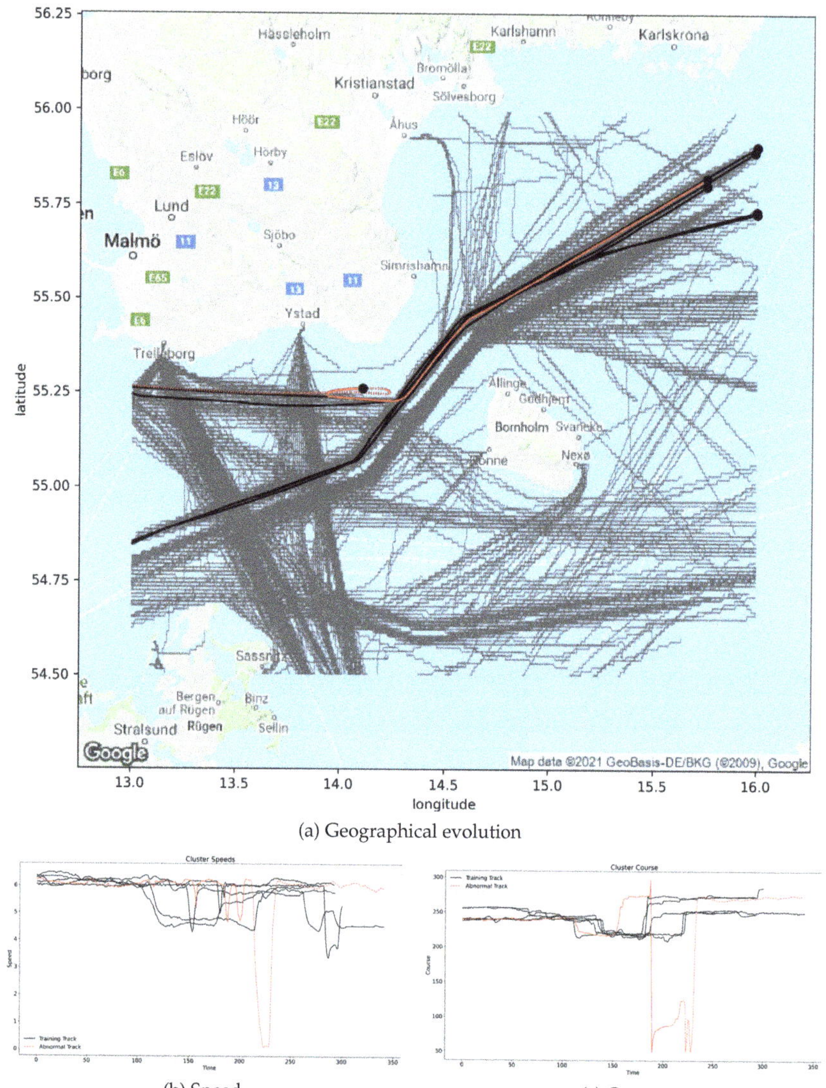

(a) Geographical evolution

(b) Speed

(c) Course

Figure 13. Top five most similar trajectories (black) to the abnormal trajectory in red determined by Equation (4) at time step $t = 205$. The future trajectory is shown in dashes.

The three visualizations issued from the kinetic context provide complementary information to the VTS operator. Indeed, if the the U-turn may be overlooked in Figure 13a,b, the plot of the course is clear. Another hypothesis could be that the abrupt change of direction suggests a turn toward a harbor, but this is dismissed by the fact that the vessel is still close to the shipping lane.

For comparison, we propose an interpretation of the outputs of the VRNN, the best-performing model, to justify the anomaly of the red trajectory. Note that VRNN outputs a multivariate Bernoulli distribution, which indicates if an input can or cannot be predicted/reconstructed accurately. In Figure 14, we plot the multivariate Bernoulli distribution of the VRNN model at two time steps during and after the first U-turn. During the U-turn, the speed and course are not reconstructed accurately. After the U-turn, the model recovers and is able to reconstruct accurately both the speed and the course. This means that at $t = 205$, the trajectory will still be flagged as abnormal because it was at $t = 189$. Thus, the operator needs to manually evaluate several previous updates, to understand the context that led the model to flag the anomaly. Another solution is to assess the situation based on the visualization of the different dimensions of the trajectory, as in Figure 13, but without the support of the black trajectories. This complicates the task, as one would expect—for example, pilot boats performing U-turns close to a shipping lane but not commercial vessels. Again, the operator needs to look for more information before making a decision.

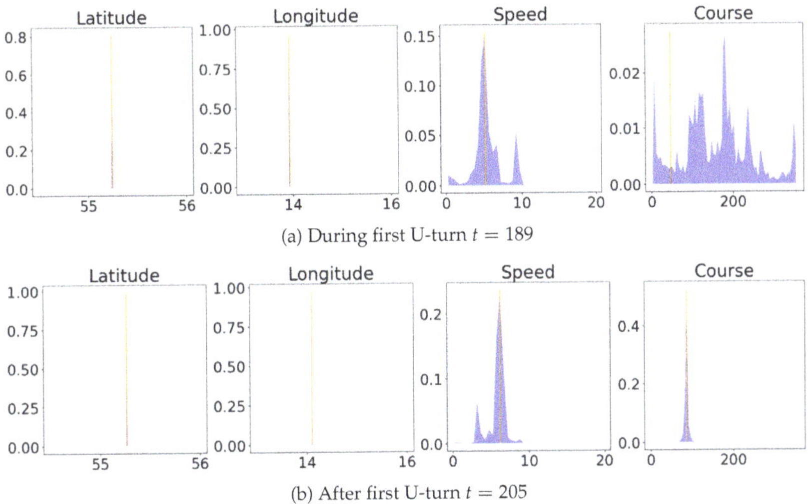

Figure 14. Multivariate Bernoulli distribution (blue) for the VRNN output at two time steps observed during and after the first U-turn. True values are indicated by orange lines.

5. Conclusions

In this work, we have presented and made public two large hand-annotated AIS traffic datasets for abnormal maritime behavior detection that we have created. One dataset contains all traffic AIS data around Sjælland Island during November 2021. The second dataset contains all AIS trajectories around Bornholm Island during December 2021, with the full annotation for the data of 13 December 2021, with events expected to be of interest to VTS operators. In addition, we also have proposed an abnormality detection algorithm for maritime trajectories based on two-step clustering, for which we proposed a novel kinematic similarity measure based on DTW. The two separate steps allow the clustering to better focus on the kinematic behavior expected in a certain geographical position. This disentanglement of positional and kinematic features results in better descriptions of behavioral patterns and clusters with well-defined and unique kinematic behaviors. These behavioral clusters and kinematic similarity measures can be used to provide context to the VTS operator, to accept or reject the algorithm prediction. We evaluated our proposed abnormality detection method on the annotated data of 13 December 2021 of the Bornholm dataset. Although our proposed model achieved a lower area under the ROC than the VRNN model, it had a clear advantage, in terms of runtime and interpretability, over current deep learning methods. In future work, we aim to compute the proposed similarity measures with neural networks and to utilize deep models for reconstruction-based outlier detection that also align trajectories in the latent space, to preserve a certain level of interpretability.

A limitation of our method is that it has been designed for and tested mostly on near-coastal traffic with a large variety of different maritime behaviors. We speculate that application in the open ocean with more extreme weather differences would result in separate clusters for each weather profile. We leave this question for future work.

Author Contributions: Conceptualization, K.V.O.; methodology, K.V.O. and A.B.; software, K.V.O. and A.B.; formal analysis, K.V.O.; investigation, K.V.O.; resources, S.H.; data curation, K.V.O. and S.H.; original draft preparation, K.V.O.; review and editing, A.B., M.C.K., R.J., A.N.C. and L.H.C.; visualization, K.V.O.; supervision, A.B., M.C.K., R.J., A.N.C. and L.H.C.; funding acquisition, R.J. and L.H.C. All authors have read and agreed to the published version of the manuscript.

Funding: This work was financially supported by the Danish Ministry of Defence Acquisition and Logistics Organisation, grant no. 4600005159. Visual Intelligence publications are financially supported by the Research Council of Norway, through its Centre for Research-based Innovation funding scheme (grant no. 309439), and Consortium Partners.

Data Availability Statement: The data used in this paper are made freely available at https://data. dtu.dk/collections/AIS_Trajectories_from_Danish_Waters_for_Abnormal_Behavior_Detection/6287 841, (accessed on 20 October 2023).

Acknowledgments: The datasets used in this paper and the visualization tools to identify abnormal behavior were provided by Terma A/S (Lystrup, Denmark).

Conflicts of Interest: The authors declare no conflict of interest. The funders had no role in the design of the study; in the collection, analyses, or interpretation of data; in the writing of the manuscript; or in the decision to publish the results.

Abbreviations

The following abbreviations are used in this manuscript:

AIS	Automatic Identification System
MMSI	Maritime Mobile Service Identity
GPS	Global Positioning System
VTS	Vessel Traffic Service
LOF	Local Outlier Factor
ROI	Region of Interest
DTW	Dynamic Time Warping
LCSS	Longest Common SubSequence
DP	Douglas–Peucker algorithm
RVAE	Recurrent Variational AutoEncoder
VRNN	Variational Recurrent Neural Network
AH	Average Haversine
KNN	K-Nearest Neighbors
AUC	Area Under the receiver operating characteristic Curve

References

1. IMO. *About IMO*; International Maritime Organization: London, UK, 2020. Available online: http://www.imo.org/en/About/Pages/Default.aspx (accessed on 20 October 2023).
2. Asariotis, R.; Benamara, H.; Lavelle, J.; Premti, A. Maritime Piracy. Part I: An Overview of Trends, Costs and Trade-Related Implications . UNCTAD 2014. Available online: https://eprints.soton.ac.uk/368254/ (accessed on 20 October 2023).
3. Lebedev, A.O.; Lebedeva, M.P.; Butsanets, A.A. Could the accident of "Ever Given" have been avoided in the Suez Canal? *J. Phys. Conf. Ser.* **2021**, *2061*, 12127. [CrossRef]
4. European Maritime Safety Agency. *Annual Overview of Marine Casualties and Incidents*; Technical Report; European Maritime Safety Agency: Lisbon, Portugal, 2022.
5. Long, T.; Widjaja, S.; Wirajuda, H.; Juwana, S. Approaches to combatting illegal, unreported and unregulated fishing. *Nat. Food* **2020**, *1*, 389–391. [CrossRef]
6. Ljungqvist, M. Confirmed Sabotage at Nord Stream . (In Swedish) Available online: https://www.aklagare.se/nyheter-press/pressmeddelanden/2022/november/bekraftat-sabotage-vid-nord-stream/ (accessed on 20 October 2023).
7. International Maritime Organization (IMO). *International Convention for the Safety of Life at Sea (SOLAS), Chapter V: Safety of Navigation, Regulation 19*; International Maritime Organization (IMO): London, UK, 1998.
8. MarineTraffic. A Day in Numbers. MarineTraffic Blog. Available online: https://www.marinetraffic.com/blog/a-day-in-numbers/ (accessed on 20 October 2023).

9. Pallotta, G.; Vespe, M.; Bryan, K. Traffic knowledge discovery from AIS data. In Proceedings of the 16th International Conference on Information Fusion, IEEE, Istanbul, Turkey, 9–12 July 2013.

10. Liu, B.; De Souza, E.N.; Matwin, S.; Sydow, M. Knowledge-based clustering of ship trajectories using density-based approach. In Proceedings of the 2014 IEEE International Conference on Big Data, IEEE Big Data 2014, Washington, DC, USA, 27–30 October 2014; pp. 603–608.

11. Zhao, L.; Shi, G. A trajectory clustering method based on Douglas-Peucker compression and density for marine traffic pattern recognition. *Ocean Eng.* **2019**, *172*, 456–467. [CrossRef]

12. Yang, J.; Liu, Y.; Ma, L.; Ji, C. Maritime traffic flow clustering analysis by density based trajectory clustering with noise. *Ocean Eng.* **2022**, *249*, 111001. [CrossRef]

13. Murray, B.; Perera, L.P. An AIS-based deep learning framework for regional ship behavior prediction. *Reliab. Eng. Syst. Saf.* **2021**, *215*, 107819. [CrossRef]

14. Pallotta, G.; Jousselme, A.L. Data-driven detection and context-based classification of maritime anomalies. In Proceedings of the 2015 18th International Conference on Information Fusion (Fusion), IEEE, Washington, DC, USA, 6–9 July 2015.

15. Nguyen, D.; Vadaine, R.; Hajduch, G.; Garello, R.; Fablet, R. GeoTrackNet-A Maritime Anomaly Detector using Probabilistic Neural Network Representation of AIS Tracks and A Contrario Detection. *IEEE Trans. Intell. Transp. Syst.* **2021**, *23*, 5655–5667. [CrossRef]

16. Riveiro, M.; Pallotta, G.; Vespe, M. Maritime anomaly detection: A review. *Wiley Interdiscip. Rev. Data Min. Knowl. Discov.* **2018**, *8*, e1266. [CrossRef]

17. Stach, T.; Kinkel, Y.; Constapel, M.; Burmeister, H.C. Maritime Anomaly Detection for Vessel Traffic Services: A Survey. *J. Mar. Sci. Eng.* **2023**, *11*, 1174. [CrossRef]

18. Endsley, M.R. From Here to Autonomy: Lessons Learned From Human—Automation Research. *Hum. Factors* **2017**, *59*, 5–27. [CrossRef]

19. Wang, L.; Chen, P.; Chen, L.; Mou, J. Ship AIS Trajectory Clustering: An HDBSCAN-Based Approach. *J. Mar. Sci. Eng.* **2021**, *9*, 566. [CrossRef]

20. Liu, B.; de Souza, E.N.; Hilliard, C.; Matwin, S. Ship movement anomaly detection using specialized distance measures. In Proceedings of the 2015 18th International Conference on Information Fusion (Fusion), IEEE, Washington, DC, USA, 6–9 July 2015.

21. Hu, J.; Kaur, K.; Lin, H.; Wang, X.; Hassan, M.M.; Razzak, I.; Hammoudeh, M. Intelligent Anomaly Detection of Trajectories for IoT Empowered Maritime Transportation Systems. *IEEE Trans. Intell. Transp. Syst.* **2022**, *24*, 2382–2391. [CrossRef]

22. Liu, H.; Liu, Y.; Li, B.; Qi, Z.; Rizvi, J.; Liu, H.; Liu, Y.; Li, B.; Qi, Z. Ship Abnormal Behavior Detection Method Based on Optimized GRU Network. *J. Mar. Sci. Eng.* **2022**, *10*, 249. [CrossRef]

23. Li, J.; Liu, J.; Zhang, X.; Li, X.; Wang, J.; Wu, Z. A Novel Hybrid Approach for Detecting Abnormal Vessel Behavior in Maritime Traffic. In Proceedings of the 2023 7th International Conference on Transportation Information and Safety (ICTIS), Xi'an, China, 4–6 August 2023; pp. 1–7.

24. Widyantara, I.M.O.; Hartawan, I.P.N.; Karyawati, A.A.I.N.E.; Er, N.I.; Artana, K.B. Automatic identification system-based trajectory clustering framework to identify vessel movement pattern. *Iaes Int. J. Artif. Intell.* **2023**, *12*, 1–11. [CrossRef]

25. Breunig, M.M.; Kriegel, H.P.; Ng, R.T.; Sander, J. LOF: Identifying Density-Based Local Outliers. In Proceedings of the 2000 ACM SIGMOD International Conference on Management of Data, SIGMOD '00, Dallas, TX, USA, 16–18 May 2000; pp. 93–104.

26. Larsen, M.S. Russian 'Ghost Ships' Are Turning the Seabed into a Future Battlefield, 2023. Available online: https://foreignpolicy.com/2023/05/02/russia-europe-denmark-spy-surveillance-ships-seabed-cables/ (accessed on 20 October 2023).

27. Laxhammar, R.; Falkman, G. Inductive conformal anomaly detection for sequential detection of anomalous sub-trajectories. *Ann. Math. Artif. Intell.* **2015**, *74*, 67–94. [CrossRef]

28. Zhen, R.; Jin, Y.; Hu, Q.; Shao, Z.; Nikitakos, N. Maritime Anomaly Detection within Coastal Waters Based on Vessel Trajectory Clustering and Naïve Bayes Classifier. *J. Navig.* **2017**, *70*, 648–670. [CrossRef]

29. Klaas, G.; De Vries, D.; Van Someren, M. Machine learning for vessel trajectories using compression, alignments and domain knowledge. *Expert Syst. Appl.* **2012**, *39*, 13426–13439.

30. Douglas, D.H.; Peucker, T.K. Algorithms for the Reduction of the Number of Points Required to Represent a Digitized Line or its Caricature. In *Classics in Cartography: Reflections on Influential Articles from Cartographica*; John Wiley & Sons, Ltd.: Hoboken, NJ, USA, 2011; pp. 15–28.

31. Pallotta, G.; Vespe, M.; Bryan, K. Vessel Pattern Knowledge Discovery from AIS Data: A Framework for Anomaly Detection and Route Prediction. *Entropy* **2013**, *15*, 2218–2245. [CrossRef]

32. Ester, M.; Kriegel, H.p.; Sander, J.; Xu, X. A density-based algorithm for discovering clusters in large spatial databases with noise. In Proceedings of the 4th International Conference on Knowledge Discovery and Data Mining, Portland, OR, USA, 2–4 August 1996.

33. Luo, S.; Zeng, W.; Sun, B. Contrastive Learning for Graph-Based Vessel Trajectory Similarity Computation. *J. Mar. Sci. Eng.* **2023**, *11*, 1840. [CrossRef]

34. Zhao, L.; Shi, G. Maritime Anomaly Detection using Density-based Clustering and Recurrent Neural Network. *J. Navig.* **2019**, *72*, 894–916. [CrossRef]

35. Shamos, M.; Preparata, F. Computational Geometry An Introduction. In *Computational Geometry an Introduction*, 1st ed.; Schneider, F., Gries, D., Eds.; Springer: New York, NY, USA, 1985; Chapter 5, p. 223.

36. Nanni, M.; Pedreschi, D. Time-focused clustering of trajectories of moving objects. *J. Intell. Inf. Syst.* **2006**, *27*, 267–289. [CrossRef]

37. Olesen, K.V.; Christensen, A.N.; Hørlück, S.; Clemmensen, L.K.H. AIS Trajectories from Danish Waters for Abnormal Behavior Detection. 2022. Available online: https://data.dtu.dk/collections/AIS_Trajectories_from_Danish_Waters_for_Abnormal_Behavior_Detection/6287841 (accessed on 20 October 2023).
38. Soefartsstyrelsen. Historical AIS Data. Available online: https://dma.dk/safety-at-sea/navigational-information/ais-data (accessed on 20 October 2023).
39. Satopaa, V.; Albrecht, J.; Irwin, D.; Raghavan, B. Finding a "Kneedle" in a Haystack: Detecting Knee Points in System Behavior. In Proceedings of the 31st International Conference on Distributed Computing Systems Workshops, Minneapolis, MI, USA, 20–24 June 2011.
40. Rousseeuw, P.J. Silhouettes: A graphical aid to the interpretation and validation of cluster analysis. *J. Comput. Appl. Math.* **1987**, *20*, 53–65. [CrossRef]

Journal of
Marine Science and Engineering

Article

AIS Data Manipulation in the Illicit Global Oil Trade

Andrej Androjna [1], Ivica Pavić [2], Lucjan Gucma [3], Peter Vidmar [1] and Marko Perkovič [1,*]

[1] Maritime Department, Faculty of Maritime Studies and Transport Portorož, University of Ljubljana, 6320 Portorož, Slovenia; andrej.androjna@fpp.uni-lj.si (A.A.); peter.vidmar@fpp.uni-lj.si (P.V.)
[2] Maritime Department, Faculty of Maritime Studies Split, University of Split, 21000 Split, Croatia; ipavic71@pfst.hr
[3] Department of Marine Traffic Engineering, Maritime University of Szczecin, 70-500 Szczecin, Poland; l.gucma@am.szczecin.pl
* Correspondence: marko.perkovic@fpp.uni-lj.si

Abstract: This article takes a close look at the landscape of global navigation satellite system (GNSS) spoofing. It is well known that automated identification system (AIS) spoofing can be used for electronic warfare to conceal military activities in sensitive sea areas; however, recent events suggest that there is a similar interest of spoofing AIS signals for commercial purposes. The shipping industry is currently experiencing an unprecedented period of deceptive practices by tanker operators seeking to evade sanctions. Last year's announcement of a price cap on Russian crude oil and a new ban on Western companies insuring Russian cargoes is setting the stage for an increase in illegal activity. Our research team identified and documented the AIS position falsification by tankers transporting Russian crude oil in closed ship-to-ship (STS) oil transfers. The identification of the falsified positions is based on the repeated instances of discrepancies between AIS location suggestions and satellite radar imagery indications. Using the data methods at our disposal, we reconstructed the true movements of certain tankers and encountered some surprising behavior. These false ship positions make it clear that we need effective tools and strategies to ensure the reliability and robustness of AISs.

Keywords: automatic identification system (AIS); tankers; falsification; spoofing and jamming

Citation: Androjna, A.; Pavić, I.; Gucma, L.; Vidmar, P.; Perkovič, M. AIS Data Manipulation in the Illicit Global Oil Trade. *J. Mar. Sci. Eng.* **2024**, *12*, 6. https://doi.org/10.3390/jmse12010006

Academic Editor: Dejan Brkić

Received: 19 November 2023
Revised: 9 December 2023
Accepted: 12 December 2023
Published: 19 December 2023

1. Introduction

In spite of the ban on imports of Russian crude oil and products and the ban on ships calling at certain ports, a number of shipping companies are supporting Russian exporters. While it must be recognized that there is a fundamental clash between nations and industries, the growing ghost fleet of substandard ships carrying sanctioned Russian goods around the world's seas is certainly a danger at sea: many of these vessels are not seaworthy, operate with incompetent crews, and have already been involved in enough accidents to raise some degree of alarm. All of this is possible because the bans on imports of Russian crude oil and ships entering certain ports apply only to certain countries [1,2]. Yet, some shipping companies can continue to transport Russian oil without facing consequences as a ghost fleet of unregistered vessels (that do not officially exist) is growing, and it is difficult to track down such vessels. The ghost fleet can transport these goods undetected because they are not registered with any country's maritime authority and are not subject to any regulations or inspections. These vessels undermine the effectiveness of sanctions by allowing sanctioned goods to travel around the world undetected. They also pose a security risk to global shipping, given the lack of inspections. Finally, they create an environment in which illegal activities, such as smuggling, can flourish. Addressing these risks requires coordinated efforts by governments and international organizations. For now, we must set aside the debate over the logic and efficacy of sanctions themselves. One approach is to increase the surveillance and monitoring of maritime activities to detect suspicious vessels.

Another is to impose stricter regulations on shipping companies and force them to provide more detailed information about their vessels and cargoes. In addition, countries can work together to enforce sanctions more effectively by sharing information and coordinating their efforts.

It should be noted that cargo transfers do not occur in ports, but at sea, either at an anchorage outside territorial waters where the water depth permits this, or further out at sea where ships either drift with the current and wind or move slowly, engaging in a process called ship-to-ship transfer (STS transfer), which is commonly used in the shipping industry to transfer large quantities of oil or other cargo between vessels. STS transfers can be performed using a variety of methods, including hoses, pipelines, and pumps [3,4]. However, if performed improperly, this process poses risks, such as oil spills and accidents. As a result, strict regulations and safety protocols are in place to ensure that such transfers are conducted safely and efficiently, and this is even more important as the demands for oil and other commodities continue to grow and STS transfers are likely to continue to play an important role in moving goods around the world.

Most commonly, STS operations are used to handle crude oil, petroleum products, and liquified gas [5]. The STS operations of the International Convention for the Prevention of Pollution from Ships (MARPOL) Annex I's cargo within territorial waters (TTWs) and the exclusive economic zone (EEZ) of a MARPOL Contracting Party are reported in advance to the competent authority of the coastal state, while coastal state authorities may require notifications for MARPOL Annex II and other cargoes. All other issues are subject to local regulations [6]. STS transfers require careful planning and coordination to ensure the safety of the vessels involved and to prevent oil spills, accidents, or damage to the ships or cargo. Planning considers factors, such as the limits of the transfer area, environmental constraints, weather conditions, sea state, traffic density, good holding conditions, ships' characteristics, and compliance with international and local regulations [7–9].

This operation requires compliance with various legal and technical requirements established by the International Maritime Organization (IMO) and other relevant organizations. IMO Resolution MEPC 186(59)/2009 amended the MARPOL Annex I by adding Chapter 8 to provide a common international framework addressing applications, preparations of STS operation plans, persons in overall advisory control, retention of records on board, and notifications for operations within the TTW and EEZ [8]. In addition, the publication Ship to Ship Transfer Guide for Petroleum, Chemicals and Liquefied Gasses, developed by maritime industry organizations, provides guidance and recommendations for the planning and actions of all responsible persons involved in the operation [5].

Since it is a complex operation that can lead to accidents at sea, it is necessary to conduct a safety and risk assessment. Ventikos and Stavrou [4] described some recent developments, including the identification of risk factors and their roles in risk assessments. They applied a fuzzy inference system (FIS) for the risk assessment [9] and a process failure mode and effects analysis (PFMEA) in combination with FIS methodology to assess and evaluate risk scenarios of STS transfer operations [10]. Sultana et al. conducted a comparative analysis between a system theoretical process analysis (STPA) and hazard and operability studies (HAZOP) for liquefied natural gas (LNG) STS transfer systems and concluded that the STPA technique provided effective systematic guidance and recommendations for safety requirements [11]. These studies demonstrate the importance of risk assessments for STS operations.

Depending on the legal basis, STS transfer can be divided into two categories: legitimate and illicit. Legitimate transfer is an operation conducted in accordance with IMO regulations, technical guidelines of relevant organizations, and national regulations within the framework of regular (legal) maritime trade. In contrast, illicit transfer is conducted to disguise the violation of sanctions or embargoes, smuggle illegal cargo, or, simply, engage in any illegal maritime trade. There are numerous reported cases of sanction violations committed through illicit STS operations [12–15]. In the two piracy hotspots (Southeast Asia and West Africa), illicit STS is also known to have been performed as part of piracy and armed robbery operations aimed at stealing oil cargoes [16,17]. In recent developments, STS transfer is becoming one of the oil smuggling techniques used worldwide, basically outside the traditional trade routes. The Israeli company (one should always consider potential biases of private company results, given the fraught diplomatic conditions in the world) Windward's Maritime AI^{TM} platform identified four suspicious operational patterns in Russian oil smuggling. These patterns are aimless journey, checkpoint avoidance, floating storage and ID, and location tampering, which allows vessels to disguise their locations and conduct illicit STS operations that usually occur in the mid-North Atlantic [18].

The patterns show the ability to conduct illegal activities to evade sanctions. According to the UK P&I Circular 01/22, there are some techniques used to evade sanctions, such as manipulating AIS data, changing the vessels' physical appearance, falsifying ship and/or cargo documents, and multiple STS operations [15]. Of particular interest to our research is AIS data manipulation used to evade sanctions through illicit STS operations.

Although AIS is not a system for preventing illegal activities at sea, AIS data deviations can be a great help in detecting such activities. Originally, AIS was intended as a communication system to improve the safety of navigation, exchanging navigational and other data between ships and the shore. At present, the data from AISs are not only used for safety of navigation, but also for maritime safety and security purposes, such as vessel tracking; route and fishing activity monitoring; maritime risks and trend analyses; accident investigations; waterway planning; management and maintenance; scientific, environmental, and ecological purposes; and the prevention of illegal fishing, piracy, armed robbery, etc. [19–26]. In connection with the expansion of the use of AIS data, it should be mentioned that research has recently been conducted on the use of AIS data in the area of critical infrastructure protection. Soldi et al. analyzed how information from different sources, in particular, AIS and satellites (i.e., synthetic aperture radar—SAR), could be fused to detect anomalous and suspicious behavior and identify threats to critical underwater infrastructure [27].

The wide application of AIS was adopted because of its advantages. Numerous design-related disadvantages led to AIS vulnerabilities. AIS is an insecure open broadcast system [28] that transmits on two dedicated maritime public VHF frequencies [29]. AIS data are not encrypted and there are no mechanisms to perform authentication, timing, and validity checks [28–30]. Therefore, AIS stations are vulnerable to spoofing, hijacking, and availability disruption [31]. Various solutions have been proposed in the context of improving the protection and integrity of AIS data. In this sense, Iphar et al. described the use and misuse of AIS data and proposed a data integrity assessment method to detect anomalies in AIS messages. They also presented a risk assessment that individualized these messages and evaluated the risk values according to the different message characteristics [32,33]. AIS signal reception is one of the disadvantages that can be exploited in illegal activities. Salmon et al. analyzed the problems with the reception of AIS signals and introduced the concept of black holes for maritime areas where AIS signals could not be received by coastal stations [34]. Malicious actors can manipulate the data from AISs and transmit deceptive or false data [28].

The UK P&I identified three techniques of AIS data manipulation for the purpose of evading sanctions. These techniques are switching off the AIS, GPS/GNSS manipulation, and AIS misuse [15]. Technically, switching off an AIS is the simplest technique. During STS transfer, the AIS is switched off. According to Regulation V/19.2.4.7 of the International Convention for the Safety of Life at Sea (SOLAS), an AIS must be in continuous operation [35], while according to IMO Resolution A.1106(29), switching off an AIS is allowed only when its operation can endanger the safety or security of the ship or when a security incident is imminent [36]. Since STS transfer does not threaten the safety and security of the ship, turning off the AIS is not allowed. It is important to note that the maritime community is making additional efforts to discourage switching off AISs. In this regard, the Baltic and International Maritime Council (BIMCO) have developed an AIS switch-off clause for time and voyage charter parties that allows shipowners and charterers to terminate contracts with contractors who switch off AISs for improper reasons [37].

In our previous articles [38,39], we scrutinized the landscape of AISs as an important source of information for maritime situational awareness (MSA), highlighting its vulnerabilities, challenges, and cybersecurity in relation to safe navigation and shipping. Complemented by a case study of a specific spoofing event near Elba in December 2019 [39], which confirmed that a typical maritime AIS could easily be spoofed and, in the case of this study of suspicious STS oil transfers, generate a false position or even no position, we confirmed, once again, that AIS messages could be spoofed, disrupted, deliberately faked, or simply shut down.

The paper is structured as follows: Section 2 presents the methodology; Section 3 presents AIS vulnerabilities and results made explicit through case studies. Section 4 presents the discussion and Section 5 the conclusions.

2. Methods

2.1. Automatic Identification System

At present, the easiest way to track ship traffic is through the AIS system, as all large ships are equipped with a radio system that automatically transmits dynamic and static information about the ship. The position of the ship, its speed, and even its rate of turn are transmitted very frequently; a normal ship, while underway transmits at least 6 positions per minute. When maneuvering, the frequency increases by threefold, and when the ship is at anchor or in port, it transmits the data once every three minutes. Although the rate of turn (ROT) is an AIS function, this value is not displayed correctly in some messages, and therefore may only be used sometimes. The data quality is better for large vessels displacing over 50,000 tons, which must be equipped with an ROT indicator that provides more reliable reports on the rate of turn transmitted by the AIS system. A coastal AIS base station can also interrogate the ship's AIS transmitter to speed up the transmission, up to every two seconds, which is especially useful when the ship is maneuvering or when ships at anchor are exposed to bad weather. So, when we track a ship on its course, we generally have a position at least every 72 m (without taking into account the GNSS positioning performance and possible jamming or spoofing activities), and the exact data are given by Equation (1) and Figure 1 presenting a plot of all the ship distances traveled between consecutive measurements.

$$S(\triangle T, V) = V \cdot \frac{1852}{3600} \cdot \Delta T(V) \left\|\begin{array}{l} \Delta T(V) = 10s \; for \; V < 14kn \\ \Delta T(V) = 6s \; for \; 14 \leq V < 23kn \\ \Delta T(V) = 2s \; for \; V \geq 23kn \end{array}\right. \tag{1}$$

where ΔT is the AIS reporting interval and V is the ship's velocity (speed over ground).

Of course, all this is true only if the ship is within the range of a shore base station and if the VHF link operates at a high performance, but this is not always the case. The expectations for integrity, availability, and reliability were unfortunately not met by the AIS developers or could only be explained by its anti-collision short-range nature, where integrity and reliability were not the basic requirements. For VHF propagation, the range depends mainly on the heights of the transmitter and receiver antennas. The expected range between the AIS base station with an antenna 49 m above sea level and the ship station with an antenna height of 25 m (smaller vessels) corresponds to 30 NM according to Equation (2):

$$D = 2.5\left(\sqrt{h_1}\right) + \sqrt{h_2}, \; D = 2.5\left(\sqrt{49\,m} + \sqrt{25\,m}\right) = 30\,NM \tag{2}$$

Some coastal stations are located on very high hills; so, an AIS base station at an altitude of 1089 m can track a larger ship at a distance of 100 NM, as seen in the following table.

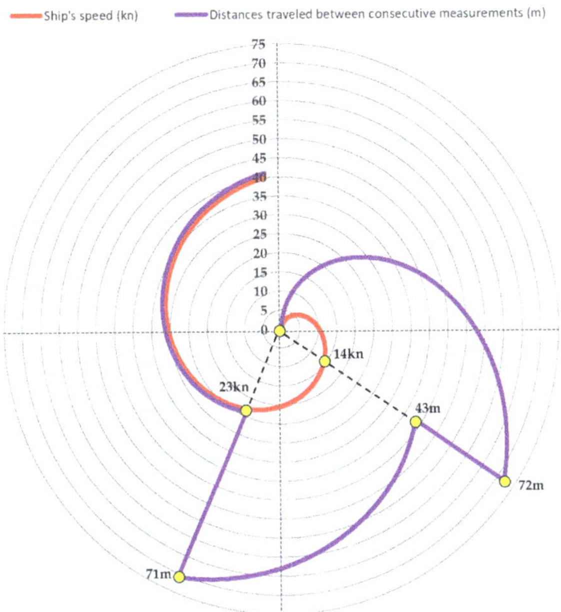

Figure 1. Distances traveled between successive AIS messages.

Occasionally, distances can be much larger due to the effects of atmospheric ducts on AIS propagation [40,41]; these situations should normally be relatively rare, but with the heavy evaporation of the water surface, this is more common over the ocean in summer. A very illustrative picture of the possible atmospheric ducts can be seen in Figure 2, where the range of AIS stations in the Black Sea area is very high. However, considering the relatively low height of all base stations' VHF antennas around Costanța, it is difficult to believe that all stations can receive signals from smaller vessels located in the Sea of Azov.

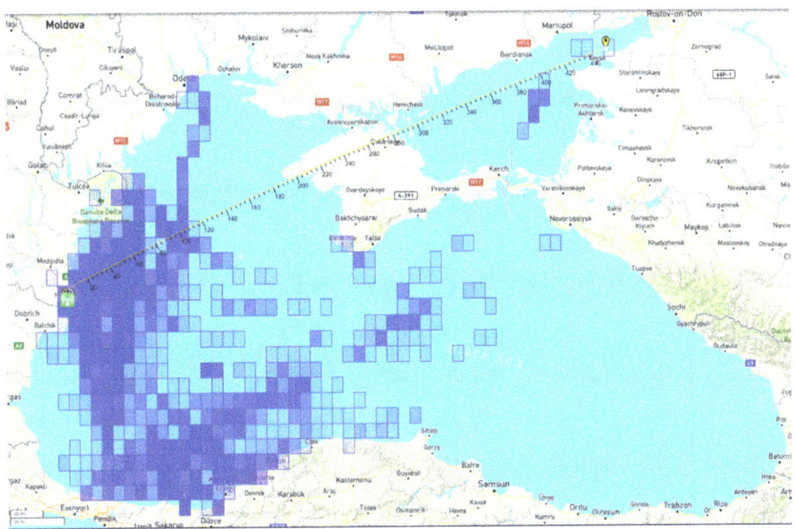

Figure 2. AIS range of Costanța base station (9 March 2023); source: produced from Marine Traffic.

Using the range method from at least two base stations, it was possible to determine the positions of suspect vessels located at distances beyond the range of the terrestrial AIS system.

An example of an exceptional AIS range is shown in the following figure, depicting the average and maximum receiving distances of three AIS stations during the period from October 2022 to 2023. All three stations are located near Costanța at similar altitudes, namely, Costanța Station (h = 41 m), Constanta Kalimbassieris Maritime Station (h = 20 m), and Offshore Costanța Station (h = 10 m). Figure 3 shows the extreme range of the AIS system on 9 March 2023, when it exceeds 450 nautical miles. One can speak of atmospheric disturbances, but other stations in the western part of the Black Sea do not have such a range. Nor can one imagine special conditions in March, known as the time of the summer solstice. All three stations are located at low altitudes; their realistic range, according to Table 1, is not more than 40 NM, even for communication with a large merchant ship, where the AIS antenna can be placed at a height of 60 m above sea level.

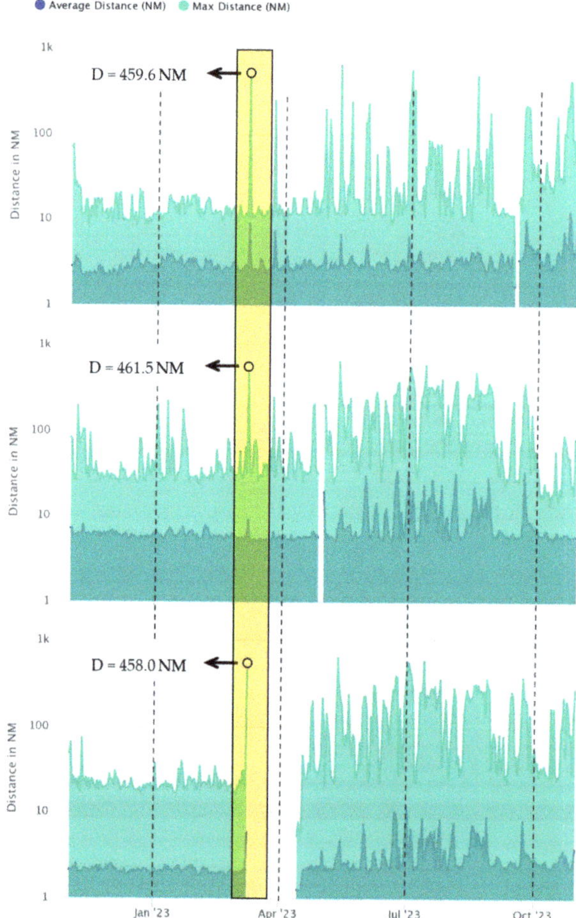

Figure 3. Average and maximum receiving distances of three AIS stations: Costanța Station (h = 41 m), Constanta Kalimbassieris Maritime Station (h = 20 m), Offshore Costanta Station (h = 10 m); source: produced from Marine Traffic.

Table 1. The range of AIS signals in NM.

Ship station antenna height (m)	Base Station Antenna Height (m)																																	
	4	9	16	25	36	49	64	81	100	121	144	169	196	225	256	289	324	361	400	441	484	529	576	625	676	729	784	841	900	961	1024	1089	1156	1225
4	10.0	12.5	15.0	17.5	20.0	22.5	25.0	27.5	30.0	32.5	35.0	37.5	40.0	42.5	45.0	47.5	50.0	52.5	55.0	57.5	60.0	62.5	65.0	67.5	70.0	72.5	75.0	77.5	80.0	82.5	85.0	87.5	90.0	92.5
9	12.5	15.0	17.5	20.0	22.5	25.0	27.5	30.0	32.5	35.0	37.5	40.0	42.5	45.0	47.5	50.0	52.5	55.0	57.5	60.0	62.5	65.0	67.5	70.0	72.5	75.0	77.5	80.0	82.5	85.0	87.5	90.0	92.5	95.0
16	15.0	17.5	20.0	22.5	25.0	27.5	30.0	32.5	35.0	37.5	40.0	42.5	45.0	47.5	50.0	52.5	55.0	57.5	60.0	62.5	65.0	67.5	70.0	72.5	75.0	77.5	80.0	82.5	85.0	87.5	90.0	92.5	95.0	97.5
25	17.5	20.0	22.5	25.0	27.5	30.0	32.5	35.0	37.5	40.0	42.5	45.0	47.5	50.0	52.5	55.0	57.5	60.0	62.5	65.0	67.5	70.0	72.5	75.0	77.5	80.0	82.5	85.0	87.5	90.0	92.5	95.0	97.5	100.0
36	20.0	22.5	25.0	27.5	30.0	32.5	35.0	37.5	40.0	42.5	45.0	47.5	50.0	52.5	55.0	57.5	60.0	62.5	65.0	67.5	70.0	72.5	75.0	77.5	80.0	82.5	85.0	87.5	90.0	92.5	95.0	97.5	100.0	102.5
49	22.5	25.0	27.5	30.0	32.5	35.0	37.5	40.0	42.5	45.0	47.5	50.0	52.5	55.0	57.5	60.0	62.5	65.0	67.5	70.0	72.5	75.0	77.5	80.0	82.5	85.0	87.5	90.0	92.5	95.0	97.5	100.0	102.5	105.0
64	25.0	27.5	30.0	32.5	35.0	37.5	40.0	42.5	45.0	47.5	50.0	52.5	55.0	57.5	60.0	62.5	65.0	67.5	70.0	72.5	75.0	77.5	80.0	82.5	85.0	87.5	90.0	92.5	95.0	97.5	100.0	102.5	105.0	107.5
81	27.5	30.0	32.5	35.0	37.5	40.0	42.5	45.0	47.5	50.0	52.5	55.0	57.5	60.0	62.5	65.0	67.5	70.0	72.5	75.0	77.5	80.0	82.5	85.0	87.5	90.0	92.5	95.0	97.5	100.0	102.5	105.0	107.5	110.0

Thus, it can be concluded that a vessel presenting as 450 NM from Costanța is actually quite close to Costanța, since there are no other stations in neighboring Black Sea countries receiving a signal from this vessel, and the other stations are at least 50 NM away from the base stations near Costanța. The method presented for identifying a vessel that is falsifying its position is simple and robust, but a full identification requires an AIS VHF signal information (VSI) string containing the TOA record and RSI information, i.e., the time of arrival and received signal strength.

At present, an AIS is an indispensable system for any vessel traffic service and vessel traffic monitoring center. Each dedicated maritime traffic monitoring application has the ability to set various warning systems, such as automatically displaying all vessels within the traffic zone that are in close proximity to each other or adding an AIS vessel-type filter (80 to 84) to display only tankers, so that only vessels that may be conducting STS operations are displayed. Of course, such a system can also be used to send warnings in the event of near collisions, the loss of an AIS signal, or when a vessel is entering or leaving the traffic zone. Sudden ship movements can be indicated by the ship triggering an alarm at unrealistically high speeds or by a ship coming ashore. Operators or maritime authorities have numerous means at their disposal to ensure the high quality of traffic monitoring, which is essential for safety at sea and the protection of the marine environment.

Various overview charts can be created from the large amount of AIS data, such as displaying major shipping lanes, classifying traffic by vessel type, traffic density, forecasting shipping routes [42], etc. The Figure 4 show the monitoring of shipping routes in the wake of the war between Ukraine and Russia. A quick look at the maritime traffic already shows a change in the shipping routes (marked with yellow dashed circles and ellipse), with tankers from Russia heading toward Costanța (an area off Costanța outside territorial waters) where STS operations are taking place. A very large amount of cargo is also diverted to inland waterways (marked higher on the map).

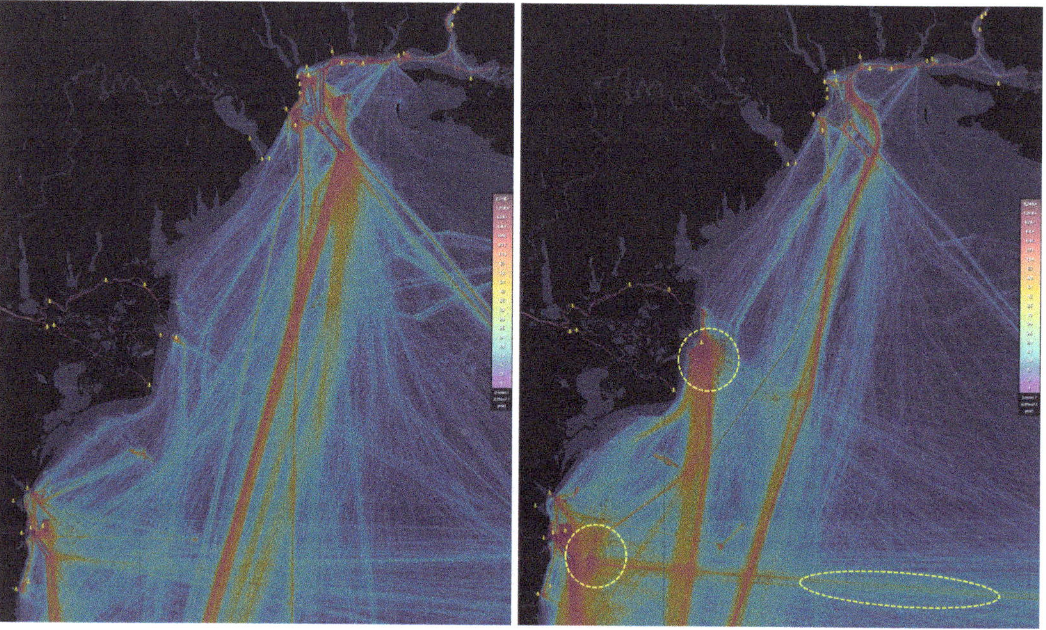

Figure 4. Comparison of traffic density during the years 2021 and 2022; source: produced from Marine Traffic.

Of course, it is also possible to monitor ships outside the radio horizon. The signal from the AIS is broadcast in all directions, including into space, where it is received by AIS onboard satellites, providing a much wider and homogeneous coverage, which is of great value for the overall assessment of maritime traffic. There are three points to note: there is a cost associated with transmitting the data to the shore station; the temporal coverage is much less than a land-based AIS system; and the satellite AIS (S-AIS) does not provide a report from the ship at a specific time interval. Therefore, there is also another complementary, but mandatory, long-range identification and tracking (LRIT) system that requires the ship to send a position report every six hours. Each flag state is required to establish or select an LRIT data center to directly collect LRIT reporting the data from a ship entitled to fly its flag. The LRIT information is always available to the ship's flag state in accordance with the data distribution plan developed by IMO to meet the flag state's requirements. The IMO data distribution plan meets the flag state requirements and is linked to the routing rules. The International LRIT Data Exchange provides the other flag states with valid access to the LRIT data of the ships concerned [43]. LRIT and AISs have many differences, but primarily, AIS is used to avoid collisions, while LRIT is used to monitor specific vessels of interest to the government. These fundamental differences affect both how the self-reporting systems work and how often the self-reports are required. For example, for collision avoidance, only vessels in close proximity need to know a vessel's position; therefore, STS VHF transponders are sufficient. In the case of LRIT, the global nature of the system appears to be the deciding factor in adopting a satellite-based system for transmitting LRIT messages. Unfortunately, we were not able to obtain S-AIS or LRIT data. The latter are of particular interest, as it is currently unknown to what extent LRIT is resistant to spoofing.

2.2. Space-Based Observation and Ship Detection Practices

As for non-cooperative observation systems, the main tools used for maritime surveillance are optical cameras, infrared cameras, and radar. These can be deployed from land, ships, aircrafts, or satellites. Each type of sensor and platform has its strengths and weaknesses in terms of features, such as spatial resolution, update rate, range, coverage, etc. Satellite-based sensors have the particular advantage of remote access, global coverage, regular updates, and comprehensive data collection, making them the only viable option in some scenarios and the most economical in others. The use of satellite imagery is therefore an essential tool for locating ships at sea. In particular, satellite-based radar imagery, usually in the form of synthetic aperture radar (SAR), is very popular for monitoring ships at sea; ships can be detected relatively easily, even through clouds or in the absence of daylight [44,45]. However, interest in the capabilities of optical imaging for maritime surveillance has increased greatly in recent years, perhaps largely due to the growing number of optical imaging satellites. Another technology useful for light detection is the visible infrared imaging radiometer suite (VIIRS) generaly used for detecting ships at night based on light intensity [46]. A detailed literature review of the methods used for detecting and classifying ships using space-based platforms can be found in [47]. There are several satellite platforms for ship detection purposes; but, in this study, we focused on the Sentinel 1 (SAR) and Sentinel 2 (optical) platforms of the European Space Agency's Copernicus program, as they provided accurate quality data and were freely available. To incorporate our scent-based research into further studies, we could also be provided with data from ESA and third-party missions for which we have already applied. This would further improve the ability to accurately represent the ship's position and thus reduce the possibility of the falsification of the presented data on the ship's position. Ship detection with the Sentinel 1 mission (C-band radar) falls into the category of non-cooperative systems and enables the detection of ships that do not have AISs or other vessel monitoring systems on board, e.g., the VMS of fishing vessels. With two platforms (Sentinels 1A and 1B), the system has a revisit time of only a few days, which is a great opportunity for ship detection. The most suitable mode for ship detection is the interferometric wide angle (IW), which covers an extended area with a width of 250 km at a spatial resolution of 20×22 m and ensures the detection of larger ships, such as shuttle tankers, in almost all weather conditions. The most suitable polarization for object detection at sea is VH (Vertical transmit-horizontal receive). The Sentinel application platform (SNAP) was used to analyze the SAR images in this study. SNAP is a pixel-based algorithm that uses a systematic method to identify and eliminate false alarms from objects on the sea surface. The constant false alarm rate (CFAR) method was used to determine the threshold for an object to be indicated as a vessel at sea.

The Sentinel 1-SAR workflow for ship detection consists of three steps: pre-processing (removing thermal noise, loading a new orbit file, performing the ternary correction, creating a subset, and masking the land area), object detection with the built-in CFAR algorithm, and finally extracting the ships' positions.

The Copernicus Sentinel-2 missions of 2 identical satellites provide free multispectral images with a maximum update rate of 5 days and a resolution of up to 10 m (4 bands out of 13). Both satellites, Sentinels 2A and 2B, are in the same orbit at a mean altitude of 786 km and are separated by 180 degrees, which is the basis for the systematic coverage of all continental land areas (including inland waters) between 56° South and 82.8° North latitudes. More importantly, for ship tracking, it also covers all coastal waters up to 20 km from the coast, all islands with an area larger than 100 km^2, all EU islands, all enclosed seas, and the entire Mediterranean. With an orbital swath of 290 km, optical images can not only be used for ship detection purposes, but also for identification, i.e., target classification, which is quite a challenging task and is well described in [48]. A multispectral approach based on optical learning can be used both on the high seas and in harbors and eliminates the need to store a vector map of coastlines.

3. Results of the Case Studies

Since the beginning of the war in Ukraine began, Russian ships have been banned from docking in European Union ports; however, Russian oil and oil derivatives have managed to circumvent the sanctions and reach Europe. The maritime world is probably facing an unprecedented period of fraudulent shipping practices by tanker operators trying to circumvent the sanctions. Inspired and encouraged by the research of Bergman [49], Windward [50], and Savchuk [51], we systematically investigated how Russian oil has circumvented EU sanctions to reach European ports.

European Union sanctions prohibit Russian ships from docking in EU ports. Nevertheless, new shipping activities were observed off the Romanian coast in the spring of 2022. The 20-year-old Liberian-flagged crude oil tanker New Legend anchored off the coast of Costanța on 24 April 2022. The ship did not move for several months and served as a storage tanker for the transshipment of oil of Russian origin; other tankers soon appeared on the scene. Figure 5 shows the spatial distributions of vessel traffic in 2021 (left) and 2022 (right), with the location of the transshipment clearly visible.

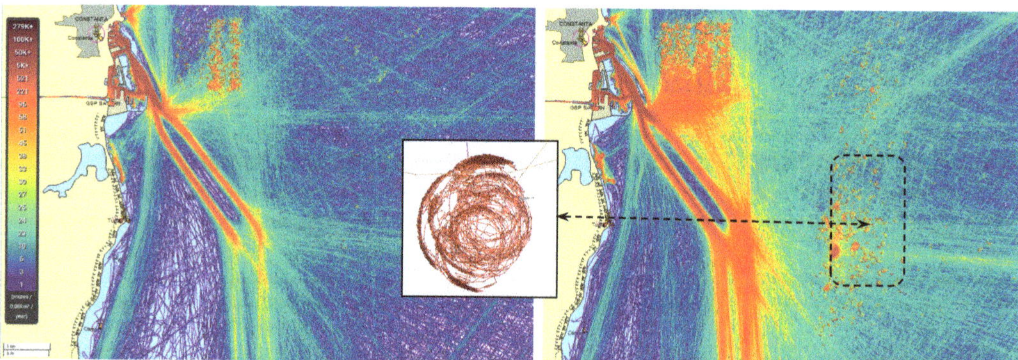

Figure 5. AIS traces of STS operations near the port of Costanța; source: produced from Marine Traffic.

The middle part of Figure 5 shows the track of the New England, which lay at anchor for several months, during which time it received cargo from a number of smaller Russian tankers. The transshipment activity in the first two weeks is shown in the Table 2. Laden Russian tankers called at New England and then sailed back to the area of Novorossiysk or though the Azov Sea up to the port of Svetlyy Yar on the Volga River. With a carrying capacity of almost 160,000 tons, New England can store cargo of more than 20 VF-class tankers. After ten tankers had discharged their cargo to the storage tanker, two tankers, Beks Swan and Kimolos Warrior, took on the cargo, the latter discharging onto Nordic Cosmos in the Laconian Gulf on 20 May 2022. Beks Swan did not sail into the Black Sea last year, but regularly took on cargo in three Russian Baltic ports. At this point, it should be noted that, both in the areas off Costanța and in the Laconian Gulf, where STS operations were conducted even closer to the coast, the tankers were also assisted by local tugboats when approaching the storage tankers, suggesting that the authorities were aware of all the activities taking place. On May 5, an additional transshipment also occurred on the coast off Costanța, with the storage tanker Haifa Adumello taking on the cargo of the tanker VF-8.

Table 2. First transshipment operations off the coast of Costanța.

Date	Offloading Vessel	Loading Vessel	AIS	Satellite
24 April 2022	VF-4		On	
25 April 2022	VF-5		On	
28 April 2022	VF-21		On	
1 May 2022	VF-18		On	
1 May 2022	VF-13		On	
1 May 2022	VF-6		Off	
2 May 2022	VF-12		On	
3 May 2022	VF-4		On	
3 May 2022	Kapitan Permyakov		On	
5 May 2022	VF-21		On	
6 May 2022		Beks Swan	Off	
7 May 2022	VF-5	Beks Swan	Off	Sentinel-2
7 May 2022		Kimolos Warrior	Off	

Two interesting cases are those of Beks Swan and the tanker VF-5, which were simultaneously involved in a cargo transfer, where both switched off their AISs, the larger tanker for 24 h. Figure 6 (Sentinel-2 L2A true-color image from 7 May 2022) shows in yellow dashed circles both tankers moored to New Legend.

To confirm the deliberate deactivation of the vessel's AIS system, the reception of the AIS base stations must first be checked. However, it could be assumed that there was a deliberate deactivation, as in this case, all other vessels in the vicinity of New England were seen by the AIS signals. To confirm the AIS coverage of the STS site, we provided the AIS reception statistics for the Mediterranean and Black Sea compiled by the European Maritime Safety Agency (EMSA), where the Mediterranean AIS Regional Exchange System (MAREΣ) was available [52]. The estimated monthly AIS coverage of the MAREΣ is represented by layers, as shown in Figure 7.

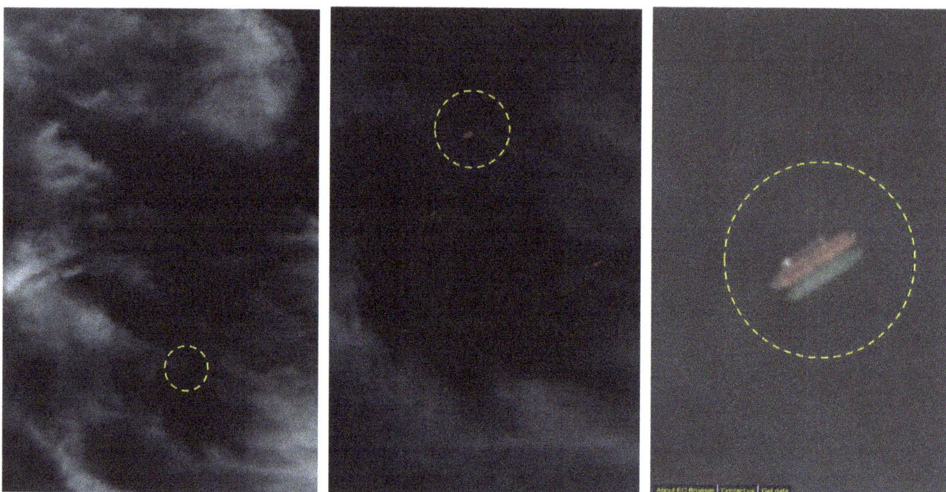

Figure 6. Satellite optical image, Sentinel-2 (7 May 2022, 09:08:20). The large ship in the middle is the storage tanker New England, on the port side is the Russian tanker VF-5, and on the starboard side is the tanker Beks Swan flying the flag of the Marshall Islands.

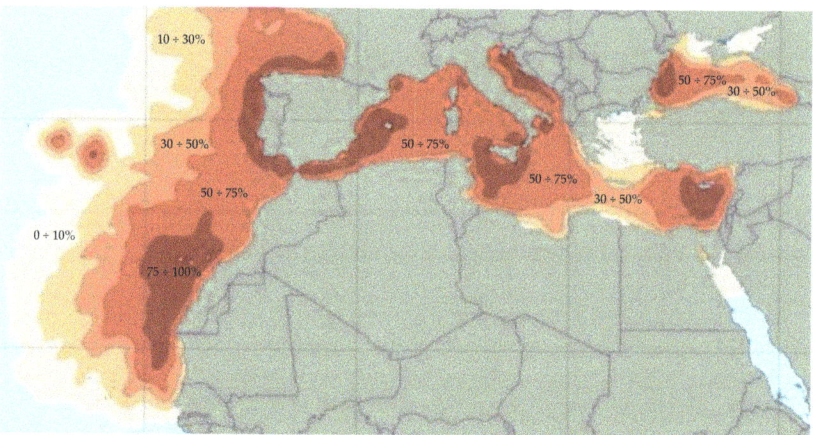

Figure 7. The estimated monthly AIS coverage of the MAREΣ 2022 [52].

Each layer was assigned to an area and corresponded to a probability range for the reception of AIS information from the vessels. The following probability ranges were evident: 0–10%, 10–30%, 30–50%, 50–75%, and 75–100%. The area in which STS operations were conducted off the coast of Costanța was well covered, with a detection probability between 75 and 100%.

Figure 8 shows the turnover of all VF tankers and some transshipment tankers deployed from May to September 2023. It can be seen that Russian oil traffic is also routed to ports within the EU.

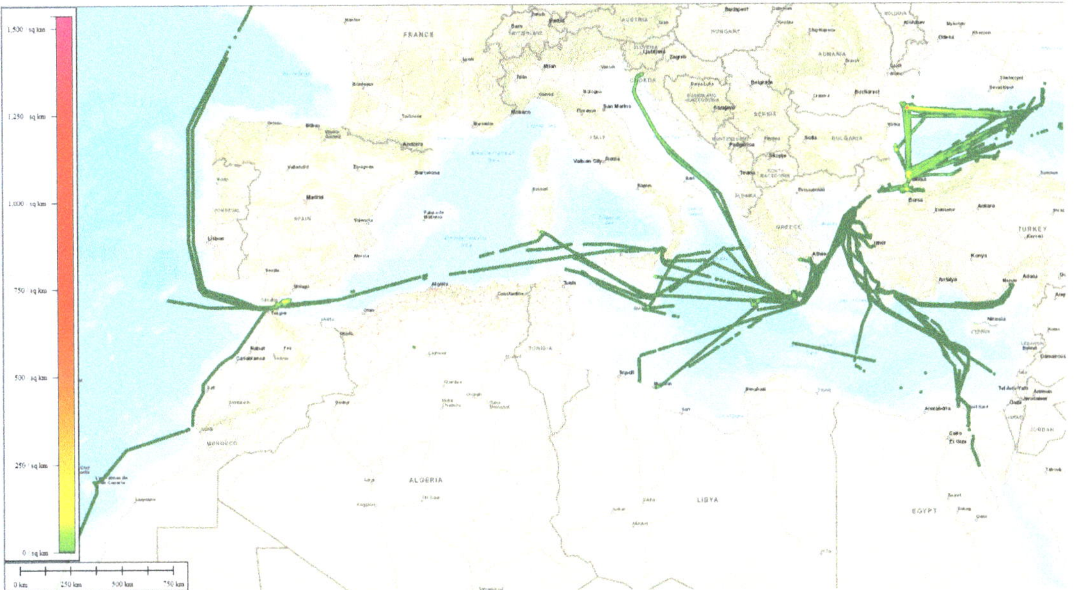

Figure 8. The Black Sea oil STS transfer activities and its transit from May to September 2022.

During the investigation, we were also able to follow the transshipment from start to finish. On 2 July 2022, New Legend transferred its cargo to the tanker San Sebastian in 434 min (Figure 9). The STS operations were monitored by the two satellite platforms SAR and Optical.

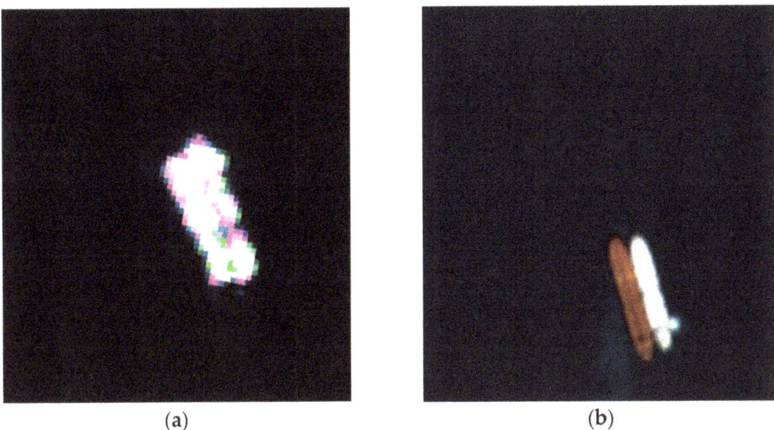

(a)   (b)

Figure 9. Tankers New Legend and San Sebastian on 2 July 2022: (**a**) radar image as it appears on Sentinel-1; (**b**) optical image as it appears on Sentinel-2.

The complexity of tracking tankers involved in the Russian oil business is illustrated in Figure 10, which shows the route of the tanker Kriti Future over several months, calling at Russian ports, the STS site in Greece, and several other ports, many of which are in EU countries.

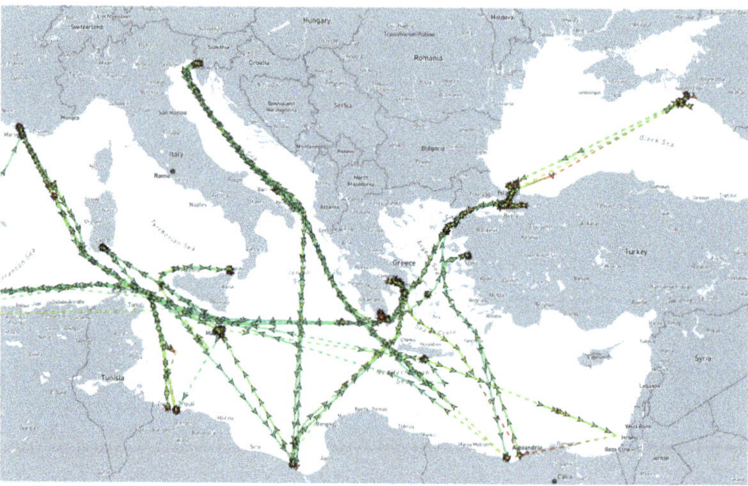

Figure 10. Trajectory of Kriti Future tanker; source: adapted from Marine Traffic.

Like Bergman [49], we found that Russian-flagged tankers often switched off their transponders, making it difficult to track their movements. The misuse of AIS collision avoidance signals jeopardizes maritime safety by increasing the risk of collisions, oil spills, and other serious accidents. A combination of satellite imagery and AIS data from the tankers shows that, while the tankers reported their position in the western Black Sea, sometimes for weeks, their true position was near the Kerch Strait, 400 miles to the northeast. Many of these tankers are substandard and do not comply with international maritime safety regulations. This significantly increases the risk of dangerous accidents and oil spills, with disastrous consequences for seafarers, coastal states, and the marine environment.

The example in Figure 11 shows another STS operation with Russian oil. However, here, the tanker did not want to reveal its actual position, so this ship did not switch off the AIS system and became a so-called "dark tanker". It falsified its position and displayed it in a different location than where it actually was, performing a spoofing activity that could be very dangerous. We became aware of this vessel because, according to the AIS system, it almost collided with another vessel in the STS area (Laconian Bay), which also reduced its speed, as such a vessel appeared on the electronic chart display and information system (ECDIS) and radio detection and ranging (RADAR) screens, but not on the actual horizon. The tanker Turba had intercepted Strea as she was heading into the sunset and must have been surprised by the alerts from her navigation system. Turba's rapid course change was also very unusual, apparently performing an impossible maneuver.

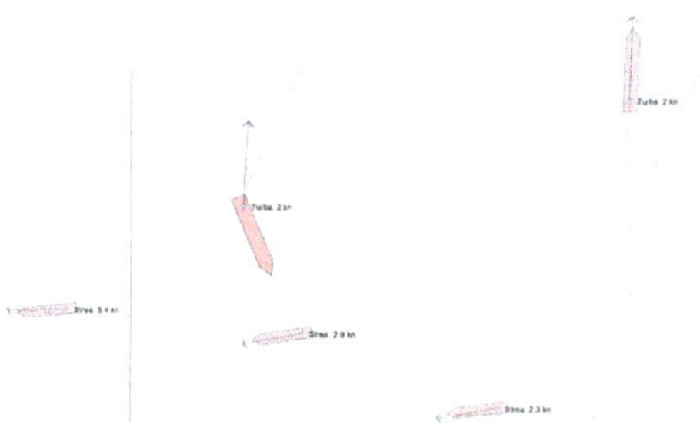

Figure 11. A look-a-like near miss with a false AIS position.

Sentinel-1 images were available and captured the situation just minutes before the apparent near miss. Using SNAP application, we processed the image and exported the detected vessels to the GIS application, into which we also imported the AIS data. It immediately became clear that there was no Turba tanker in the vicinity of the Strea vessel, which was actually performing an STS operation with the Simba tanker, i.e., Turba was showing an incorrect position; at this time, the offset was 6 NM, as shown in Figure 12. Furthermore, the movement and orientation of the vessel did not match the movement of the Simba tanker. We also checked the optical images, and the following day, 20 September 2023, we observed the same situation: Turba was still falsifying its AIS transmission. Such activities are very dangerous and the maritime authority should significantly increase its vigilance and sanction such culprits.

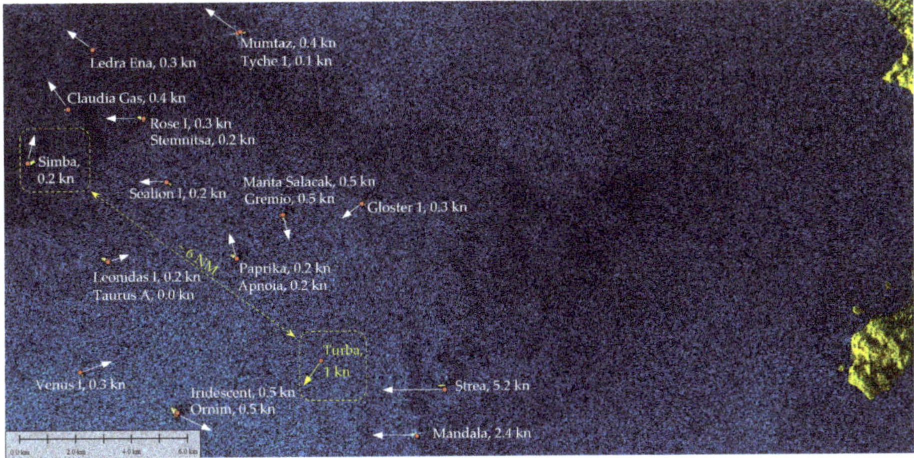

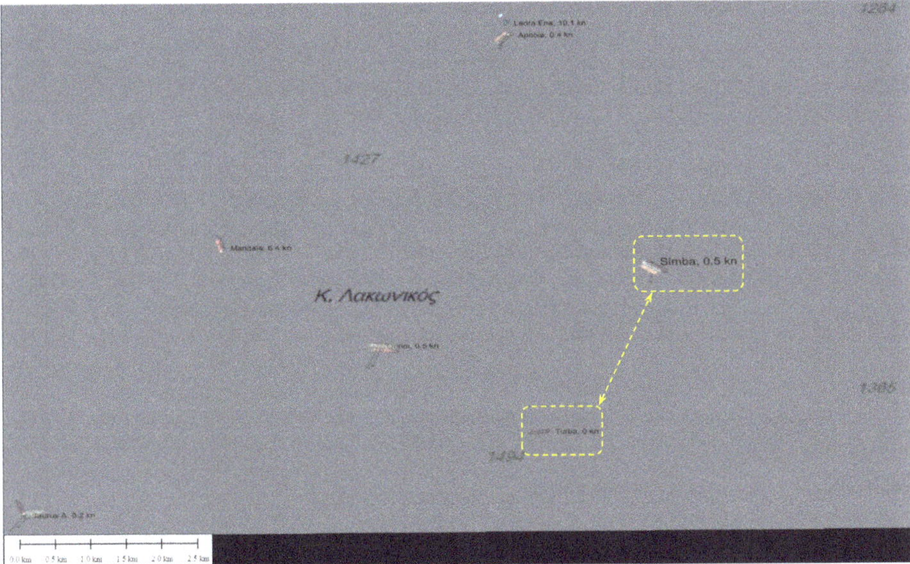

Figure 12. SAR image from 19 September 2023 and optical image from 20 September 2023 showing the AIS spoofing of the tanker Turba during an STS operation in a busy area. The SAR image as well as the optical image shows that the Turba tanker is located next to the Simba tanker while the AIS image shows a large offset.

Previously, on 14 September 2023, the tanker Simba had taken on cargo at the familiar offshore location of Constanţa from the tanker New Trust, which had taken on cargo from the tanker VF-4 at the same time. Simba had communicated with New Trust with the help of the two tugs Vulkan and Dynamax. It should be emphasized once again that these operations occurred off the coast of Constanţa, albeit outside territorial waters, but with the help of port resources. The situation is similar off the coast of Malta. In the Gulf of Laconia, STS operations are conducted in the immediate vicinity of the coast. The main point in this regard is that such dangerous and illegal activities are more common than we can express at this point.

4. Discussion

The international community is making great efforts to curb the trade in Russian crude oil. Last December, the Price Cap Coalition (EU, G7, and Australia) responded to the ongoing Russian invasion of Ukraine by capping the price of Russian seaborne oil sold to global markets. Companies based in the coalition countries are currently allowed to provide services that support the sale of Russian oil, including shipping, insurance, and trade finance, but only if the price paid to Russia does not exceed USD 60 per barrel. The goal of the coalition is to reduce Russian revenues received from oil sales while ensuring the uninterrupted flow of Russian oil to global markets, thus preventing a negative supply shock that can have short-term negative consequences for the rest of the world [53]. However, a report by the Centre for Research on Energy and Clean Air (CREA) shows that, following Moscow's invasion of Ukraine, some countries have increased their imports of Russian oil and have processed ("laundered") it into products that are sold to countries that have imposed sanctions on Russian oil, the so-called "laundromat" countries [54]. Our research confirms that Russian oil is finding its way to countries at limited prices in the form of diesel, jet fuel, and petrol.

The data from CREA [54] are remarkable: in the 12 months following the invasion of Ukraine, the EU spent USD 19.3 billion on these Russian-origin products, followed by Australia with USD 8.74 billion, the United States with USD 7.21 billion, the United Kingdom with USD 5.46 billion, and Japan with USD 5 billion.

It goes without saying that the armed conflict in Ukraine has triggered a multitude of sanctions affecting the industry. In addition to the suspicion of STS oil transfers mentioned in the article, there is also a new type of fraud that has been introduced to the global community: vessel identity laundering (VIL). According to Tsakiris [55], this is a novel tactic whereby one or more vessels assume a different identity on an AIS in order to provide "dirty" vessels (i.e., those associated with illegal activities) with a "clean" identity, and where at least one vessel in this operation assumes an identity obtained through IMO number fraud (Unmasked-vessel identity laundering). With the sale of older tankers increasing, there is a possibility that this can soon increase. One hopes that serious ship owners want to avoid being involved in illegal transactions. From the points of view of security and environmental risks, the emergence of these shadow ships and the countries willing to host them is worrying for all those involved in the maritime industry. There is little to discourage their activities. Yet, the shadow fleet is almost entirely composed of ships that are past their prime and on their way to the scrapyard, and this is a growing and increasingly dangerous set of circumstances. These vessels sail under flags of convenience, which are subject to less stringent legislations and allow for lower maintenance standards. This has led to the fleets and ships engaged in legal traffic being exposed to a greater risk of accidents. Accidents involving shadow tankers have increased worldwide, such as fires, groundings, and pollution. In 2022, there were eight groundings, collisions, or near misses involving tankers carrying sanctioned crude oil or oil products [1]. One example from 2023 is the m/v Pablo, which was part of the world's growing fleet of shadow tankers that caught fire, exploded, and completely burnt down off the coast of Malaysia, not far from the busy Singapore Strait. Fortunately, the tanker was not loaded with oil, so the environmental damage caused was limited [2] and therefore did not attract public attention. If the ship had been carrying oil, there would have been an environmental disaster with about 600,000 barrels being spilled. To this day, the wreckage of the ship remains untouched where the accident occurred because no one is willing to send out a salvage crew and because no one knows who will pay for it. So, the case remains unsolved and is a typical example of what happens when there is an environmental disaster [56].

The analysis shows that the falsification of the AIS led to a tripling in the size of the so-called dark fleet that transported illegal oil from Russia around the world. By tracking ships during this study, it was discovered that many shadow tankers changed their names and registries, some virtually after every voyage, their ownership remaining a mystery.

The data provided by Geollect [57] show that, since 14 May 2023, the data from commercial vessels' AISs are remotely spoofed to create the impression of a 65 km Z symbol on the Black Sea, as seen on the open source tracking software (Figure 13). The tracks that created the image indicate high ship speeds of 102 knots (188 km per hour), which is another indication that it is fake. The perpetrator behind the spoofing may have used high-frequency signals to mimic a real signal with ease, causing the ship's signal to display false information.

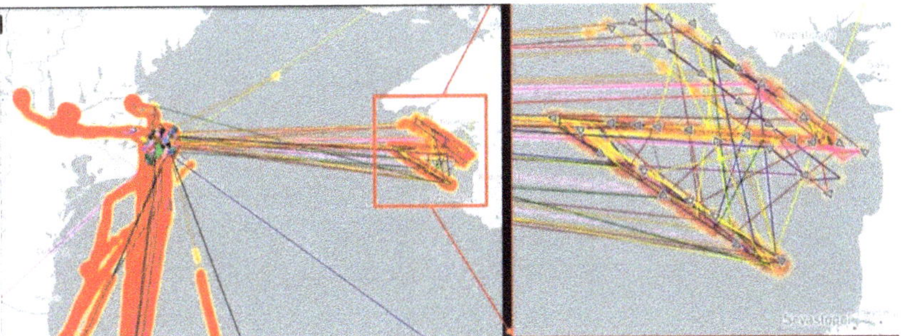

Figure 13. AIS spoofing data projecting a Z symbol on the Black Sea during 14–21 May 2023; source: Geollect [57].

5. Conclusions

Geopolitics and the maritime ecosystem are interdependent and complicated, and maritime risks are constantly evolving. This article shows how much things have changed in unpredictable ways over just the last 18 months. This research was conducted using predictive analytics and an artificial intelligence platform. Without these tools, it would be almost impossible to keep up with the maritime risks occurring at present due to the flood of data and their complexity. But, to also improve the quality of the input data, it will be necessary in future studies to plan, combine, and compare the data from several different platforms, e.g., the ESA oil-trading analytics monitor (OTAM) and ship detection systems, to determine the ship's position more accurately.

With the looming global recession and economic sanctions from the West, combined with the political and security situations in the neighboring eastern countries, we can expect to see an increase in fraudulent maritime practices, particularly a combination of illegal activities, GNSS manipulation, and secret STS meetings. Most likely, new nodes will continue to emerge to disguise illegal activities, while the old ones will be less effective.

The false-position data sent by the tanker Beks Swan demonstrates the importance of developing robust automated systems to detect and flag the fraudulent use of AISs. Otherwise, it will not be long before we can no longer trust AISs. The international community has taken decisive action to restrict the sale of Russian crude oil to fund the armed conflict, but these measures will be ineffective if AIS falsifications go undetected. This article showed that false positions could be detected with the data methods currently available. We can be confident that the increasing automation of these methods will soon mean that any ship falsifying its position will only end up attracting attention to its illegal activities, which are also linked to the complex and multi-layered shadow fleet. The impact of Russian sanctions has meant that the number of these vessels has doubled in the last 18 months. This is a worrying development that threatens the world fleet and the environment. With collaborative action and cooperation from all stakeholders, substandard ships can be prevented from threatening our maritime industry and our environment. Yet, to date, no action has been taken. Legislations need to be strengthened so that the standards of the individual countries that issue ships with certificates for the transport of dangerous goods (oil) are complied with. The fundamental problem is that, where profits are involved, political will tends to be compromised.

It is very difficult to understand that, on the one hand, we are still encountering ships in the maritime world that should have been consigned to the scrapyard long ago, while on the other hand, the maritime industry is making every effort to address the increasing cyber risks by building cyber security systems into ship designs that enables strong protection against threats while ensuring compliance with the new regulations of the International Association of Classification Societies (IACS) Unified Requirements (URs) E27 and E26 [58], which complement the IMO cyber security regulations that have been in force since 1 January 2021.

In our view, the deployment of multiple public and private satellite navigation constellations that do not rely on signal transmissions should provide the shipping industry with much-needed improved protection against criminal activity at sea in the foreseeable future.

Author Contributions: Conceptualization, A.A., I.P. and M.P.; methodology, M.P.; software, M.P. and L.G.; validation, A.A., M.P., I.P. and L.G.; formal analysis, P.V. and L.G.; investigation, A.A., I.P. and M.P.; resources, A.A. and M.P.; data curation, P.V.; writing—original draft preparation, A.A., I.P., P.V. and M.P.; writing—review and editing, A.A., L.G. and M.P.; visualization, A.A.; supervision, M.P. and L.G.; project administration, P.V.; funding acquisition, P.V. All authors have read and agreed to the published version of the manuscript.

Funding: This research received no external funding.

Institutional Review Board Statement: Not applicable.

Informed Consent Statement: Not applicable.

Data Availability Statement: The data for this analysis were obtained from the Maritime Administration Office in Slovenia and the ELMAN S.r.l in Italy.

Acknowledgments: We thank Elman S.r.l for their assistance with the analysis of the AIS data, UL FPP postgraduate student Migel Mehlmauer and US proofreader Rick Harsch.

Conflicts of Interest: The authors declare no conflict of interest.

Abbreviations

AIS	Automated Identification System
BIMCO	Baltic and International Maritime Council
CFAR	Constant False Alarm Rate
CREA	Centre for Research on Energy and Clean Air
ECDIS	Electronic Chart Display and Information System
EEZ	Exclusive Economic Zone
EMSA	European Maritime Safety Agency
EU	European Union
ESA	European Space Agency
FIS	Fuzzy Interference System
GIS	Geographic Information System
GNSS	Global Navigation Satellite System
GPS	Global Positioning System
HAZOP	Hazard and Operability Process Analysis
IACS	International Association of Classification Societies
IMO	International Maritime Organization

IW	Interferometric Wide
LNG	Liquefied Natural Gas
LRIT	Long-Range Identification and Tracking
MAREΣ	Mediterranean AIS Regional Exchange System
MARPOL	International Convention for the Prevention of Pollution from Ships
MMSI	Maritime Mobile Service Identity
MSA	Maritime Situational Awareness
NM	Nautical Mile
OTAM	Oil-Trading Analytics Monitor
PFMEA	Process Failure Mode and Effect Analysis
RADAR	Radio Detection and Ranging
ROT	Rate Of Turn
RSI	Received Signal Strength
SAR	Synthetic Aperture Radar
S-AIS	Satellite AIS
SNAP	Sentinel Application Platform
SOLAS	International Convention for the Safety of Life at Sea
STPA	System Theoretical Process Analysis
STS	Ship To Ship
TOA	Time Of Arrival
TTWs	Territorial Waters
URs	Unified Requirements
VH	Vertical Transmit–Horizontal Receive
VHF	Very High Frequency
VIIRS	Visible Infrared Radiometer Suite
VIL	Vessel Identity Laundering
VMS	Vessel Monitoring System
VSI	VHF Signal Information

References

1. Insight: Oil Spills and Near Misses: More Ghost Tankers Ship Sanctioned Fuel. Available online: https://www.reuters.com/business/autos-transportation/oil-spills-near-misses-more-ghost-tankers-ship-sanctioned-fuel-2023-03-23/ (accessed on 23 September 2023).
2. The Diplomat: Southeast Asian States Need to Tackle the Dangerous Shadow Tanker Activities in Their Waters. Available online: https://thediplomat.com/2023/09/ (accessed on 23 September 2023).
3. ICS/OCIMF. *Ship to Ship Transfer Guide (Petroleum)*, 4th ed.; Witherby Publishing Group: Livingston, UK, 2005.
4. Ventikos, N.P.; Stavrou, D.I. Ship to Ship (STS) Transfer of Cargo: Latest Developments and Operational Risk Assessment. *SPOUDAI J. Econ. Bus.* **2013**, *63*, 172–180.
5. Tokić, T.; Frančić, V.; Hasanspahić, N.; Rudan, I. Training Requirements for LNG Ship-to-Ship Transfer. *Pomor. Zbornik.* **2021**, *60*, 49–63. [CrossRef]
6. CDI/ICS/OCIMF/SIGTTO. *Ship to Ship Transfer Guide for Petroleum, Chemicals and Liquefied Gases*, 1st ed.; Witherby Seamanship: Livingston, UK, 2013; pp. 13–15. ISBN 9781856095945.
7. Shipowners Security for Small and Specialist Vessels, Buletin, Issue Date 27/01/2015, Ship to Ship Oil Transfer Operations: We Would Like to Advise Members of Claims Arising from Poor Cargo Practices Being Adopted on Board Tankers. Available online: https://www.shipownersclub.com/media/2015/01/Ship-to-Ship-oil-transfer-operations.pdf (accessed on 17 April 2023).
8. IMO Resolution MEPC.186(59). *Amendments to the Annex of the Protocol of 1978 Relating to the International Convention for the Prevention of Pollution from Ships, 1973*; IMO Publishing: London, UK, 2009.
9. Stavrou, D.I.; Ventikos, N.P. Ship to Ship Transfer of Cargo Operations: Risk Assessment Applying a Fuzzy Inference System. *J. Risk Anal. Crisis Response* **2014**, *4*, 214–227. [CrossRef]
10. Stavrou, D.I.; Ventikos, N.P. Risk evaluation of Ship-to-Ship transfer of cargo operations by applying PFMEA and FIS. In Proceedings of the 2015 Annual Reliability and Maintainability Symposium (RAMS), Palm Harbor, FL, USA, 26–29 January 2015; pp. 1–7. [CrossRef]
11. Sultana, S.; Okoh, P.; Haugen, S.; Vinnem, J.E. Hazard analysis: Application of STPA to ship-to-ship transfer of LNG. *J. Loss Prev. Process Ind.* **2019**, *60*, 241–252. [CrossRef]
12. Suspicion of Illegal Ship-to-Ship Transfers of Goods by North Korea-Related Vessels. Available online: https://www.mofa.go.jp/fp/nsp/page4e_000757.html (accessed on 22 April 2023).
13. Sanctions and STS Transfers—Legal Risks. Available online: https://www.skuld.com/topics/legal/sanctions/sanctions-and-sts-transfers--legal-risks/ (accessed on 22 April 2023).

14. United Nations Security Council. S/2021/777. Letter Dated 3 September 2021 from the Panel of Experts Established Pursuant to Resolution 1874 (2009) Addressed to the President of the Security Council. Available online: https://www.securitycouncilreport.org/atf/cf/%7B65BFCF9B-6D27-4E9C-8CD3-CF6E4FF96FF9%7D/S_2021_777_E.pdf (accessed on 22 April 2023).

15. UKPANDI. Circular 01/22 Sanctions—Recent Deceptive Practices. Available online: https://www.ukpandi.com/news-and-resources/circulars/2022/circular-0122-sanctions-recent-deceptive-practices/ (accessed on 22 April 2023).

16. ReCAAP ISC. *Regional Guide to Counter Piracy and Armed Robbery against Ships in Asia*; ReCAAP ISC: Singapore, 2016; p. 5. Available online: https://www.american-club.com/files/files/Regional_Guide_to_Counter_Piracy_and_Armed_Robbery_Against_Ships_in_Asia.pdf (accessed on 22 April 2023).

17. Kamal-Deen, A. The Anatomy of Gulf of Guinea Piracy. *Nav. War Coll. Rev.* **2015**, *68*, 93–118. Available online: https://digital-commons.usnwc.edu/nwc-review/vol68/iss1/7/ (accessed on 22 April 2023).

18. Windward. High Sea Russian Oil Transfers Are Far from the Only Smuggling Method. Available online: https://windward.ai/blog/high-sea-russian-oil-transfers-are-far-from-the-only-smuggling-method/ (accessed on 18 April 2023).

19. Goudossis, A.; Katsikas, S.K. Towards a secure automatic identification system (AIS). *J. Mar. Sci. Technol.* **2019**, *24*, 410–423. [CrossRef]

20. Ramin, A.; Masnawi, A.; Shaharudin, A. Prediction of Marine Traffic Density Using Different Time Series Model from AIS Dana of Port Klang and Straits of Malacca. *Trans. Marit. Sci.* **2020**, *2*, 217–223. [CrossRef]

21. Eriksen, T.; Greidanus, H.; Delaney, C. Metrics and provider-based results for completeness and temporal resolution of satellite-based AIS services. *Mar. Policy* **2018**, *93*, 80–92. [CrossRef]

22. Natale, F.; Gibin, M.; Alessandrini, A.; Vespe, M.; Paulrud, A. Mapping fishing effort through AIS data. *PLoS ONE* **2015**, *10*, e0130746. [CrossRef]

23. Pallotta, G.; Vespe, M.; Bryan, K. Vessel Pattern Knowledge Discovery from AIS Data: A Framework for Anomaly Detection and Route Prediction. *Entropy* **2013**, *15*, 2218–2245. [CrossRef]

24. Perkovic, M.; Twrdy, E.; Harsch, R.; Vidmar, P.; Gucma, M. Technological Advances and Efforts to Reduce Piracy. *TransNav* **2012**, *6*, 203–206.

25. Lee, E.S.; Mokashi, A.J.; Moon, S.J.; Kim, G.S. The Maturity of Automatic Identification Systems (AIS) and Its Implications for Innovation. *J. Mar. Sci. Eng.* **2019**, *7*, 287. [CrossRef]

26. Fournier, M.; Hilliard, R.C.; Rezaee, S.; Pelot, R. Past, present, and future of the satellite-based automatic identification system: Areas of applications (2004–2016). *WMU J. Marit. Aff.* **2018**, *17*, 311–345. [CrossRef]

27. Soldi, G.; Gaglione, D.; Ramponi, S.; Forti, N.; D'Afflisio, E.; Kowalski, P. Monitoring of Critical Undersea Infrastructures: The Nord Stream and Other Recent Case Studies. *IEEE Aerosp. Electron. Syst. Mag.* **2023**, *38*, 4–24. [CrossRef]

28. IALA Guideline G1082. An Overview of AIS, Edition 2.1. IALA-AISM. 2016. Available online: https://www.iala-aism.org/product/g1082/ (accessed on 20 April 2023).

29. Kessler, G.C. Protected AIS: A Demonstration of Capability Scheme to Provide Authentication and Message Integrity. *TransNav* **2020**, *14*, 279–286. [CrossRef]

30. Caprolu, M.; Di Pietro, R.; Raponi, S.; Sciancalepore, S.; Tedeschi, P. Vessels Cybersecurity: Issues, Challenges, and the Road Ahead. *IEEE Commun. Mag.* **2020**, *58*, 90–96. [CrossRef]

31. Balduzzi, M.; Wilhoit, K.; Pasta, A. A Security Evaluation of AIS. *Trend Micro Res. Pap.* **2018**, 1–9. [CrossRef]

32. Iphar, C.; Napoli, A.; Ray, C. An expert-based method for the risk assessment of anomalous maritime transportation data. *Appl. Ocean Res.* **2020**, *104*, 102337. [CrossRef]

33. Iphar, C.; Ray, C.; Napoli, A. Uses and Misuses of the Automatic Identification System. In Proceedings of the IEEE OCEANS 2019, Marseille, France, 17–20 June 2019. [CrossRef]

34. Salmon, L.; Ray, C.; Claramunt, C. Continuous detection of black holes for moving objects at sea. In Proceedings of the 7th ACM SIGSPATIAL International Workshop on GeoStreaming, Burlingame, CA, USA, 31 October–3 November 2016. IWGS '16. [CrossRef]

35. IMO. *The International Convention for Safety of Life at Sea (Consolidated Edition 2020)*; IMO Publishing: London, UK, 2020; pp. 15, 34. ISBN 9789280116908.

36. IMO Resolution A.1106(29). *Revised Guidelines for the Onboard Operational Use of Shipborne Automatic Identification Systems (AIS)*; IMO Publishing: London, UK, 2015.

37. BIMCO. AIS Switch Off Clause 2021. Available online: https://www.bimco.org/contracts-and-clauses/bimco-clauses/current/ais_switch_off_clause_2021 (accessed on 22 April 2023).

38. Androjna, A.; Brcko, T.; Pavic, I.; Greidanus, H. Assessing Cyber Challenges of Maritime Navigation. *J. Mar. Sci. Eng.* **2020**, *8*, 776. [CrossRef]

39. Androjna, A.; Perkovič, M.; Pavic, I.; Mišković, J. AIS Data Vulnerability Indicated by a Spoofing Case-Study. *Appl. Sci.* **2021**, *11*, 5015. [CrossRef]

40. Han, J.; Wu, J.; Zhang, L.; Wang, H.; Zhu, Q.; Zhang, C.; Zhao, H.; Zhang, S. A Classifying-Inversion Method of Offshore Atmospheric Duct Parameters Using AIS Data Based on Artificial Intelligence. *Remote Sens.* **2022**, *14*, 3197. [CrossRef]

41. Tang, W.; Cha, H.; Wei, M.; Tian, B. The effect of atmospheric ducts on the propagation of AIS signals. *Aust. J. Electr. Electron. Eng.* **2019**, *16*, 111–116. [CrossRef]

42. Sigillo, L.; Marzilli, A.; Moretti, D.; Grassucci, E.; Greco, C.; Comminiello, D. Sailing the Seaformer: A Transformer-Based Model for Vessel Route Forecasting. In Proceedings of the 2023 IEEE 33rd International Workshop on Machine Learning for Signal Processing (MLSP), Rome, Italy, 17–20 September 2023; pp. 1–6. [CrossRef]
43. Cheng, Y. Satellite-based AIS and its Comparison with LRIT. *TransNav* **2014**, *8*, 183–187. [CrossRef]
44. Zhang, T.; Zeng, T.; Zhang, X. Synthetic Aperture Radar (SAR) Meets Deep Learning. *Remote Sens.* **2023**, *15*, 303. [CrossRef]
45. Passah, A.; Sur, S.N.; Abraham, A.; Kandar, D. Synthetic Aperture Radar image analysis based on deep learning: A review of a decade of research. *Eng. Appl. Artif. Intell.* **2023**, *1*, 123. [CrossRef]
46. Marzuki, M.I.; Rahmania, R.; Kusumaningrum, P.D.; Akhwady, R.; Sianturi, D.S.; Firdaus, Y.; Sufyan, A.; Hatori, C.A.; Chandra, H. Fishing boat detection using Sentinel-1 validated with VIIRS Data. *IOP Conf. Ser. Earth Environ. Sci.* **2021**, *925*, 012058. [CrossRef]
47. Kanjir, U.; Greidanus, H.; Oštir, K. Vessel detection and classification from spaceborne optical images: A literature survey. *Remote Sens. Environ.* **2018**, *207*, 1–26. [CrossRef]
48. Ciocarlan, A.; Stoian, A. Ship Detection in Sentinel 2 Multi-Spectral Images with Self-Supervised Learning. *Remote Sens.* **2021**, *13*, 4255. [CrossRef]
49. Russian Tanker Falsifies AIS Data, Hides Likely Activity around Malta and Cyprus. Available online: https://skytruth.org/2022/12/russian-tanker-falsifies-ais-data-hides-likely-activity-around-malta-and-cyprus/?fbclid=IwAR1865fkOeoxP5h0arBYaSgtPQZw4AOXEvkr1ntMQ1IXTOJwqjf8U_iS6gk (accessed on 15 December 2022).
50. The Impact of Russia's Year-Long Invasion on the Maritime Ecosystem & Global Economy. Available online: https://www.hellenicshippingnews.com/the-impact-of-russias-year-long-invasion-on-the-maritime-ecosystem-global-economy/ (accessed on 15 February 2023).
51. How Russian oil Evades EU Sanctions and Land in European Ports. Available online: https://www.rferl.org/a/russia-ukraine-eu-oil-sanctions-shipping/32025726.html (accessed on 20 September 2022).
52. European Maritime Safety Agency. 19th Mediterranean AIS Expert Working Group Meeting—Report. EMSA Ref. Ares(2023)183539-11/01/2023. Available online: https://www.emsa.europa.eu/ssn-main/258-other-ssn-initiatives/4885-19th-mediterranean-expert-working-group-meeting.html (accessed on 15 December 2022).
53. Johnson, S.; Rachel, L.; Wolfram, C. Theory of Price Caps on Non-Renewable Resources. National Bureau of Economic Research. 2023. Available online: https://www.nber.org/system/files/working_papers/w31347/w31347.pdf (accessed on 23 September 2023).
54. The Laundromat: How the Price Cap Coalition Whitewashes Russian Oil in Third Countries. Available online: https://energyandcleanair.org/wp/wp-content/uploads/2023/04/CREA_The-Laundromat_How-the-price-cap-coalition-whitewashes-Russian-oil-in-third-countries.pdf (accessed on 8 May 2023).
55. Tsakiris, I. Bringing the 'Dark' Fleet into the Light. *IUMI EYE* **2023**, *40*, 15. Available online: https://iumi.com/news/iumi-eye-newsletter-march-2023/bringing-the-dark-fleet-into-the-light (accessed on 20 March 2023).
56. Michelle Wiese Bockmann: Shifty Shades of Grey: Dark Fleet Shipping Sanctioned Oil around the World. Available online: https://www.youtube.com/watch?v=-DH9XkwN3tY (accessed on 23 September 2023).
57. Geollect: Ship AIS Data Spoofed to Draw Pro-War Russian Z Symbol in Black Sea. Available online: https://lnkd.in/eug5qJVz (accessed on 25 May 2023).
58. Jarle Coll Blomhoff: Building Strong Cyber Security into Ship Design. Available online: https://www.dnv.com/expert-story/maritime-impact/building-strong-cyber-security-into-ship-design.html (accessed on 12 October 2023).

MDPI AG
Grosspeteranlage 5
4052 Basel
Switzerland
Tel.: +41 61 683 77 34

Journal of Marine Science and Engineering Editorial Office
E-mail: jmse@mdpi.com
www.mdpi.com/journal/jmse

www.ingramcontent.com/pod-product-compliance
Lightning Source LLC
LaVergne TN
LVHW071543170726
843515LV00008B/2204